BREAKING AND DISSIPATION OF OCEAN SURFACE WAVES

Over the past 15 years, our understanding of the breaking and dissipation of ocean waves has undergone a dramatic leap forward. From a phenomenon which was very poorly understood, it has emerged as a process whose physics is clarified and quantified. This book presents the state of the art of our current understanding of the breaking and dissipation of ocean waves.

Wave breaking is an intermittent random process, very fast by comparison with other processes in the wave system. Distribution of wave breaking on the water surface is not continuous, but its role in maintaining the energy balance within the continuous wind–wave field is critical. Such breaking represents one of the most interesting and challenging problems for fluid mechanics and physical oceanography. Ocean wave breaking also plays the primary role in the air–sea exchange of momentum, mass and heat, and it is of significant importance for ocean remote sensing, coastal and ocean engineering, navigation and other practical applications. This book outlines the state of wave-breaking research and presents the main outstanding problems. Non-breaking dissipation and non-dissipative effects of breaking waves, their influence on the atmospheric boundary layer and on upper-ocean mixing, including extreme weather conditions, are also discussed.

Breaking and Dissipation of Ocean Surface Waves is a valuable resource for anyone interested in this topic: researchers, modellers, forecasters, engineers and graduate students in physical oceanography, meteorology and ocean engineering.

ALEXANDER BABANIN is a Professor in the Faculty of Engineering and Industrial Sciences at Swinburne University of Technology, Melbourne, Australia. His bachelor's degree in Physics and master's degree in Physical Oceanography are from the Faculty of Physics at Lomonosov Moscow State University, and his Ph.D. on Spectral Characteristics of Surface Wind Wave Fields is from the Marine Hydrophysical Institute in Sevastopol. His research interests involve wind-generated waves, air–sea interaction and ocean turbulence, including dynamics of surface ocean waves, wave breaking and dissipation, air–sea boundary layer, extreme oceanic conditions, wave statistics, ocean mixing and remote sensing.

BREAKING AND DISSIPATION OF OCEAN SURFACE WAVES

ALEXANDER V. BABANIN
Swinburne University of Technology,
Melbourne, Australia

CAMBRIDGE
UNIVERSITY PRESS

University Printing House, Cambridge CB2 8BS, United Kingdom

One Liberty Plaza, 20th Floor, New York, NY 10006, USA

477 Williamstown Road, Port Melbourne, VIC 3207, Australia

314-321, 3rd Floor, Plot 3, Splendor Forum, Jasola District Centre, New Delhi-110025, India

79 Anson Road, #06-04/06, Singapore 079906

Cambridge University Press is part of the University of Cambridge.

It furthers the University's mission by disseminating knowledge in the pursuit of education, learning and research at the highest international levels of excellence.

www.cambridge.org
Information on this title: www.cambridge.org/9781108454773

First published 2011
First paperback edition 2018

A catalogue record for this publication is available from the British Library

Library of Congress Cataloging in Publication data
Babanin, Alexander V., 1960-
Breaking and dissipation of ocean surface waves / Alexander V. Babanin.
p. cm.
ISBN 978-1-107-00158-9 (Hardback)
1. Ocean waves–Measurement. 2. Ocean waves–Simulation methods. I. Title.
GC205.B34 2011
551.46′3–dc22

2011011866

ISBN 978-1-107-00158-9 Hardback
ISBN 978-1-108-45477-3 Paperback

This book is dedicated to my parents Vladimir and Vera
to whom I am indebted.

Contents

Preface

Wave breaking is a fascinating object to watch and to research. Wind-generated waves are the most prominent feature of the ocean surface, and so are breaking waves manifested by sporadic whitecaps on the wavy surfaces of oceans and lakes. The breaking is so apparent that one does not need to be an oceanographer to have an opinion on what it is and how it works.

Such breaking, however, represents one of the most interesting and most challenging problems for both fluid mechanics and physical oceanography. It is an intermittent random process, very fast by comparison with other processes in the wave system. The distribution of wave breaking on the water surface is not continuous, but its role in maintaining the energy balance within the continuous wind–wave field is critical. Ocean wave breaking also plays the primary role in the air–sea exchange of momentum, mass and heat, and it is of significant importance for practical applications such as ocean remote sensing, coastal and ocean engineering, and navigation among others.

Understanding such wave breaking, predicting its behaviour, the breaking rates and breaking strength, and even an ability to describe its onset have been hindered for decades by the strong nonlinearity of the process, together with its irregular and ferocious nature. Lately, knowledge of the breaking process has significantly advanced, and this book is an attempt to summarise the facts into a consistent, albeit still incomplete, picture of the phenomenon.

In the book, a variety of definitions relating to wave breaking are discussed and formulated, and methods for breaking detection and measurement are examined. Experimental, observational, numerical, analytical and statistical approaches and their outcomes are reviewed. The present state of wave-breaking research and knowledge is analysed and the main outstanding problems are outlined.

Most attention is dedicated to the research of wave-breaking probability and severity, which lead to the second most important topic of the book – wave energy dissipation. Existing theories, measurements and applications of spectral dissipation due to wave breaking are reviewed. Non-breaking dissipation and non-dissipative effects of breaking are also discussed. Finally, roles of breaking in the atmospheric boundary layer, in air–sea interactions and in upper-ocean mixing, including extreme weather conditions, are analysed.

At the end of the book, many questions that are still unanswered are listed. To find the missing answers, the means are available, and in a way this is a matter for concentrated effort. Except, perhaps, for extreme weather conditions. We know that it is not only the waves and wave breaking, but also other processes in the atmosphere and upper ocean that experience changes in dynamic regime under hurricane-like wind force conditions, but robust instrumentation, observational techniques, physical concepts and analytical theories able to be applied to this are still under development.

The book is based on the review of a substantial amount of literature on wave-breaking-related topics which has accumulated over the years, as well as on my own contributions. It has been both interesting and instructive to see how the outcomes of studies conducted in the 1960s and even earlier are still relevant today, or are sometimes rediscovered.

I am most grateful to all of my colleagues over many years of collaborative wave-breaking research; material from our joint papers forms a significant part of this book. Among them, Yuri Soloviev, Ian Young, Mark Donelan and Dmitry Chalikov must receive a special mention. Their influence and contribution to my research and understanding of the phenomenon are hard to overestimate.

I am also thankful to partners from the wave-research community, particularly from The WISE Group, endless discussions with whom helped me to shape my current views on wave breaking and associated phenomena. It is hard, if not impossible to mention all the names over all those years, and here I will only acknowledge colleagues who directly assisted through discussions, advice, by providing their yet unpublished results, or in some other way in the course of preparation of this book: Fabrice Ardhuin, Sergei Badulin, Alexander Benilov, Roman Bortkovskii, Guilimette Cauliez, Luigi Cavaleri, Emanuel Coleho, Vladimir Dulov, Jean-Francois Filipot, Sebastien Fouques, Alina Galchenko, Andrey Ganopolski, Johannes Gemmrich, Isaac Ginis, Changlong Guan, Tetsu Hara, Tom Herbers, Alessandro Iafrati, Vladimir Irisov, Alastair Jenkins, Cristian Kharif, Jessica Kleiss, Harald Krogstad, Vladimir Kudryavtsev, Petar Liovic, Anne Karin Magnusson, Vladimir Makin, Richard Manasseh, Paul Martin, Jason McConochie, Ken Melville, Alexey Mironov, Nobuhito Mori, Jose Carlos Nieto-Borge, Miguel Onorato, Efim Pelinovskii, Will Phillips, Bill Plant, Andrey Pleskachevsky, Andrei Pushkarev, Fangli Qiao, Yevgenii Rastigejev, Torsten Retzlaff, Jim Richman, Erick Rogers, Aron Roland, Eugene Sharkov, Lev Shemer, Victor Shrira, Igor Shugan, Alex Soloviev, Marec Srokosz, Miky Stiassnie, Sergei Suslov, Jim Thomson, Pavel Tkalich, Alessandro Toffoli, Hendrik Tolman, Yulia Troitskaya, Karsten Trulsen, Anatoli Vakhguelt, Andre van der Westhuijsen, David Wang, Takuji Waseda.

The role of the editor of *Acta Physica Slovaca* Vladimir Buzek is also pivotal. It was his suggestion that I should write a review paper on wave breaking for his journal, which has culminated in this book on breaking, dissipation and related topics.

The last, but most significant, is my wife Anna and son Andrei whose encouragement and patience must be acknowledged. For many months, family nights and weekends have been replaced with a book on wave breaking, and without their support this book would never have been written. Anna also provided most valuable technical assistance with the text, figures, copyright and permission issues, and even with wave hunting in the search for those photos that are now included in the Introduction and on the book cover.

1
Introduction

Wind-generated waves are the most prominent feature of the ocean surface. As much as the oceans cover a major part of our planet, the waves cover all of the oceans. If there is any object in oceanography that does not need too much of a general introduction, it is the surface waves generated by the wind.

Being such a conspicuous entity, these waves, however, represent one of the most complex physical phenomena of nature. Three major processes are responsible for wave evolution in general, with many more whose significance varies depending on conditions (such as wave-bottom interaction which is only noticeable in shallow areas). The first process is energy and momentum input from the wind. The waves are generated by turbulent wind, and the turbulence is most important both for their initial creation and for subsequent growth (e.g. Miles', 1957; Miles, 1959, 1960; Phillips', 1957; Janssen, 1994, 2004; Belcher & Hunt, 1993; Belcher & Hunt, 1998; Kudryavtsev *et al.*, 2001, among many others). There is, however, no fixed theory of turbulence to begin with. Experimentalists have to deal with tiny turbulent fluctuations of air which are of the order of 10^{-5}–10^{-6} of the mean atmospheric pressure and which must be measured very close to the water surface, typically below the wave crests (e.g. Donelan *et al.*, 2005). The wind input process is quite slow and it takes hours of wind forcing (thousands of wave periods) and tens and hundreds of kilometres of fetch for waves to grow to a considerable height.

The second process is weak, resonant, nonlinear interactions within the wave system which can only be neglected for infinitesimal waves. For most of its existence, the wind–wave can be regarded as almost sinusoidal (i.e. linear), but its very weak mean nonlinearity (i.e. deviation of its shape from the sinusoid) is generally believed to define the wave's evolution. This is due to such waves, unlike linear sinusoids, exchanging energy when they cross-path. They cross-path because waves of different scales (i.e. different frequencies and wavelengths) propagate with different velocities, and also because waves tend to propagate at a range of angles with respect to the mean wind direction. Such weak interactions appear to be of principal importance. The longer (and higher) the waves are, the faster they move, and therefore the visibly dominant waves move with speeds close to the wind speed. This means that they virtually move in the still air, there is almost no wind for them. If they are still obviously wind generated, how does the wind produce such waves? The answer that is most commonly accepted now, is that the wind pumps energy mostly into shorter (high-frequency) and slowly moving waves which then transfer this energy across the continuous spectrum of waves of all scales towards longer (lower-frequency) components thus allowing those to grow – by means of nonlinear interactions. So, this small nonlinearity plays a large role in developing wind–waves as we know them. Analytically, to account for this sort of interaction the theoreticians have to solve relevant equations of hydrodynamics with accuracy down to expansion terms of the third order (e.g. Hasselmann, 1962; Zakharov, 1968; Hasselmann *et al.*, 1994; Badulin *et al.*, 2005). Experimentally, such interactions could not have been studied directly because of a great number of technical difficulties, one of which is the slowness of the process, thousands of wave periods being its time scale. Here, we would also like to mention that there are alternative approaches to explaining the evolution of long wind-generated waves.

The third most important process that drives wave evolution is the wave energy dissipation. Common experience tells us that wind-generated waves, no matter how strong the wind and how long its duration and wave fetch, do not grow beyond a certain limit. In the absence of mainland in the Southern Ocean, high continuous westerly winds are free to run the waves around the globe and thus provide conditions of unlimited wind–wave forcing and growth. Yet, the significant wave height (height of one third of the highest waves) rarely goes beyond 10 m. Individual waves of some 30 m are occasionally reported (e.g. Liu *et al.*, 2007), but these are very seldom and would certainly be the ultimate limit for wind-generated waves on the planet. Therefore, there is a process that controls the wave growth from above, and that is wave dissipation.

1.1 Wave breaking: the process that controls wave energy dissipation

There are a number of physical mechanisms in the oceanic and atmospheric boundary layers, other than breaking, that contribute to wave energy dissipation (e.g. Babanin, 2006; Ardhuin *et al.*, 2009a), but once wave breaking occurs it is the most significant sink for energy. In well-developed deep-water wind-forced waves, it is believed that breaking accounts for more than 80% of dissipation. Wave energy is proportional to the wave height squared, and therefore a sudden reduction of wave height during breaking by, for example, two times, signifies a four-times' reduction in energy. Obviously, provided there is a sufficient number of waves breaking, such a dissipation mechanism is much more efficient compared to viscosity, to the interaction of waves with winds, currents, background turbulence and to other ways of gradual decline. The energy lost to breaking is spent on injecting turbulence and bubbles under the ocean interface, emitting spray into the air, and thus wave breaking, and wind-generated waves in general, play a very significant role in negotiating the exchange of momentum, heat and gases between the atmosphere and the ocean.

Breaking happens very rapidly, it only lasts a fraction of the wave period (Bonmarin, 1989; Rapp & Melville, 1990; Babanin *et al.*, 2010a), but the wave may indeed lose more than half of its height (Liu & Babanin, 2004). Thus, the wave energy slowly accumulated under wind action and through nonlinear transfer over thousands of wave periods is suddenly released in the space of less than one period. Obviously, this process, the breaking-in-progress process of wave collapse, is a highly nonlinear mechanism of very rapid transfer of wave energy and momentum to other motions. So far, there are no adequate mathematical and physical descriptions of such a process.

Conceptually, however, the physics of wave collapse is completely different from the physics leading to breaking onset. While collapse is driven, to a greater extent, by gravity and inertia of the moving water mass and, to a lesser extent, by hydrodynamic forces, breaking onset occurs mostly due to the dynamics of wave motion in the water. Approaching breaking onset by a background wave is also very rapid, and also happens in the space of one wave period (e.g. Bonmarin, 1989; Babanin *et al.*, 2007a, 2009a, 2010a), but it should be considered separately from the following wave collapse. Essentially, the breaking process consists of two different sets of physics – one leading to breaking and another driving the wave breaking once it has started. These are not entirely disconnected, however, and the outcome of breaking collapse appears to 'remember' the 'input' that made a wave break. This will be discussed in more detail in Section 7.3.2.

The distinct difference between whitecapping dissipation and other processes involved in wave evolution is also determined by the fact that not every wave breaks whereas every wave experiences continuous energy input from the wind and continuous nonlinear energy exchange with other components of a continuous wave spectrum. A typical picture of a wavy surface under moderately strong wind conditions is shown in Figure 1.1. Waves of all scales, forming a continuous spectrum in terms of wave periods and lengths, exist simultaneously and run concurrently with different phase speeds, riding on top of each other or intercepting momentarily in different directions. All of these waves are subject to wind input and nonlinear exchange, but as is seen in Figure 1.1 just a small fraction of them break. Only under very strong winds does the rate of breaking crests reach 50% or more, but normally it is well below 10% (Babanin *et al.*, 2001). This means that, on average, it is every 20th or even every 50th wave that breaks, and this is sufficient to hold the energy balance in the wave system where every single wave gains energy one way or another. In the continuous time-space environment of a continuous wave spectrum and continuous physical processes, random breaking, which is intermittent in time and does not cover the surface uniformly, appears to control the equilibrium and ultimately wave growth. There is evidence that coverage of the ocean surface with breaking has a fractal dimension rather than being a two-dimensional surface (Zaslavskii & Sharkov, 1987), and this fact provides further mathematical complications if a description of this phenomenon is attempted by means of hydrodynamics or statistics.

It is important to mention at this stage that the three major processes, wind energy input, energy redistribution due to nonlinear interactions and energy dissipation, are closely coupled, affect each other and are equally important in wave evolution. Obviously, there would be no waves if they were not generated by the wind, but the wind input mechanism alone cannot explain evolution to any extent. As soon as the waves grow beyond the infinitesimal stage, nonlinear interactions begin to play an important role, and soon after, once individual steeper waves start to break, whitecapping dissipation assumes its responsibility as the balance holder.

Figure 1.1 Wind–wave pattern at a moderate wind. Waves of all scales are present simultaneously, representing a continuous wave spectrum. Only a small fraction of them are breaking at each scale. Gulf of St Vincent, Indian Ocean, August 2010

So-called whitecapping dissipation is the dissipation due to wave breaking, but it is not always that waves form whitecaps when they break (i.e. so-called micro-breaking discussed in Section 2.8 below). Since such a notion contradicts the general intuitive perception of wave breaking, we first have to answer a question: what do we call wave breaking?

1.2 Concept of wave breaking

Definitions pertaining to different physical and mathematical aspects of the wave breaking process will be formulated in Chapter 2. Here, we would like to discuss a common concept of breaking – that is what is a wave-breaking event and how is it generally perceived?

In Figure 1.2, a linear harmonic sinusoidal wave (sometimes called Airy wave), a Stokes wave, and an incipient breaker of the same height and length, i.e. all waves of the same average steepness, are compared graphically. This figure tests our ability to describe nonlinear behaviour of waves theoretically. The Stokes wave is a perturbation solution of hydrodynamic equations, assuming that the steepness of the waves is small. Obviously, although this traditional approach does produce a nonlinear wave shape, it does not look like anything close to a breaking wave as we perceive it (dash-dotted line). Needless to say the steepness of a breaking wave can hardly be expected to be small.

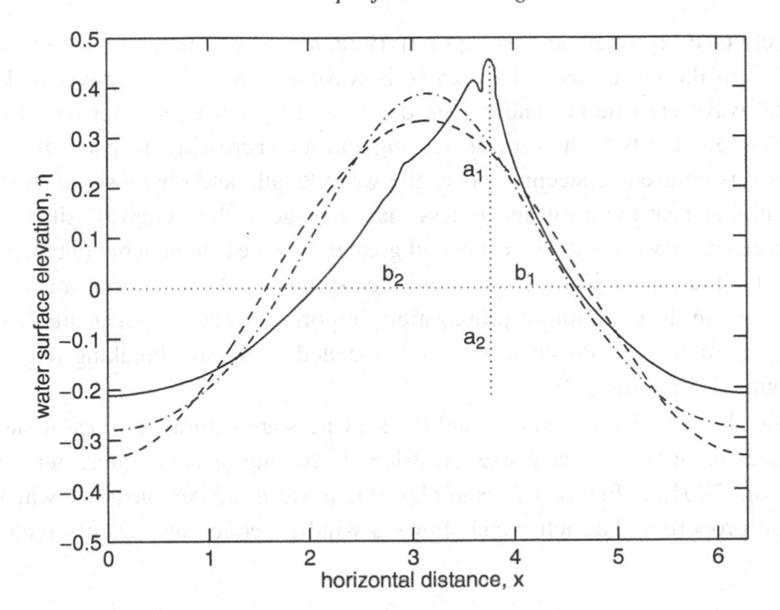

Figure 1.2 Geometric definition of wave skewness and asymmetry. The wave propagates from left to right. b_1 and b_2 are horizontal distances from the breaker crests to the zero-upcrossing and -downcrossing respectively. a_1 and a_2 are the breaker crest height and trough depth respectively. Solid line – numerically simulated incipient breaker (skewness is $S_k = 1.15$, asymmetry is $A_s = -0.51$). Dashed line – harmonic wave of the same wavelength and wave height ($S_k = A_s = 0$). Dash-dotted line – nonlinear wave of the same length and height obtained by means of perturbation theory ($S_k = 0.39$, $A_s = 0$). Dotted lines – mean (zero) water level (horizontal) and line drawn from the breaker crest down to the level of its trough (vertical). Figure is reproduced from Babanin *et al.* (2010a) with permission

An incipient breaker shown in Figure 1.2 (solid line) is produced numerically by means of the Chalikov–Sheinin model (hereinafter CS model (Chalikov & Sheinin, 1998; Chalikov & Sheinin, 2005)) which can simulate propagation of two-dimensional waves by means of solving nonlinear equations of hydrodynamics explicitly. The shape of such a wave is very asymmetric, with respect to both vertical and horizontal axes, and even visually the wave looks unstable.

Instability is a key word in the breaking process. The wave that we interpret as the incipient breaker in Figure 1.2 cannot keep propagating as it is: it will either relax back to a less steep, skewed and asymmetric shape, or collapse. We will define the steepness, skewness and asymmetry (with respect to the vertical axis) as

$$\epsilon = ak = \pi \frac{H}{\lambda}, \tag{1.1}$$

$$S_k = \frac{a_1}{a_2} - 1, \tag{1.2}$$

$$A_s = \frac{b_1}{b_2} - 1, \tag{1.3}$$

respectively (see Figure 1.2 and its caption). Here, a is wave amplitude, H is wave height ($a = H/2$ in the linear case), $k = 2\pi/\lambda$ is wavenumber and λ is wavelength, a_1 and a_2 are the wave crest height and trough depth, and b_1 and b_2 are horizontal distances from the breaker crests to the zero-upcrossing and -downcrossing, respectively. Thus, the steepness ϵ is an average steepness over the wave length, and obviously, local steepness is much higher near the crest and is less than average at the trough. Positive skewness $S_k > 0$ represents a wave with a crest height greater than the trough depth (a typical surface wave outside the capillary range), and negative asymmetry $A_s < 0$ corresponds to a wave tilted forward in the direction of propagation. Importantly here, experimentally observed negative asymmetry A_s has been broadly associated with wave breaking (e.g. Caulliez, 2002; Young & Babanin, 2006a).

Intrinsically, both the asymmetry and the skewness are natural features of steep deep-water waves regardless of their size, crest length, forcing or generation source (see e.g. Soares *et al.*, 2004). In Figure 1.3, examples of real waves are demonstrated which exhibit both these properties. The left panel shows a wind-generated and wind-forced wave of

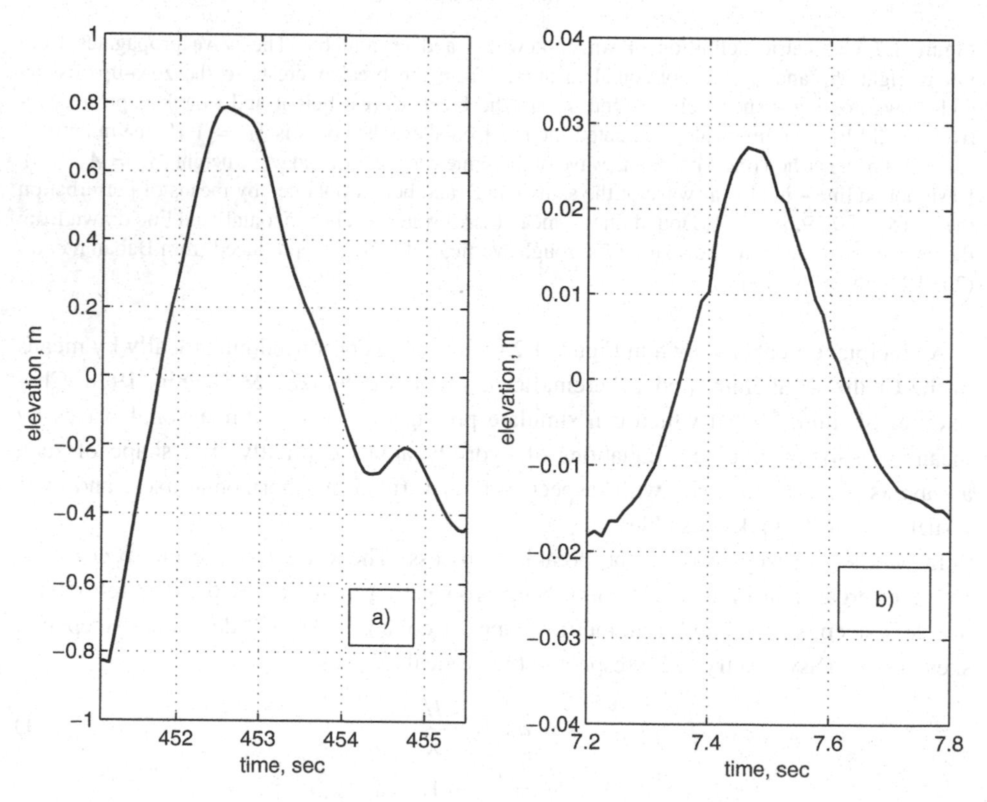

Figure 1.3 Real waves exhibiting both skewness and asymmetry. The waves propagate from right to left. a) Field wave measured in the Black Sea; b) laboratory (two-dimensional) wave measured in ASIST

height 1.6 m measured in a natural directional wave field in the Black Sea. The right panel shows a freely propagating (very small) two-dimensional wave of height 4 cm without wind forcing. This wave was mechanically generated in the Air Sea Interaction Salt water Tank (ASIST) of the University of Miami.

Once the skewness is non-zero and the amplitude a is not clearly determined, a definition of the wave steepness in terms of ak becomes ambiguous. Therefore, unless otherwise specified, the steepness will be expressed in terms of wave height $H = a_1 + a_2$ rather than wave amplitude a, as $\epsilon = Hk/2$. In these terms, steepness $\epsilon = 0.335$ of the wave shown in the figure far exceeds the limits of a perturbation analysis.

The dashed line in Figure 1.2 represents a steep sinusoidal wave ($S_k = A_s = 0$). Such a wave will immediately transform itself into a Stokes wave (e.g. Chalikov & Sheinin, 2005). The steep Stokes wave in the figure (dash-dotted line) is highly skewed ($S_k = 0.39$), but remains symmetric (i.e. $A_s = 0$). The incipient breaker in Figure 1.2 ($S_k = 1.15$, $A_s = -0.51$) was produced by the CS model, in a simulation which commenced with a monochromatic wave of $\epsilon = 0.25$. Such a wave profile looks visually realistic for a breaker and corresponds to, or even exceeds, experimental values of skewness and asymmetry for breaking waves previously observed (e.g. up to $A_s = -0.5$ instantaneously in Caulliez (2002) or $A_s = -0.2$ on average in Young & Babanin (2006a)). It is worth noting that the steepness of the individual wave has grown very significantly at the point of breaking: from $\epsilon = 0.25$ initially to $\epsilon = 0.335$.

When collapsing, the wave shape becomes singular at least at some points along the wave profile (i.e. space derivatives of the surface profile have discontinuities). This stage of wave subsistence is called breaking. Breaking of large waves produces a substantial amount of whitecapping, but smaller waves, the micro-breakers, do not generate whitecaps or bubbles and lose their energy directly to the turbulence.

Examples of various breaking and non-breaking waves are shown in Figures 1.4–1.8. Figures 1.4–1.5 show deep-water waves. The swell in Figure 1.4 are former wind–waves which have left the storm region where they were generated. They most closely conform to our intuitive concept of what the ideal wave should look like: uniform and long-crested, with crests marching parallel to each other. Their steepness is low and they do not break until they reach a shore.

Wind-forced waves hardly resemble this ideal picture. They look random and chaotic, they are multi-scale and directional, and they break. In Figure 1.5 a deep-water breaker is shown whose height is in excess of 20 m.

In Figure 1.6 (see also the cover image), waves approaching finite depths and, ultimately, the surf zone are pictured. In finite depths, waves break more frequently. Possible reasons are two-fold. Mainly, the waves break for the same inherent reason as in deep water, but they do it more often because the bottom-limited waves are steeper on average. Another fraction of waves break due to direct interaction with the bottom; this fraction grows as the waters become shallower (see Babanin *et al.*, 2001, for more details).

If deep-water waves enter very shallow environments, as shown in Figure 1.7, all of them will break and ultimately lose their entire energy to interaction with the bottom, to

Figure 1.4 Ocean swells are gently sloped former wind–waves propagated outside the storm area. They have low steepness and do not break. Image © iStockphoto.com/Jason Ganser

Figure 1.5 Large wave breaking in deep water. Distance from the mean water level to the lowest deck of the Fulmar Platform in the North Sea is 21 m. The photo is courtesy of George Forristal

Figure 1.6 Deep-water waves approach finite depths, grow steep and break. Hallett Cove, South Australia. Photo from Anna Babanina

Figure 1.7 Shallow water breaking. 100% of waves coming into the shallows break and lose all of their energy. This photo was purchased from http://www.istockphoto.com/

Figure 1.8 Wave ripples. Such waves are steep and break frequently, but do not form whitecapping. Port Phillip Bay, Victoria. Photo from Anna Babanina

sediment transport, to production of turbulence, bubbles and droplets, to mean currents and to generation of a small amount of waves reflected back into the ocean.

In the close-up picture in Figure 1.8, the waves can still be breaking even though it is now not possible to spot them visually. Short (in terms of wavelength) and small (in terms of wave height) ripples can nevertheless be quite steep. Such micro-breakers do not generate whitecapping, but demonstrate all the other singular surface features and irreversibly lose a significant part of their height and dissipate their energy (see e.g. Jessup *et al.*, 1997a).

At very strong wind-forcing conditions, wave-breaking behaviour is different yet again, and even the definition of breaking needs to be adjusted. As seen in Figure 1.9, taken from an aeroplane in Hurricane Isabel, the air–water interface is now smeared, the atmospheric boundary layer being full of droplets (spray) and the water-side boundary layer is filled

Figure 1.9 Aerial photo of the ocean surface at extreme wind forcing (Hurricane Isabel). The photo is courtesy of Will Drennan

with bubbles. The distinct interface is effectively replaced by a two-phase medium and the notion of wave shape and its singularity become vague.

While breaking due to inherent hydrodynamic reasons still takes place, the wind is now capable of instigating the process. Additionally, the wind blows away the steep wave crests. This latter event breaks the surface, creates surface singularities and reduces the wave energy, but cannot be treated as conventional breaking. Wave breaking and wave energy dissipation in such extreme conditions are poorly understood even in a phenomenological sense.

Thus, in order to avoid ambiguity, the wave-breaking process requires more specific definitions before further discussions are conducted. Therefore, the notion of the breaking onset, a classification of the wave-breaking phases, definitions of the breaking probability and breaking severity, types of breaking waves including the concept of micro-breaking, and breaking criteria will be discussed next in Chapter 2.

2

Definitions for wave breaking

Following the intuitively familiar concept of wave breaking discussed in Section 1.2, a variety of phenomenological definitions can be formulated. Wikipedia suggests such a definition, applicable across the range of wave processes, including water waves as well as electromagnetic waves, waves in plasmas and in other physical media:

"In physics, a breaking wave is a wave whose amplitude reaches a critical level at which some process can suddenly start to occur that causes large amounts of wave energy to be dissipated. At this point, simple physical models describing the dynamics of the wave will often become invalid, particularly those which assume linear behavior" (http://en.wikipedia.org/wiki/Wave_breaking).

The Glossary of Meteorology of the American Meteorological Society defines the breaking of ocean surface waves more specifically:

"A complex phenomenon in which the surface of the wave folds or rolls over and intersects itself. In the process it may mix (entrain) air into the water and generate turbulence. The causes of wave breaking are various, for example, through the wave steepening as it approaches a beach, through an interaction with other waves in deep water, or through the input of energy from the wind causing the wave to steepen and become unstable" (http://amsglossary.allenpress.com/glossary/browse?s=w&p=11).

As discussed above, a more explicit physical, yet only mathematical definition of the wave-breaking phenomenon is hardly possible. Wave breaking can occur due to a number of different causes which will result in different appearances of the wave breaker, different physics of wave energy dissipation in the course of breaking and different outcomes in terms of the impact on the wave field, on the subsurface water layer and even on the solid bottom in finite-depth environments. While breaking, a wave and the associated underwater motion go through a number of different stages, with different dynamics, different surface, acoustic, void-fraction, optical and other signatures. It is these stages that we will classify in this chapter which will help us to avoid ambiguity throughout the rest of the book. We will also define here the main quantitative characteristics used to describe the frequency of occurrence of wave breaking and the strength of breaking, and will describe breaking types and breaking criteria.

2.1 Breaking onset

Figure 1.2 demonstrates an incipient wave breaker modelled by means of a CS model (solid line). Visually and intuitively, the shape of the breaker appears quite realistic, and therefore it is instructive to review the model's definition of breaking onset. A numerical model cannot operate by means of phenomenological definitions, and obviously the inception of breaking had to be explicitly defined in mathematical terms.

In numerical simulations, a wave is regarded as breaking if the water surface becomes vertical at any point (Babanin *et al.*, 2007a, 2009a, 2010a). The criterion for terminating the model run was defined by the first appearance of a non-single value of surface in the interval $x = (0, L)$:

$$x(i+1) < x(i), \quad i = 1, 2, 3 \ldots, N-1, \tag{2.1}$$

where N is the number of points on the wave profile over its length L.

This definition is further illustrated in Figure 2.1, also simulated by means of the CS model. Here, development of a very steep harmonic wave with initial steepness $\epsilon = ak = 0.32$, is shown in terms of dimensionless time (in the horizontal scale, 2π corresponds to the wave period). This is effectively a rapidly developing breaker, as in the two-dimensional CS model such waves break within one period.

The model has obvious limitations in simulating the final stages of incipient breaking and was stopped when the water surface became vertical at any point. Strictly speaking, this geometrical property of the surface can be used as a physical definition of breaking onset. In numerical simulations it was noticed that local steepness can be very large, but the carrier wave can still recover to a non-breaking state. If, however, a negative slope appears locally, the wave never returns to a non-breaking scenario because the water volume intersecting the vertical line tends to collapse. That is, after the moment when criterion (2.1) has been reached, the solution never returns to stability: the volume of fluid crossing the vertical $x(i)$

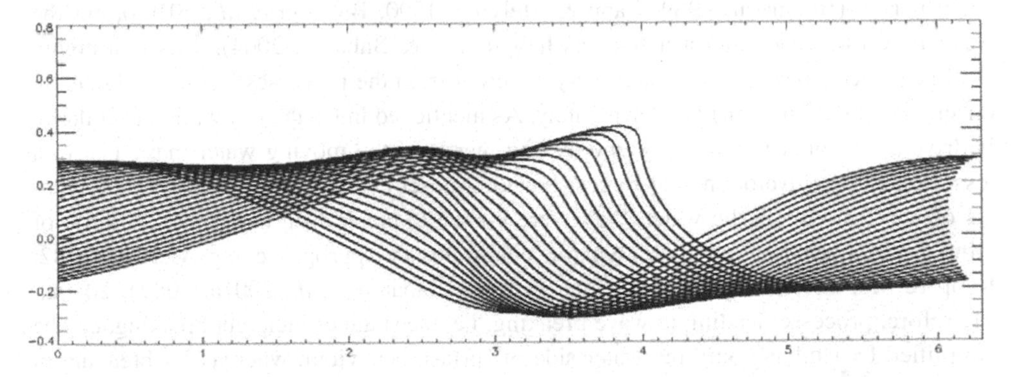

Figure 2.1 Numerical simulation of a steep wave evolving towards breaking, as predicted by the CS model. The wave propagates from left to right. Chalikov (2007, personal communication)

rapidly increases until breaking occurs. Apparently, the same considerations are applicable to physical waves too.

At present, the concepts of incipient breaking and breaking onset are poorly defined and even ambiguous. Traditionally, the initial phases of a breaker-in-progress are treated as incipient breaking. As an example, let us consider how 'near-breaking' was defined by Caulliez (2002). In this paper, surface elevations were recorded, differentiated, and the wave was regarded as a 'near-breaker' if its slope exceeded 0.586 any time between two subsequent zero-downcrossings. This criterion is an estimate of the highest slope that a Stokes wave can reach (Longuet-Higgins & Fox, 1977). But if this slope is exceeded, then the wave is not about to break – it is already breaking. This is not an incipient breaker, but represents breaking in progress. The features and physics of breaking-in-progress, however, may be very different to that of incipient breaking (Section 2.2). Thus, investigation of geometric, kinematic, dynamic and other properties of breaking-in-progress, such as whitecapping, void fraction, acoustic noise emitted by bubbles etc., will be of little assistance if we are seeking to understand the causes of breaking and breaking onset.

In this book, as in Babanin *et al.* (2007a, 2009a, 2010a), we suggest that breaking onset is defined as an instantaneous state of wave dynamics where a wave has already reached its limiting-stability state, but has not yet started the irreversible breaking process characterised by rapid dissipation of wave energy. That is, breaking onset is the ultimate point at which the wave dynamics caused by initial instabilities is still valid. This definition allows identification of the onset and, once the location of breaking onset can be predicted, allows measurement of the physical properties of such waves. The state of breaking onset is instantaneous, unlike the stages of incipient breaking and breaking in progress (see Sections 2.2 and 2.4). The latter can be further subdivided into stages with different properties and different dynamics (Rapp & Melville, 1990; Liu & Babanin, 2004).

2.2 Breaking in progress

Beyond the point of onset, breaking occurs very rapidly, lasting only a fraction of the wave period (Bonmarin, 1989; Rapp & Melville, 1990; Babanin *et al.*, 2010a), but the wave may lose more than a half of its height (Liu & Babanin, 2004). This is a highly nonlinear mechanism, conceptually very different from the processes leading to breaking onset, and should be considered separately. As mentioned in Chapter 1, while the collapse is driven, to a greater extent, by gravity and inertia of the moving water mass and, to a lesser extent, by hydrodynamic forces, breaking onset occurs mostly due to the dynamics of wave motion in the water. Waves are known to break even in the total absence of wind forcing, provided that hydrodynamic conditions are appropriate (e.g. Melville, 1982; Rapp & Melville, 1990; Brown & Jensen, 2001; Babanin *et al.*, 2007a, 2009a, 2010a). Therefore, processes leading to wave breaking, i.e. the stage of incipient breaking, can be simplified by studying only the water side of surface behaviour, whereas for breaking in progress, the air–sea interaction part, such as whitecapping (e.g. Guan *et al.*, 2007), void fraction (e.g. Gemmrich & Farmer, 2004), work against buoyancy forces (e.g. Melville *et al.*, 1992) and wind–wave momentum/energy exchange (e.g. Babanin *et al.*, 2007b) are of essential importance.

The pre- and post-breaking physics are not entirely disconnected, and the outcome of breaking collapse appears to 'remember' the 'input' which made the wave break. This will be discussed in more detail in Chapter 7. Here, we would like to emphasise that among wave-breaking definitions, breaking in progress needs to be considered separately and can be further subdivided into distinctly different phases.

For classification of wave-breaking phases, we will follow the logic suggested by Liu & Babanin (2004). They envisaged and quantified in terms of relative duration a single breaking event as passing through several distinct stages from both the external appearance of breaking and the underlying physics involved. These stages are incipient breaking, developing breaking, subsiding breaking and residual breaking. The first stage leads to breaking onset as described in Section 2.1. Developing breaking and subsiding breaking are different phases of breaking in progress. Residual breaking is a follow-up dynamic impact of the breaking event, rather than wave breaking as such and it will be discussed in Section 2.3.

Liu & Babanin (2004) aimed at testing the Liu (1993) breaking-detection approach based on the wavelet technique, by means of field data (see Section 3.7). Wavelet analysis indicated wave breaking if the downward acceleration, obtained from a surface-elevation series, exceeded a predetermined threshold value. The data were obtained under a variety of wind–wave conditions in deep water in the Black Sea and in a finite-depth environment in Lake George, Australia. Both data sets included synchronised time series of surface elevations and wave-breaking marks. Both had been extensively used to study breaking statistics for different wave spectra and in different environmental conditions, and therefore detailed results of the analyses were available for comparison (Babanin, 1995; Babanin & Soloviev, 1998a,b; Banner *et al.*, 2000; Babanin *et al.*, 2001).

In the Black Sea, wave breaking was detected and marked visually (Babanin, 1995), and in Lake George detection was conducted by acoustic means (Babanin *et al.*, 2001). Overall statistics of the number of breaking events and their frequency of occurrence matched the outcomes of the wavelet technique very well. Detailed examination of the results, however, indicated essential differences. The wavelet approach, along with the measurements, was generally successful in capturing breaking-wave events on many occasions, although on some other occasions one of them failed to detect a breaker while the other indicated that breaking had occurred. This is not unexpected as both measurements, i.e. detection of breaking by visual or acoustic means and the theoretical wavelet approach, should be anticipated as relevant to different phases of wave breaking.

At the incipient stage, the wave begins to find its continuous surface becoming difficult to sustain, so it is about to break, but is not breaking yet. An incipient breaker does not have whitecapping coverage as the breaking crest does not turn over. This is how Phillips *et al.* (2001) described development of the crest breaking:

"A single breaking event is generally initiated at some point on the wave crest and spreads laterally so that its average length is of order half its ultimate length, the width of the broken patch".

Before this broken patch starts developing, the wavelet method will already be detecting the breaking because the downward acceleration will exceed the threshold value while closing the breaking onset. The visual technique implemented at the Black Sea and the acoustic technique implemented at Lake George, however, will not detect a breaking. The two experimental techniques effectively make use of the occurrence of whitecapping as the basic measurement point and, since there is no whitecapping, i.e. no 'broken patch', they will fail to detect the wave crest as a breaking crest.

The developing stage is characterised by the lateral spread of breaking with a whitecapping appearance for the crest to pass over the measurement point, so a developing breaker should be readily detected using whitecap-oriented measurements. But the developing breaker also exhibits an increase in wave front steepness before it subsides. Rapp & Melville (1990) in their subsection 3.4 defined front steepness as the ratio of crest-to-front-zero-crossing height to crest-to-front-zero-crossing length and showed that this is larger, compared to the incipient breaking front steepness, for both spilling and plunging breakers. As shown by Liu & Babanin (2004), even though the front steepness is not unambiguously linked to the maximal instantaneous downward acceleration, this is an indication that the over-limiting acceleration values may persist through the developing stage, and thus the developing breaker will be detected by the wavelet method as well.

The relaxing or subsiding stage of breaking has not received as much attention in the literature as developing breaking. Therefore it is not quite clear, for example, what will happen to the breaking crest once it has reached its maximal length according to Phillips *et al.* (2001) or when the front steepness of breakers, described by Rapp & Melville (1990), will start to decrease. But at some stage it will start to decrease. For example, in his observations of plunging breakers (see Section 2.8 about breaking types), Bonmarin (1989) described the so-called 'splash phenomenon' due to interaction of the breaking-water jet with the surface in front:

> "The elevation of the splash of water can rise as high as the original plunging crest. When several successive splash-up cycles occur, gradual decrease of the potential energy from one cycle to the next is observed".

In Figure 2.2, a segment of a Black Sea record is plotted which shows surface elevations with wave breaking marked by visual observations of whitecaps (dots) and by means of the wavelet analysis (vertical bars). For the Black Sea waves shown, the mean front steepness was 0.045. The second and the third breakers picked up by the visual method have front steepness values of 0.052 and 0.075 which is greater than the mean steepness as one can intuitively expect for a breaking wave. The first breaker, however, which was detected visually because a whitecapping crest propagated past the measuring wave probe, has a front steepness of 0.011, well below mean wave steepness. Clearly, this broken wave, which still carries a whitecapping patch, is not expected to be detected by the wavelet method based on the acceleration criterion.

Figure 2.3 further illustrates the necessity for subdivision of the breaking process into the three phases. It shows the properties of individual waves in the range of frequencies

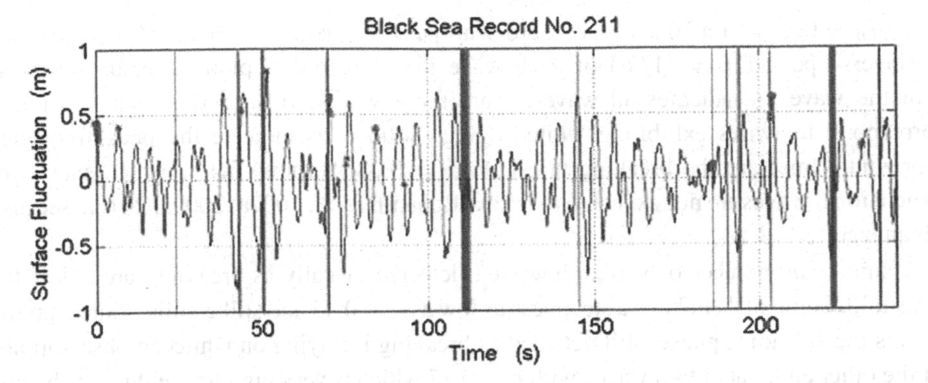

Figure 2.2 Segment of a Black Sea record. Surface elevations with wave breaking marked by visual observation of whitecaps (dots) and by means of the wavelet analysis (vertical bars). Figure is reproduced from Liu & Babanin (2004) (copyright of Copernicus Publications on behalf of the European Geosciences Union)

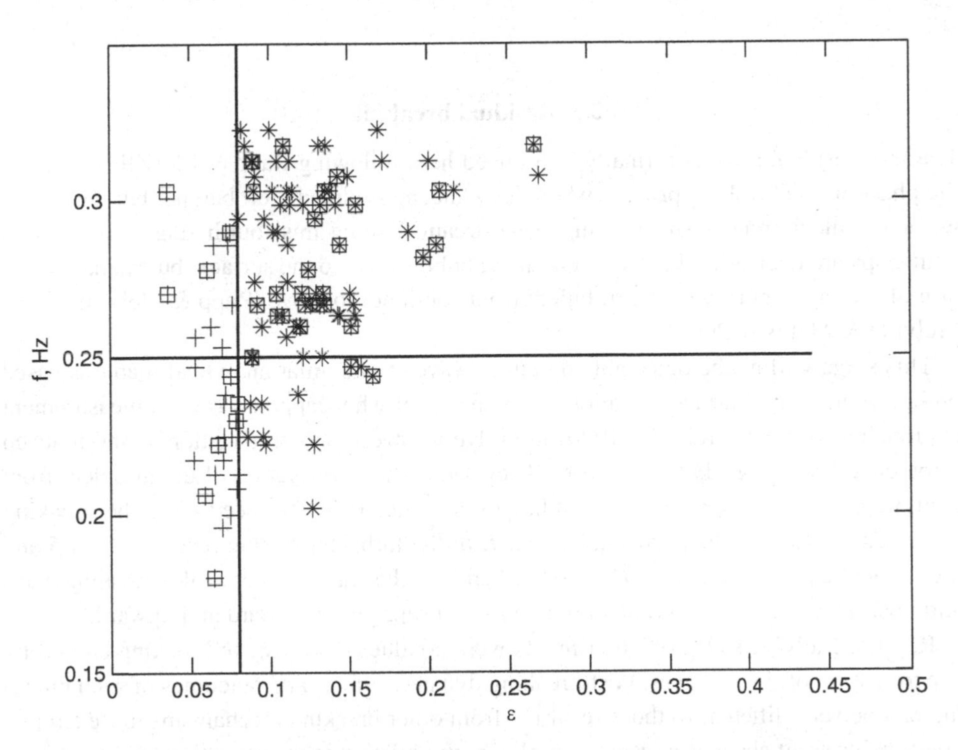

Figure 2.3 Properties of individual waves in the range of frequencies $f = f_p \pm 0.3 f_p$ of a Black Sea record with $f_p = 0.25$ Hz (Rec. 244, Table 5.1). Frequency (inverse period) f versus steepness ϵ. + – all waves, $*$ – those waves with $\epsilon > 0.08$, boxes – those waves exhibiting whitecapping. Solid lines show peak frequency $f_p = 0.25$ Hz (horizontal) and the breaking threshold $\epsilon = 0.08$ (vertical)

$f = f_p \pm 0.3 f_p$ of a Black Sea record with peak frequency $f_p = 0.25$ Hz. Frequency f (inverse period $f = 1/T$) of each wave in the record is plotted versus steepness ϵ of the wave; $+$ indicates all waves, $*$ are those waves with $\epsilon > 0.08$, squared boxes correspond to waves exhibiting whitecapping. Solid lines indicate the peak frequency (horizontal line) and the breaking threshold $\epsilon = 0.08$ (vertical) below which even two-dimensional waves are not expected to break (Babanin *et al.*, 2007a, 2009a, 2010a, see also Chapter 5).

A significant number of waves, however, detected visually as breaking, are below the threshold. Some of them have a steepness as low as $\epsilon = 0.03$ and still exhibit whitecapping. This is the subsiding phase, still detected as breaking if relying on whitecap observations. At the other end, out of two waves with $\epsilon \approx 0.27$ which is very high for field waves by any standards, one wave does not exhibit whitecaps and another does, that is, the first one is on its way up to limiting steepness (incipient breaker) and another is on its way down while collapsing (developing breaker). In Figure 8.6, out of 6347 individual waves considered 54 were in this high-steepness range ($\epsilon = 0.25$–0.3), half were breaking and half were not. These observations highlight uncertainties and ambiguities in existing definitions of breaking rates and the need for classification of the wave-breaking phases.

2.3 Residual breaking

The last, residual stage is formally introduced here following Rapp & Melville (1990) as the phase of the breaking process when the whitecap is already left behind, but the underwater turbulent front is still moving downstream. During this fourth stage of breaking, whitecaps are decreasing in size as entrained bubbles rise to the surface, but spatial evolution of mixing continues as the turbulent front continues to move (Rapp & Melville, 1990; Melville & Matusov, 2002).

This stage will not be detectable by either wavelet or similar analytical methods based on interpretation of surface elevation, or by means of whitecapping-oriented measurement approaches. Rapp & Melville (1990) used dye to investigate propagation of the induced turbulence following a breaking event. They found that propagation of the turbulent front continues for many wave periods, although at a much slower speed once the breaking wave has passed by. The horizontal extent L of the turbulent mixing reaches $kL \approx 5$ and the vertical extent D reaches $kD \approx -1$, where k is the wavenumber of the breaking wave and positive values of the turbulent plume mean propagation forward and upward.

Rapp & Melville's (1990) experiments were conducted with wave breaking caused by superposition of linear waves. Post-breaking dynamic effects and outcomes of such breaking can be very different to those resulting from other breaking mechanisms, for example, breaking brought about as a result of nonlinear modulation in the wave train or even superposition of nonlinear waves. Therefore, it is actually not clear whether the residual stage is a general feature of wave breaking and will persist in cases of breaking other than those caused by linear superposition.

The residual-breaking process should be important for upper-ocean mixing (see Section 9.2.1). As far as the wave field and boundary-layer interactions are concerned, however, the dynamic impact of the breaking event by this stage is already spent: the wave has already lost the associated momentum and energy to turbulence and mean currents, the air bubbles have been injected under the surface and the droplets suspended into the air, and flow-separation and other impacts on the atmospheric boundary layer are exhausted. The wave as such has already left the point of the underwater turbulent front and even if there is interaction of surface waves with this enhanced sub-surface turbulence, it is subsequent waves in the wave train that are now involved.

2.4 Classification of wave-breaking phases

Thus, following Liu & Babanin (2004) we classify the wave-breaking process into four stages: incipient breaking, developing breaking, subsiding breaking and residual breaking. As discussed in Section 2.2, we expect both whitecapping-based breaking measurements and analytical approaches based on limiting breaking-onset criteria to detect the same breaking events only at the developing stage of breaking phases. The incipient breakers, for example, will be detected by the wavelet method and will not be detected by measurements, and the subsiding breakers, on the other hand, will be detected by measurements whereas the wavelet method will fail to pick them up (Liu & Babanin, 2004).

The relative durations of the different breaking phases are not immediately clear, especially those of incipient and subsiding phases, as these are the least investigated. If the latter two phases are comparable in terms of duration, then the breaking count, i.e. total number of breakers detected on the basis of measurements and those obtained by the wavelet approach will be the same, even though there will be no 100% match of waves indicated as breakers. This is checked in Figure 2.4.

Two kinds of counts are shown here: total breaking counts are designated with open triangles and counts of perfect matchings are designated by triangles with a circle inscribed inside. The Lake George finite-depth cases are shown by triangles with vertex points up, and the Black Sea deep-water cases are shown with triangular vertex points down.

The straight line through points of total breaking counts in the figure is the one-to-one correlation line. The points follow the line with a 0.94 correlation coefficient and a 0.19 deviation index (DI) defined as

$$DI = \frac{1}{M} \sum \frac{|N_{\text{wavelet}} - N_{\text{measured}}|}{N_{\text{measured}}} \tag{2.2}$$

where M is the total number of records available and N_{measured} and N_{wavelet} are the breaking cases detected by measurements and by the wavelet method respectively.

Thus, the wavelet method effectively predicts the same breaking statistics as the measurements, based on single-point observations of whitecaps or related acoustic noise. This implies that the incipient and the subsiding phases of wave breaking have approximately the same relative duration.

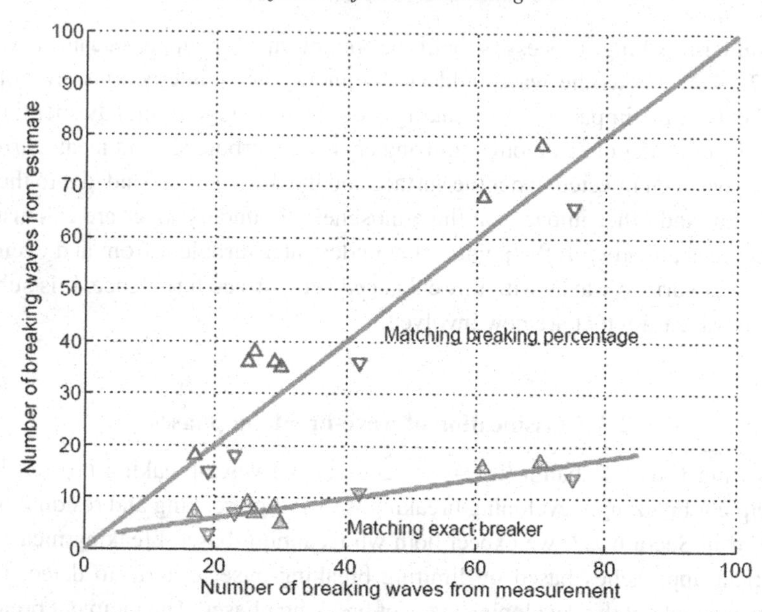

Figure 2.4 Comparison of breaking detection quantities obtained by means of the wavelet analysis (vertical scale) and observations of whitecaps (horizontal scale). The total breaking counts (by either of the methods) are designated with open triangles and counts of perfect matchings are designated by triangles with a circle inscribed inside. The Lake George finite-depth cases are shown by triangles with vertex points up, and the Black Sea deep-water cases are shown with triangular vertex points down. The two lines are the one-to-one correlation line (upper line) and the best linear fit to the exact match data points (lower line). The figure is reproduced from Liu & Babanin (2004) (copyright of Copernicus Publications on behalf of the European Geosciences Union)

The straight line through the triangles with circles is the best linear fit to the data. The data points are the number of perfect matchings we can really amass. In terms of percentages, the results range from a low of 16.7% to a high of 41.2% with an average of 28.4% of perfect matching cases. This outcome reveals the relative duration of the developing breaking phase where both analytical and observational methods are expected to detect the same events.

The slope of the lower line is nearly $\frac{1}{3}$ of the total-count curve. The total count in fact only includes two phases (i.e. incipient breaking plus developing breaking in the case of the wavelet method and developing breaking plus subsiding breaking in the case of whitecapping observations). This means that the second breaking stage lasts for approximately half the time compared to either incipient or subsiding phases. Thus, the developing phase is the shortest (fastest) stage and only accounts for some 20% of the wave breaking duration (and even less if the residual stage is included). Note that our present interpretation of Figure 2.4 is somewhat different to the conclusions reached by Liu & Babanin (2004).

As seen in Figure 2.4, there are small differences between wave-breaking detection techniques, and therefore the breaking phases in deep water (Black Sea) and finite depth (Lake George). Wave breaking in finite depths does exhibit some peculiarities which will be discussed in Chapter 5. These peculiarities, however, are not of a principal nature and here, quite likely, the minute differences observed are technical rather than physical.

Indeed, the wavelet analysis somewhat overestimates the total breaking percentage measured at Lake George and slightly underestimates that measured at the Black Sea compared to the one-to-one line. At Lake George, the acoustic method used only allowed detection of the dominant breakers in $\pm 30\%$ vicinity of the spectral peak (Babanin *et al.*, 2001), and therefore the wavelet method, whose performance is not limited by the spectral peak band, may have an extra number of breakers contributing to the total statistics. At the Black Sea, contrary to this, the breakers were detected visually and did not have an upper-frequency bound except that of the physical capability of the observer to actually see small whitecaps. Therefore, the wavelet detection method may have failed for small breakers whose profile was not sampled well enough but whose whitecapping was detected by the observer, and this could lead to underestimation of the deep-sea breaking rates. Details of the experimental techniques are described in Chapter 3.

Another possible contributor to the observed deep-water–finite-depth variation can be actual physical dissimilarity in relative durations of the breaking phases in the two environments. Such dissimilarity, if it exists, should be further investigated and cannot be addressed with certainty here. With some certainty, however, we can say that there appears to be no difference in duration of the developing-breaking phase between the deep water and finite depths. On the lower line, the Black Sea and Lake George points are scattered evenly which means that the second-phase duration does not depend on whether breaking occurs in deep or in finite-depth water.

The relative duration of the fourth, residual-breaking phase, with respect to the other three phases, is much longer. While the total duration of the first three breaking stages only accounts for a fraction of the wave period, the residual phase lasts for many wave periods (see Section 2.3). It is essentially a different, post-breaking process of dissipation of turbulence generated by the wave-breaking impact that occurs during the active phase of a breaking event.

Alternative classifications of the wave-breaking phases, with different physics applicable at the different phases, are also available in the literature. For plunging breaking, Bonmarin (1989) classified these phases as the overturning phenomenon, the splash-up phenomenon and degenerated forward flow. Essentially these phases are similar to the second, third and fourth (residual) phases introduced here. Even though Bonmarin (1989) also indicates his overturning phase as a development towards the breaking point, in terms of this chapter it is the breaking in progress, as the overturning phase is characterised by developing negative asymmetry A_s, which in our classification happens after breaking onset (see Section 5.1.1).

According to Bonmarin (1989), the characteristic timescale for the transition of the steep progressive wave to the breaking point is of the order of $T/10$ where T is the wave period.

In the context of the above estimates for the breaking-in-progress stage, this would mean that the entire breaking process (excluding the residual stage) lasts for half of a wave period.

Bonmarin (1989) also suggested the interesting analogy of plunging breaking with turbulent flow: the overturning phase would compare with laminar flow, the breaking as such to the transient stage and the afterbreaking to fully developed turbulence.

> "If this comparison is adopted,... the experimental procedures and the theoretical and numerical approaches have to be adapted to these three different phases".

We should point out that the water motion in the overturning crest is far from being laminar (e.g. Gemmrich, 2010), but overall the suggested analogy is quite attractive and promising.

2.5 Breaking probability (frequency of occurrence)

Breaking probability, or as it is often called breaking rates or frequency of breaking occurrence, is one of the most important statistical characteristics of wave fields that contain breaking events. Since wave breaking is the main sink of energy in such fields, but not every wave is breaking, the breaking probability, together with the breaking severity (Section 2.7), provide a means of statistical description of wave energy dissipation and other dynamic impacts caused by breaking.

In order to achieve this description, an understanding of the distribution of breaking probability, as well as breaking severity across the spectrum is needed. In other words, we need to be able to predict how frequently waves of different scales will break and how much energy they will lose in a breaking event. It is also most important to understand the physics that controls breaking rates and severity, and to quantify and parameterise the dependence of the breaking occurrence and severity on wave development and other environmental characteristics. These will be the main topics of Chapters 5 and 6; here we will suggest the main definitions.

Previous authors have used various characteristics and parameters to describe breaking statistics or probabilities. Wu (1979) provided a summary of whitecap coverage statistics based on field observations by Monahan (1971) and Toba & Chaen (1973) who analysed photographs of the water surface. Longuet-Higgins & Smith (1983) detected 'jumps' in the rate of change of the elevation signal related to the passage of breaking crests. Weissman *et al.* (1984) analysed changes in spectral energy in the 18–32 Hz frequency band to detect breaking events. In his Loch Ness measurements, Thorpe (1992) associated large time-derivatives of the signal strength in sonograph records with the occurrence of breaking wave crests. Ding & Farmer (1994) used an array of four hydrophones to detect and track breaking waves and thus determine their duration, velocity and spacing in terms of 'active acoustic coverage.' Gemmrich & Farmer (1999) detected breaking waves from a buoy on the basis of air entrainment within breaking waves measured by changes in the electrical conductivity of the water at fixed depths just below the surface.

More details on the measurement of breaking probability and statistics will be provided in Chapter 3. It is evident that thresholds underlying most of these methods are empirical and there is no standardisation amongst them. Besides, in terms of breaking rates some of the statistics appear ambiguous. The same whitecap coverage, for example, can be achieved at different frequencies of occurence of breaking events, depending on whether the events (waves) are large or small and whether the severity is strong or weak.

Here, following Babanin (1995), Banner *et al.* (2000) and Babanin *et al.* (2001), we define the breaking probability b_T for the dominant waves as the mean passage rate past a fixed point of dominant wave-breaking events per dominant wave period T_d or, in other words, the percentage of breaking crests n within a sequence of N wave crests:

$$b_T = \frac{nT_d}{t} = \frac{n}{N} \tag{2.3}$$

where $t = NT_d$ is the duration of the wave record. Another definition of breaking probability for dominant waves is also possible (2.36), and is in fact of greater practical significance, but it will be introduced and justified in Section 2.7 because its meaning cannot be fully appreciated before the issues of breaking strength are discussed.

The nondimensional quantity b_T is expressed in terms of the main temporal scale T_d of the wave field, making it convenient for comparisons and analysis. The definition, however, is not as simple as it first seems. Because of the continuous-spectrum nature of wind-generated waves, even the notion of dominant waves needs clarification. Otherwise, any spectral distribution of breaking rates defined this way will be ambiguous, i.e. $b_T(f)$ dependence on frequency f:

$$b_T(f) = \frac{n(f)T}{t} = \frac{n(f)}{N(f)}. \tag{2.4}$$

Indeed, counting crests of waves for a specific period $T = 1/f$ makes no physical sense as such a count will return a zero value.

Therefore, any spectral characteristic of breaking probability $b_T(f)$, including breaking of dominant waves with peak frequency f_p, from the very beginning implies the use of a spectral band $f \pm \Delta f$ such that wave crests with frequencies (periods) within this band, i.e. individual waves with frequencies of

$$f - \Delta f \le f \le f + \Delta f \tag{2.5}$$

can be counted. In Banner *et al.* (2000) and Babanin *et al.* (2001), the dominant waves were assumed to have frequencies within ±30% of the vicinity of the spectral peak, i.e.

$$f = f_p \pm \Delta f_p = f_p \pm 0.3 f_p, \tag{2.6}$$

following the width of the peak enhancement region of the JONSWAP formulation for the frequency spectrum $F(f)$ (Hasselmann *et al.*, 1973):

$$F(f) = \alpha g^2 (2\pi)^{-4} f^{-5} \exp\left[-\frac{5}{4}\left(\frac{f}{f_p}\right)^{-4}\right] \cdot \gamma^{\exp\left[-\frac{(f-f_p)^2}{2\sigma^2 f_p^2}\right]}. \tag{2.7}$$

Here, α is the level of equilibrium interval (tail) of the spectrum, g is the gravitational constant, γ is the peak enhancement factor, and σ is the width of the spectral peak.

In the original JONSWAP formulation, γ, the right width σ_{right} and the left width σ_{left} were chosen to be constant, but in reality α, σ and γ depend on the wave development stage and in general vary significantly (e.g. Donelan *et al.*, 1985; Babanin & Soloviev, 1998a).

Based on comparisons of the count for dominant breaking conducted by two independent acoustic techniques, the dominant frequency band was later reassigned as

$$\Delta f_p = \pm 0.35 f_p \tag{2.8}$$

(Manasseh *et al.*, 2006). Other frequency bands, such as $\Delta f = \pm 10\%$ are also often employed (e.g. Banner *et al.*, 2002; Babanin & Young, 2005; Manasseh *et al.*, 2006; Babanin *et al.*, 2007c). This narrower band is usually needed when breaking probabilities are defined for frequencies other than the spectral peak. Away from the peak, the spectral density decays very rapidly which would make a count of the crests and the relative importance of the breaking waves very uncertain if the spectral band were too broad.

Is there a physical meaning for the spectral band implied in the breaking-probability definition (2.3)–(2.5)? For the dominant waves, such a physical meaning can certainly be pointed out. That is, the width of the spectral peak defines the characteristics of the groups of dominant waves, and the wave-breaking frequency of occurrence depends on these groups.

Indeed, for a narrow-banded spectrum, which the spectral peak of wind-generated waves is, the width of the peak is related to modulational properties in the train of dominant waves. This relation was investigated statistically by Longuet-Higgins (1984). He suggested a nondimensional bandwidth parameter ν to describe such modulational properties:

$$\nu^2 = \frac{m_2 m_0}{m_1^2} - 1 \tag{2.9}$$

where m_i is the spectral moment of order i and in the general case is defined as

$$m_i = \int_0^\infty f^i F(f) df. \tag{2.10}$$

Longuet-Higgins (1984) found that integral limits in (2.10) have to be set to

$$\Delta f_p = \pm 0.5 f_p \tag{2.11}$$

in order to explain experimentally observed properties of wave groups. This limit is somewhat larger than the width of the JONSWAP peak enhancement or the $\Delta f = \pm 30$–35% in the breaking statistics of dominant waves, but it is reasonably close.

Once they exist, the groups of dominant waves have a verified association with breaking probability. Donelan *et al.* (1972) reported observing several consecutive waves breaking at the peak of the group envelope before sufficient energy was lost from the group. In another open-ocean study, Holthuijsen & Herbers (1986) found a significant influence of wave groups on wave breaking. When breaking occurred, the position of the first breaker in a group was slightly ahead of the centre of the group. They concluded that the overall fraction of breaking occurring within wave groups was close to 70%.

Support for the connection between frequency bandwidth, wave groups and breaking occurrence also comes from hydrodynamic, rather than just statistical studies of modulational instabilities of wave trains. Theoretical approaches, started by Zakharov (1966, 1967, 1968), Benjamin & Feir (1967) and Longuet-Higgins & Cokelet (1978) (see Yuen & Lake (1982) for an essential review, but progress on this topic is continuing), numerical models (e.g. Dold & Peregrine, 1986; Chalikov & Sheinin, 1998, 2005) and laboratory experiments (e.g. Babanin *et al.*, 2007b, 2009a, 2010a; Stiassnie *et al.*, 2007) all point out that conditions for the onset of breaking, i.e. a very steep individual wave, can result from the evolution of nonlinear wave groups. Initial conditions for such an evolution consist of steep monochromatic waves with sidebands, i.e. this also involves some characteristic bandwidth defined by the sidebands.

Interpretation of the applicability of this kind of modulational instability to field waves has been a subject for debate for quite some time. First of all, the field waves have a continuous spectrum and therefore the notion of primary waves and sidebands is uncertain: waves at every frequency can be treated as both primary waves and sideband disturbances. There have been attempts to draw analogies between the monochromatic wave trains with sidebands and spectral waves. It has been shown (Onorato *et al.*, 2001; Janssen, 2003) that the modulational properties in a spectral system of nonlinear waves depend on the ratio of wave steepness $\epsilon = ak_0$ to spectral bandwidth $\Delta\omega/\omega_0$, where k_0 and ω_0 are some characteristic wavenumber and radian frequency, respectively ($\omega = 2\pi f$), a is the mean amplitude at this wavenumber, $\Delta\omega$ is a characteristic width of the spectral peak, $\Delta\nu$ is a dimensionless characteristic depth, and N_m is the number of waves in the modulation:

$$M_I = \frac{\epsilon}{\Delta\omega/\omega_0} = \frac{\epsilon}{\Delta\nu} = \epsilon N_m. \tag{2.12}$$

This same ratio, if the properties of the primary wave and sidebands are used, was shown to be important in the original studies of instabilities of weakly modulated trains of monochromatic carrier waves of small amplitude. Here, we will denote this ratio as M_I (Modulational Index). The physical applicability and interpretation of the analogy is still a subject for discussion. Note, however, that even if applicable in the case of a continuous spectrum, definition (2.12) has a physical meaning for the modulational properties of dominant waves only, as there are no characteristic k_0, ω_0 and $\Delta\omega$ away from the spectral peak.

The second major uncertainty is the essentially two-dimensional nature of the hydrodynamic modulational instabilities discussed. The theories, numerical simulations and laboratory experiments mentioned above are two-dimensional or quasi-two-dimensional. Against direct extrapolation of the outcomes of these studies into field conditions were the known experimental and theoretical results on limitations which the nonlinear modulational mechanism has in broadband, and particularly in three-dimensional fields (e.g. Brown & Jensen, 2001; Onorato *et al.*, 2002, 2009a,b; Waseda *et al.*, 2009a). Eventually evidence appeared that showed that in directional wave fields it is active (Babanin *et al.*, 2010a, 2011a). This will be further discussed in Section 5.3.3.

Thus, a concrete physical mechanism (or a number of mechanisms) that connects the breaking probability of dominant waves with the bandwidth in the definitions (2.3) and (2.6) is not fully certain, but links between breaking rates and the wave groups, and therefore breaking rates and the spectral-peak bandwidth, are apparent. This is not the case, however, with the bandwidths of spectral bins other than the spectral peak in (2.4) and (2.5). The wave process at those scales is not narrow banded, therefore artificially imposed bandwidths have no apparent meaning in terms of wave groups and modulation of wave trains of respective scales. While the choice of bandwidth for higher-frequency parts of the spectrum $F(f)$ is usually done by analogy with the spectral-peak region, the physical analogy is not applicable in this case, therefore defining the breaking probabilities in the same terms appears to be a matter of convenience rather than physics (see Babanin & Young, 2005; Babanin *et al.*, 2007c, for a discussion of differences in the physics of breaking at the spectral peak and in the higher-frequency region). These issues will be discussed in more detail in Sections 5.3.1 and 5.3.2.

Another potential complication of definitions (2.3)–(2.5) must also be clearly stated. When breaking probabilities according to these definitions are estimated experimentally, the number of wave crests is counted. For each determined breaker, the frequency f (period T) of the wave is also extracted, by zero-crossing analysis (e.g. Manasseh *et al.*, 2006), the riding wave removal method (Schulz, 2009) or by some other means. Thus the total number of breaking waves $n(f)$ is found for each frequency. The total number of expected waves at a frequency is

$$N(f) = t/T = t \cdot f \tag{2.13}$$

and expressions (2.3)–(2.4) can then be used. As previously discussed in this section, in practice, in order to obtain $n(f)$ and $b_T(f)$, the wave frequencies are effectively discretised into bands $f \pm \Delta f$ (2.5) because $n(f)$ and $b_T(f)$ are not spectral densities but statistical quantities, and there are no exact matches between measured wave periods and a given $T = 1/f$.

It is clear, however, that if the waves of any given period $T = 1/f$ are counted by the zero-crossings or other means in a wave record of duration t, the resulting count $N_c(f)$ will be less than the nominal reference count $N(f)$ given by (2.13), because in real seas, waves of periods other than $1/f$ will occupy some part of the duration t. In terms of the breaking-probability definition, it would not matter if the ratios $N_c(f)/N(f)$ were constant across the frequency and therefore the crest counts were simply proportional to the reference count. This is, however, not the case.

Banner *et al.* (2002) demonstrated that if the bandwidth $\Delta f = \pm 0.3 f_c$ were chosen for a central frequency f_c in their experiment, $N_c(f)/N(f)$ was about 0.65 at the spectral peak ($f_c = f_p$) and gradually decreased for higher frequencies ($f_c > f_p$), asymptoting to 0.2 at $f_c/f_p > 2$. The ratio also depended on the choice of bandwidth, i.e. it would be different, for example, for $\Delta f = \pm 0.1 f_c$. Therefore, to avoid this additional uncertainty, the breaking probability b_T used in this book will be based on the reference count (2.13) rather than on an actual count of the total number of wave crests in each frequency bin, unless stated otherwise.

2.6 Dispersion relationship

The dispersion relationship, that is the relationship between the temporal (i.e. wave period/frequency) and spatial (i.e. wavelength/wavenumber) scales is a general property of surface water waves rather than a wave-breaking characteristic. It has to be introduced among definitions here, however, because it will frequently be mentioned and used throughout this book.

In a general case of a weakly nonlinear deep-water unidirectional modulated wave train, it can be written as

$$\omega^2 = gk\left(1 + \frac{1}{2}a^2k^2 + \frac{a_{xx}}{8k^2a}\right)^2 \tag{2.14}$$

(e.g. Yuen & Lake, 1982). In deriving (2.14), the steepness $\epsilon = ak$ (1.1) was assumed to be a small parameter, and therefore subsequent perturbation terms of higher orders of ϵ are not shown. The first term on the right characterises the linear frequency dispersion of surface waves, and the second term describes nonlinear correction to the dispersion due to finite amplitude. The last term in (2.14), where a_{xx} is the second derivative of wave amplitude a with respect to spatial coordinate x, comes from assuming that wave frequency, wavenumber and amplitude are slowly varying functions in space and time. It is only relevant for nonlinearly modulated wave trains and is usually omitted.

For wave trains of constant frequency, wavenumber and amplitude, the phase speed c is

$$c = \frac{\lambda}{T} = \frac{\omega}{k} = \sqrt{\frac{g}{k}\left(1 + \frac{1}{2}a^2k^2\right)}. \tag{2.15}$$

When waves can be treated as linear, that is their steepness (1.1) is small

$$\epsilon = ak \to 0, \tag{2.16}$$

then the linear dispersion relationship is simply

$$\omega^2 = gk \tag{2.17}$$

and the linear phase speed is

$$c = \sqrt{\frac{g}{k}} = \frac{g}{\omega}. \tag{2.18}$$

As seen in (2.18), even linear waves with different frequencies/wavenumbers propagate with different phase speeds. For nonlinear waves, phase speed (2.15) additionally depends on wave amplitude/steepness. These properties of surface waves are broadly used in wave-breaking experimental techniques to achieve frequency or amplitude focusing of waves of different scales at a particular point in space/time, in order to make these waves break.

The phase speed $c(\omega)$ of waves is different to the group velocity $c_g(\omega)$ which characterises the speed of propagation of wave energy. The difference is significant, and for linear waves in deep water the ratio is

$$c_g = \frac{d\omega}{dk} = \frac{1}{2}\sqrt{\frac{g}{k}} = \frac{1}{2}c = \frac{1}{2}\frac{g}{\omega}. \tag{2.19}$$

Group velocity is another most important characteristic to be employed in wave-breaking studies as it describes the speed of wave envelopes (e.g. Yefimov & Babanin, 1991). The dominant waves exist as wave groups whose elevation envelope decays away from the peak magnitude close to the centre of the group. Thus, the individual waves propagate through the group at a relative speed $\frac{1}{2}c$ and correspondingly change their height as they propagate. Close to the centre of the group their height/steepness is largest, and that is where they most frequently break.

2.7 Breaking severity

Breaking severity is another most important characteristic of wave breaking. By breaking severity, or breaking strength, we mean the amount of energy lost in an individual breaking event. Note that in the spectral wave environment the breaking of a wave of a certain scale can mean or involve loss of energy, or sometimes gain of energy by waves of other scales. Here, we will be interested in defining this property in such a way that, taken together with the breaking probability from Section 2.5, they will determine the wave energy dissipation rate. This way, if both the breaking probability and breaking severity can be measured experimentally, whitecapping dissipation can be estimated experimentally including the spectral distribution of the dissipation. So far, spectral dissipation functions employed in wave models remain, to a large extent, speculative formulations.

Note also that in the case of breaking severity a breaking event does not comprise a breaking wave only, but rather the wave group which the breaking waves are propagating through. Such detail is essential and will be discussed in the current section. As with the breaking probabilities in Section 2.5, however, there are further uncertainties even within such a simple and general definition. In the case of breaking severity, these uncertainties are even deeper due to the fact that this property has received far less attention, and the energy loss in the course of breaking is a much less studied process compared to the phenomena leading to breaking occurrence.

Difficulties in studying breaking severity are indeed serious. While hydrodynamic theories, even simplified versions of them such as those based on irrotational wave motion and assumptions of small initial amplitudes and disturbances, can be employed to investigate the probability of a breaking onset, they are clearly not applicable once the onset happens. On their own, no such theories can describe in detail the process of wave collapse and the associated energy loss in the generation of turbulence, entraining the air bubbles and pockets into the water and emitting spray into the air. In the absence of a theory, the numerical modelling of breaking strength suffers obvious limitations too. With no analytical theory to model, we have to simulate basic equations for two-phase flows, including water filled with bubbles and air filled with droplets. Computationally, these are very demanding models, but they have advanced very significantly over the past decade (see Section 7.2).

Thus, applications which involve breaking severity have to rely on experimental data. These are quite sparse and difficult to obtain as well, and the difficulties are many. Most importantly, in order to estimate the loss of energy, contact measurements are usually needed. That is, at the very least, the wave energy loss has to be estimated by direct measurement immediately before and immediately after a breaking event. This is a complicated exercise even in the laboratory because it involves a controlled breaking at a particular location, between the wave probes.

While possible in principle (e.g. Rapp & Melville, 1990; Pierson *et al.*, 1992; Meza *et al.*, 2000; Manasseh *et al.*, 2006; Babanin *et al.*, 2007a, 2009a, 2010a), the controlled-breaking approach has many limitations. First of all, controlled breaking is often achieved by artificial means, for instance through focusing many linear waves of different scales by using their different phase speeds. It is quite likely that this is not the way, or at least not the only way that breaking is controlled in general, and breaking severity in particular in natural environments (Babanin *et al.*, 2009a, 2010a). Secondly, keeping controlled breaking between two particular probes significantly limits the variety of wave properties and environmental dependencies that can be investigated. And finally, some processes leading to wave breaking, and in particular the most important mechanism of modulational instability of nonlinear waves trains, may be impaired in three-dimensional natural field circumstances, which fact further depletes the generality of conclusions achieved in two-dimensional wave tanks. See Chapter 6 for more details.

In the field, estimating the breaking strength by means of measurement of wave-breaking events before and after breaking is a challenge. Breaking is random, sporadic and infrequent, and the probability of it starting and ending right between two deployed wave gauges is very low. To achieve this, long measurement series would have to be taken using an extensive spatial array of wave probes, concurrent with detailed monitoring and recording of information on wave breaking, i.e. the start, the end and the direction of propagation of the breaking waves through the array. Depending on the instrumentation employed, a significant complication of contact *in situ* measurements is the destructive power of wave breaking which can damage the equipment, particularly as the most interesting breaking events are the most severe. Significantly promising in this regard, both in terms of measuring the waves losing their energy as they are breaking and in terms of being detached from the direct wave-breaking impact, is the stereo-imaging technique which allows us to video-record three-dimensional surface elevations over an area, with high resolution in time and space (Gallego *et al.*, 2008).

While many remote-sensing techniques have been developed to investigate wave breaking, these are mostly intent on detecting the breaking events and measuring breaking probabilities rather than the energy losses (see Section 3.6). In this regard, the bubble-detection method of Manasseh *et al.* (2006) is also significantly promising. The size of the bubbles formed in the course of breaking can be determined by passive acoustic means, i.e. by recording the noise produced by a breaking wave and quantifying the characteristics of this noise. Essentially, bubble size appears to be connected to the strength of breaking. Thus, the new acoustic technique, once it is fully validated and calibrated, has the potential to allow measurements of breaking severity by means of underwater hydrophones. The hydrophones are very small devices, easy to deploy, cheap and therefore cost-effective even if lost in a field experiment. Their power consumption is low which would allow long-term deployment and observations with a power supply based on batteries. Also, importantly, because the sound attenuation in water is relatively low, the hydrophones can be deployed well below the surface, even at the bottom in the case of finite-depth field sites, thus avoiding the destructive impact of wave-breaking events (see Babanin *et al.*, 2001; Young *et al.*, 2005; Manasseh *et al.*, 2006).

In the meantime, reliable and comprehensive understanding of breaking severity is the principal issue for all applications involving dissipation of wave energy, which thus range from engineering problems to general topics of air–sea interaction, extending as far as climate modelling. Indeed, in order to estimate the rate of energy loss from the wave field, whether it is then converted into energy of impact on engineering structures or into mixing of the upper ocean, it is necessary to know how frequently the waves break and how much energy they lose in a breaking event. If we denote the breaking severity, i.e. the mean energy loss due to breaking of waves of a particular scale f as $E_s(f)$, then the amount of energy dissipated per wave crest at this scale $D(f)$ is:

$$D(f) = b_T(f)E_s(f). \tag{2.20}$$

In terms of the spectral dissipation function S_{ds} employed in wave forecast models, the dissipation rates per unit of time are:

$$S_{ds}(f) = \frac{n(f)E_s(f)}{t} = \frac{n(f)}{N(f)}\frac{N(f)}{t}E_s(f) = \frac{D(f)}{T}. \tag{2.21}$$

This practical definition can be used to estimate the spectral distribution of dissipation by experimental means (Manasseh *et al.*, 2006; Babanin *et al.*, 2007c). It has, however, the same obvious issues with the spectral band (2.5), as does the practical definition of breaking probability in Section 2.5, because, again, $n(f)$ and $E_s(f)$ in the right-hand side are not spectral densities but are statistical quantities.

The wave energy per unit area is (e.g. Young, 1999)

$$E(f) = \frac{1}{8}\rho_w g H(f)^2 \tag{2.22}$$

where ρ_w is the density of water, and the normalised potential wave energy can be expressed simply in terms of the wave height $H(f)$

$$E(f) = H(f)^2. \tag{2.23}$$

Then, for generality and for parameterisation purposes, it is convenient to represent the severity in nondimensional terms as a fraction s of this energy which is lost to breaking:

$$E_s(f) = s(f)E(f) = s(f)H(f)^2, \tag{2.24}$$

that is

$$S_{ds}(f) = \frac{s(f)n(f)H(f)^2}{t}. \tag{2.25}$$

In linear waves, the depth-integrated kinetic energy $Q(f)$

$$Q(f) = \rho_w\big(u(f)^2 + w(f)^2\big) \tag{2.26}$$

equals the potential energy (2.22) (here, u and w are horizontal and vertical components of the surface orbital velocity respectively), and the total energy can easily be estimated. In strongly nonlinear waves, this is not necessarily true and therefore (2.22)–(2.25) and further expressions in terms of wave height H should be applied to wave trains measured before and after the rapid transient processes of breaking occur. This is how experiments intended for the measurement of breaking severity are usually designed (e.g. Rapp & Melville, 1990; Manasseh *et al.*, 2006; Babanin *et al.*, 2009a, 2010a), but to avoid ambiguity this should be explicitly mentioned.

The severity coefficient s, or simply breaking severity or breaking strength hereinafter, is often treated as a fixed fraction, for instance

$$s = 50\% \tag{2.27}$$

of the energy that a wave had before breaking (this is approximately the estimate that was obtained, for example, in laboratory experiments by Xu *et al.* (1986)). Such an estimate, however, is not general due to four main reasons, and the resulting deviations from this mean estimate can be enormous.

The first reason is the cause of the wave breaking, or in other words the means by which wave breaking was achieved in the laboratory. These means are many. For example, waves can be made to break using an artificial obstacle or a submerged shoal, and this is a robust practical way to investigate relevant wave-breaking properties and phenomena (e.g. Ramberg & Bartholomew, 1982; Manasseh *et al.*, 2006; Calabresea *et al.*, 2008). If natural deep-water processes leading to breaking are simulated, there is still a variety of possibilities: superposition of linear waves achieved through use of frequency dispersion (e.g. Longuet-Higgins, 1974; Rapp & Melville, 1990; Meza *et al.*, 2000, see (2.18)), superposition of nonlinear waves through amplitude dispersion (e.g. Donelan, 1978; Pierson *et al.*, 1992, (see 2.15)) and evolution of nonlinear wave groups (e.g. Melville, 1982; Babanin *et al.*, 2007a, 2009a, 2010a). The latter evolution to breaking may exhibit essentially different characteristics in directional wave fields (Onorato *et al.*, 2009a,b; Waseda *et al.*, 2009a; Babanin *et al.*, 2011a). In the case of wind forcing, wave breaking may be affected, and this effect is of principal significance as far as breaking strength is concerned (Babanin *et al.*, 2009a, 2010a; Galchenko *et al.*, 2010). Wave breaking and the severity of short spectral waves to a large extent are caused or at least impacted by the modulations brought about by longer waves (e.g. Donelan, 2001; Babanin & Young, 2005; Babanin *et al.*, 2007c). Breaking of laboratory waves can also be made random if the waves are, for example, wind-generated (e.g. Xu *et al.*, 1986; Hwang *et al.*, 1989), in which case the physical mechanism leading to the breaking is one or a combination of the above.

Whatever laboratory practice is chosen, breaking severity can now be estimated. Relevant differences between estimates due to different breaking mechanisms, however, appear to be of a very essential nature (Babanin *et al.*, 2009a, 2010a, see also Section 7.3.2) which puts the notion of a roughly constant breaking strength in serious doubt. How applicable this notion is in the field is another very important question. Field breaking is a combination of the various mechanisms, although some may prove less frequent and less significant; linear superposition is likely to be among the latter (Babanin *et al.*, 2009a, 2010a, see also Chapter 10). In any case, breaking severity seen in Figure 2.3, based on Black Sea measurements, is very different from 50%. In this figure, waves which display whitecapping, i.e. breaking waves, are shown squared for the spectral-peak frequency band. Their steepness ranges from $\epsilon = ak = 0.27$ down to $ak = 0.03$. If the ratio signifies a loss of wave height in the course of breaking, then it amounts to nine times and translates into energy loss

$$\frac{E_{\text{after}}}{E_{\text{before}}} = \frac{\epsilon^2_{\text{after}}}{\epsilon^2_{\text{before}}} = 0.012 \tag{2.28}$$

or, according to (2.23),

$$E_s = H^2_{\text{before}} - H^2_{\text{after}} = \frac{4}{k^2}\left(\epsilon^2_{\text{before}} - \epsilon^2_{\text{after}}\right) = 0.988 H^2_{\text{before}}. \tag{2.29}$$

If, however, at breaking onset the steepness were even higher (i.e. $ak = \frac{H}{2}k = 0.44$ as measured by Brown & Jensen (2001) for linear-superposition breaking and by Babanin *et al.* (2007a, 2010a) for modulational-instability breaking), then the energy loss is even greater:

$$E_s = H^2_{\text{before}} - H^2_{\text{after}} = 0.995 H^2_{\text{before}}. \tag{2.30}$$

Above, E_{before}, ϵ_{before}, H_{before} are respectively the wave energy, wave steepness and wave height immediately before breaking, and correspondingly E_{after}, ϵ_{after}, H_{after} immediately after breaking.

Estimates (2.28)–(2.30) should be treated with caution because the wavelength changes somewhat in the course of breaking (e.g. Babanin *et al.*, 2007a, 2009a, 2010a) and therefore the change of steepness ϵ in (2.28)–(2.30) depends on the change of wavenumber k too, not only on the reduction of the wave amplitude a. Besides, as mentioned above, the expression for total wave energy in terms of wave height (2.23) is strictly valid for linear waves and can be biased in the case of strongly nonlinear waves which the breakers are. Also, individual waves change their height as they propagate through the group even without breaking.

In any case, however, the breaking strength exhibited by an individual wave in the Black Sea is more like

$$s = 99\% \tag{2.31}$$

rather than 50% in the laboratory experiment (2.27). While seemingly extreme, result (2.31), however, is not unprecedented: for example, in his investigations Bonmarin (1989) gave examples of almost complete dissipation of potential energy of plunging breakers, "as if a wave crest had never existed at this location".

This discussion brings us to the second reason why the (2.27)-like constant-fraction estimate of severity coefficient s for a single breaking event should not be general. While in the laboratory wave energy (2.24) can be treated in terms of a single wave which is breaking, in the field the change of wave height of an individual wave is an ambiguous property. As was mentioned in Section 2.6, the dominant waves exist as wave groups, and individual waves propagate through the group at a relative speed $\frac{1}{2}c$ and correspondingly change their height as they propagate.

Dominant waves usually break close to the peak of the group envelope (e.g. Donelan *et al.*, 1972; Holthuijsen & Herbers, 1986; Babanin, 1995). This means that once a wave has started breaking at the top of a group, its height is decreasing due to both the breaking and its moving towards the front face of the group. The latter reduction would occur regardless of whether the wave was breaking or not.

Since the active phase of wave breaking lasts a fraction of a wave period (e.g. Rapp & Melville, 1990), and the wave group typically consists of 7–10 waves (e.g. Longuet-Higgins, 1984), then a wave propagating at relative speed $\frac{1}{2}c$ will start and finish breaking within the group. It is logical and more accurate, therefore, to estimate the breaking strength in terms of group breaking severity s_g, by integrating the wave energy loss over the group:

$$E_{sg} = s_g \frac{8}{\lambda_g} \int_0^{\lambda_g} \eta(x)^2 dx. \tag{2.32}$$

Here, E_{sg} is the energy loss from the entire group, λ_g is the length of the group, and the surface elevations $\eta(x)$ are integrated along a distance x within the group before the breaking started. The factor of 8 is included in order to have wave energy expressed in terms of wave height H, rather than amplitude a, for consistency with definitions (2.23)–(2.30). Note that there is no frequency scale f explicitly mentioned here, because the apparent wave grouping is a property of dominant waves only. Therefore, in the spectral sense, definition (2.32) is applicable to the frequency band of the spectral peak which according to (2.5)–(2.8) and (2.11) is in the range $\Delta f_p = \pm(0.3{-}0.5) f_p$.

With such a definition, the whitecapping dissipation function should be estimated as

$$S_{ds} = \frac{n_{\text{group}} E_{sg}}{t}, \tag{2.33}$$

where n_{group} means the number of wave groups where breaking has occurred. Values of severity s_g estimated this way will essentially be lower than the severity measured for individual waves because the energy loss effect is smeared over all waves in the group.

In the case of a single breaker, they can be approximately related to severity s through the number n_g of waves in the group:

$$s = n_g s_g. \tag{2.34}$$

The number n_{group} is different to the number n of breaking waves in (2.4) and (2.21) because typically there is more than one individual wave breaking within a wave group in close proximity in time and space (e.g. Donelan *et al.*, 1972; Babanin *et al.*, 2009a, 2010a; Galchenko *et al.*, 2010, among many others). As has already been mentioned, field observations by Donelan *et al.* (1972) revealed several consecutive waves breaking at the peak of the group envelope. This was interpreted in a quasi-linear sense: that is, subsequent waves propagate with their phase speeds within the group according to (2.19), exceed a limiting height/steepness as they approach the peak of the group and keep breaking until sufficient energy is lost from the group. Babanin *et al.* (2009a, 2010a), in a laboratory experiment with two-dimensional wave breaking due to modulational instability, found that the number of one-after-another breaking waves in a wave group can be one or more, depending also on wind forcing and breaking severity. They showed that subsequently breaking waves are strongly coupled if the breaking strength is large enough.

Therefore, subsequent breaking of a number of waves on the top of the wave group is effectively a single breaking event in the dynamics of groups. Although different individual waves are breaking, they cannot be separated in order to measure the wave-group energy loss (2.32) for each of the breakers. Thus, a typical scenario for a breaking event is

$$\begin{aligned} n_{\text{group}} &= 1, \\ n &\geq 1. \end{aligned} \tag{2.35}$$

In this regard, it is helpful to introduce a group-breaking probability b_{Tg} in terms of the relative number of wave groups containing one or more breaking waves coupled in such a way that they can be treated as a single breaking event:

$$b_{Tg} = \frac{n_{\text{group}} T_g}{t} = \frac{n_{\text{group}}}{N_{\text{group}}} \tag{2.36}$$

where T_g is the period of the wave group and N_{group} is the total number of wave groups in the wave record.

These issues of individual waves breaking within the group can be illustrated by means of Figures 6.2 and 6.3 in Chapter 6. Starting from Figure 6.3, we can see what happens with waves propagating through the group without breaking or with only gentle breaking. The time series are compared immediately before (solid line) and immediately after breaking (dashed line). Breaking of the four incipient breakers (highest waves in each solid-lined wave group) started and ended between the two probes. The breaking was very gentle when observed visually and the wave following this gentle incipient breaker did not break. The time necessary to travel the distance between the two probes was estimated as 1.04 s and therefore the dash-lined record was shifted back by 1.04 s in order to superpose the two wave trains.

As mentioned above, the individual waves and the group propagate with different speeds. In the case of deep-water waves of 1.5 Hz frequency shown in Figure 6.3, and in a close-to-linear scenario, according to (2.19) the relative speed of wave propagation within the group is $c_{relative} = 0.52\,\mathrm{m/s}$ and the relative position of a wave over 1.04 s will be shifted by 0.5 m. This is comparable with one wavelength of such waves, $\lambda = 0.69\,\mathrm{m}$. That is, in the absence of breaking, each wave would move approximately one position ahead, and the highest wave in the solid-lined group would become the second highest (in front of it) in the dash-lined group. Because of this, the height of the highest wave would be significantly reduced without any breaking.

Since breaking does occur, the pattern is more complicated. The breaking was quite gentle visually, and indeed its impact on the individual wave that broke (now the second wave in each dash-lined group) is not that large. It is quite significant, however, over the entire group because of, obviously, nonlinear coupling between different waves in the course of the breaking event. Other details of such a breaking impact will be discussed in Chapter 6; here we would like to highlight the fact discussed above: that is, in the case of wave breaking within a wave-group structure, estimating the breaking strength in terms of measurements of the breaking wave only is ambiguous and even misleading.

This is even more valid for strong breaking, when two or more subsequent waves break one after another. This is demonstrated in Figure 6.2. Again, wave series immediately before and after the breaking are compared. Breaking of the three incipient breakers seen in the solid-lined wave train happened (started and finished) between the two probes. The wave that is seen following the incipient breaker at the second probe also broke between the two probes. These breakings happened in a period of 1.2 s, the time required by the waves to travel the distance between the probes. Therefore, the dash-lined record is time-shifted by 1.2 s in an attempt to superimpose the two wave trains.

Now that a strong breaking has occurred, the correlation between the two time series is quite poor. The incipient breaker and wave following it practically disappeared, as well as the entire modulation. The number of waves in the segment has also changed. This is a collapse of the wave group which from this point in time and space should restart its evolution in hydrodynamic terms, from the new initial conditions. Again, further details will be examined in Chapter 6, but it is apparent that the severity can only be estimated in terms of the wave group (2.32) rather than in terms of individual waves.

The third reason for (2.27)-like constant-fraction estimates of the severity coefficient s being not general is the influence of various environmental conditions on real wave fields. Ocean waves are directional (i.e. three-dimensional), wind-forced and experience many sorts of interactions with background currents, turbulence and other phenomena that may affect the breaking strength, with or without alterations of the breaking probability. For example, Babanin *et al.* (2009a, 2010a) demonstrated that the presence of wind increases the breaking probability due to modulational instability of nonlinear wave groups, but decreases the breaking severity. This reduction extends to the severity becoming virtually zero at reasonably strong wind forcing, which effectively means that the wind may essentially impair or even erase breaking caused by the evolution of wave groups.

Thus, both breaking probability and breaking severity become functions of the wind forcing, with opposite trends, and this fact will affect breaking dissipation (2.25) and (2.33) in a way which is hard to predict at this stage:

$$\begin{aligned} b_T &= b_T(U/c), \\ s &= s(U/c). \end{aligned} \tag{2.37}$$

Here, U is some characteristic wind speed, and therefore the ratio U/c describes the wind forcing, that is, how fast the wind is with respect to the phase speed of the waves c. In any case, the notion of roughly constant breaking severity is absolutely inapplicable in such a scenario.

The fourth reason to be highlighted is the spectral nature of breaking severity in real wave fields. Except for the case of pure swell which does not break anyway, ocean waves have a continuous spectrum extending to higher and lower frequencies, and often with multiple peaks. In the realistic spectral environment, it is quite rare that a physical process is limited to a particular frequency or wavelength scale and does not have an impact across the spectrum. This is certainly true with respect to such strongly nonlinear processes as those that determine breaking severity.

An example of the spectral distribution of breaking severity is demonstrated in Figure 6.1. In Figure 6.1a, spectra of the time series of Figure 6.2 are plotted: the solid line is the pre-breaking spectrum and the dashed line is the after-breaking spectrum. Figure 6.1b is the ratio of the two spectra.

Even though the laboratory waves were generated monochromatically and sinusoidally, they are now strongly nonlinear and the first, second and third harmonics are clearly visible in their spectra, indicated by vertical solid lines (again, solid lines correspond to the pre-breaking state and dashed lines are close to the after-breaking state). The breaking severity effect (the ratio in Figure 6.1b) is distinctly spectral. While it is the main wave that is breaking (around a frequency of 2 Hz at pre-breaking), the energy is lost from all the harmonics as well. In both absolute and relative terms, most of the loss has come from the peak which was reduced by a factor of 5. With the exception of the second harmonic (reduced and shifted to 3.6 Hz), the other harmonics have almost completely disappeared. For frequencies above the spectral peak, the average ratio is approximately 1.7. Across the entire range of relevant frequencies, from the peak up to 11 Hz, the average ratio is 1.8 which translates into a total spectral loss of $s_{\text{spectral}} = 45\%$. Here, spectral severity is defined as

$$E_{s_{\text{spectral}}} = 16\int_0^\infty F_{\text{before}}(f)df - 16\int_0^\infty F_{\text{after}}(f)df = s_{\text{spectral}}16\int_0^\infty F_{\text{before}}(f)df \tag{2.38}$$

where $E_{s_{\text{spectral}}}$ is the energy loss across the entire spectrum and $F_{\text{before}}(f)$ and $F_{\text{after}}(f)$ are wave frequency spectra before and after the breaking event. The factor of 16 is introduced for consistency, as, by definition the significant wave height H_s is

$$H_s = 4\sqrt{m_0} = 4\sqrt{\int_0^\infty F(f)df}. \tag{2.39}$$

This way, the energy in (2.38) is expressed in terms of wave height, as in (2.23)–(2.32).

In the case of a spectrum being estimated over a single wave group, or in circumstances where breaking occurs in every wave group as in dedicated laboratory experiments (e.g. Babanin *et al.*, 2009a, 2010a),

$$s_{\text{spectral}} = s_{\text{group}}. \tag{2.40}$$

Otherwise,

$$s_{\text{spectral}} = \frac{n_{\text{group}}}{N_{\text{group}}} s_{\text{group}}. \tag{2.41}$$

In the case of field waves, because wave groups at scales other than the spectral peak cannot be identified, definitions (2.40)–(2.41) are only relevant for the breaking of dominant waves.

Apart from the mere energy loss, downshift of the spectral energy also occurred in Figure 6.1, and the peak frequency moved from 2 Hz to less than 1.8 Hz. Both the spectral downshift in the course of breaking (e.g. Tulin & Waseda, 1999) and the spectral distribution of the energy loss due to a breaking (e.g. Pierson *et al.*, 1992; Meza *et al.*, 2000) have been reported in a number of other studies. The latter two papers, however, provide an account on the spectral pattern of breaking severity which is quite different both to Figure 6.1 (i.e. Babanin *et al.*, 2009a, 2010a) and to each other.

Meza *et al.* (2000), who studied the dissipation of energy of laboratory two-dimensional waves by means of frequency dispersion, found that the energy is lost almost entirely from the higher frequencies, whereas the spectral peak remained unchanged after breaking. Pierson *et al.* (1992), who stimulated the laboratory breaking through amplitude dispersion, obtained the opposite result: that is, most of the energy is lost from the primary wave. This main conclusion is similar to that of Babanin *et al.* (2009a, 2010a), but the spectral breaking impact in their experiment is still different. In the experiments by Pierson *et al.* (1992), while the dominant wave loses energy, some components of the spectrum actually gain energy. This outcome is quite physical because breaking is known to be associated with the generation of short waves, whether by means of the production of parasitic capillary waves on the front face of the breaker (e.g. Crapper, 1970; Ebuchi *et al.*, 1987) or because of a plunging jet impacting the surface (e.g. Hwang, 2007).

This brings us to another issue relevant to breaking severity in a spectral environment. It is well known that field waves do not necessarily break at the spectral peak (dominant breaking), but in fact breaking occurs frequently or even more frequently at smaller scales (e.g. Babanin, 1995; Banner *et al.*, 2002; Melville & Matusov, 2002; Babanin & Young, 2005; Hwang, 2007; Mironov & Dulov, 2008; Babanin *et al.*, 2007c). Here, by smaller scales we mean frequencies/wavenumbers of the spectrum tail, away from the spectral peak, that is, waves that are relatively shorter than the dominant waves rather than just short waves in terms of some dimensional limit of length. As discussed above, such waves do not form groups, at least not in the narrow-banded wave-group sense. Therefore, the problem of breaking severity distributed across the group is not relevant for them.

Instead, other issues which are non-existent in the case of dominant breaking become important.

These waves can potentially be breaking because of two reasons. First of all, one would expect them to break due to inherent physical processes that lead surface waves to break in any scenario. Secondly, their breaking is induced by the longer waves. This is either the straining action of longer waves (Longuet-Higgins & Stewart, 1960; Phillips, 1963; Donelan, 2001; Donelan *et al.*, 2010) or forced breaking (Babanin & Young, 2005; Young & Babanin, 2006a). The straining action takes place because of the modulation of short wave trains by the underlying large waves which causes the short-wave steepness to increase at the forward faces of longer waves, resulting in their frequent breaking. Forced breaking is the breaking of short waves in a wake or on the top of large-wave breaking triggered by the large breaker. One way or another, but similarly to the subsequent breaking of dominant waves on top of their groups described above, sequential or simultaneous breaking of short waves is expected in this case on top of longer waves. This sequence is linked to the phase of the longer waves (i.e. it happens on the front face, at the top or in the wake of the long wave). By analysing this phase link, Filipot (2010, private communication) concluded that the induced breaking dominates at frequencies $f > 3f_p$.

If these short breakers (wavelengths λ_{short}, frequencies f_{short}) are also coupled in a nonlinear sense, then, similarly to definition (2.32) for groups of dominant waves, the short-breaker severity s_{short} needs to be defined in terms of a value averaged over the period of the dominant wave:

$$E_{s_{\text{short}}} = s_{\text{short}} \frac{8}{\lambda_{\text{long}}} \int_0^{\lambda_{\text{long}}} \eta_{\text{short}}(x)^2 dx. \tag{2.42}$$

Here, $E_{s_{\text{short}}}(f)$ is the energy lost due to the breaking of short waves, λ_{long} is the length of the modulating underlying wave, and $\eta_{\text{short}}(x)$ are surface elevations just before the breaking started, but with the low-frequency/wavenumber oscillations of the mean surface ($\lambda > \lambda_{\text{short}}, f < f_{\text{short}}$) filtered out. Wavelength λ_{long} is also not a constant value of, for example, dominant waves with frequency f_p. There has to be

$$\begin{aligned} \lambda_{\text{long}}(f_{\text{long}}) &\gg \lambda_{\text{short}}(f_{\text{short}}), \\ f_{\text{long}} &\ll f_{\text{short}}, \end{aligned} \tag{2.43}$$

but other than that λ_{long} and f_{long} can take on any scale.

We should note that the severity of breaking of short waves may depend on their scale f_{short} and therefore $E_{s_{\text{short}}}$ can be treated as a spectral function $E_{s_{\text{short}}}(f)$. Like the breaking probability of Section 2.5, in practical terms frequency f implies some spectral band $f \pm \Delta f$ (2.5) such that surface elevations $\eta_{\text{short}}(x)$ in (2.43) are bandpassed into this band.

The outcome of induced breaking is the so-called cumulative effect (Babanin & Young, 2005; Young & Babanin, 2006a; Babanin *et al.*, 2007c, see also Section 7.3.4), accumulation of breaking energy losses at smaller scales which are coupled with the behaviour of longer waves. Further from the spectral peak, the cumulative energy loss grows and tends to dominate over the inherent breaking severity. This is a purely spectral effect which is always present regardless of whether or not the dominant waves break.

Therefore, as far as the spectral impact of the breaking severity is concerned (i.e. spectral distribution of the energy loss in a single breaking event), there appears to be a broad range of possibilities, uncertainties and ambiguities. Depending on the physics leading up to the wave breaking, energy may or may not be lost from the primary breaking wave. In all scenarios, energy loss is endured across the entire spectrum, certainly at scales smaller than that of the primary breaker. These smaller scales may be harmonics of the primary wave or free waves (e.g. Meza *et al.*, 2000; Young & Babanin, 2006a), but in any case their breaking occurrence and breaking strength are strongly coupled with the breaking and severity of the primary waves. Additionally, the wave energy is downshifted as a result of the breaking, at least if the breaking is strong enough (Babanin *et al.*, 2009a, 2010a). This means that part of the energy, that which is lost from the primary breaking scale, is not in fact removed from the wave system. Finally, some scales in the wave spectrum, distant from the scale of the primary breaker, may actually acquire energy as a result of breaking. These scales are both below and above the scale of the breaker (Pierson *et al.*, 1992). Overall, the (2.27)-like notion of a fixed or even a reasonably approximate average value for the severity s is a gross simplification or is simply inapplicable in a spectral environment.

In summary, we have provided four definitions for breaking severity (2.24), (2.32), (2.38) and (2.42). While this may seem confusing and even discouraging, there is an apparent order which should allow for measurement and quantifying of the breaking strength. The basic definition of the energy lost by a single isolated breaker (2.24), although the most obvious, has perhaps the least practical significance and can only be employed in refined conditions when breaking is due to superposition of linear or nonlinear modes, or to a subsurface obstacle, mostly in the laboratory.

Equation (2.38) is the most general definition in spectral terms. It accounts for the spectral impact of a breaking, which can include energy losses, energy gains and energy exchange (shifting) between different wave scales in the spectrum. Since the spectrum, however, is a resolution of waves, often nonlinear, into linear modes, the spectral distribution of breaking strength can signify both the actual energy lost by free shorter waves and a reduction in nonlinearity of the primary waves (bound harmonics). Also, the attenuation of the short waves in the wake of large breaking can occur without breaking, for example, because of interactions of the short waves with intensive turbulence in the trace of the large breaker (e.g. Banner *et al.*, 1989). In physical space, this loss will be attributed to the large breaker, or to the group where this breaker occurred, and in the spectral sense it will be placed at high frequencies/wavenumbers even though the waves at those scales may not break. Obviously, the spectral definition of breaking severity (2.38) can be rewritten for any spectral band $f_1 < f < f_2$ if there are reasons to believe that the loss of energy is restricted to this band, or if such a band is of dedicated interest:

$$\frac{1}{16} E_{s_{\text{spectral}}}(f_1, f_2) = \int_{f_1}^{f_2} F_{\text{before}}(f)df - \int_{f_1}^{f_2} F_{\text{after}}(f)df = s_{\text{spectral}} \int_{f_1}^{f_2} F_{\text{before}}(f)df. \tag{2.44}$$

Note that in the formally introduced definition (2.44) the primary breaker, which originally caused the energy loss, does not necessarily have to belong to the frequency range $f_1 < f < f_2$.

Definitions (2.32) and (2.42) describe the breaking severity measured in physical space. The first accounts for all energy losses that take place due to a breaking within a wave group. As discussed above, these are breakings of dominant waves, although the spectral impact may be distributed across all wave scales. The second definition relates specifically to energy losses of waves short compared to the spectral peak f_p according to (2.43). The most important significance of these definitions is that, together with measurements of the breaking probability (2.36) and (2.4), they allow us to estimate experimentally the spectral dissipation function (2.21). As already mentioned, this function has so far been the most elusive and speculative property of wave modelling and forecasting (see e.g. Young & Babanin, 2006a; Babanin *et al.*, 2007c; Babanin & van der Westhuysen, 2008; Babanin *et al.*, 2010c). It will be discussed in detail in Chapter 7.

2.8 Types of breaking waves: plunging, spilling and micro-breaking

Breaking waves are usually subdivided into three types: plunging, spilling and micro-breaking (e.g. Weissman *et al.*, 1984). More detailed and complicated classifications are also available. For example, Griffin (1984) introduces a collapsing type as a limiting case of the plunging breaker and a surging-breaker type over a sloping beach.

The plunging breaker is the most commonly perceived picture of a breaking wave: at breaking onset, the crest curves forward and forms a plunging jet that impacts and penetrates the water surface in front of the wave, entrains air and turbulence deep under the surface, and potentially can trap an air pocket between the jet and the former front face of the wave, which will then disintegrate into large bubbles with corresponding consequences in terms of gas exchange across the interface and particularly in terms of acoustic noise produced by the event.

The spilling type is a less dramatic, but somewhat more frequent (e.g. Katsaros & Atakturk, 1992) kind of breaking when the crest destabilises and collapses, spilling the water over the front slope of the wave. Price (1970, 1971) suggested a perturbation theory leading to the spilling breaking. Qiao & Duncan (2001) used particle image velocimetry (PIV) to investigate development of spilling breakers in the laboratory. They found that, first, a bulge is formed which, however, does not result in formation of a jet as in the plunging breaker described above. Rather, the bulge starts to slide down the front face of the wave.

In many regards, the dynamics of the spilling breaker is similar to having multiple smaller jets impacting the water surface; these jets also lead to formation of bubbles and generation of turbulence. Essential also, as far as turbulence and dissipation are concerned, are the shear stresses between the sliding-down bulge and the wave orbital motion. Plunging breaking is more energetic in terms of energy/momentum loss from the waves and correspondingly the energy/momentum transferred to the ocean, as well as in terms of turbulence generation, bubble penetration depth and gas exchange across the surface, but as said above the latter is more frequent. For the dynamics of plunging and spilling breakers in shallow waters see, for example, Janssen (1986b).

The micro-breaking phenomenon deserves to be mentioned separately among other wave-breaking definitions, not so much because it is a different kind of breaking, but rather in order to avoid confusion and to state that the dynamics of micro-breaking should be basically the same as that of regular breaking. Different is the external signature of such breaking, which is that micro-breaking does not produce whitecapping and therefore does not exhibit whitecapping-related acoustic, optical and other signatures.

The term 'whitecapping' is often used interchangeably with the term wave breaking, and even the dissipation function employed in spectral wave modelling is routinely called whitecapping dissipation. A significant part of the spectral distribution of such dissipation, however, corresponds to the scales where waves break without producing whitecaps. These are short gravity waves, short in absolute rather than relative terms, whose breaking intensity is too weak to warrant air entrainment visible as whitecapping (or too weak to overcome the surface tension at the wave crest and form a jet (Tulin & Landrini, 2001)). Investigators of such micro-scale breaking point out that this phenomenon is in fact much more widespread than are the whitecaps (e.g. Jessup *et al.*, 1997a). According to Tulin & Landrini (2001), these are waves of

$$\begin{aligned} \lambda &\leq 25\,\mathrm{cm}, \\ f &\geq 2.5\,\mathrm{Hz}, \end{aligned} \tag{2.45}$$

i.e. in most cases this is a major part of the wave spectrum.

Since micro-breaking is not visible, new means have had to be developed in order to detect and to be able to quantify its breaking rates and severity. Katsaros & Atakturk (1992) used high-resolution video recording for this purpose, which involved a significant manual effort to process the data and obtain the statistics. Jessup *et al.* (1997a) employed infrared imagery that allowed automatic data processing. The idea is based on the existence of a 'skin' thermal layer at the ocean surface. The depth of this layer is of the order of 0.1 mm, and the top of the layer is a few tenths of a degree Celsius cooler than the bottom. Micro-breaking disrupts the skin layer and exposes the water with bulk upper-ocean temperature to the surface where the temperature differences between the micro-breaking wake and the background surface can be observed in the infrared-light range.

Jessup *et al.* (1997a) write:

> "The conceptual model we present, which explains our infrared observations, suggests that thermal detection of microscale wave breaking may serve as a de facto definition of the phenomenon itself".

In other words, as much as visually observed whitecapping can be treated as a de facto definition of breaking in the common sense, regardless of the physics behind the phenomenon, the infrared signature defines micro-breaking. We agree this is true, with a small addition: the disruption of the thermal skin layer identifies the micro-breaking provided that the breaking does not generate whitecaps. Obviously, large breakers, which do produce whitecapping, also break the skin layer (Jessup *et al.*, 1997b).

Lowen & Siddiqui (2006) compared three methods of detecting micro-breaking: the wave slope, the areal extent of the thermal wake and the variance of the vorticity in the micro-wave crest region, based on empirical threshold values. The comparison showed that the vorticity method is the most accurate, although it is obviously a difficult method to apply in field conditions. Infrared remote-sensing techniques intended to deal with micro-breaking will be discussed in more detail in Section 3.6.

Tulin & Landrini (2001) investigated micro-breaking by means of numerical simulations of the breaking onset with account taken of capillary effects, and by means of radar observations of micro-breakers in the laboratory. They provide interesting insights into the hydrodynamics of formation and propagation of micro-breaking waves. According to them, surface tension has a significant impact on the breaking of waves shorter than 2 m, to such an extent that for waves characterised by (2.45) no jet is produced at the crest. Instead,

"a forward facing bulge growing out of the wave crest is formed... Unlike energetic breakers where the shape continually evolves in transient fashion,... the microbreaker can propagate for a considerable distance without significant change of shape". Tulin & Landrini (2001) note that for very short waves (sub-micro-breakers) the bulge separates from the main flow "forming a cup on top of the wave, as noted by Ebuchi *et al.* (1987)".

2.9 Criteria for breaking onset

Wave breaking obviously happens when the water surface loses stability and collapses. What is the physics leading to this collapse has been one of the main questions for wave-breaking studies, and is one of the main topics of this book. The understanding of this physics has been elusive, but dealing with breaking is necessary across a very broad range of oceanographic applications as described throughout the book. Therefore, in the course of wave-breaking investigations, various criteria which supposedly indicate a breaking onset and therefore may help to detect it have been employed. They may or may not assist in the understanding of the physics, and may not even rely on the physics as such, but they at least need to be outlined at this stage in order to explain various experimental, analytical, statistical, probabilistic and numerical methods described later in this book.

Wave-breaking criteria are typically broadly classified into three categories: geometric, kinematic and dynamic criteria (e.g. Wu & Nepf, 2002). Most commonly employed are those that follow from the Stokes limiting steepness (Stokes, 1880):

$$H/\lambda = 1/7 = 0.142. \tag{2.46}$$

If converted into the nowadays commonly used steepness parameter ϵ (1.1), this criterion reads:

$$\epsilon_{\text{limiting}} = (ak)_{\text{limiting}} = 0.443. \tag{2.47}$$

In a wave of such limiting steepness, the crest takes the form of a

$$\theta_{\text{limiting}} = 120^\circ \tag{2.48}$$

corner flow (see also McCowan, 1894; Packham, 1952; Lenau, 1966; Price, 1970, 1971).

The Stokes limit can also be translated into kinematic limit, i.e. the surface wave orbital velocity becomes greater than the wave's phase speed:

$$u_{\text{orbital}} = a\omega = c, \tag{2.49}$$

or into downward surface acceleration, i.e. a dynamic limit of

$$a_{\text{downward}} = \frac{1}{2}g. \tag{2.50}$$

Because of their significance for wave-breaking studies, these criteria have been extensively revisited in the modern literature, both in theoretical and experimental studies, and we refer the reader to Longuet-Higgins *et al.* (1963), Longuet-Higgins (1969a, 1974), Brown & Jensen (2001), Stansell & MacFarlane (2002), Wu & Nepf (2002), Babanin *et al.* (2007a) and Toffoli *et al.* (2010a), among many others.

For two-dimensional waves, Brown & Jensen (2001) confirmed the steepness (2.47) as the breaking limit for wave breaking induced by linear wave focusing, and Babanin *et al.* (2007a) for breaking due to modulational instability. For directional waves, Toffoli *et al.* (2010a) investigated the probability distribution of the steepness of individual waves in three-dimensional wave fields, based on two different field data sets (Black Sea and Indian Ocean, including waves generated by tropical cyclones), and two sets obtained in different directional wave tanks (Marintek's ocean basin in Trondheim, Norway and the directional tank of the Kinoshita Laboratory/Rheem Laboratory of the University of Tokyo, Japan) and found a threshold for the ultimate steepness as

$$\frac{H_{\text{front}}k}{2} = 0.55 \tag{2.51}$$

for the front steepness, and

$$\frac{H_{\text{rear}}k}{2} = 0.45 \tag{2.52}$$

for the rear steepness. Here, account was taken of the fact that the front and rear troughs of breaking waves were not symmetric (for example, Figures 1.3, 5.5, 5.14, see also Soares *et al.*, 2004), i.e. wave height H_{front} is estimated as the vertical distance between the crest and the front trough and H_{rear} is for the rear trough. Since the probability function has a cutoff at the threshold value, i.e. does not extend into the higher values of steepness, this result should be interpreted as the ultimate steepness beyond which the directional waves will certainly break.

Babanin *et al.* (2011a), based on measurements of pre-breaking three-dimensional waves, showed that their steepness at the breaking onset is more like

$$\frac{Hk}{2} = 0.46 - 48. \tag{2.53}$$

They argued that the higher ultimate steepness values observed by Toffoli *et al.* (2010a) perhaps relate to transient waves, i.e. those already breaking. That is, short-crested directional waves start breaking if the steepness (2.53) is reached, but in the course of breaking they can achieve even higher steepness (2.51) at the front face. The limiting steepness (2.53) for directional waves is only slightly higher than that of (2.47) for two-dimensional wave trains; this issue will be further discussed in Chapter 5.

Stansell & MacFarlane (2002) specifically investigated the kinematic criterion (2.49) in a dedicated laboratory experiment. Since three different interpretations of what is the wave's phase velocity are possible, they verified the criterion with respect to all three definitions:

> "The first definition, based on the equivalent linear waves, is constant over the wavelength of the wave. The second, based on partial Hilbert transforms of the surface elevation data, is local in space and time giving instantaneous values at all space and time measurements. The third, based on the speed of the position of the crest maximum, is local in time but not in space".

The orbital velocity at the crest of breaking and non-breaking waves was measured by means of a PIV system, and breaking in an intended location was achieved through linear focusing.

The conclusions were not overly supportive of the classical criterion. The ratio of orbital velocity to phase velocity was found to be, at most, 0.81 in the plunging breakers and 0.95 in a spilling breaker. In this regard, we would like to make two comments. First of all, this finding means that when criterion (2.49) is satisfied, the waves are definitely breaking, and therefore experimental detection techniques, and statistical and theoretical approaches based on this criterion will underestimate rather than overestimate breaking probabilities. Secondly, linear focusing is most likely to be a mechanism of secondary importance when it comes to natural field wave breaking (see Chapters 4–6, 10). The breaking caused by instabilities of nonlinear wave trains appears to be a feasible common mechanism for field breaking, and its physics is quite different from that of breaking because of linear superposition. For such a breaking onset, brought about by modulational instability, the experiments and findings of Stansell & MacFarlane (2002) need to be revisited.

For such modulational breaking, an unconventional interpretation of the kinematic criterion was given by Tulin & Landrini (2001). In a comprehensive overview of wave-breaking investigations conducted by their research group over a period of more than 15 years, the authors, in particular, provided analytical derivation and experimental validation of their version of this criterion:

> "... upon passing through the peak of a modulation group, when the orbital velocity at the wave crest, u_c, exceeds the wave group velocity $d\omega/dk = c_g$, then the wave crest and trough both rise, the front face steepens, the wave crest sharpens, and eventually a jet forms at the crest, leading finally to splashing and a breakdown of the wave".

Thus, the criterion is stated for wave breaking due to modulational instability rather than that due to focusing (see Chapters 4 and 5 about the evolution of nonlinear wave groups to breaking). This does not signify the ratio of orbital and phase velocities at the point of breaking onset, as (2.49), but rather a point of no return: i.e. if the orbital velocity at the crest exceeds such a value, it may still be increasing (and perhaps even reach values comparable with those given by (2.49)), but inevitably, the wave will eventually break:

$$u_{\text{orbital}} = \frac{d\omega}{dk} = c_g. \tag{2.54}$$

If the crest-particle velocity is below (2.54)-value, the wave will not proceed to breaking. According to Tulin & Landrini (2001), the criterion is true not only for deep-water waves, for which the ratio of phase and group velocities is determined by (2.19), but also for modulating trains in finite depths where this ratio is smaller.

Analytically, the criterion is justified by Tulin & Landrini (2001) through considering the propagation of Stokes waves through a wave group. In the case of a uniform wave train, the trajectory of the crest is a horizontal line. In the case of the crest passing through the peak of a concave envelope, it must decelerate which, for gravity waves, is impossible unless values of the crest orbital velocities are below those defined by (2.54). This convincing derivation is supported by substantial experimental observations outlined further by Tulin & Landrini (2001). Potentially, this criterion can serve to separate observed wave breakings occurring due to different physical causes, e.g. modulational-instability breaking from directional-focusing breaking, which is not connected with the modulation of wave trains (Fochesato *et al.*, 2007).

Since most wave measurements are conducted as time series of the surface elevation, where the wave period (frequency) rather than wavelength (wavenumber) is known, the Stokes steepness (2.47) can be further converted into its height/period version. This is usually done by means of a linear dispersion relationship (see 2.17). Then, the criterion suggests that the waves will break if (Ramberg *et al.*, 1985; Ramberg & Griffin, 1987)

$$H \geq 0.027 g T^2. \tag{2.55}$$

A variety of other criteria, both related and unrelated to the Stokes limit, have also been proposed. Among the geometric limiters, Longuet-Higgins & Fox's (1977) maximal inclination of the surface, which the Stokes wave can reach before breaking, should be mentioned:

$$\theta_{\text{critical}} = 30.37^\circ, \tag{2.56}$$

as it has been employed extensively in experimental studies (see Section 3.3).

Other widely used geometric properties of the pre-breaking wave shape are skewness S_k (1.2) and A_s (1.3) (statistically, they are the third moment of the surface elevation and the third moment of the Hilbert transform of surface elevation, respectively). They were introduced by Kjeldsen & Myrhaug (1978), along with front and rear crest steepness. These are empirical, rather than theoretically justified criteria based on the common perception of the breaking wave as one with a sharp crest and front face leaning forward (see Section 1.2 and Figure 1.2). Quantitative criteria for the asymmetry, for example, were suggested by Kjeldsen & Myrhaug (1980), Myrhaug & Kjeldsen (1986) and Kjeldsen (1990) based on field observations, by Bonmarin (1989) and Griffin *et al.* (1996) from laboratory experiments, and by Tulin & Landrini (2001) through numerical simulations. In addition, Kjeldsen & Myrhaug (1980) found that the front trough of the incipient breaker is shallower compared to the rear trough, which is a persistent feature of wave breaking due to modulational instability, observed in the laboratory (Babanin *et al.*, 2010a) and produced by means of fully nonlinear numerical simulations (Dyachenko & Zakharov, 2005, see also Chapter 4).

Tulin & Landrini (2001), however, point out that deformation of the wave shape prior to breaking

"has to be viewed as a consequence of the breaking process and not the cause of it",

and we fully agree with this. In particular, recent investigations of breaking onset brought about by modulational instability of wave trains showed that the asymmetry at the breaking point is a rapidly changing characteristic, with a value close to zero and increasingly becoming negative. Thus, it would be difficult to employ asymmetry as a certain geometric breaking criterion. Skewness, on the other hand, indeed exhibits a robust asymptotic trend to a limiting value of

$$Sk_{\text{limiting}} = 1 \tag{2.57}$$

for two-dimensional waves (Babanin *et al.*, 2007a, 2009a, 2010a) and

$$Sk_{\text{limiting}} \approx 0.7 \tag{2.58}$$

for three-dimensional waves (Babanin *et al.*, 2011a).

Returning to the dynamic criteria, the most significant difficulty of applying limiting downward-acceleration (2.50) to real sea waves is the fact that the natural wave fields are multi-scaled and therefore the surface elevation at any point and any instant of time is a superposition of an unlimited number of wave components. Therefore, among other dynamic criteria, we should mention the downward acceleration derived by Longuet-Higgins (1985) for a complicated sea, rather than for a monochromatic Stokes wave:

$$a_{\text{downward}} = g. \tag{2.59}$$

As a general approach, the acceleration criterion is treated in the sense that the wave surface will break when its downward acceleration exceeds a limiting fraction γ, which is a tuning parameter, i.e.

$$a_{\text{downward}} = a\omega^2 > \gamma g. \tag{2.60}$$

When it is applied to a monochromatic wave, this approach is straightforward even though there are uncertainties about the value of γ. In theoretical/statistical studies of the limiting-steepness Stokes wave, it has generally been assumed that $\gamma = 0.5$ ((2.50), Longuet-Higgins *et al.*, 1963; Snyder *et al.*, 1983). The latter paper also verified this limit by observing the breaking of dominant waves in the field. Field observations of Ochi & Tsai

(1983) (see their interpretation by Srokosz (1986)) for broadband breaking and laboratory studies of Hwang *et al.* (1989) showed that γ is closer to 0.4, and some other field measurements (e.g. Holthuijsen & Herbers, 1986; Liu & Babanin, 2004) further indicated that the value of γ could be even lower.

For more breaking-onset properties and criteria, we refer the reader to a near-comprehensive table of the possible wave-breaking threshold variables in Snyder & Kennedy (1983). In this paper, in addition to those already mentioned here, such characteristics as vertical velocity gradient, horizontal velocity divergence, surface curvature and vertical acceleration gradient are included. Bonmarin (1989) used 37 geometric parameters or coefficients, and an additional 8 parameters in terms of potential energy, to characterise the breaking wave.

This brief summary of breaking-onset indicators would be incomplete without mentioning a set of criteria that came from numerical simulations of the evolution of nonlinear wave groups with imposed modulation (Banner & Tian, 1998; Song & Banner, 2002; Banner & Peirson, 2007). In general terms, these criteria can be regarded as dynamic, i.e. the rate of change of the local mean wave energy density/the rate of change of the momentum flux averaged over half a wavelength/the local average mean energy flux to the energy maximum in the wave group. Such criteria can hardly be employed in a practical sense and even their verification in refined laboratory conditions proved intangible and inconclusive (i.e. Section 4.2 in Babanin, 2009).

Overall, we would like to say that breaking-onset criteria have proved very useful and successful in wave-breaking studies of all kinds, but once the physics of breaking is understood better, their significance will diminish or even eventually become redundant, with the definite exception of the Stokes steepness limit (2.47) and related geometric, kinematic and dynamic features (2.48)–(2.50). As we will try to show later (Chapters 4, 5), it appears that waves break, not because of some particular physical mechanism, but rather because in the course of their evolution they reach this limiting steepness and the water surface becomes unstable, regardless of the processes behind the steepness growth. The same growth mechanism, if it does not bring the wave height up to the Stokes limit (or its modification in the case of directional waves (see Toffoli *et al.*, 2010a; Babanin *et al.*, 2011a, and Section 5.3.3)), will not make the wave break. Therefore, probably, no single criterion, except the limiting steepness, can be a robust breaking predictor in all circumstances. Mechanisms that can lead to such a steep wave occurrence can be few, with modulational instability likely to be largely responsible for the breaking of dominant waves.

2.10 Radiative transfer equation

The radiative transfer equation (RTE) plays an important role in the context of wave-breaking and breaking-dissipation studies, their intentions and motivations. It will not be explicitly used in this book for derivations or modelling, but it will be referred to, and therefore needs to be mentioned and described among the other relevant definitions relating to the wave-breaking process.

Since it was first proposed by Hasselmann (1960), RTE has been widely used in scientific studies and practical applications relating to the evolution of wind-generated waves. In water of finite depth, this equation takes the form (Komen *et al.*, 1994)

$$\left(\frac{\partial}{\partial t} + (\mathbf{c}_g + \mathbf{U_c}) \cdot \frac{\partial}{\partial \mathbf{x}} - \nabla\Omega \cdot \frac{\partial}{\partial \mathbf{k}}\right)\frac{\Phi}{\omega} = S_{in} + S_{nl} + S_{ds} + S_{bf} \qquad (2.61)$$

where the left-hand side represents the evolution of the wave action density Φ/ω as a result of the physical processes of atmospheric input from the wind S_{in}, nonlinear interactions of various orders within the spectrum S_{nl}, dissipation due to 'whitecapping'/breaking S_{ds} (2.25), and decay due to bottom friction S_{bf}. Function $\Phi(f, k, \theta)$ is the directional wave spectrum, i.e. θ is the direction of wave propagation. All the source terms are also spectra along the wavenumber–frequency-direction, as well as functions of other parameters. The term $\mathbf{U_c}$ is the surface current and $\Omega(\mathbf{k})$ is the Doppler-shifted frequency $\Omega(\mathbf{k}) = \omega(k) + \mathbf{k} \cdot \mathbf{U_c}$.

The dissipation-due-to-breaking source (sink) term S_{ds} (2.25) was introduced and defined above (Section 2.7) and will be mentioned many times throughout this book. While the waves would not be generated without the wind input S_{in} in the first place, the other terms in (2.61) are not of the secondary importance they may seem. Without S_{ds}, the wind-generated waves would keep growing which obviously does not happen. As soon as the waves pass the infinitesimally small stage, the dissipation switches on. The dissipation due to breaking does not turn on until the spectral threshold value is overcome (see Sections 5.2, 5.3.2), but once it is active it becomes as significant as the wind input and the other two general terms mentioned in (2.61) as far as wave evolution is concerned. Many more less general source/sink terms in RTE are possible, but the four introduced in (2.61) are invariably important in the finite-depth wave environment, and the first three in deep water.

Knowledge of the terms other than dissipation, based on either experimental or analytical (or both) approaches, is incomplete but still rational (see Komen *et al.*, 1994 and The WISE Group, 2007, in general and Donelan *et al.* (2006) and Babanin *et al.* (2007b) on recent developments on the wind input S_{in}). In contrast, understanding of the dissipation term remains poor. A number of theoretical and conjectural approaches have been attempted to predict the spectral dissipation function, but none of these have been validated experimentally. It is generally assumed that S_{ds} is a function of the wave spectrum Φ:

$$S_{ds}(f, k, \theta) \sim \Phi(f, k, \theta)^n, \qquad (2.62)$$

but there is no agreement even on such basic ground as to whether the spectral dissipation $S_{ds}(f, k, \theta)$ is linear in terms of the spectrum $\Phi(f, k, \theta)$ or not, i.e whether $n = 1$ or $n > 1$. On the other hand, such experimentally known features of wave-breaking dissipation as threshold behaviour, cumulation of dissipation at smaller scales and dependence of dissipation on the wind at extreme-forcing conditions have not been accounted for in present-day dissipation terms in any way (see Section 7.4). See Chapters 5, 6 and 7 for more detailed discussions on the dissipation-term issues and and The WISE Group (2007) for an up-to-date discussion of the entire topic of state-of-the-art wave modelling.

3

Detection and measurement of wave breaking

Measurements and even detection of wave breaking are challenging tasks, particularly if carried out by unattended devices in the open ocean. As a result, there are vast amounts of wave records accumulated, and most of them contain breaking waves, but information about the breaking cannot be extracted. There is an obvious need for methods and instrumentation to directly detect breaking events and measure their properties, and for the analytical means and criteria to identify the breaking waves in existing time series of surface elevations.

Until recently, visual observations were arguably the only reliable means of breaking detection. These are based on viewing and quantifying information on whitecaps produced by breaking waves and are obviously biased towards large breakers (see Section 2.8). Over the past two decades, more technological methods have become available, both contact type and by remote sensing of the ocean surface or subsurface. These utilise acoustic (passive and active), optical (both visible and infrared range), reflective and other properties of breakers which distinguish them from the more homogeneous background wave field. Without giving a comprehensive review, we will mention here passive acoustic techniques based on air bubbles ringing while being created during air entrainment (e.g. Lowen & Melville, 1991a; Ding & Farmer, 1994; Babanin *et al.*, 2001; Manasseh *et al.*, 2006), sonar observations of bubble clouds produced by breaking wind–waves (e.g. Thorpe, 1992), aerial imaging (e.g. Melville & Matusov, 2002; Kleiss & Melville, 2011, 2010), infrared remote sensing of breaking waves (e.g. Jessup *et al.*, 1997a), radar observations of microwave backscatter from breakers (e.g. Jessup *et al.*, 1990; Lowen & Melville, 1991a; Smith *et al.*, 1996; Phillips *et al.*, 2001), and conductivity measurements of the void fraction produced by breakers (e.g. Lammarre & Melville, 1992; Gemmrich & Farmer, 1999), amongst others. New analytical methods of detecting the breaking in wave records have also become available, for example, by means of Hilbert transform (e.g. Huang *et al.*, 1998) or wavelet analysis (e.g. Liu & Babanin, 2004) applied to the wave series.

This section gives a brief review of wave-breaking detection and estimation techniques. We start from visual observations of whitecapping which have been a traditional way of investigating the phenomenon and remain a valid and highly efficient method today. Contact measurements of the breaking properties provide a means for direct estimates, which, unlike visual observations, are not subject to human error. These are subdivided into two sections dedicated to laboratory techniques and to their counterparts applicable in field experiments where, as far as wave breaking is concerned, requirements for *in situ* devices are very different from the laboratory. Remote sensing approaches are outlined next. Of these, the acoustic methods are singled out into a separate section because, in our view, they are the most advanced and most promising with regard to breaking studies. Finally, analytical techniques are reviewed briefly. These include a section on detecting breaking events in existing surface-elevation time series and a section on analyses of wave-breaking statistics based on some initial assumptions. Both approaches appeal to outliers, either in wave records or in probability distributions, which would not have taken place if the waves were not breaking.

It should be mentioned that the majority of the wave-breaking measurement and analysis methods are intended for studies of breaking rates or probability, whereas breaking severity measurement methods are underdeveloped if not marginal. In this regard, almost all experimental techniques mentioned above and described in more detail below show promise, but have never been sufficiently elaborated for the purpose of estimating severity. The most direct way of estimating breaking-energy loss is of course measuring the wave train immediately before and immediately after a breaking event (see Section 2.7), but even in the laboratory this is problematic. Among the other methods, acoustic (Section 3.5) and infrared (Section 3.6) techniques have a proven record of being able to qualitatively distinguish breaking events of different strengths, but even these are still to be advanced into a quantitative calibration stage. In the meantime, as mentioned in Section 2.7, knowledge and understanding of breaking severity remain quite poor, but are as important as those of breaking probability. Therefore, it is worth emphasising at this stage that the lack of progress in the breaking-strength topic is not only due to the need for relevant dedicated studies, but to a great extent is hampered by the absence of robust and reliable experimental methodologies.

3.1 Early observations of wave breaking, and measurements of whitecap coverage of ocean surface

In this section, whitecap-coverage observations and parameterisations are described and analysed, dating from historical to modern. In this context, definitions of the characteristic features and properties of the atmospheric boundary layer are also discussed.

Early documented scientific observations of wave breaking date back to the late 1940s (Munk, 1947) and were followed by a number of research attempts which intensified in the 1960s (i.e. Blanchard, 1963; Gathman & Trent, 1968; Cardone, 1969; Monahan, 1969, 1971; Monahan & Zietlow, 1969). They usually dealt with some identification of the whitecap coverage of the ocean surface, usually by means of photography and without actual collocated measurements of the waves. Wave measurement techniques were still under development in those days, and data logging and computer facilities for recording, storage and processing long time series of surface elevations were not available. Therefore, the whitecap measurements were interpreted and parameterised in terms of wave-generating winds. Some observations (i.e. Munk, 1947) were quantified as the number of foam patches per unit area thus allowing their relation to the breaking probability (see Section 2.5). The majority of the results were expressed as a percentage of the whitecap coverage of the surface, which property depends on a combination of the breaking probability, the breaking strength and lifetime of the whitecap foam. The first two properties combined relate to the overall dissipation due to wave breaking (2.20), where contributions of the breaking rates (Section 2.5) and breaking strength (Section 2.7), however, cannot be separated. The whitecapping lifetime depends on environmental conditions, such as water temperature, salinity and surfactants, and technically speaking it is not a characteristic of wave breaking. Therefore, the whitecap coverage bears essential uncertainties if treated as a property of wave breaking or a quantitative feature of wave energy dissipation.

Parameterisation of breaking dependences, including the whitecap surface coverage, in terms of the surface-wind speed is of course a reasonable approach as the waves are generated by the wind and many characteristic properties of the wave field ultimately correlate with the wind. Because of the very large density difference between the air and the water, however, the wind impact on breaking is mostly indirect. That is, the wind slowly pumps energy into the waves, which gradually grow under the wind action, and as their steepness increases, the hydrodynamic mechanisms (rather than wind forcing) lead some waves to break. This will be discussed in detail in Chapters 5 and 6. Here, we will mention that the capacity of the wind to stimulate or even affect the breaking onset as such is marginal, unless the wind forcing is very strong (i.e. $U/c > 10$, see Babanin *et al.*, 2009a, 2010a), and the physics of wave-breaking parameterisations should be expressed in terms of the wave, rather than the wind properties (e.g. Babanin, 1995; Felizardo & Melville, 1995; Banner *et al.*, 2000; Babanin *et al.*, 2001).

The last two studies used Black Sea (Section 3.7) and Lake George (Section 3.5) field data, combined with measurements from a small shallow lake and data points from the Southern Ocean, which together made up a very diverse data set, which argued that relating the breaking probability to the wind speed provided a reasonable correlation within each individual data set, but when the diverse data were combined, these correlations essentially degraded. In the combined data set, with a broad range of dominant wave lengths (10–300 m) and wind speeds (5–20 m/s), there was no correlation between the breaking rates and the wind speed, whereas the dependence of the breaking probability on the properties of the wave field remained.

Keeping this understanding in mind, it is interesting to revisit the early results. They clearly exhibited many features of wave-breaking behaviour that are being 'rediscovered' years later, and they set many standards to follow. For example, Munk (1947) concluded that there was no breaking at wind speeds $U < 6\,\mathrm{m/s}$ and Gathman & Trent (1968) found no whitecaps at $U < 3.1\,\mathrm{m/s}$. Similarly, Blanchard (1963) demonstrated a significant increase in whitecap coverage at $U > 3\,\mathrm{m/s}$ and Monahan (1969) observed that the fractional coverage was very small, less than 0.1% for light winds, and only started to grow and depend on the surface wind at $U > 4\,\mathrm{m/s}$. Such results are consistent with later observations (e.g. Donelan & Pierson, 1987; Bortkovskii, 1997, 1998; Babanin *et al.*, 2005, see also Section 5.3.4), but most importantly agree with the recently found threshold behaviour of breaking onset. These days, the threshold is expressed in terms of the characteristic steepness of the wave field rather than in terms of the wind speed, which makes a difference in cases of complicated wave fields, or wave fields ambiguously coupled with the local wind (Banner *et al.*, 2000; Babanin *et al.*, 2001, 2007a). See, however, discussions of the low-wind breaking threshold in Sections 5.3.4 and 9.1.1.

Furthermore, we can mention that Cardone (1969) suggested an analytical model for whitecap-coverage dependence on the wind by assuming that the whitecaps are a manifestation of wave-energy dissipation which balances the energy input to the waves from the wind. This idea was to be followed by a great many studies in subsequent decades (e.g. Wu, 1979; Zhao & Toba, 2001; Guan *et al.*, 2007).

The apparent quantitative differences between the early observations were not unexpected for both technical and physical reasons. On the technical side, variations in the accuracy and methodology of wind measurements were pointed out by the authors of early studies themselves (Monahan, 1971). The shortcomings of the manual techniques of processing photographic images became clearer once digital methods were developed (see, e.g. Stramska & Petelski, 2003). For example, the manual methods did not take into account the geometry difference between the near-field and far-field in oblique photography, and as a result the whitecapping coverage of the former was overestimated and of the latter underestimated. This would bring about discrepancies between observations taken at different angles of incidence and introduce scatter and bias, particularly significant at low wind speeds where the presence or absence of a breaking event in the near-field could essentially distort the overall statistics.

Physical processes that can affect the size and duration of the persistence of whitecaps in different environmental circumstances are very many, and they could certainly contribute to the quantitative disagreements too. The early researchers were aware of at least some of these, and here we refer to a recent update by Stramska & Petelski (2003):

"For example, wind history, local hydrodynamic conditions such as currents and swell, directionality of the wave field, presence of biological surfactants, and variations of the water temperature and atmospheric stability all can contribute to variability in the W versus U_{10} relationship"

(see also Bortkovskii, 1997, 1998).

The envelope of dependence of the fractional whitecap coverage W on wind speed U_{10} at 10 m height above mean sea level (i.e. experimental fit through the highest-coverage values at various wind speeds) was suggested by Monahan (1971) for ocean whitecapping as follows:

$$W = 0.00135 U_{10}^{3.4}. \tag{3.1}$$

Monahan (1971) reviewed the earlier observations and found, as we have just discussed, that available parameterisations were qualitatively similar, but quantitative deviations were essential.

The earlier observations concentrated on whitecap coverage, rather than on more detailed and coupled-with-waves characteristics of the wave breaking, out of necessity, because of the lack of experimental techniques and data handling capacities. Later, studies of such coverage acquired importance in its own right due to the development of methods of remote sensing of the ocean surface (e.g. Reul & Chapron, 2003; Anguelova & Webster, 2006, see also Section 3.6). Such methods either rely directly on measurements of whitecap coverage in order to estimate, for example, the surface winds by means of (7.4)-like dependences (e.g. Wu, 1969, 1979; Monahan *et al.*, 1981), or have to take such coverage into account as the whitecaps are up to 10 times brighter than the water surface (e.g. Moore *et al.*, 1997; Reul & Chapron, 2003) and therefore alter the observed ocean colour and other remote-sensing properties of the ocean (e.g. Monahan & O'Muircheartaigh, 1986; Ester & Arnone, 1994; Gordon, 1997; Sharkov, 2007). In this regard, it is worth mentioning that albedo due to whitecapping essentially depends on the sea-water temperature (Bortkovskii, 1997).

In view of this importance, experimental dependences of W versus U_{10} continue to be revisited, and a larger number of new dependences, or modifications to existing dependences have been proposed. In a recent paper, based on a combined empirical–statistical–analytical approach, Yuan *et al.* (2009) suggested an upper envelope for such dependences. For any given wind speed, such an envelope would describe whitecap coverage at infinite wave fetch under infinite wind duration. The vast literature on this topic cannot be reviewed in the current section and we refer the reader to books by Monahan & MacNiocaill (1986), Sharkov (2007) and most recent updates by Zhao & Toba (2001), Stramska & Petelski (2003), Anguelova & Webster (2006), Guan *et al.* (2007) and Yuan *et al.* (2009) and respective references in these publications. Here, we would like to highlight the studies most relevant for the discussions in this book.

Wu (1979) further reviewed available observations of whitecap coverage (7.4) almost 10 years after the first such review by Monahan (1971) and at that stage suggested, based on some semi-theoretical/semi-empirical argument, that the data are not inconsistent with such dependence:

$$W = \alpha U_{10}^{3.75}, \tag{3.2}$$

where coefficient $\alpha = 1.30$–2.90 varies mainly as a function of the stability conditions of the atmospheric boundary layer. Stramska & Petelski (2003) introduced a wind-speed threshold into this kind of dependence:

$$W = \begin{cases} 4.18 \cdot 10^{-5}(U_{10} - 4.93)^3 & \text{if } U_{10} \geq 4.93\,\text{m/s}, \\ 0 & \text{if } U_{10} < 4.93\,\text{m/s}. \end{cases} \tag{3.3}$$

If the wind is below the threshold, no breaking/whitecapping is observed, otherwise the whitecap coverage W depends on the excess of wind speed above its threshold value, rather than on the wind speed itself. The threshold was determined as

$$U_{10_{\text{threshold}}} = 4.93\,\text{m/s}. \tag{3.4}$$

Such threshold-like dependences are consistent with parameterisations of the breaking probability (see Chapter 5), at least for straightforward wave-development conditions when the threshold wind speed can be related to the threshold characteristic steepness in the wave field (see also Section 5.3.4).

When using the power of 3 in (3.3), Stramska & Petelski (2003) followed the approach prevailing in the 1990s (e.g. Monahan & Lu, 1990; Monahan, 1993). While in the earlier parameterisations of

$$W \sim U_{10}^{p}, \tag{3.5}$$

the power p was obtained by means of fitting the experimental data points, the latter approach assumes and enforces $p = 3$. The semi-theoretical reasoning behind such an assumption is the same as that of Wu (1979) whose conclusion, however, was $p = 3.75$. Therefore, it is worth briefly outlining the argument, and reviewing it somewhat in the light of more recent knowledge.

The physical argument states, by definition, that the total energy flux from the wind to the waves, and therefore the total wave energy dissipation rate in a quasi-stationary case is approximately:

$$S_{ds} = \tau U, \tag{3.6}$$

where

$$\tau = \rho_a u_*^2 = \rho_a C_D U_{10}^2 \tag{3.7}$$

is the wind stress or the momentum flux from the air to the water, U is some characteristic velocity of energy propagation in the low atmospheric boundary layer, u_* is the so-called friction velocity, ρ_a is the air density, and

$$C_D = \frac{u_*^2}{U_{10}^2}, \tag{3.8}$$

is the drag coefficient, a parameter introduced for convenience for converting the U_{10} wind into u_*. Thus, according to (3.6)–(3.7), the dissipation is proportional to some wind speed cubed, but the precise nature of this proportion depends on what the characteristic speed U is.

If $U = u_*$ (e.g. Wu, 1979; Soloviev *et al.*, 1988; Agrawal *et al.*, 1992; Melville, 1994), then

$$S_{ds} = \rho_a u_*^3 = \rho_a C_D^{3/2} U_{10}^3. \tag{3.9}$$

If $U = U_{10}$ (e.g. Demchenko, 1993; Bister & Emanuel, 1998), we obtain

$$S_{ds} = \rho_a u_*^2 U_{10} = \rho_a C_D U_{10}^3. \tag{3.10}$$

Obviously, a conclusion on speed U depends on which model of the boundary layer is employed, and other options are possible between the two extremes mentioned, including more complicated characteristic speeds defined by integral properties of the sheared boundary-layer air flow (e.g. Kudryavtsev *et al.*, 2001).

Treating the whitecap coverage W (3.5) as a direct indicator, or even as a property directly proportional to the dissipation S_{ds} (3.9) and (3.10), however, is a significant oversimplification which is quite poorly justified. While the $S_{ds} \sim U^3$ relationship for steady wind–wave fields is true, it does not necessarily imply that $W \sim U^3$, even if the dependence of the bubble lifetime on environmental conditions does not play its role. Dominant waves, if they break, provide a major contribution to whitecapping. For mature seas, however, they do not break (e.g. Banner *et al.*, 2000; Dulov *et al.*, 2002), and for developing seas a combined effect of frequency of their occurrence and severity of their breaking is not necessarily proportional to U^3 (e.g. Babanin *et al.*, 2009a, 2010a). Also there is multiple evidence that the bulk of energy flux to the waves is supported by short waves (e.g. Terray *et al.*, 1996) which bulk is also lost through dissipation at high frequencies (e.g. Babanin *et al.*, 2007c). Such short-wave breaking has a strong connection with the dominant waves, rather than directly with the wind, and the energy-containing waves appear to strongly modulate the respective whitecapping which is physically located near the dominant crests. At the spectral end of very short waves, furthermore, their breaking produces little or no whitecapping (see Section 2.8).

With this in mind, let us review the theoretical argument based on the assumption that coverage W is proportional to the wind energy flux (3.6). If the drag coefficient

$$C_D = \text{constant}, \tag{3.11}$$

we arrive at $p = 3$ assumed in (3.3) and similar dependences. It is, however, not constant. There is a great variety of experimental parameterisations of C_D versus U, and none of them suggests that the drag coefficient is independent of wind speed. Wu (1979), based on Wu (1969) and Garratt (1977), used

$$C_D \sim U_{10}^{1/2} \tag{3.12}$$

and obtained $p = 3.75$ in (3.2) and (3.9). Most experimental dependences are much steeper than (3.12). We refer to Babanin & Makin (2008) for a recent review and to their 'ideal-condition' parameterisation of

$$C_D = 1.92 \cdot 10^{-7} U_{10}^3 + 0.000953. \tag{3.13}$$

In any case, (3.3) and similar approximations where $p = 3$ was assumed, cannot be expected to remain general. Whichever power p is chosen or fitted, however, deviations from fit are still observed at high wind speeds (e.g. Monahan, 1971; Wu, 1979; Stramska & Petelski, 2003, and others), particularly as in a very strong wind the sea drag C_D indeed saturates (see Section 9.1.3). The explanation of such an effect seems obvious. At high wind speeds, whitecap coverage can no longer characterise the balance (3.6) between the wind input and wave dissipation. As Munk (1947) puts it, for 'fresh breeze' "spume tends to be blown from the breaking wave crests". According to Munk (1947), the fresh breeze is winds in excess of 8.7 m/s. The blown spume should create an enhanced foam coverage, in addition to the coverage due to the strength of the wave breaking, and the respective bias of the W versus U dependences. Therefore, any such parameterisations should be treated with caution for applications in extreme weather conditions and even at reasonably strong winds (see also Section 9.1.2).

This brings us to further cautious considerations regarding the validity of the interpretations of whitecap coverage as a characteristic for wave energy dissipation. Bortkovskii & Kuznetsov (1977) and Bondur & Sharkov (1982) and Bortkovskii (1987a) introduced a classification of whitecap systems that we will describe by using a direct quote from a later book of one of the authors (Sharkov, 2007):

"... specific form of foam systems in optical images allows us to confidently identify at least two classes (types) of foam formations: (1) 'wave crest foam' (i.e. 'whitecaps'), the so-called 'short-living form' (i.e. 'dynamic foam') of foam activity with a lifetime of units of seconds; and (2) spotty structures (or 'foam streaks'), 'static foam' (or 'residual foam') with a lifetime of about 10 seconds to several minutes... At wind velocities higher than 15 m/s there arises a special class of stable foam systems: the thread-like systems caused by capture of air bubbles by Langmuir vortices (i.e. Langmuir circulation)"

(e.g. Langmuir, 1938; Craik & Leibovich, 1976; Smith, 1992, 1998; McWilliams *et al.*, 1997; Melville *et al.*, 1998; Phillips, 2003, 2005; Thorpe, 2004; Sullivan & McWilliams, 2010).

Within this classification, Bortkovskii & Kuznetsov (1977) produced parameterisations both for the total whitecap coverage W:

$$W = 0.0166U_{10}^{2.25} \tag{3.14}$$

and for the ratio of active whitecapping W_A and passive whitecapping W_B:

$$\frac{W_B}{W_A} = \begin{cases} 0.255U_{20} - 1.99 & \text{if } U_{20} > 9\,\text{m/s}, \\ 0 & \text{if } U_{20} \leq 9\,\text{m/s}. \end{cases} \tag{3.15}$$

Here, $W = W_A + W_B$, and U_{20} is the wind speed at 20 m height. Bortkovskii & Kuznetsov (1977) also provided an interesting observation that

$$\frac{W_B}{W_A} \sim u_*, \tag{3.16}$$

but concluded that the exact dependence is much more complicated than this.

Bortkovskii (1987a) further extended this kind of observation and proposed a variety of statistics for active and passive stages of whitecapping. In particular, he connected the duration t_A of the active stage with wind speed:

$$t_A = 0.296U_{10} + 0.593 \tag{3.17}$$

and also suggested a connection between wind speed and velocity of propagation of the breaking fronts. In view of the present understanding of breaking processes, these dependences on the wind are most likely indirect, that is they reflect changes that occurred in the wave system, such as downshifting of the wave energy under persistent action of the wind. This way, for higher winds the waves will be longer and faster moving, and even if the relative breaking duration (in terms of the wave period) does not change, the absolute duration (3.17) will grow. If so, dependence (3.17) should be used with caution, as, for the same wind speed U_{10}, under-developed or over-developed (by comparison with the average stage of development in measurements of Bortkovskii (1987a)) estimates based on (3.17) will be biased.

Bortkovskii (1987a) also produced a number of other statistics and dependences which are worth noting. Among them, dependence of the duration t_B of the existence of passive whitecapping on water temperature, and long-wind extent of passive whitecaps as a function of temperature and wind speed. For active whitecaps, he obtained the rate of lateral spreading of the active whitecaps versus U_{10}, an expression for the geometrical proportion of the longitudinal and lateral whitecap dimensions, and histograms of these dimensions and of their product (active whitecap area). For this area, he suggested separate dependences for the tropical regions and for the Southern Ocean and concluded that the number of whitecaps per unit area does not depend on the wind in the range of speeds $10\,\mathrm{m/s} \leq U_{10} \leq 20\,\mathrm{m/s}$, among other interesting findings.

Coming back to the classification of whitecaps into active and passive, they are generally accepted now, as was re-introduced later by Monahan (1993), to be named stage A and stage B whitecaps respectively. It is apparent that, even if both stages are connected with the breaking probability and breaking strength, the diffusive stage B depends to a great extent on the lifetime of the residual bubble clouds. This lifetime, as was already known from earlier observations, is an environmental characteristic rather than a property of wave breaking or dissipation. Monahan (1971) compared diffusion rates of oceanic whitecaps with those in fresh water (Monahan, 1969) and found that the fresh-water whitecap area decays approximately 1.5 times faster. Thus, an essential variation of the whitecap coverage, in similar meteorological conditions and similar wave fields, is expected for water bodies with different salinity.

It is interesting to note that there is considerable controversy in the literature regarding further roles played by salinity in this regard. While some researchers find that the bubbles are smaller in salt water (e.g. Haines & Johnson, 1995), others claim they are not (e.g. Cartmill & Su, 1993; Wu, 2000). Wu (2000), in particular, found rather that,

"a greater volume of air is entrained in salt than in freshwater to generate many more bubbles".

If this was true, salt-water breaking would have to be more severe which is difficult to justify physically.

From early observations, it was also known that the lifetime of whitecap foaming L_t depends on the water temperature (Miyake & Abe, 1948). Laboratory experiments of Miyake & Abe (1948) produced a rather strong dependence:

$$L_t \sim \exp(-T_w/25) \tag{3.18}$$

where T_w is the water temperature. Mention of the temperature effect is scattered around later studies (Wu, 1979; Monahan & MacNiocaill, 1986; Monahan & O'Muircheartaigh, 1986; Bortkovskii, 1987a,b, 1997; Wu, 1988; Bortkovskii & Novak, 1993; Stramska &

Petelski, 2003), but a systematic and conclusive effort to investigate this is yet to be completed. For instance, Bortkovskii (1997) confirms the trend indicated by (3.18), that the lifetime of foaming reduces as the temperature increases, but at the same time claims that whitecap coverage increases at higher temperatures. This creates an additional uncertainty about coverage W in different meteorological conditions, and in this regard it is also necessary to point out what appears to be a persistent confusion about the dependence of W on atmospheric stability.

Variations of dependence of W versus U due to atmospheric stability were foreshadowed in very early studies (Monahan, 1971), and were even quantified to some extent later (e.g. Wu, 1979) and are still highlighted now (e.g. Stramska & Petelski, 2003). It is hard to justify physically, however, such a connection between the stability of the atmospheric boundary layer and the whitecap coverage of the ocean surface. Wave breaking, whitecap coverage and lifetime of the foam are essentially water, not air properties. We see two possible explanations of the correlations between whitecaps and stability, which are observable.

First of all, the unstable atmospheric conditions (which signify water being warmer than the lower atmosphere) affect the wind profile. For the steady low boundary layer such a profile is described as

$$U(z) = \frac{u_*}{\kappa}\left[ln\frac{z}{z_0} - \psi\left(\frac{z}{L}\right)\right] \tag{3.19}$$

where z is the vertical coordinate, $\kappa = 0.42$ is the von Karman constant, ψ is a function of the Monin–Obukhov length scale L (we refer the reader to the literature on atmospheric physics for more details (e.g. Komen *et al.*, 1994)). Under stable and neutral conditions, the profile is logarithmic and U_{10} wind in (3.5) and τ stress in (3.7), or other relevant wind characteristics are easily found. In unstable conditions, there can be significant deviations from the logarithmic profile because of function ψ. These are quite difficult to take into account appropriately, hence bias occurs in estimates of U_{10}, u_*, τ and in the corresponding dependences. In this case, some correlations between the whitecap coverage and atmospheric conditions can be observed which are, however, due to the bias of surface wind estimates rather than due to dependence of W on stratification of the air.

The second possible explanation is a misinterpretation of some data. For example, Wu (1979) analysed whitecap-coverage data measured independently by Monahan (1971) and by Toba & Chaen (1973) and demonstrated a clear separation of the data obtained in stable, neutral and unstable conditions. In both data sets, for the same wind speeds, whitecap coverage on average was higher when the atmosphere was stable, lower when the stratification was neutral and even lower in unstable circumstances. Since, however, a significant range of surface water temperatures was involved (in Monahan (1971) it was from 17.5°C to 30.55°C), and the surface temperature trend was not removed, it would affect the conclusions. Ordering the stratification as stable, neutral and unstable would correspond, on average, to increasing the water temperature, and therefore to decreasing whitecap coverage according to (3.18). This is exactly the trend observed by Monahan (1971) and attributed to atmospheric stability without first ruling out the surface-temperature dependence.

Further uncertainties of (3.5)-like parameterisations of coverage W in terms of wind speed U, which may prove essential in some circumstances, have to be highlighted. Again, early researchers were already aware of these. Monahan (1971) mentions variations of whitecap coverage as a function of wave fetch and wind duration. These days, this is translated as dependence of W on the second parameter, U/c, which characterises inverse wave age (stage of wave development or wind forcing) (e.g. Zhao & Toba, 2001; Stramska & Petelski, 2003; Guan *et al.*, 2007). Indeed, this should be expected based on the apparent physical argument. As discussed above, the whitecap coverage depends on breaking probability and breaking strength. Both of these depend on wave age (see e.g. Katsaros & Atakturk, 1992; Banner *et al.*, 2000, for the breaking probability). In a laboratory experiment Babanin *et al.* (2009a, 2010a) demonstrated that, for stronger forcing U/c, the breaking probability is higher, whereas the breaking severity is weaker. Thus, the trends are opposite, and therefore their combined effect on whitecap coverage may not be clearly pronounced, but obviously variations of magnitude of W, at the same wind speed U, but for different development stages U/c, can be expected.

Another essential uncertainty, also pointed out by the early researchers, is due to the presence of biological surfactants. Monahan (1971) mentions

> "variations in sea water surface tension caused by the occasional presence of organic films (Abe *et al.*, 1963; Garrett, 1967)".

Such 'occasional presence' may have a significant impact on regional variations of whitecap-coverage dependence, because of different biological activity in an ocean region. Stramska & Petelski (2003), for example, conducted observations in a polar area, where the amount of dissolved organic material is different to the waters of traditional observations, and concluded that some of the historical relationships for W-versus-U would overestimate their observations by a factor as large as 8.

To conclude this brief overview of the early observations of whitecaps, we have to say that, while studies of this property of wave breaking have now spanned more than six decades, many uncertainties still remain in parameterisations of whitecap coverage. Given that the importance of this oceanic characteristic grows as the means of remote sensing of the ocean develop, it is important to realise these uncertainties and address them. Some of them are linked with gaps in our understanding of the wave-breaking process, and some are not. Reliable experimental means for whitecapping detection, monitoring and quantifying are certainly available (see Section 3.6).

3.2 Traditional means (visual observations)

When capacities for measuring and storing surface elevation data became available, further research of wave breaking concentrated on the hydrodynamic characteristics of breaking, rather than on the connections with the wind as in Section 3.1. As far as the wave-breaking part of the measurement is concerned, still, for many years visual observations remained the most reliable, or sometimes the only means of detecting breaking events. These included manual tagging of breaking waves in the surface elevation records (Weissman *et al.*, 1984; Holthuijsen & Herbers, 1986; Stolte, 1992, 1994; Babanin, 1995) or video-taping of the surface, accompanied by collocated measurements of the waves (Katsaros & Atakturk, 1992; Babanin *et al.*, 2001). In the laboratory experiments or field observations, tagging can be performed in real time while observing recorded breaking waves and marking them electronically. In the case of video-taping, such tagging can be done by means of repeated watching of the recorded videos, which reduces the possibility of human error.

In any case, the observations deal with visible whitecapping, that is with the breaking-in-progress stage of the process (see Section 2.2). Therefore, they mostly concentrate on studies of breaking probability (see Section 2.5) and dependences of the breaking occurrence on wave, wind and other environmental characteristics.

The first major observation of the kind was by Holthuijsen & Herbers (1986). In many regards, it was a breakthrough in experimental studies of wave breaking, particularly in field conditions. One of the authors was watching a wave buoy and

"triggered a radio signal each time an active whitecap was seen to pass under the buoy. The signal was recorded synchronously with the buoy signal on one tape in an onshore station, thus identifying breaking waves with an 'on-off' signal".

Then wave-by-wave analysis of individual waves was conducted by means of the zero-crossing method. Probability and statistics of wave height, crest height, wave period, steepness and asymmetry of the breaking waves were studied for the first time and dependence of the breaking probability on wind speed was considered.

In particular, it was concluded that

"the breaking occurs at wave steepness values much less than the theoretically expected steepness of a limiting wave" (2.47).

This issue has been discussed in Section 2.1 above, and we now know that measurements of a wave that already exhibits whitecaps cannot provide estimates of the limiting steepness at breaking onset.

Other outcomes of the papers, for example, conclusions on the connection of breaking occurrence with the structure of wave groups, remain valuable up-to-date knowledge on wave breaking. Holthuijsen & Herbers (1986) concluded that the probability of a wave group having at least one breaker was higher for longer wave groups. In their observations, almost 100% of long groups, i.e. those consisting of seven waves, produced a breaker. Overall, 69% of all breaking waves happened within a wave group. Those were breaking close to the centre of wave groups, slightly ahead of this centre. As will be discussed later, such observations are consistent with the present understanding of the two-phase behaviour of wave breaking (see Young & Babanin, 2006a, Chapters 5 and 6) and with the asymmetric shape of nonlinear wave groups due to modulational instability (Shemer *et al.*,

2002), which is expected to be the major mechanism responsible for wave breaking in deep water (see Chapter 5).

Babanin (1995) conducted wave measurements and wave-breaking observations in the field and the laboratory. In the field, the wave probes were resistance staff gauges that recorded the elevations and buoys that recorded the surface acceleration (Babanin *et al.*, 1993). In all cases, breaking detection was conducted by means of visual observation and synchronous electronic labelling of breaking events.

As in Holthuijsen & Herbers (1986), it was concluded that the mean steepness of the breaking waves is well below the limiting steepness (2.47) and that the waves usually break close to the centre of wave groups. Unlike Holthuijsen & Herbers (1986), however, it was found that all the breakers displayed a well-pronounced negative front-to-back asymmetry (1.3) and that a dominant breaker at the top of wave group envelopes is usually accompanied by one or sometimes two subsequent breaking dominant crests. In any case, as mentioned above, both observations deal with the waves already breaking rather than with breaking onset, and as is now well-understood, in the course of such breaking a wave can go through a broad range of steepnesses, which will be gradually decreasing after having reached some maximum, and of asymmetries which will tend to oscillate (e.g. Babanin *et al.*, 2007a, 2009a, 2010a, 2011a, see also Section 2.1 and Chapter 5).

Acceleration measurements by Babanin (1995) demonstrated a very strong asymmetry of the time series of vertical acceleration in breaking events. While in general oscillations of the acceleration the signal is quite symmetric, in the course of breaking the downward acceleration could reach a value of (2.50). This is consistent with theoretical expectations for monochromatic limiting waves (e.g. Longuet-Higgins *et al.*, 1963) and with observations of waves that can be treated as quasi-monochromatic (e.g. Liu & Babanin, 2004), but it is significantly less than (2.59) which could be expected for a complicated sea (Longuet-Higgins, 1985).

Babanin (1995) marked conditionally large and small breaking events by electronic tags of different duration which allowed him to research their statistics separately. Together with measurements of both surface elevation and surface acceleration in breaking waves, this helped to reveal new features of the wave breaking properties across the spectrum. The large breakers were interpreted as those of dominant waves belonging to the spectral peak region, and small breakings were related to the equilibrium interval of the spectrum.

Breaking of energetic waves amounted to approximately one third of the total number of breakers which is about half of the percentage measured by Holthuijsen & Herbers (1986). When the same data were used for further analysis, this ratio was found to depend on the wave development stage (Banner *et al.*, 2000), so this and the above-mentioned differences between the Holthuijsen & Herbers (1986) and Babanin (1995) observations can perhaps be attributed to the wave age of the observed sea states. The latter study also showed that, while average breaking rates did depend on wind speed, variations of the dominant breaking occurrence actually correlated with the running variance of surface elevations. Since such a wave variance is mostly determined by the contributions from the spectral peak, this finding demonstrated that properties of the waves themselves, i.e. hydrodynamics rather than the wind, ultimately control the breaking process of dominant waves.

The breaking of small waves, on the other hand, appears to correlate with the running variance of the acceleration. The spectrum of the acceleration is dominated by high-frequency contributions into the variance (e.g. Babanin *et al.*, 1993). Thus, the correlation signifies a connection of the small-scale breaking with variations of the level of equilibrium interval in the wave spectra and points to the two-phase behaviour of wave breaking which will be an important discussion issue in breaking dissipation (Section 7.3.4).

Thus, Babanin (1995) concluded that the short-term equilibrium of the wave spectrum tail (i.e. level α in the spectrum shape formulation (2.7)) was supported by changes in the frequency of breaking occurrence of waves that belong to this tail (see also Section 8.2). Breaking of the dominant waves controls the energy in the spectral peak region, that is the enhancement γ of the peak (2.7). Overall breaking rates were also found to depend on the spectral width which was formulated as

$$\nu = \frac{m_0}{f_p F(f_p)} \tag{3.20}$$

following Belberov *et al.* (1983).

Observations similar to those by Holthuijsen & Herbers (1986) and Babanin (1995) have obvious limitations due to human error when marking rapid and suddenly happening events, and due to the physical ability of the observer to only resolve large enough events. Holthuijsen & Herbers (1986)

> "watched the buoy from either a nearby observation tower (~100 m from the buoy) or from a nearby ship (~50 m from the buoy)".

Their estimate is that they could see whitecaps of the minimum size of approximately 15 cm. Babanin (1995) watched the whitecaps occurring either inside an array of wave staffs or on a drifting buoy. In the first case, the observer was located 16 m directly above the array and could resolve any breaking that produced whitecaps. In the case of the buoy, the observer was watching this from a drifting research vessel, some 100 m away, and had a clear view of the whitecaps down to the size of the buoy (less than a metre (Babanin *et al.*, 1993)).

Such limitations can be overcome to some extent by high-resolution video taping of the wave surface. Having followed such an approach, Katsaros & Atakturk (1992) reported quite unique statistics, including that of micro-breaking. Contrary to many expectations (see Section 2.8), micro-breaking rates in their observations did not appear higher than those for plunging and spilling breaking. This counter-intuitive conclusion can perhaps still be explained by the difficulty of visual resolution of micro-breakers, even in the course of repeated video observations. An important finding of Katsaros & Atakturk (1992) is that, while the overall amount of plunging/spilling breaking depends on average on the background mean wind speed, the amount of micro-breaking does not. This can now be explained by the cumulative effect of wave breaking, that is by the fact that small-scale breaking is affected by the breaking of larger waves or modulation due to the larger waves. In such a case, the cumulative effect at high frequencies will be such that the induced breaking of the short waves will dominate the breaking caused for inherent reasons. Therefore, breaking rates at small enough scales should be fully determined by the behaviour of the longer waves in the system, and will be poorly correlated with the background wind (Babanin & Young, 2005; Babanin *et al.*, 2007c, see also Section 7.3.4). Another important conclusion of Katsaros & Atakturk (1992) is that, if the breaking statistics is investigated in terms of the wave-generating wind, it will depend on two parameters: the wind speed and the wind forcing u_*/c_p (here, c_p is the phase speed of waves at the spectral peak f_p).

Other examples of visual observations and manual tagging of breaking events include investigations by Weissman *et al.* (1984) and Stolte (1992, 1994) where the tags were used to develop *ad hoc* criteria in order to then computerise the subsequent data processing. These and other contact measurements of wave breaking, based on assumed trial-and-error criteria, will be reviewed in Section 3.3.

Thus, the traditional visual observations gave a significant boost to understanding, description and parameterisations of breaking rates in different wind–wave fields. This is, however, a very laborious kind of study involving an observer who marks and then counts breaking events by visually monitoring wave probes, where the surface elevations are recorded, and waiting for whitecapping to occur. While the visual observations do not suffer from uncertainties due to empirical thresholds set in 'automated' wave-breaking detection techniques described later in this chapter, they are nevertheless subject to human error and are too manually intensive and time consuming to be broadly employed in modern wave research.

3.3 Contact measurements

Contact measurements of wave breaking imply instrumentation placed at the air–sea interface where the breaking actually happens. What should they measure however? There are vast volumes of records by wave probes sensing surface elevations, velocities and accelerations that have accumulated over decades; most of them undoubtedly had breaking waves passing over and recorded, but they are impossible to identify without some criteria which would distinguish breaking from non-breaking waves.

A significant number of clever and elaborate attempts and approaches have been tried in this regard. Since the most obvious feature of breaking is discontinuity of just about every geometrical, physical and even chemical property at the ocean surface, many studies aimed to develop an *ad hoc* criterion by means of trial and error to detect such a discontinuity. Others have relied on known physical limiters, i.e. limiting Stokes steepness (2.47), limiting acceleration (2.60) or limiting orbital velocity (2.49).

Contact measurements of wave breaking is an area of research extensively explored and developed over a period of some 30 years, and it is impossible to mention even briefly all relevant studies. Here, we will suggest a review of a small selection of examples in an attempt to illustrate the ongoing experimental effort on the phenomenon which has been resisting such investigations and still remains difficult to approach experimentally.

The field study by Thorpe & Humphries (1980) used a breaker detection scheme based on the jump of the surface slope associated with the leading edge of the breaking region. Following the tradition of whitecap observations, they were seeking a correlation between wind speed and breaking probability. The data clearly indicated such a dependence, but no quantitative relationship was provided.

A similar breaking detection technique was developed by Longuet-Higgins & Smith (1983) who suggested a physical-limiter criterion; this was followed by many studies based on this criterion. The authors reasoned that

"if a breaking wave passes the recorder, we expect a sudden jump in surface elevation"

which has to be detectable because of large values of the time t derivative of surface elevation $\partial\eta/\partial t$. To detect the jumps, a wire wave probe was deployed on a specially designed buoy. The buoy was tested first in the laboratory and then applied in the field to look for the derivative

$$R = \frac{\Delta\eta}{\Delta t} = c\frac{\Delta\eta}{\Delta x} = c\tan\theta \tag{3.21}$$

where the discretisation Δ is determined by the sampling frequency of the time series, and the limit is defined by the maximal possible inclination of the surface (2.56). The authors argued that if lower values of R are set as the criterion for the jump-detection circuit, breaking in progress will be identified, but caution should be exercised not to have the circuit respond to steep, but non-breaking waves.

Longuet-Higgins & Smith (1983) applied their technique in the field, and measurements were conducted for wind speeds from 1 m/s to 14 m/s. At the maximal wind speed, the breaking rates found were of the order of 1% which is consistent with later field observations (e.g. Babanin *et al.*, 2001). Interestingly, however, they found that histograms of measured surface jumps in breaking waves did not depend on the actual selection of the critical value for R (provided, of course, it was high enough). Measured values of R often exceeded what would signify the value in the maximal theoretical ratio of

$$\frac{R}{c} = 0.586 \tag{3.22}$$

which follows from (3.21) and (2.56).

Detection of the surface-elevation jumps in accordance with the technique defined by (3.21)–(3.22) was employed for wind-generated waves by Xu *et al.* (1986) and Caulliez (2002) in the laboratory. Xu *et al.* (1986) measured the probability of breaking occurrence, breaking height and breaking duration, their dependence on the wind and their interconnections. The data disintegrated into two distinct groups, the dominant breakers and the second group

"probably associated with either the breaking of small waves riding on long waves or other profile irregularities".

This is a feature that is becoming familiar and has already been mentioned a few times before – two-phase behaviour of wave breaking in a spectral environment (see Chapters 5 and 6 for more details). Xu *et al.* (1986) described statistics and dependences for the two groups separately. For the dominant breakers, they found an average steepness of $ak = 0.375$. This will be discussed later, since the detection was based on ratio R/c exceeding the limit (3.22); this is most likely the steepness of waves already breaking rather than a criterion for incipient breaking, as it was interpreted.

Caulliez (2002) found that breaking waves detected in the wave records this way exhibit a self-similar shape with very strong negative asymmetry of $A_s \approx 0.5$ (1.3). Like Longuet-Higgins & Smith (1983) and Xu *et al.* (1986), Caulliez (2002) also observed the ratio R/c (3.22) exceeding the maximal theoretical value of 0.586. On average, the measured maximal slopes of wavefronts were 45° which is

"much larger than 30°, the value predicted for the highest Stokes wave" (i.e. 2.56).

Longuet-Higgins & Smith (1983), Xu *et al.* (1986) and Caulliez (2002) interpreted the fact that they found $\frac{R}{c} > 0.586$ as measuring 'the waves that are just about to break', 'not actually breaking', 'near-breaking', or in other words incipient breaking or breaking onset (see Section 2.1). Here, we would disagree with the interpretation. If θ_{critical} (2.56) defines the maximal possible inclination of the surface for steady waves, then inclination angles of $\theta > \theta_{\text{critical}}$ should signify the surface that is unstable and is already breaking. As the review of pictures of breaking waves in Chapter 1 shows, local inclination of the surface in the course of breaking can reach almost any angle and perhaps even 90° in a vertically plunging jet. The fact that histograms measured by Longuet-Higgins & Smith (1983) did not depend on the critical value for R supports such a conjecture: raising the level of R did not remove the breakers in progress and did not eventually limit the statistics to only the incipient breakers because the breakers in progress exhibited surface steepness, locally, higher than the limiting Stokes-wave steepness.

Therefore, the techniques developed by Longuet-Higgins & Smith (1983), Xu *et al.* (1986) and Caulliez (2002) appear to be an excellent practical tool to identify and measure properties and statistics of breaking in progress, rather than those for breaking onset. This supposition is further supported by the experimental results of Babanin *et al.* (2007a, 2009a, 2010a) who found that at the breaking onset a wave is nearly symmetric with respect to the vertical, but it inevitably starts leaning forward (develops a progressing negative asymmetry (1.3)) as it begins and continues to break, in accordance with measurements of the breaking-wave shape by Caulliez (2002).

Another example of using physical limiters to detect breaking events is a laboratory study on wind-generated waves by Hwang *et al.* (1989). The authors employed two limiting criteria for the water waves in order to detect breaking, i.e. a geometric criterion (2.56) and a kinematic criterion (2.49), in a comprehensive study intended to investigate breaking probability, breaking duration and lengthscale, breaking phase with respect to the wave shape, breaking severity and the probability of multiple subsequent breaking within one wave group. The instantaneous local values for the surface slope and the orbital velocity were obtained from the time series of surface elevation by means of a Hilbert transform (see Section 3.7).

Breaking regions were found to be geometrically similar, with the duration of their persistence and spatial extent being proportional to the scale of breaking waves. This observation gave good experimental support to the dissipation theory by Hasselmann (1974) which has formed the basis for dissipation terms employed in spectral models up to now, and has only started to be challenged recently, in the light of new experimental evidence (see Chapter 7).

Interesting observations were made with respect to the breaking phase, which challenged the unambiguity of the breaking criteria themselves. According to the authors,

> "... based on the geometric criterion, the breaking inception is on the downwind side of the wave crest, while the kinematic criterion indicates inception on the upwind side. This delicate difference, although restricted to a narrow region of $\pm 10^\circ$ of the wave crest demonstrates that the breaking phenomena described by these two criteria, as well as by other threshold variables, are not exactly the same".

Depending on the wind speed, a trend was observed for the breaking to move closer to the crest at higher wind speeds. Since for higher wind speeds Babanin *et al.* (2009a, 2010a) observed depleted severity, it may be that the phase of the wave-breaking region relative to the wave crest is linked to the breaking strength.

The probability of breaking, breaking duration and breaking length were investigated in terms of wind speed. Repeated breaking within wave groups was also found to correlate with the wind, but the probability of a repetition in these laboratory observations was quite low compared to field observations (e.g. Donelan *et al.*, 1972; Holthuijsen & Herbers, 1986; Babanin, 1995). Such a probability was about 20% at 16 m/s wind and only around 10% at 7 m/s. Current understanding of repeated breaking indicates that it is primarily due to the behaviour of nonlinear wave groups, i.e. hydrodynamics, rather than the wind. The strong wind, however, can influence wave grouping (through slowing down growth of the modulational instability) and correspondingly the breaking and reduce the breaking severity, and thus decrease or even diminish the probability of multiple subsequent breaking (Babanin *et al.*, 2009a, 2010a; Galchenko *et al.*, 2010). For laboratory waves, which are always young, wind forcing is usually quite strong even at moderate wind speeds and thus could have affected the observed statistics of repeated breaking.

Estimates of breaking severity, because they are rare, are most instructive. Hwang *et al.* (1989) evaluated the loss of potential energy in terms of ratio $\Delta H/H_{\text{before}}$ where $\Delta H = H_{\text{before}} - H_{\text{after}}$. They found that it is

$$\frac{\Delta H}{H_{\text{before}}} = 1 - \frac{H_{\text{after}}}{H_{\text{before}}} = 30\% \pm 3\%. \tag{3.23}$$

If translated into energy, this gives

$$s = 1 - \frac{H^2_{\text{after}}}{H^2_{\text{before}}} \approx 50\%. \tag{3.24}$$

Significantly lower were the losses of kinetic energy Q (2.26):

$$s_{\text{kinetic}} = \frac{\Delta Q}{Q_{\text{before}}} = 20\text{–}25\%. \tag{3.25}$$

Such a difference is qualitatively consistent with recent observations. Young & Babanin (2006a), when comparing spectra of pre-breaking and post-breaking waves, found some 40% of energy lost in the peak region of the surface-elevation power spectrum in the case of dominant breaking, while the peak region of the velocity spectrum measured below the wave troughs remained virtually unaltered.

An interesting account of a contact laboratory experiment with wind-generated waves was provided by Leikin *et al.* (1995). It was dedicated to measurements of wave asymmetry A_s (1.3), and although breaking detection was not conducted, discussion of these results is relevant here.

Leikin *et al.* (1995) found a strong correlation of the asymmetry A_s with inverse wave age u_*/c_p, whereas correlations of skewness S_k (1.2) and steepness ϵ (1.1) with wind forcing were poor. Thus, they supposed that

"the vertical asymmetry of waves is caused by direct wind forcing".

At the same time, they conducted bispectral analysis which revealed strong nonlinear coupling between the main wave and its harmonics. Therefore, Leikin *et al.* (1995) also concluded that the observed wave system can be treated as mainly the dominant component with its harmonics propagating at the same speed. Normally, such a system would have non-zero skewness S_k, but not, on average, non-zero asymmetry A_s. Therefore, the phase shift between the main component and the harmonics was considered and found to increase with u_*/c_p, and thus explained the growth of vertical asymmetry.

The two conclusions, i.e. the direct influence of the wind on wave shape and this shape being a result of bound nonlinear harmonics, seem contradictory. The first process signifies a strong air–sea interaction, whereas the second appears a purely hydrodynamic phenomenon. The influence of the wind could perhaps be considered to be responsible for the overall growth of the steepness and consequently the nonlinearity, including bound harmonics, but such straightforward reasoning cannot explain the growing phase shift between the carrier wave and the harmonics, and most importantly, there was in fact no correlated growth of wave steepness as a function of wind forcing.

We would suggest a different explanation of the trend observed by Leikin *et al.* (1995). At high wind forcing there is indeed a strong connection between the wind input into waves and the asymmetry (Agnon *et al.*, 2005). This connection, however, does not bring about negative average asymmetry, it only causes correlated oscillations of the input and A_s. In our view, what makes the asymmetry non-zero on average (in the formulation of Leikin *et al.* (1995) it is positive rather than negative for waves leaning forward) is wave breaking.

The range of wind speeds involved in the experiment was $u_* = 0.27\text{–}1.71$ m/s which, if converted into corresponding winds at a standard 10 m height using (3.19), gives $U_{10} \approx$ 8–48 m/s. These signify some very high wind speeds. Agnon *et al.* (2005) found that for strongly forced and highly nonlinear wind–waves, the wind–wave energy transfer, which depends on the mean wind speed, and the wave asymmetry are correlated, and they both oscillate with the same period. Thus, the connection of magnitude of the asymmetry with the wind speed observed by Leikin *et al.* (1995) can be explained through the mechanism of air–sea coupling.

The waves, however, have to become nonlinear first, and this comes through the gradual growth of the wave steepness as a result of wind action. According to Agnon *et al.* (2005), the steepness ϵ and skewness S_k also exhibit oscillations correlated with the wind speed, but no direct connection of ϵ and S_k was observed by Leikin *et al.* (1995).

This discrepancy can be explained by the growth of wave-breaking occurrence that accompanies increasing wind speeds. In nonlinear wave trains, steepness, skewness and asymmetry all oscillate due to modulational instability (Babanin *et al.*, 2009a, 2010a). The mean value of A_s remains close to zero unless a wave breaks. The wave always breaks when it starts leaning forward, i.e. its steepness and skewness are maximal and the asymmetry $A_s \approx 0$. In the course of breaking such a wave exhibits $A_s < 0$ and at this stage the oscillations of the asymmetry are interrupted. Therefore, if many wave-breaking events take place within a wave record, the average asymmetry over such a record will deviate towards negative values, whereas maximal values of ϵ and S_k are not affected (see also Section 5.3.3 and Eqs. (7.18)–(7.19)).

It is also instructive to notice that values of asymmetry A_s in Leikin *et al.* (1995) tend to saturate at $u_*/c_p \geq\sim 1.2$ (their Figure 1a). Since the phase speeds of the laboratory waves measured were $c_p \leq\sim 1$ m/s, this translates into the saturation at wind speeds

$$U_{10} \approx 34\,\mathrm{m/s} \tag{3.26}$$

which is consistent with recent observations both in the field (Powell *et al.*, 2003; Jarosz *et al.*, 2007) and the laboratory (Donelan *et al.*, 2004) for the saturation of sea drag (3.8) at wind speed (9.12), and for the change of regime of gas transfer (9.50). Kudryavtsev & Makin (2007) investigated the former case and explicitly connected it to wave breaking.

Returning back to the issue of contact measurements of wave breaking, two good examples of trial-and-error methods are papers by Weissman *et al.* (1984) and Stolte (1992). As mentioned in Section 3.2, both studies used visual observations in order to develop an *ad hoc* breaking criterion and further process wave-breaking records automatically.

Weissman *et al.* (1984) were measuring the surface elevations with very thin wire probes and were detecting the breaking waves based on spectral energy in the 18–32 Hz frequency band. A wave was regarded as breaking if the energy density exceeded some threshold level in the vicinity of the crest of a respective wave group. The technique was applied in a fetch-limited field experiment, and the breaking statistics at ~6 m/s wind was considered. Again, it was found that the mean steepness of breaking waves is well below the Stokes criterion (2.47). While temporal coverage of the breaking was only 1.2%, the relative high-frequency energy in those events was ten times that, i.e. 12%. Such a difference may bear significant implications for interpretations of radar and other remote-sensing techniques that rely on short-scale waves.

Stolte (1992) high-passed wave records at 1 Hz cut-off frequency and used visual observations to deduce empirical criteria: e.g. +2.5 cm-jump over 0.31 s to mark the start of a breaker and −1.25 cm over 0.016 s for the breaker's end. Very interesting statistics for the breaker velocity, height, length, momentum, spectral distribution of the breaking momentum and parameterisations of the total and relative numbers of breaking were obtained. Stolte (1992) also provided an experimental dependence of breaking severity on wind speed, which is quite a unique result.

The contact means for detecting wave breaking and quantifying the breaking physics do not have to be restricted to only measuring surface elevations or other surface properties. Lammarre & Melville (1992), Su & Cartmill (1992) and Gemmrich & Farmer (1999), for example, developed a technique based on conductivity measurements below the surface. The breaking causes air-entrainment into the water column which is accompanied by a reduction of electrical conductivity. Effectively, the void fraction in the water is detected which signifies and quantifies breaking events, their probabilities, severity and other properties. Gemmrich & Farmer (1999) defined a wave as the breaking event if the air fraction exceeds 8% in accordance with the theory of Longuet-Higgins & Turner (1974). The method is quite sensitive and accurate, and when used in field experiments allows us to obtain the spectral distribution of breaking probability (Banner *et al.*, 2002).

Thus, the trial-and-error criteria based on interpretation of one or another kind of discontinuity/jump caused by the breaking can bring very fruitful and useful outcomes and should not be underestimated. Some of them, however, are very involved and leave room for uncertainties if they are to be repeated. For example, Stolte (1992) admits that his criteria had to be modified for more- and for less-intensively breaking wave fields. With respect to the Weissman *et al.* (1984) technique, Katsaros & Atakturk (1992) wrote:

> "Since the threshold, i.e. the absolute value of the measured band energy varies with wind speed (stress), gustiness, underlying long waves, currents etc., it had to be determined individually for each run".

Gemmrich & Farmer (1999), who used the void-fraction criterion, pointed out:

> "While this criterion utilizes a well-defined property of all breaking waves except microbreaking, the definition of a suitable threshold again seems arbitrary and depends on the precise depth of the measurement as well as other factors that we cannot control, such as the vertical gradient of air fraction".

3.4 Laboratory measurements in deterministic wave fields

Laboratory measurements may involve both contact and remote-sensing methods and deal with either random (for example, wind-generated as in Section 3.3) or deterministic wave fields. The principal difference of the latter, with respect to field observations, is that in such controlled repeatable experiments the location of breaking is known. If so, the necessary measurements of wave breaking can be planned without having to detect the breaking event first. These can therefore be fine measurements which either require precise positioning of measuring devices or employ delicate high-precision instrumentation, impossible to transport and deploy in the field, for example, particle image velocimetry (PIV).

In the content of this section, methodological and phenomenological differences between directional and shortcrested waves will also be discussed. Another topic in this section, which has been a discussion issue lately, is estimates of the wave-breaking dissipation based on the dimensional argument which involves speed of breaker propagation.

Rather than just concentrating on passive observations and measurements, in the laboratory an experimental effort can create conditions to simulate breaking by means of a designated physical mechanism, for example by means of superposition of linear waves achieved through use of frequency dispersion (e.g. Cummins, 1962; Davis & Zarnik, 1964; Longuet-Higgins, 1974; Rapp & Melville, 1990; Griffin *et al.*, 1996; Meza *et al.*, 2000), or a superposition of nonlinear waves through amplitude dispersion (e.g. Donelan, 1978; Pierson *et al.*, 1992), evolution of nonlinear wave groups (e.g. Melville, 1982; Babanin *et al.*, 2007a, 2009a, 2010a), or simply by artificial means such as concentration of wave energy because of converging channel walls (Van Dorn & Pazan, 1975; Ramberg *et al.*, 1985; Ramberg & Griffin, 1987) or over an obstacle or a submerged shoal (e.g. Ramberg & Bartholomew, 1982; Manasseh *et al.*, 2006; Calabresea *et al.*, 2008).

Conditions can also be created to exaggerate some wave-breaking features and extend the breaking time in order to study the phenomenon in greater detail. Care must be taken to properly scale down the observed and extended features, to clearly realise limits of applicability of the modelled conditions, but the potential rewards of such laboratory efforts can be very significant.

In this regard, an experiment by Duncan (1981) should be highlighted whose results were broadly implemented and whose ideas stimulated very many related experiments, applications and studies (e.g. Phillips, 1985; Melville, 1994; Phillips *et al.*, 2001; Melville & Matusov, 2002; Gemmrich *et al.*, 2008, among many others). In the experiment, a steady breaker was produced by means of towing a submerged hydrofoil with phase speed of the breaking wave. The profile of the breaking waves, velocity distributions, turbulent wake and other properties of the steady breaker were measured and investigated.

Of particular importance for subsequent studies and applications was Duncan's (1981) conclusion that the energy dissipation rate in such a breaker can be described by a single independent variable, the wave's phase speed c. Thus, the rate of energy loss per unit length of the breaking front is proportional to:

$$S_{ds}(c) \sim \frac{\rho_w}{g} c^5. \tag{3.27}$$

Since in the linear or quasi-linear sense the phase speed translates into frequency/wavenumber, such a conclusion signifies the existence of a universal spectral function for wave energy dissipation, and attempts have been undertaken to formulate and quantify such a dissipation term (Melville, 1994; Phillips *et al.*, 2001; Melville & Matusov, 2002; Gemmrich *et al.*, 2008, among others).

In reality, however, breaking is principally unsteady, and this brings about a proportionality coefficient broadly called the breaking parameter b_{br}. Measurements within realistic unsteady-breaking conditions in the laboratory and in the field have seen b_{br} varying by some two to four orders of magnitude (see Section 7.4 for more details). Some studies pointed out its dependence on wave steepness ak (e.g. Melville & Matusov, 2002; Drazen *et al.*, 2008), others argue that it is not the average steepness (1.1), but rather local variations of the steepness, that is wave slope, crest-to-wavelength ratio, along with other characteristics such as density of the whitecapping foam, relative orbital velocity (with respect to the phase speed) that form such a breaking parameter b_{br} (Gemmrich *et al.*, 2008).

If we identify a range of open ocean-sea-lake peak frequencies very broadly as $f_p = 0.1$–0.3 Hz, then the range of dissipation rates at the spectral peak $S_{ds} \approx 9{\cdot}10^5$–$4{\cdot}10^3$ according to (3.27), which is only two orders of magnitude. That is, the predicted range of change of the dissipation based on the Duncan (1981) hypothesis is comparable or even less than the range that can be inflicted by uncertainties brought about by the 'proportionality coefficient' b_{br}. Given the fact that hypothesis (3.27) has been widely exploited in the wave-dissipation literature lately, it should be emphasised that its practical significance appears in fact rather low in the light of the enormous variability of this proportionality coefficient.

Besides, the connection of dissipation in a breaking wave with its phase speed is clearly not applicable in the case of induced breaking, i.e. the breaking of short waves caused by large waves, whereas such cumulative dissipation tends to dominate at high frequencies (Babanin & Young, 2005; Babanin *et al.*, 2007c). Therefore, even if the proportionality coefficient in (3.27) could be adequately quantified, the dissipation formulation of this kind, strictly speaking, is only applicable at the spectral peak region, where the cumulative effect is negligible. The cumulative effect is also absent below the peak, but clearly the dissipation (3.27) is not suitable there as it would have to keep growing towards longer waves, which contradicts common sense.

Generation of deterministic unsteady breaking in the laboratory has been most frequently attempted through focusing the wave energy by using wave frequency dispersion. The waves are generated mechanically with frequency being linearly decreased and thus the group velocity increased (2.19). As a result, linear superposition of waves in a predetermined location is achieved (although at the later stages, because of the increasing average steepness in the converging wave packet, some essential nonlinear interactions take place (Brown & Jensen, 2001)). The technique was originally suggested in ship design testing (Cummins, 1962; Davis & Zarnik, 1964) and was introduced into the broad research of wave breaking by Longuet-Higgins (1974), since when it has been employed in numerous laboratory experiments.

The most comprehensive study of this kind is that by Rapp & Melville (1990). Laser Doppler velocimetry was used to obtain two components of the velocity in the breaking region, but the main wave-measuring probes were simple surface-piercing resistance wires. Since the method allows positioning of the breaking location quite precisely, energy losses and other relevant information concerning the breaking can be obtained with only two wave gauges. The technique deals with groups of waves, rather than an individual wave that is breaking, and it was found

> "advantageous to consider the momentum flux and energy flux loss from the wave packet as opposed to considering individual waves"

(see also discussion in Section 2.7).

In Section 2.7 and throughout the text, the results of Rapp & Melville (1990) are mentioned and discussed with regard to a number of relevant topics. This comprehensive and thorough study represents a benchmark in terms of methodology of laboratory investigations of wave breaking and in terms of the physics of breaking which is caused by the linear superposition of a number of waves. Some features of such breaking are general, but some are distinctly different if compared with the physics of breaking that occurs, for example, as a result of nonlinear evolution of wave groups (Babanin *et al.*, 2010a).

What needs to be additionally highlighted in this section is the technique of flow visualisation of the breaking region devised by Rapp & Melville (1990) (in this regard, see also Janssen (1986b)). Rapp & Melville (1990) performed photography on the propagation and diffusion of dye initially floated on the surface. It was a precursor to modern flow-visualisation means for wave motion, such as PIV (e.g. Melville *et al.*, 2002; Qiao & Duncan, 2001; Grue & Jensen, 2006; Oh *et al.*, 2008; Babanin & Haus, 2009), but many results of Rapp & Melville (1990) on subsurface mixing due to a breaking event, such as temporal and spatial measures of the mixing, their connection with wave properties and momentum losses in the course of the breaking, remain state of the art to date.

With respect to studying the early stages of wave breaking, dye-visualisation in fact proved superior in some senses. The PIV techniques rely on projected laser sheets which illuminate a plane in the wave flume seeded with small neutral-buoyancy particles. Sequences of images, illuminated at high rate, are taken by a sensitive camera and allow detection of small displacements of the particles. Thus, two components of the water velocity can be obtained with high frequency resolution (and modern PIV techniques in some flows can obtain all three components).

The light, however, is scattered by air bubbles, which are many in the course of breaking in progress, particularly in the case of plunging breakers. Thus, the measurements need to be done outside of the aerated region. Melville *et al.* (2002), for example, essentially repeated the experimental setup of Rapp & Melville (1990) in order to investigate by PIV details of the velocity and vorticity fields due to breaking. They, however, had to start their measurements three wave periods after breaking onset, to allow for the large bubbles to surface. This points out limitations in the PIV technique. First of all, the most intensive phase of breaking is unavailable to investigation. Secondly, PIV instrumentation cannot attend to the dynamics of flow with void fraction, and this is an important dynamics as a large proportion of the breaking-wave energy is dissipated due to work against the bubble-buoyancy forces (up to 50% in Lammarre & Melville (1991, 1992) and at least 14% according to a more recent study of Blenkinsopp & Chaplin (2007)).

The linear-superposition technique was also employed by Nepf *et al.* (1998) and Wu & Nepf (2002) in their laboratory comparison of two- and three-dimensional wave breaking. The experiment was conducted in a wave basin where water fronts were generated by 13 independently programmed and driven paddles. If moved concurrently, the paddles would create a long-crested wave, but the crest could have been tapered laterally, and this was done by applying a cosine window to the signal which controlled the motion of the set of paddles. While the method of converging a wave packet due to phase dispersion was the same as in other tests which used such a superposition, the tapering allowed creation of short-crested waves and thus investigated the influence of the crest's three-dimensionality on wave breaking.

Here, we would like to point out that short-crestedness and directionality of the waves are often confused and used interchangeably in the literature, although this is not the same feature. Indeed, if 2D-Fourier-like analysis is applied to two-dimensional wavy surfaces, in the case of short-crested waves the outcome will be an angular distribution of wave energy. Such an angular distribution, however, will not be a δ-function even if the short-crested waves are strictly unidirectional. The opposite is also true: that is the superposition of long-crested waves having come from different directions will be decomposed by the 2D Fourier transform into a finite-width directional spectrum. This is because Fourier analysis, when applied formally, will treat the former situation also as a superposition of long-crested harmonic waves which they are not, and therefore the wave energy placed by such analysis at oblique angles is just a noise in the Fourier sense due to the short-crestedness.

This may seem to be an abstract mathematical argument, but it has a principal physical consequence for wave breaking. Modulational instability, which can lead to wave breaking, still exists in two-dimensional wave trains with three-dimensional wave crests, even though the breaking onset is set back from steepness $\epsilon = 0.29$ in strictly 2D waves to $\epsilon = 0.44$ in 2D waves with 3D wave crests (Melville, 1982; Babanin *et al.*, 2009a, 2010a). The proper directionality, however, can influence the modulational instability and therefore the wave breaking for this reason, if the directional spreading of the wave field is broad enough (Onorato *et al.*, 2009a,b; Waseda *et al.*, 2009a; Babanin *et al.*, 2011a).

In this context, Nepf *et al.* (1998) and Wu & Nepf (2002) conducted experiments with short-crested rather than directional waves. The three-dimensional structure of the crest was created by either focusing or diffracting the waves generated separately by the 13 paddles. Surface elevations were measured by an array of six wave gauges deployed on a carriage which could traverse the array to a measurement point near the wave-breaking location. An extensive set of geometric (wave steepness and shape), kinematic (2.49) and dynamic (up-frequency spectral energy shift) criteria were investigated.

It was found that the imposed directionality of the crest could either increase (focusing waves) or decrease (diffracting waves) the wave steepness at breaking onset (as opposed to the lateral modulational instability of the wave crest which increases the steepness at the incipient breaking (e.g. Melville, 1982)). Shape parameters at the breaking were altered in a similar manner. The kinematic and dynamic criteria, on the contrary, were not affected by the crest directionality. Since the dynamic criterion considered is sensitive and difficult to estimate, particularly in the field, it was concluded that the kinematic criterion (2.49) is the most robust property to describe the breaking onset.

Special attention was paid to the severity of breaking in the circumstances. The severity was defined in terms of wave groups (2.32). It was found that

> "spatially focusing and diffracting wave packets lost 34% and 18% of their energy, respectively, as a result of plunging breakers and lost 12% and 9%, respectively, as a result of spilling breakers. Comparable two-dimensional breakers with the same spectral shape lost 16% for plunging and 12% for spilling".

The breaking strength and its spectral distribution in the case of breaking caused by linear focusing, considered by Nepf *et al.* (1998) and Wu & Nepf (2002), is very different to the severity distribution in the breaking event caused by amplitude dispersion (Pierson *et al.*, 1992) or by nonlinear modulation (Babanin *et al.*, 2009a, 2010a). The wave superpositions, whether these are due to frequency or amplitude dispersion, concentrate wave energy and thus create a steep wave which then becomes unstable and breaks. A very steep wave can be brought about by a completely different physical mechanism, modulational instability of wave trains. Such instability leads to formation of nonlinear wave groups, within which rapid instantaneous concentration of wave energy occurs at some fetch. The location of such an event, which results in wave breaking if the concentration leads to formation of a steep enough wave, is repeatable in deterministic experiments, but it is not as precise as that due to the superpositions of waves discussed above. Breaking happens at a particular phase of the nonlinear group, close to the top of the group whose envelope is not symmetric, and since the group usually comprises a non-integer number of carrier waves, the exact position of the breaking oscillates (Melville, 1982; Babanin *et al.*, 2007a, 2009a, 2010a).

The oscillation, however, is quasi-periodic and measurements of the breaking onset or of other phases of breaking events can be conducted by a simple wave probe through recoding a number of breaking events at the location (Babanin *et al.*, 2007a, 2009a, 2010a). Since it is believed that the modulational instability may be one of the primary mechanisms responsible for wave breaking in ocean wave fields, later in this book significant attention will be paid to laboratory experiments dealing with breaking due to such instability (Chapters 5, 6).

3.5 Acoustic methods

Underwater ambient acoustic noise is generated through a number of possible sources such as precipitation, formation of bubbles in a saturated condition (e.g. Blanchard & Woodcock, 1957), biological sources (e.g. Chitre *et al.*, 2006), or breaking waves. In most circumstances, the latter is by far the dominant source (e.g. Kerman, 1988, 1992; Farmer & Vagle, 1988; Thorpe, 1992; Felizardo & Melville, 1995; Bass & Hey, 1997; Tkalich & Chan, 2002; Manasseh *et al.*, 2006, among many others). Here, we will describe the relation of ambient sound to wave breaking following Manasseh *et al.* (2006) and

Babanin *et al.* (2001) (see also Babanin *et al.* (2007b) and Section 8.3 for additional discussions). In the current section, detailed description of the Lake George field experiment will also be given. This complex experiment (Young *et al.*, 2005) is relevant for the acoustic methods of the present section and will also be referred to throughout the book with respect to other topics.

The ambient sound level in the ocean at a given frequency may vary by 20 dB, increasing with the wind speed (e.g. Knudsen *et al.*, 1948; Wenz, 1962; Kerman, 1988, 1992; Ding & Farmer, 1994). Wind and wave effects are most marked in the 0.1 kHz–10 kHz band. The general mechanisms of sound creation in this band are understood, although their interrelationships are not. Wind pumps energy into the wave spectrum, causing wave growth which can lead to breaking. The whitecapping from a breaker creates bubbles near the surface, and bubbles emit sound. However, it is known that the wind dependence is indirect. In most situations, it is the hydrodynamic evolution of the waves that determines whether breaking occurs (e.g. Babanin *et al.*, 2001, 2009a, 2010a). The wave-breaking bubbles can either generate the sound themselves as described below, or can transform pressure fluctuations in the air into acoustic noise in the water (Didenkulov, 1992).

Subdividing the 0.1 kHz–10 kHz band allows more detailed explanations. In general, it is above 0.5 kHz that the wind-dependent component of the sound spectrum dominates (Wenz, 1962). Furthermore, Bass & Hey (1997) and Babanin *et al.* (2001) showed that the sound spectrograms due to breakers become evident above 0.5 kHz. Theoretical work (e.g. Meyer, 1989) suggests that the bubble-formation process dominates the acoustic spectrum at frequencies greater than 0.5 kHz. From the basics of bubble acoustics briefly described below, 0.5–10 kHz corresponds to the natural emissions of millimetre-sized bubbles at near-surface depths. Frequencies around 0.1–0.5 kHz are likely to be produced by bubble clouds, not individual bubbles (e.g. Prosperetti, 1988; Lu *et al.*, 1990; Tkalich & Chan, 2002).

It has been well known since the time of Rayleigh (1917) that individual bubbles oscillate volumetrically with a natural frequency that depends on their size (see Leighton, 1994, for a review), suggesting an obvious application to instruments analysing bubbly flows. The simple-harmonic solution to the Rayleigh–Plesset equation describing bubble-acoustic oscillations shows that a single bubble's natural frequency is inversely related to bubble size, according to

$$\omega_0 = \sqrt{\frac{3\gamma_0 P_0}{\rho_w}}\,\frac{1}{R_0}, \tag{3.28}$$

(Minnaert, 1933) where ω_0 is the sound radian frequency, γ_0 is the ratio of the specific heats of the gas, P_0 is the absolute liquid pressure and R_0 is the equivalent spherical radius of the bubble. If the number of bubbles is assumed infinite, continuum approximations based on (3.28) permit overall acoustic properties of a bubbly cloud to be calculated (e.g. Commander & Prosperetti, 1989; Duraiswami *et al.*, 1998). The acoustic properties of bubbles have been the basis of several oceanographic instruments (e.g. Phelps *et al.*,

1996; Terrill & Melville, 2000) as well as industrial instruments (Duraiswami *et al.*, 1998; Boyd & Varley, 2001; Manasseh *et al.*, 2001) although none are in widespread use. Most systems measure bubble-size distributions, relying on an active principle. Sound is sent into the water and the attenuation or reflection of the resulting signals is interpreted to infer the bubble-size distribution.

However, bubbles also passively emit sound at their natural frequency, that is without being forced by an external sound field. As a bubble detaches from its parent body of gas, it produces an acoustic pulse. This may be due to sudden compression of the trapped gas as the bubble pinches off (Manasseh *et al.*, 1998). This 'ringing' of the bubble may last less than 10–20 cycles; for example, for a 2 mm diameter bubble (i.e. 3 kHz natural frequency), the pulse can last less than 10 ms. While any disturbance may cause the bubble to ring, the highest-amplitude sounds are created when a bubble is pinched off (Chen *et al.*, 2003; Vazquez *et al.*, 2005, 2008) or coalesces (Manasseh *et al.*, 2008). Many bubble-creation events occur per second, during processes ranging from filling a glass to wave breaking. Although humans perceive this as a continuous noise, it is due to many discrete, brief events. An individual bubble's pulse becomes briefer as the bubbles are produced more closely to each other; and the frequency of the signal drops during the pulse, with the earliest acoustic cycles being closest to the natural frequency given by (3.28) (Manasseh, 1997). It was shown by Manasseh *et al.* (2004) that these effects may be explained by inter-bubble acoustic interactions as the system becomes more 'cloud-like'. Furthermore, sound intensity drops rapidly with distance from the bubbles, which may be considered as monopole sources (Longuet-Higgins, 1989; Leighton, 1994).

These phenomena suggest that a sufficiently short time window triggered on a signal peak often contains data specific to a single, nearby, newly-formed bubble (Manasseh *et al.*, 2001). This implies that appropriately thresholded acoustic data can generate statistics as a function of time, on both the number of bubbles produced and their size. Detection of the bubbles can be linked to the periods of the waves producing them, if the wave records are taken simultaneously, and thus can be used to study the breaking probability (see Chapter 5). Bubble size appears to be an indicator of breaking severity (see Chapter 6). Thus, a single device can be employed to obtain and study wave-breaking energy dissipation. We should note, however, that, like most of the other methods described in this chapter, it has to rely on an empirical threshold.

To summarise the above, acoustic methods of investigation of breaking can be broadly subdivided into two large groups of studies: active acoustic probing and passive acoustic techniques, and less broadly into research on sound produced by individual bubbles and bubble clouds. The latter subdivision is only less broad as far as wave-breaking applications are concerned. The physics of oscillations of individual bubbles (Medwin & Daniel, 1990; Lowen & Melville, 1991b; Manasseh, 1997; Manasseh *et al.*, 1998, 2006) and bubble clouds (Carey & Bradley, 1985; Prosperetti, 1985; Tkalich & Chan, 2002; Manasseh *et al.*, 2004) are quite different. Detection and quantification of wave-breaking effects by means of individual bubbles or their clouds, however, simply implies dealing with different spectral bands of the acoustic noise produced in the course of breaking.

The subdivision into active and passive acoustic methods, on the other hand, is essential in wave-breaking studies. There are two basic active-acoustic techniques employed in this regard. One of the methods of active probing makes bubbles or bubble clouds resonate when they are exposed to an external source of sound (e.g. Thorpe, 1992; Farmer *et al.*, 1998; Terrill & Melville, 2000; Gemmrich & Farmer, 2004). While they resonate at the same frequencies that they emit when being produced or collapse naturally, technically this means the presence of sonar in the water which has to be powered and maintained. This limits applications of the technique, particularly in extreme weather conditions, and on a long-term or even regular basis. As a result, such active acoustic methods have been rather extensively employed to investigate oceanic phenomena related to breaking such as bubble clouds, bubble size distributions and void fraction (e.g. Thorpe, 1992; Farmer *et al.*, 1998; Terrill & Melville, 2000; Gemmrich & Farmer, 2004), but not so much wave-breaking physics and statistics.

Another active acoustic method, which has been gaining momentum over the past two decades or so, is based on using the reflective properties of water heterogeneities (such as small particles of matter, bubbles or even turbulent vortices) or of the water surface. The former uses Doppler shift of the sound reflected by moving inhomogeneities to determine the motion velocity, and the latter simply monitors sound reflected from the surface to measure surface oscillations. The velocity records can be acquired as three-dimensional time series (so-called acoustic Doppler velocimeters (ADV)) or as a sequence of spatial slices of one-dimensional velocity fields (pulse-to-pulse coherent Doppler profilers) (Dopbeam, Veron & Melville, 1999). Another type of active acoustic device of this kind are acoustic Doppler current profilers (ADCP) which, if, for example, positioned at a not-very-deep bottom, can measure velocity time series at a number of points between the bottom and the surface, as well as time series of surface elevations. The ADVs, Dopbeams and ADCPs are manufactured by industry both as research and applied instruments.

Such acoustic velocimeters have been used to investigate velocity fields beneath breaking waves (Doering & Donelan, 1997; Young *et al.*, 2005), to quantify differences in kinetic energy due to wave breaking (Young & Babanin, 2006a) and to measure turbulence caused by breaking, even within the crests of breaking waves (Gemmrich & Farmer, 2004). Compared to sonars, Doppler velocimeters have been in much broader use in wave-breaking studies, both in the laboratory and in the field. On one hand, this is because of the availability of industrially produced battery-operated models of ADVs and ADCPs designed for field use. On the other hand, even laboratory high-precision cable-powered versions made their way into dedicated field experiments (Veron & Melville, 1999; Gemmrich & Farmer, 2004; Young *et al.*, 2005; Young & Babanin, 2006a).

Among breaking-detection methods, passive acoustic determination of breaking and its properties has a potential advantage. The instrumentation (hydrophones) is relatively cheap, robust and easy to maintain. The hydrophones are deployed below the surface and are solid-state devices, therefore escaping most of the destructive power of breaking waves. They have been used lately even in hurricanes, to quantify and classify them

(Wilson & Makris, 2006, 2008). Once deployed, they can be operated on a long-term or regular basis and collect ready-to-process time series.

Passive acoustic measurements have been employed in a number of field observations and laboratory experiments. Their first applications to wave-breaking studies were by Farmer & Vagle (1988) in the field and by Melville *et al.* (1988) in the laboratory. Both experiments showed that acoustic signatures of breaking waves can be used to identify breaking events. Farmer & Vagle (1988) used a single hydrophone and found that the mean distance between the breakers and the acoustic strength of the breakers depends on the wind speed.

Ding & Farmer (1994) further advanced the technique. They developed a directional array of hydrophones and a method to track individual breaking events out in the ocean. The directional array made possible measurements of the phase speed of breaking events, and showed it was related to the spectral scale of breaking waves (the wave period) and therefore to the spectral scale at which dissipation occurs. Ding & Farmer (1994) obtained interesting statistics on the frequency and spacing of breaking occurrences, on breaking duration, dimension and speed, and some temporal and directional spectral characteristics of breaking probability. They showed a number of distributions of the breaking probability as a function of event speeds and event directions (which are analogues of the wave spectrum frequency and direction), but did not attempt to relate the magnitude and shape of the distributions to the wave spectrum and thus to obtain the spectrum of energy dissipation.

Lowen & Melville (1991a) extended and summarised results of earlier laboratory studies. They used measurements of the acoustic pressure and concluded that the duration of the hydrophone signal above a background noise threshold is proportional to the breaking-wave period. Their estimates also showed that, albeit small (i.e. $\sim 10^{-8}$ of the dissipated wave energy), the acoustic energy radiated by breaking waves is proportional to the mechanical energy dissipated.

This important finding is illustrated in Figure 3.1 reproduced from Melville *et al.* (1992). The onset, impact and duration of the acoustic noise brought about by a breaking event are clearly seen. The acoustic energy emitted in the course of breaking can easily be quantified.

These results provided a possible method for measuring temporal spectral scales of breaking events, and even the dissipation related to those scales, using a single hydrophone. In a spectral environment, that would potentially provide the breaking probability (Section 2.5) and severity (Section 2.7) and ultimately the spectral distribution of the dissipation (2.20).

In Lowen & Melville (1991a) and Melville *et al.* (1992), the waves were made to break by means of dispersive focusing of wave packets, generated mechanically, with a pre-selected central frequency. The method was effectively developed for a single-wave environment, and determination of the scales and energy losses of breaking waves in a complex spectral environment was beyond the scope of the studies.

While showing a certain promise in investigating such an elusive characteristic of the breaking process as wave-energy spectral dissipation, the technique, however, proved inapplicable in the field. This is, mostly, due to the much higher levels of background ambient

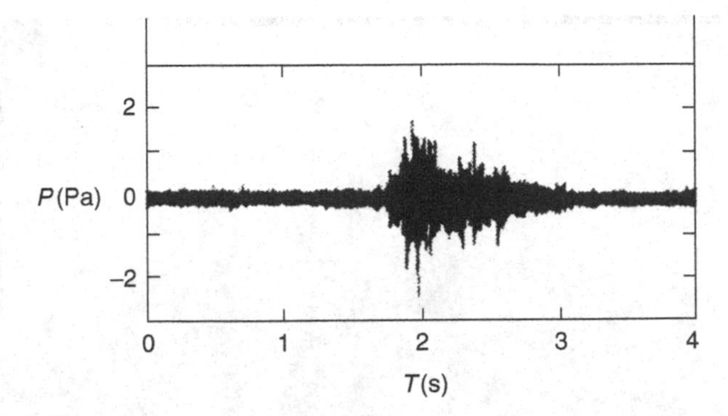

Figure 3.1 Hydrophone time series bandpass-filtered in the range 500 Hz to 10 kHz. Vertical scale is pressure in Pa, horizontal scale is time in sec. Figure is reproduced from Melville *et al.* (1992) by permission of Ken Melville

Figure 3.2 Video image of the breaker that occurred near $t = 1$ s of the one-minute record shown in Figures 3.4a,c and 3.5. Figure is reproduced from Babanin *et al.* (2001) by permission of American Geophysical Union

noise in the field which make the determination of both duration and energy of the breaking acoustic impact unfeasible (see also Babanin *et al.*, 2007b and Section 8.3).

To demonstrate this, in Figures 3.2, 3.3 and 3.4 breakers, detected by repeated viewing of video records, and synchronised wave records are shown. The data were collected during the Lake George field experiment (Babanin *et al.*, 2001; Young *et al.*, 2005). The first and last breakers of a one-minute segment of record 4 of Table 5.2 (i.e. those occurring in $t = 1$ s and $t = 55$ s) are shown in the captured video images seen in Figures 3.2 and 3.3, respectively. It is clear that these are cases where the crest of a breaking wave is passing through the wave array and over the bottom-mounted hydrophone underneath the array.

Figure 3.3 Video image of the breaker occurring near $t = 55$ s of the one-minute record shown in Figures 3.4b,d and 3.5. Figure is reproduced from Babanin *et al.* (2001) by permission of American Geophysical Union

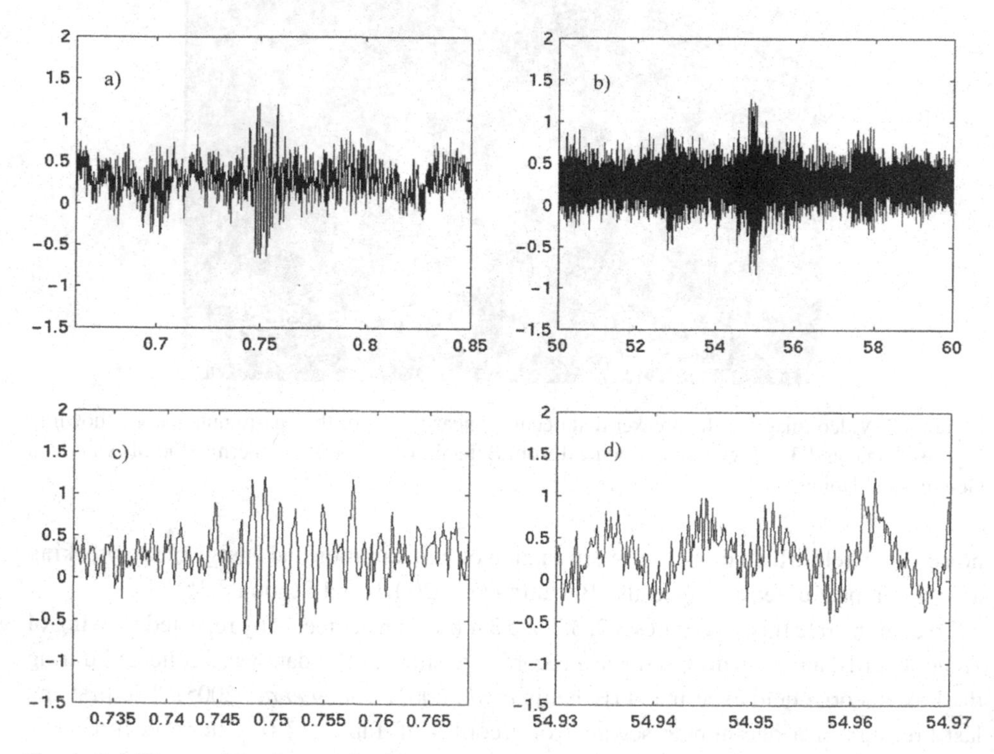

Figure 3.4 Time series of acoustic signatures of breakers shown in Figures 3.2 and 3.3. a) and c) Zoom-out and zoom-in for the breaker in Figure 3.2. b) and d) Zoom-out and zoom-in for the breaker in Figure 3.3. Figure is reproduced from Babanin *et al.* (2001) by permission of American Geophysical Union

Figures 3.4a–d plot time series of the digitised acoustic signal near $t = 1$ s and $t = 55$ s (Figures 3.4c–d are a zoom-in of Figures 3.4a–b). In both Figures 3.4a and 3.4b, there are well-defined peaks in the acoustic noise level associated with the breaking events. This is in qualitative agreement with laboratory results by Melville *et al.* (1992). In contrast to laboratory breaking waves, different fractions of energy are apparently lost by field breakers, and therefore the breaking noise impact above the background *in situ* ambient noise is not always evident in field acoustic time series. For instance, the breaking event that was observed visually at $t = 53$ s is not well defined in the time series in Figure 3.4b. It is, however, clearly seen in the corresponding acoustic noise spectrogram in Figure 3.5.

Figure 3.5 shows a spectrogram of this minute of the acoustic record. The spectrogram is a time series of consecutive spectral densities computed over 256 readings of the acoustic signal with a 128-point overlap; the segments were windowed with a Hanning window (see Babanin *et al.*, 2001, for further details). Values of the spectral density are shown using a

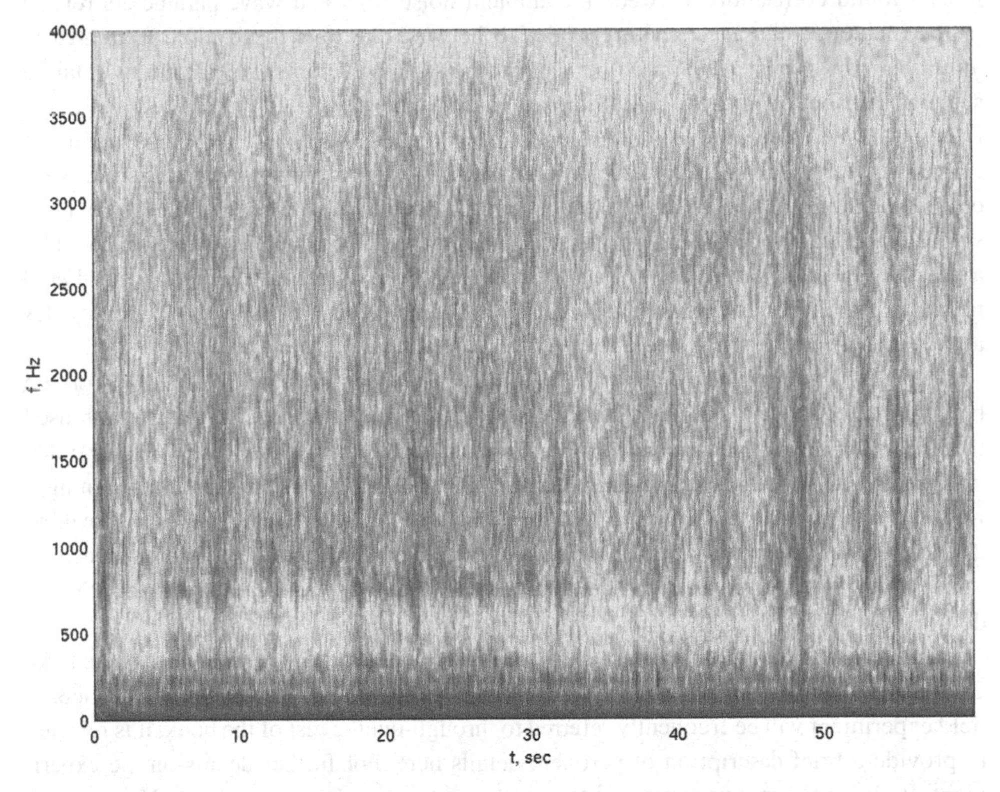

Figure 3.5 Spectrogram of one-minute record of acoustic noise recorded by a bottom-mounted hydrophone during wave record 4 of Table 5.2. Darker crests correspond to dominant waves breaking. The breaker in $t = 1$ s is depicted in Figures 3.2 and 3.4a,c, and the breaker in $t = 55$ s in Figures 3.3 and 3.4b,d. Figure is reproduced from Babanin *et al.* (2001) by permission of American Geophysical Union

logarithmic scale, with darker patches corresponding to higher values (i.e. louder recorded sound levels associated with dominant-wave breakers).

If the temporal resolution of the acoustic signal in Figures 3.4a–b is increased, as in Figures 3.4c–d, other features become apparent. For the breaker occurring in the first second (Figure 3.4c), there is an apparent alteration of the frequency of the sound: the acoustic carrier downshifts in frequency. This is not the case, however, for the breaker at $t = 55$ s (Figure 3.4d), in spite of the fact that the acoustic signature of this breaker has quite a distinct amplitude enhancement above the ambient noise. Nevertheless, both breakers are clearly seen as breaking crests in the spectrogram in Figure 3.5.

Felizardo & Melville (1995) applied the passive acoustics techniques of Lowen & Melville (1991a) and Melville *et al.* (1992) in the field, where breaking waves of all scales and various dissipation rates can be present at the same time. They argued that the dependence of ambient noise on wind is indirect, which argument signified an essential move away from the decades-long tradition of associating wave breaking with the wind, towards wave hydrodynamics which drives the physics of wave breaking as we understand it now. They indeed found correlations between the ambient noise level and wave parameters related to the incidence of wave breaking, and also between the total dissipation, estimated in a number of different ways, and the acoustic noise. No attempts were made to obtain a spectral distribution of the total dissipation. It should be mentioned that, in addition to the issue of the much larger levels of ambient noise in the field described above, extending the Lowen & Melville (1991a) and Melville *et al.* (1992) approach into the multi-scale wave environment may not be as straightforward as considering the duration of the hydrophone signal above a threshold even if it was clearly determined. The acoustic energy radiated by a breaking event and even the threshold itself (e.g. Babanin *et al.*, 2007b) can be altered because of multiple breakings nearby, or due to simultaneous breakings of different scales at the measurement spot.

A number of other passive acoustic techniques have been further developed to work on breaking detection and statistics. Bass & Hey (1997) and Babanin *et al.* (2001) both used the spectrograms of hydrophone-recorded noise to detect breaking events. As illustrated in Figures 3.2–3.5, the identification of distinct crests in the spectrograms, spanning a frequency range from 500 Hz to 4 kHz, appears to be a more reliable means of breaking detection in the complex spectral environment than the integrated ambient noise exceeding a threshold. The spectrogram method, however, can only be applied for the detection of dominant breakers.

Babanin *et al.* (2001) obtained wave-breaking data at the experimental site at Lake George near Canberra in south-eastern Australia during 1997–2000. Since the Lake George field experiment will be frequently referred to through-out the rest of the book, it is relevant to provide a brief description of pertinent details here. For further details on the experiment, its layout, instrumentation and measurements, we refer the reader to Young *et al.* (2005).

A contour map of Lake George, shown in Figure 3.6, indicates a simple bathymetry, with the bed sloping very gently toward the eastern shore of the lake. Since its bed form is

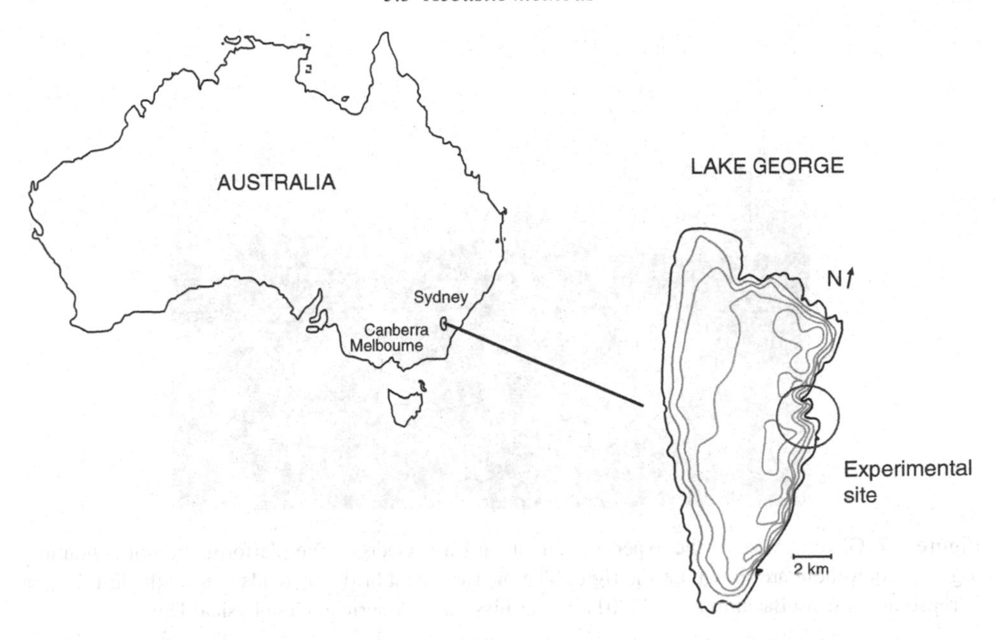

Figure 3.6 Location of the Lake George field experiment site. Figure is reproduced from Babanin *et al.* (2001) by permission of American Geophysical Union

extremely flat, Lake George is an ideal location to study *in situ* fetch-limited behaviour of wind-generated waves in a constant finite-depth environment.

The predominant wind directions at Lake George are westerly and north-westerly, and therefore the site was located near the eastern shore as seen in Figure 3.6. Under typical meteorological conditions the range of values of inverse wave age U_{10}/c_p and non-dimensional depth $k_p d$ expected at the site were $1.0 < U_{10}/c_p < 3.5$ and $0.7 < k_p d < 2.0$, respectively. At the extreme end, wind–wave records at wind forcing as strong as $U_{10}/c_p \approx 8$ were taken (Donelan *et al.*, 2006; Babanin *et al.*, 2007b). Here, k_p is the wavenumber corresponding to the spectral peak, and d is the mean water depth. For linear waves these experimental conditions are representative of intermediate-depth wind seas.

The experimental site included an observational platform with a shelter to accommodate equipment and researchers during observations (see Figure 3.7 for a general view). The platform was located 50 m from shore and was accessible via an elevated walkway. The site was equipped with a comprehensive instrumentation system for the study of finite-depth wind–wave energy sources and sinks. The measurements described here were made from a 10 m-long measurement bridge located to the side of the platform (Figure 3.7 shows a general view). Approximately half-way along the bridge, the water surface elevations were measured using an array of capacitance gauges (see also Figures 3.2, 3.3). A hydrophone was located directly below the wave probes and sensed the noise generated by individual breaking waves. The water surface around the capacitance array and hydrophone was

Figure 3.7 General view of the experimental site at Lake George. The platform and hut containing logging equipment are shown on the right. The measurement bridge extends toward the left. Figure is reproduced from Babanin *et al.* (2001) by permission of American Geophysical Union

viewed using a video camera. In addition, the records were electronically 'marked' by an observer who manually recorded visual occurrences of breaking. All the measurements were synchronised. Using this integrated system, it was possible to inter-relate the visual appearance of the water surface in any sequence of individual video frames and the acoustic signature of breaking waves at that time, together with the wave-height properties measured by the directional array at the breaking location.

An anemometer mast, accommodating three wind probes at 10 m and 5.65 m elevations above the water surface (two cup anemometers measured one-minute mean and two-second gust wind speeds and one wind vane measured wind directions), was erected 10 m from the platform beyond the end of the measurement bridge to avoid disturbing the air flow. Another anemometer mast, accommodating five wind probes at four heights closer to the surface (four cup anemometers and a wind vane), was set 6 m to the side of the bridge to ensure undisturbed airflow for these lower anemometers.

A total of 26 records of all the measured variables, relevant for studying wave breaking, taken over the period October–December 1997, were processed, and Table 5.2 summarises their parameters. Selection of records was based only on the requirement of relatively constant wind speed.

The wind speed U_{10} and wind direction were measured very close to 10 m above the water level by the uppermost wind speed and wind direction sensors on the anemometer mast. The wind probes were Aanderaa Instruments wind speed sensor 2740 and wind direction sensor 3590. Slow changes in the water depth in Lake George caused only small variations in the position of the sensors over the surface, and corresponding adjustments in the U_{10} speed were made on the basis of the wind profiles obtained with the anemometer mast.

Relatively young, strongly forced waves with $U_{10}/c_p = 3.3$–6.5 were employed for studying the wave-breaking statistics. Their spectral peak frequencies f_p ranged from 0.33 to 0.50 Hz. The wind speed U_{10} ranged from 8.5 to 19.8 m/s.

Visual detection of the dominant wave-breaking occurrence is arguably one of the most reliable methods available since it does not require any additional experimental criteria. At Lake George, most wave records were supplemented by video-taped images of the water surface surrounding the wave array. The video-taping allows repeated playback of wave records to establish the breaking statistics by visual means and to verify the results. This visual observation method was used to obtain the breaking statistics.

As mentioned above, the video records were synchronised with the surface elevation records gathered by the array of wave probes. Both the surface elevation sampling and video framing were performed at 25 Hz rate so that there was unambiguous correspondence between each wave height reading and a video image. For the video-taping a computer-controllable Panasonic model AG-7350 video recorder (VCR) was used whose time-code generator facility allowed rapid retrieval of particular segments of the recorded wave series corresponding to visually observed breaking events.

The hydrophone output was recorded on the VCR audio channel. The hydrophone had two signal gains, 20 and 40 dB. Normally, for developed wind–waves the 20 dB gain was sufficient to detect breaking waves. During data analysis, the acoustic signal was sampled at the 8 KHz rate and digitised. These synchronised time series of the acoustic signal contain potentially valuable additional information about visually observed dominant wave breakers, particularly in relation to breaker strength (see Manasseh *et al.*, 2006, and Section 6.2).

In Figure 3.5, a spectrogram of the first minute of record 4 in Table 5.2 is plotted. The dark crests across almost the entire 4 KHz frequency span in the spectrogram are associated with acoustic noise from dominant breaking waves. This was confirmed through repeated viewing of the synchronised video records. For example, the first and last breakers (near $t = 1$ s and $t = 55$ s) detected in the spectrogram in Figure 3.5 are shown in the captured video images seen in Figures 3.2 and 3.3, respectively. It is clear that these are cases when the crest of a breaking wave is passing through the wave array and over the bottom-mounted hydrophone.

After the connection between the spectrogram signatures and the visually observed dominant breakers was established, the acoustic spectrograms were used along with the video records to obtain the wave-breaking statistics (see Section 5.3.1). As a further example, Figure 3.8 shows another spectrogram of breaking occurrences during a 1 min-long segment of record 12 (Table 5.2) when the wind and waves were much weaker and only two isolated breakers occurred within the 1 min record.

Thus, the use of spectrograms rather than acoustic intensity time series is a preferred method for detection of dominant breaking events in the wave field. A similar conclusion was reached by Bass & Hey (1997), who used spectrograms to detect breakers in the natural surf zone. The acoustic time series, however, can provide useful physical insights into the breaking process, that is in order to estimate periods of individual breaking waves, the

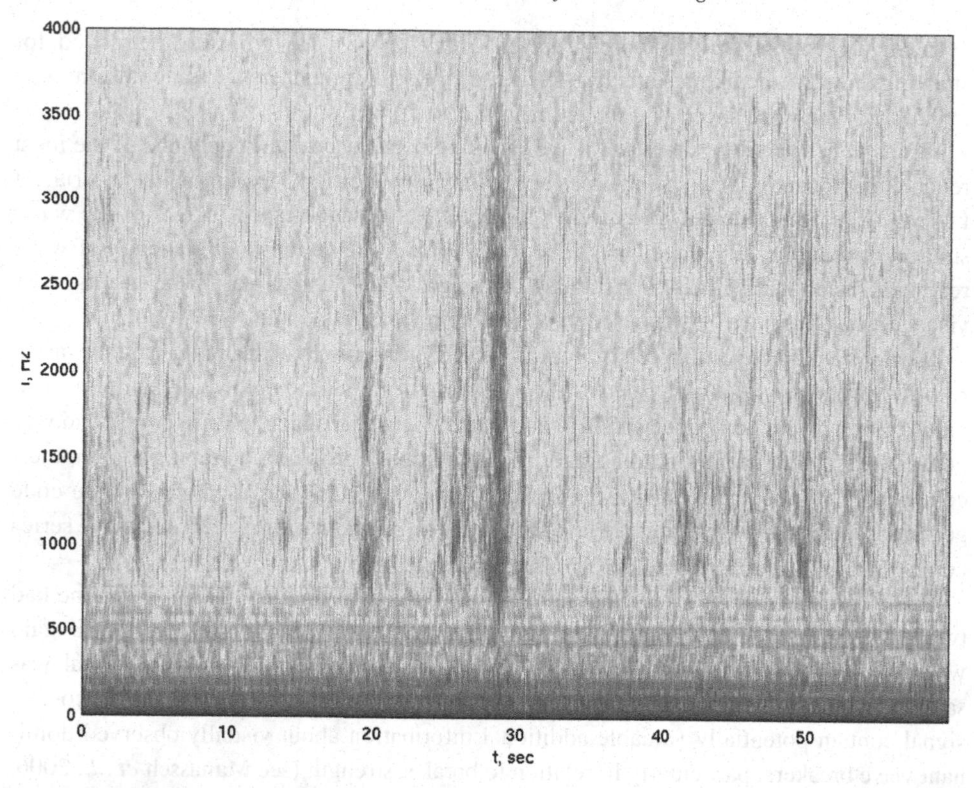

Figure 3.8 Spectrogram of one-minute record of acoustic noise recorded by a bottom-mounted hydrophone during wave record 12 of Table 5.2. Figure is reproduced from Babanin *et al.* (2001) by permission of American Geophysical Union

amount of energy lost etc., particularly in circumstances when the background ambient noise is weak as in the laboratory (Melville *et al.*, 1992).

The Lake George wave and breaking data were further used to develop another passive acoustic method, capable of detecting both dominant and small breaking events, as well as their severity (Manasseh *et al.*, 2006). As described above, individual sound pulses corresponding to the many individual bubble-formations during wave breaking events typically last only a few tens of milliseconds. For details on signal conditioning, pulse processing, determination of optimal trigger level and other elements of the analysis, we refer the reader to the original paper. Here, we will mention that with the new technique, each time a sound-level threshold was exceeded, the acoustic signal was captured over a brief window typical of a bubble-formation pulse, registering one count. Each pulse was also analysed to determine the likely bubble size generating the pulse.

Using the time series of counts and visual observations of the video record, the sound-level threshold that detected bubble-formations at a rate optimally discriminating between breaking and non-breaking waves was determined by a classification-accuracy analysis.

This diagnosis of breaking waves was found to be approximately 70–75% accurate once the optimum threshold had been determined.

The classification–accuracy analysis detailed in Manasseh *et al.* (2006) should be applied in further field experiments, covering a wider range of wind–wave environments than were available in Lake George. If in each future field experiment, synchronised surface video and underwater audio records were made with the same equipment and settings (for example, with the hydrophone the same distance below the surface and the camera the same distance above), the classification–accuracy analysis could be repeated on several data sets from a wider range of sea-state environments than studied here. If the classification accuracy of 75% is not worsened, it would suggest the discriminant recommended (in sound pressure amplitude at the hydrophone) has a universality.

The method was then used for detailed analysis of wave-breaking properties across the spectrum. To obtain wave-breaking probabilities of individual waves at different frequencies, a zero-crossing analysis was applied. From the time series of surface elevations, the period of each wave was calculated as follows. Times when the surface elevation crossed the mean or 'zero' level were noted. Two consecutive zero up-crossings were analysed and the time of the troughs preceding and following them were recorded. The difference between the trough times was taken as the period of that wave T, giving its frequency. Figure 3.9 (top panel) shows a 30 s section of the surface elevation data of record 5 (Table 5.2) used to calculate wave frequencies.

In Figure 3.9, limitations of the zero-crossing analysis at small scales are quite obvious. These limitations and other issues with the zero-crossing analysis are discussed in the Appendix of Manasseh *et al.* (2006). A superior alternative for higher-frequency breaking, which also preserves the wave shape contrary to the standard Fourier-based bandpassing procedures, would be a riding-wave removal (RWR) method (Schulz, 2009).

The synchronous passive acoustic wave-breaking data was then combined with the surface-elevation data and the zero-crossing analysis. Such bubble-detection events are shown, for the same time series, in the bottom panel of Figure 3.9. Occasional events which would correspond to negative surface elevations were excluded from the analysis to avoid possible ambiguity in detecting wave breakers when those events happened close to wave troughs. For each acoustically determined breaker, the frequency of the wave at the same time was extracted. The total number of breaking waves $n(f)$ was found for each of the calculated frequencies f.

When applied to real field data, a breaking-probability distribution could thus be obtained; this will be described in Section 5.3. This is the rate of occurrence of wave-breaking events at different wave scales. Therefore, the method provides spectral distribution of the breaking probability (2.4) rather than just the frequency of occurrence of dominant breakers (2.3) like the spectrograms. For the dominant breaking, the two techniques were compared and showed effectively the same breaking count, provided that the assumed dominant frequency band is $\Delta f_p = \pm 0.35 f_p$ (2.8). This comparison is shown in Figure 3.10.

With support from a separate laboratory experiment, the estimated bubble size was argued to be dependent on the severity of wave breaking and thus to provide information

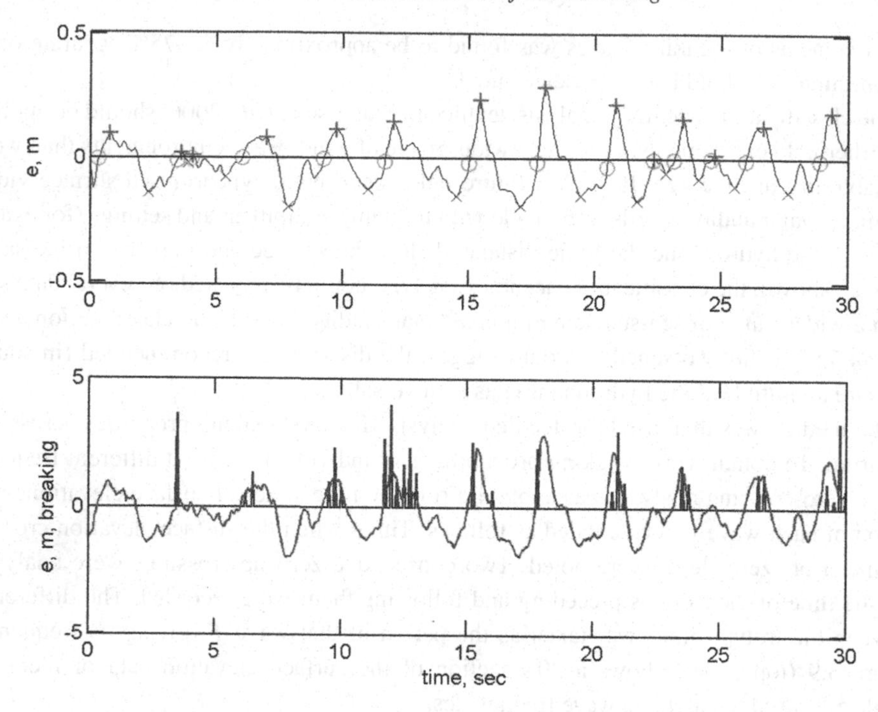

Figure 3.9 Time series of water elevation. Upper panel: o, zero crossings; +, crests; ×, troughs. Lower panel: water elevation with superimposed count of detected bubbles within 0.25 s intervals at the optimal breaking-detection discriminant. Figure is reproduced from Manasseh *et al.* (2006) © American Meteorological Society. Reprinted with permission

on the energy loss due to breaking at the measured spectral frequencies. A combination of the breaking-probability distribution and the bubble size can lead to direct estimates of the spectral distribution of wave dissipation.

In order to examine the hypothesis that the bubble size is related to breaking severity, the passive acoustic analysis was applied to data from a laboratory wave-maker. Waves with a frequency of 0.75 Hz and various amplitudes were generated in a flume of width 1.215 m at the School of Civil and Environmental Engineering of the University of Adelaide. The water depth was 225 ± 5 mm above a sandy bottom. A vertical board 45 mm wide and 150 mm deep was placed about 10 m downstream of the wavemaker; its top was 30 ± 3 mm above the mean water level so plunging breakers were forced to form over this barrier. Two capacitance probes measured the instantaneous water elevations, 640 ± 5 mm upstream of the board and 560 ± 5 mm downstream of the board. A hydrophone (Bruel & Kjaer 8103) with a diameter of 9.5 mm was mounted 60 ± 2 mm downstream of the board with its tip 55 ± 2 mm below the mean water level. The probes and hydrophone were approximately in the centre of the flume width. A typical acoustic time series is shown in Figure 3.11.

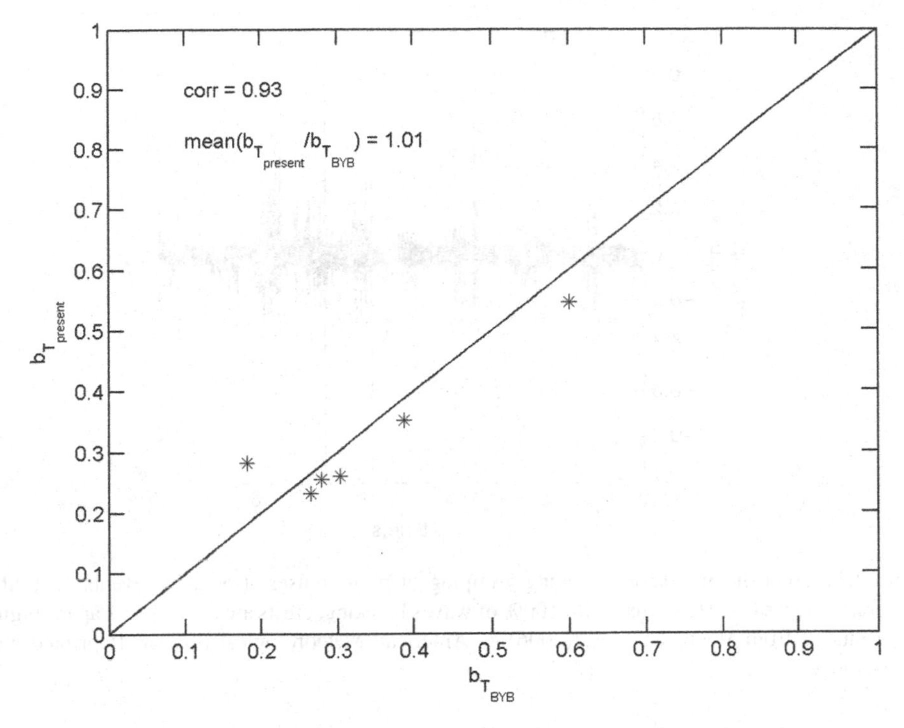

Figure 3.10 Comparison of dominant breaking probabilities (2.3) obtained by the spectrogram (denoted as 'BYB') and bubble-detection (denoted as 'present') methods. Figure is reproduced from Manasseh *et al.* (2006) © American Meteorological Society. Reprinted with permission

The acoustic signal was preamplified by a Bruel & Kjaer 2635 charge amplifier set to the hydrophone's calibration such that 1 V output represented exactly 100 Pa sound pressure amplitude. The signal was passed through a 400 Hz unity-gain high-pass filter and digitised at 40 kHz. The pulsewise processing was applied in real time on 5 min of data. Since in this laboratory testing every wave broke, the hydrophone was deliberately placed within a few cm of the bubble-formation zone, so rather than determining the optimum trigger level by classification–accuracy analysis as for field data, the criterion was simply to minimise variance in the processed data while keeping the data collection time per run reasonably brief. Typically, 500–1000 pulses were acquired.

The difference between the water elevations upstream and downstream of the wave-breaker were used to calculate the energy loss (2.24), a parameter assumed to represent the true breaking severity (Section 2.7). The results are shown in Figure 3.12, where the mean local bubble radius R_0 is shown with the 95% confidence interval calculated from the pulsewise processing. It can be seen that there is a clear, though not necessarily linear, increase in R_0 with the loss of energy by the plunging breaker. At higher wave amplitudes than those shown, breaking increasingly occurred prior to the board and between the upstream probe and the board, so those conditions could not be used for the present

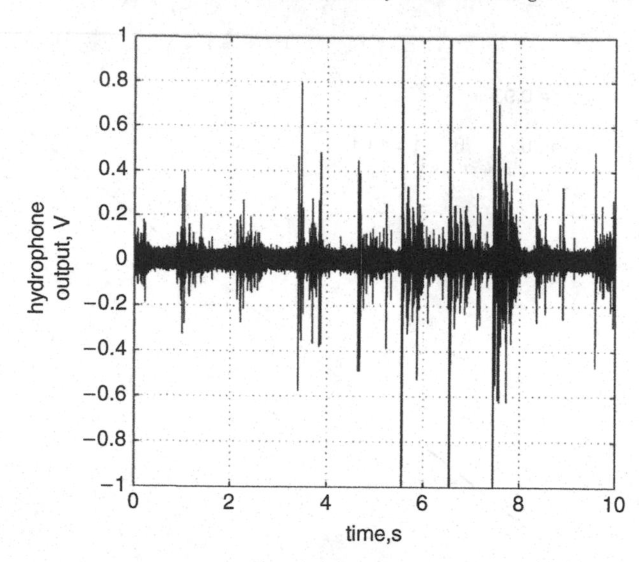

Figure 3.11 Acoustic time series showing grouping of sound pulses at breaking events. Note that wave frequency is 0.75 Hz so that, with 100% of waves breaking, crests are about 1.33 s apart. Figure is reproduced from Manasseh *et al.* (2006) © American Meteorological Society. Reprinted with permission

analysis. While this preliminary result supports the contention that R_0 can be a proxy for local breaking severity, much work is required to determine the true relationship between bubble size and breaking severity under a wider and more realistic range of breaking conditions.

Explicit calibration of the bubble size in terms of breaking strength, that is quantitative parameterisations of the wave energy loss across the spectrum in terms of the bubble size, or spectrum of the bubble sizes produced in the course of wave breaking, is still to be accomplished. Qualitatively, however, the new method signifies a tested technical means for studying spectral wave energy dissipation by non-intrusive remote-sensing passive-acoustic devices.

An itemised outline of the measurement and analysis procedure was formulated in Manasseh *et al.* (2006) as follows:

(1) A submerged hydrophone monitors sound continually;
(2) A prior statistical classification–accuracy analysis has determined a sound-pressure threshold optimally discriminating breaking from non-breaking events. When the instantaneous sound pressure exceeds this predetermined level, a very brief pulse of sound is captured, assumed to be due to a single, freshly formed bubble;
(3) The pulse frequency is rapidly measured, and translated into the bubble's radius;
(4) Running statistics on the rate of detection of bubbles and the mean bubble size during breaking events are collected;

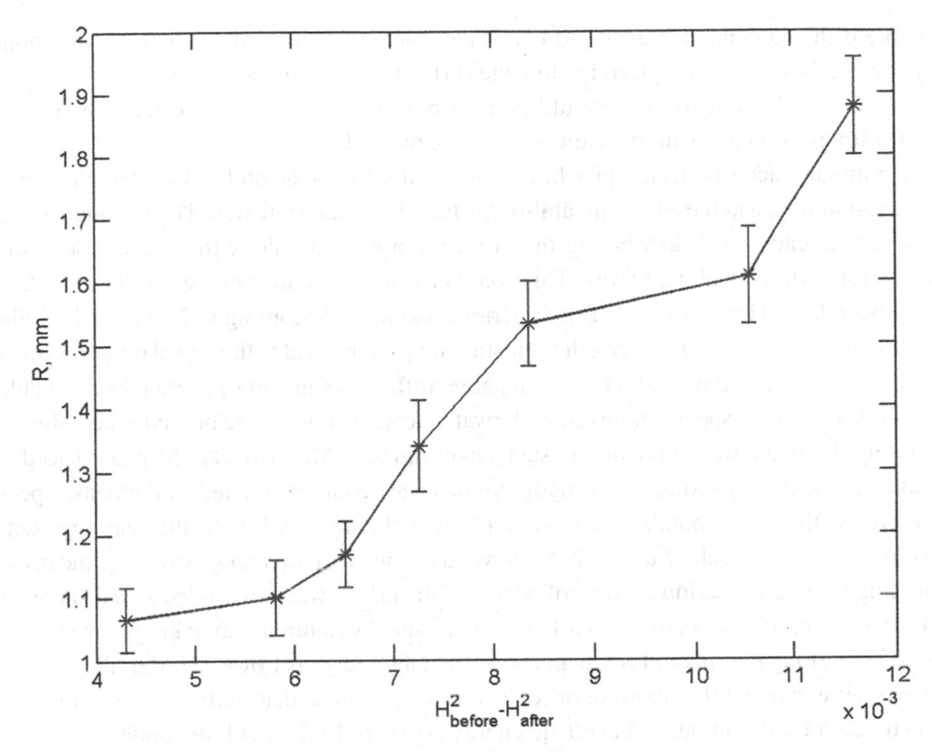

Figure 3.12 Breaking severity assumed as bubble radius R_0 versus laboratory energy loss (2.24). Vertical bars are for 95% confidence limit on R_0 (denoted as R in the figure). Figure is reproduced from Manasseh *et al.* (2006) © American Meteorological Society. Reprinted with permission

(5) Each detected bubble is linked to the synchronous wave-height record by means of a zero-crossing or RWR analysis, thus determining the period of wave breaking at the time of bubble detection and the wave-period distribution of the breaking rate (this procedure is detailed in Section 5.3 dedicated to breaking-probability analysis);
(6) From the laboratory experiments, the mean bubble size can be related to wave breaking severity;
(7) The rate of occurrence of breaking events times their severity can be used to estimate wave energy dissipation due to breaking (Sections 2.7, 6.2);
(8) The wave-period distribution of the dissipation rate is obtained (see Section 6.2).

At the time of writing, the instrumentation and equipment tested by Manasseh *et al.* (2006) only exists as a laboratory version, and its field and any industrial applications are yet to be developed. Potentially, however, the technique is very promising in this regard. Being a passive technique, it does not involve large levels of power supply and can be operated on batteries. If collocated with surface-wave measurements and intended to measure wave-breaking occurrence and severity, the device only has to record the size of bubbles and the time when they are generated. Such a sequence of double numbers, particularly given the fact that wave breaking is a relatively rare event, requires a minimal amount of electronic memory for data storage. Therefore, a field version of the device is quite feasible if the equipment is enclosed in a waterproof container, with only the hydrophone sensing head exposed, and placed below the surface away from the direct impact of breaking waves. Such an instrument would be possible to use both for dedicated experiments and for long-term observational campaigns autonomously.

In summary, acoustic techniques have consistently shown an ability to detect breaking-event signatures, and therefore the ability to study breaking statistics. The techniques are, in principle, capable of distributing the breaking probability along the spectral scales of the corresponding breaking waves. This could be done in a number of ways. The breakers could be detected and statistics of their periods obtained. According to Lowen & Melville (1991a) and Melville *et al.* (1992), the statistics are proportional to the signal duration above the threshold. While this method poses apparent difficulties in wind-generated wave fields, techniques based on spectrograms of underwater acoustic noise have proved successful for detecting dominant wave breaking in such environments. Alternatively, the phase speed of breakers, detected by a directional hydrophone array, could be related to the phase speed of waves with a corresponding frequency (Ding & Farmer, 1994). A different approach, developed by Manasseh *et al.* (2006), allows us to identify breaking waves by means of detecting bubble-formation events collocated with surface waves recorded synchronously.

Furthermore, it appears to be possible to obtain the distribution of breaking-wave energy dissipation along the spectral frequencies. In the laboratory, as Lowen & Melville (1991a) and Melville *et al.* (1992) showed, once the breaking event is detected, the energy loss can be estimated by the amount of acoustic energy radiated. In the field, the bubble-detection technique of Manasseh *et al.* (2006) demonstrates a certain promise by relating bubble size to the breaking strength.

However, no systematic studies have been done of spectral distributions of the breaking probability, and no advances have been made in obtaining an observation-based spectral distribution of the dissipation (e.g. Babanin *et al.*, 2007c). This is probably because of difficulties in applying the method of thresholding the integrated-noise-over-background, in the spectral environment of real seas with multiple breaking at various scales. In the real seas, the integrated background noise level will change, depending on wave and wind conditions, thus varying the threshold value (Ding & Farmer, 1994; Babanin *et al.*, 2007b). Even for stable wind–wave situations, the simultaneous presence of multiple wave scales can cause ambiguity in the detection of breaking events, moreover in measuring their duration or acoustic energy radiated. Babanin *et al.* (2001) synchronously detected the breakers by passive acoustics and by video recordings in the field, and showed that similar integrated noise above the background sometimes indicated a breaker and sometimes did not.

Manasseh *et al.* (2006) developed a new passive acoustic method of breaking determination using a single hydrophone. Its key difference from other methods reviewed in this section was the analysis of very brief pulses which are associated with sound emission by individual bubbles. This method shares both the advantages and some limitations of the other passive acoustic methods. It does rely on a discriminant (threshold) to trigger breaker detection and analysis and any such procedure will inevitably result in errors as discussed. However, the statistical procedures developed by Manasseh *et al.* (2006) would permit a rigorous determination of discriminants from future field data, possibly leading to a universal discriminant.

3.6 Remote sensing (radar, optical and infrared techniques)

The acoustic techniques described in Section 3.5 are based on non-intrusive measurements and therefore are themselves remote-sensing methods for studying various wave-breaking features and properties. They have been singled out into a separate section because of a number of practical advantages they appear to provide compared to other remote-sensing means, in this particular area of oceanography.

The current section is dedicated to radar and optical sensing techniques, including those based on infrared sensing. Of these, microwave radar backscattering has been applied in wave-breaking studies as often as, and sometimes together with, acoustic methods (see e.g. Melville *et al.*, 1988; Lowen & Melville, 1991a).

The radar active backscatter has been extensively used for remote sensing of the ocean surface with the purpose of detecting, monitoring and studying surface winds, ocean currents and other oceanic features. A large number of laboratory experiments have shown that the Bragg scattering mechanism, based on backscatter of microwave radiation by surface water waves with wavelengths of the order of a few cm, is the main process to be accounted for in this regard, although a variety of other contributing physical mechanisms are possible (e.g. Lowen & Melville, 1991a). The strength of the reflected signal depends on the height of the cm-waves, and because of the rapid response of the heights of such waves to environmental forcing, the backscattered signal is a good indicator of the activity and strength of the oceanographic phenomena of interest.

It was quite early by comparison with techniques other than visual observations, that it was noticed that radar backscatter could be used to study wave breaking. Pigeon (1968), Long (1974), Lewis & Olin (1980), Alpers *et al.* (1981), Kwoh & Lake (1984), Chaudhry & Moore (1984), Keller *et al.* (1986) and Trizna *et al.* (1991) all observed strong microwave return, so-called 'sea spikes' in the reflected signal, and associated it with breaking.

Breaking of very short waves is mostly induced by long waves (see Section 5.3.2). This can be either due to modulation of short waves by underlying longer waves (e.g. Longuet-Higgins & Stewart, 1960; Phillips, 1963; Donelan, 2001; Donelan *et al.*, 2010), in which case they occur on the front face of the dominant crests, or forced by larger-scale breaking (Young & Babanin, 2006a), in which case it should happen in the wake of large breaking. If the short waves break because they overcome a steepness limit like the dominant waves (i.e. Brown & Jensen, 2001; Babanin *et al.*, 2007b, 2009a, 2010a), then the Bragg scattering due to such height-saturating short waves should be locked to the phase of the dominant waves, in front of the crest or behind the crest, depending on what effect induces small-scale breaking.

Microwave backscatter indeed makes this distinction. Already in the earlier studies Alpers *et al.* (1981) and Kwoh & Lake (1984) noticed that the Bragg spikes are linked to the steep crests of breaking waves. Kwoh & Lake (1984) further concluded that Bragg scattering is not the only mechanism responsible for sea spikes at breaking crests. They distinguished the backscatter from capillary waves generated on the front face of the breaking wave, from the turbulent wake behind the crest (see also Wetzel, 1990) and from the sharp near-breaking crests themselves (see e.g. earlier work of Lyzenda *et al.*, 1983, on the topic of wedge diffraction).

Melville *et al.* (1988), Jessup *et al.* (1990), Lowen & Melville (1991a) and Melville *et al.* (1992), in the laboratory and in the field, moved from recognition of the wave-breaking contribution of radar backscatter to the question of whether these remote-sensing techniques can be used for studying the breaking statistics and physics, and the answer was positive. In this regard, combined laboratory studies of microwave backscatter and the acoustic signature of breaking are most instructive (Melville *et al.*, 1988; Lowen & Melville, 1991a; Melville *et al.*, 1992).

Such laboratory experiments showed that both reflected microwave power and radiated acoustic energy correlate with wave-energy dissipation (see also Section 3.5). As in the case of acoustic probing, this shows promise for determining the breaking strength, along with the breaking occurrence/probability. As in the acoustic case, however, this promise has so far proved difficult to realise in complicated multi-scale oceanic wave fields.

In the laboratory, the main proportion of the backscattered power appeared to precede the acoustic response to breaking. This interesting detail indicates that radar sensing can be employed for studying the incipient breaking statistics/probability, because the sound signature obviously designates the breaking in progress, the stage of active wave collapse when the whitecapping and bubbles are produced (see Sections 2.1, 2.2). Since the onset of breaking is difficult to detect because of the absence of distinct visual features, such as whitecapping, this makes microwave sensing a valuable tool for studying the initial stage of wave breaking, even if it is only in the laboratory.

Another potential application of microwave radar backscatter uses its ability to measure the Doppler spectrum of the reflected signal, and therefore orbital velocities including those at breaking onset. The latter has frequently been used as a breaking criterion ((2.49), Section 2.9) and even provides a basis for deducing a dissipation function based on observations of probability distributions of the orbital velocity, but it has not been the subject of a convincing dedicated study or comprehensively measured. In Lowen & Melville (1991a), the peak of the Doppler spectrum corresponds to a velocity within 10% of (2.49). Lamont-Smith *et al.* (2007) showed that the spectral density of this peak and its width depend on the phase velocity of breaking waves, and does not depend on the grazing angle of the radar, and they further found that the peak magnitude and the radar cross-section for breaking waves scale with the radar wavelength λ_{radar}:

$$\sim\lambda_{radar}^{1.5}. \tag{3.29}$$

Application of the technique to dedicated field studies of wave breaking followed. Jessup *et al.* (1990) found that the sea-spike contribution to the radar cross-section was $\sim u_*^3$. This suggests that the radar return can potentially be used to measure total dissipation of wave energy (i.e. 3.9).

Smith *et al.* (1996) used differences in the backscattered signal of vertical and horizontal polarisations at low grazing angles to distinguish spilling breaking events. They demonstrated a clear association of wave-breaking occurrence with the group structure of wave fields. A very interesting outcome was identification of the dispersion relationship of the breaking waves. Smith *et al.* (1996) observed developing seas, and for the young seas the dominant waves are expected to actively break (Banner *et al.*, 2000). It was found, however, that the dominant breaking frequency is about 25% above the spectrum peak frequency. Even higher levels of upshifting were reported as a result of dedicated wave-breaking microwave radar observations by Stevens *et al.* (1999). This is consistent with the observed shrinkage of individual waves prior to breaking brought about by the modulational instability, and the corresponding upshift of the wave frequency (Liu & Babanin, 2004; Babanin *et al.*, 2007b, 2009a, 2010a, see also Section 5.2).

Hwang (2007) investigated the correlation between dominant breaking and short-scale breaking based on both radar and acoustic observations. He argued that the radar-sensing techniques often process a very large number of breaking events and from these statistics an average length scale of the breakers in wave systems emerges. Based on sea-spike observations, it is about one order of magnitude shorter than the dominant wavelength of the wave field. Such evidence is very important, particularly because, as discussed previously, the microwave backscatter can account for wave breaking not producing whitecapping, which is a challenge, if not impossible for other wave-breaking detection techniques (with the exception of the infrared sensing method as discussed later in this section). This observation is consistent with some derivations for the dissipation function (Hwang & Wang, 2004).

Another potential remote-sensing technique for estimating breaking occurrences was suggested by Challenor & Srokosz (1984) and Srokosz (1986) for satellite altimeters. Indeed, Srokosz (1986) proposed a statistical model to connect the breaking probability with the fourth moment of the spectrum m_4 (see (3.47) and Section 3.8), and Challenor & Srokosz (1984) showed how m_4 may be estimated from the altimeter radar return. Satellite altimeter missions have now been flown, almost continuously, for some 25 years (see e.g. Zieger *et al.*, 2009), and thus such a technique would allow us to obtain the global distribution of the breaking probability and even its long-term trends, both global and regional, if they exist. Like many other remote methodologies outlined in this section, however, this interesting idea is still awaiting its implementation in practical studies of wave breaking.

A paper by Phillips *et al.* (2001) will be the last one mentioned in this radar-probing sub-section because it effectively links two major remote-sensing techniques, the radar backscatter methods which have been discussed above and optical observations of breaking which will be discussed next. It undertook measurements of $\Lambda(c)$, the average length of the breaking front per unit area per unit speed interval introduced by Phillips (1985). Combined with the Duncan (1981) hypothesis (3.27), measurements of $\Lambda(c)$ can provide the total energy dissipation:

$$S_{ds} = \int \frac{b_{br}\rho_w}{g} c^5 \Lambda(c) dc. \tag{3.30}$$

In Phillips *et al.* (2001), the radar measurements were conducted with a very high resolution and the moving sea spikes were clearly distinguishable over background backscatter. Many interesting observations were made. Moderate wind-speed conditions were similar to those observed in the acoustic field experiments by Ding & Farmer (1994) (see Section 3.5). Therefore, comparisons of the acoustic sensing and radar backscatter techniques could again be made, particularly as in both experiments the spectral distribution of the breaking occurrence was measured in terms of breaker phase speeds.

Phillips *et al.* (2001) found that their overall breaking statistics was close to that of Ding & Farmer (1994), but the phase speeds of breakers were on average lower (i.e. frequencies higher). While acoustic measurements did detect breaking waves propagating with a spectral peak speed c_p, the radar measurements did not. The fastest radar-measured breakers would propagate with the speed of some $0.6c_p$. As discussed earlier, the microwave backscattering technique identifies breaking events at or close to breaking onset, whereas the acoustic signature signifies wave breaking well in progress. Thus, the results reported by Phillips *et al.* (2001) are consistent with the notion also mentioned a number of times previously that the length (and therefore phase speed) of waves decreases immediately prior to breaking, if the breaking takes place as a consequence of modulational instability. Therefore, the overall number of incipient breakers and developed breakers should be statistically the same averaged value, but the statistically average spectral signature of such breakers, if they are caused by the instability of nonlinear wave groups, should not. This will be discussed in more detail in Section 5.2 (see also Babanin *et al.*, 2007b, 2009a, 2010a).

Thorough quantitative measurements of breaking fronts, such as the actual form of the phase-speed distribution of $\Lambda(c)$, were performed by Phillips *et al.* (2001) (see also Sharkov, 2007, for a review). This kind of measurement proved valuable for obtaining breaking statistics and breaking probability (see Section 2.5 for definitions). Estimates of the dissipation rates based on (3.30), however, as discussed in Section 3.4, seem to be of limited value, particularly away from the spectral peak.

Apart from its scientific merit, the paper by Phillips *et al.* (2001) also provides an important link with optical remote-sensing studies which concentrate on measuring $\Lambda(c)$. Advancements in computer power and storage in recent decades have made it possible to automatically process large numbers of digital video images of the water surface. The propagating whitecaps can be identified and distinguished from the background darker water surface, and therefore the statistics of breaking fronts, their propagation speeds and other geometric, kinematic and dynamic characteristics of the areas covered with whitecaps can be investigated by means of analysing video sequences in much the same fashion as radar imaging. Depending on the observational setup (e.g. aircraft or a close-by experimental tower), video imaging can be significantly more accurate because of the much smaller footprints of the video pixels compared to the radar cross-sections. The effects of bubbles entrained by breaking waves on the optical scattering and backscattering coefficients were investigated by Terrill *et al.* (2001).

Melville & Matusov (2002) conducted video recording of wave-breaking fronts with the purpose of obtaining both the breaking statistics and the dissipation (3.30). Video imaging was conducted from an aircraft. Statistical estimates for the wave breaking and related ocean-mixing characteristics, as well as for the breaking parameter b_{br} were obtained. In the traditional framework of relating the breaking properties directly to the wind, it was found that the distribution of $\Lambda(c)$ is proportional to

$$\Lambda(c) = 3.3 \cdot 10^{-4} \left(\frac{U_{10}}{10}\right)^3 \exp(-0.64c). \tag{3.31}$$

(note that the fit is quantitative and is not dimensionally consistent, that is, the dependence will only provide a value of Λ in m if the wind and phase speeds substituted are in m/s). It was also found that the main impact of the wave-breaking mixing effect comes from short waves which is consistent with observations that most of the wave energy/momentum input, which is supported by the shorter waves in the spectrum, is lost locally (e.g. Donelan, 1998). This work has now essentially been extended and has even included the directional distribution of $\Lambda(c, \theta)$ (Kleiss & Melville, 2011, 2010).

Dulov *et al.* (2002) performed the video recording from an observational tower, combined with collocated records of waves by means of a wave array. Their study was intent on investigation of whitecapping coverage and was mentioned in Section 3.1. It highlighted the cumulative effect of wave breaking and clearly demonstrated the induced nature of short-scale breaking. Overall, however, this kind of optical observation of breaking dynamics is far from being routine or even common.

Among the optical means of remote-sensing, infrared methods of probing the breaking waves are also most promising (Jessup *et al.*, 1997a,b; Jessup & Phadnis, 2005). Here, we quote Jessup *et al.* (1997a):

> "Under most circumstances, a net upward heat flux from the ocean occurs primarily by molecular conduction through a thermal boundary layer, or skin layer, at the ocean surface. As a result, the "skin temperature" at the top of this layer is a few tenths of a degree Celsius cooler than the bulk temperature immediately below the skin layer (Katsaros, 1980; Robinson *et al.*, 1984)".

Wave breaking disrupts the skin layer, exposes the bulk waters to the surface, and their temperature can readily be detected by the infrared-sensing devices. Thus, the wave-breaking process can be quantified in terms of the occurrence of breaking events and areal coverage of the turbulent breaking wake. In this way, the infrared technique can be employed to study not only the breaking statistics and probability, but also all sorts of turbulent subsurface mixing and air–sea interactions due to the breaking, such as heat and gas exchanges (e.g. Jessup *et al.*, 1997b).

Infrared sensing can also be used to investigate remotely the most elusive property of wave breaking, the breaking strength (Section 2.7 and Chapter 6). This strength determines the extent and intensity of the turbulent wake, and therefore the recovery rate of the skin layer as the breaking wake subsides and the surface cools, and correlates with the dissipation rate due to the breaking event (Jessup *et al.*, 1997b).

Thus, the infrared-sensing probing is an all-in-one technique for investigation of wave breaking and its consequences. It allows us to detect breaking occurrence, to investigate breaking probability and statistics, to study breaking severity, turbulent mixing and air–sea exchanges due to the breaking. Needless to say, it can be used to obtain dynamic properties of wave breaking, such as $\Lambda(c)$ ((3.30), Jessup & Phadnis, 2005). It is also applicable to micro-breaking (to which most of the other breaking-study techniques are not), and this is basically how it started.

So why do we still, more than 10 years after infrared sensing was introduced, refer to this technique as promising and having a great potential? This question is equally applicable to the other remote-sensing methods described in this section, and to some extent to the acoustic techniques of Section 3.5. In this regard, it is worth quoting Melville *et al.* (1988) who investigated both acoustic and microwave-radar signatures of breaking waves and wrote:

"Our results imply that these remote sensing techniques ultimately may be used to measure the dynamics of breaking waves, and are not restricted simply to obtaining the statistics and kinematics of breaking".

More than 20 years after these words, the remote-sensing methods mentioned have still not made breaking statistics, kinematics and dynamics well understood or even well described.

In this regard, it is perhaps not the researchers who developed the remote-sensing methods of studying breaking, and not the engineers who design the technical means of sensing who should bear sole responsibility. The methods are well developed, proved, tested and validated, and the instrumentation, for example infrared cameras, are quite sensitive, reliable and no longer very expensive. In our view, it is now the physicists who have to share the blame for the lack of progress. Once there is a clearer understanding of the physics, of what oceanic properties and features are to be sought and measured, the modern technical capabilities, including remote-sensing techniques, will be able to effectively investigate them, address issues and deliver results.

3.7 Analytical methods of detecting breaking events in surface elevation records

The analytical methods for detecting breaking events in surface elevation records, which are the subject of this section, could potentially provide a powerful means for studying wave breaking, or at the very least breaking statistics and probability. There are vast amounts of wave records accumulated over decades of wave observations and measurements, and most of them undoubtedly have the breaking events embedded.

In this section, the Black Sea field experiment is described, since its data will be broadly employed here and throughout the book. Also, conversion of the acceleration-based breaking criteria from deep water to shallow water waves is done analytically and has shown convincing agreement with observations.

The analytical techniques for breaking detection do exist, but like almost everything else related to breaking they have to rely on empirical criteria and therefore are semi-empirical rather than strictly theoretical methods. The theoretical approaches have to be able to deal with nonlinear and highly non-stationary, non-homogeneous and sporadic events, which the breaking waves are, in continuous wave time/space series and aim at detecting discontinuities or spikes within the time series or their derivatives such that they can be attributed to breaking.

A comprehensive overview of the methods for non-stationary analyses is provided by Huang *et al.* (1998). They consider the spectrograms, the wavelet analysis, the Wigner–Ville distribution, the evolutionary spectrum, the empirical orthogonal function expansion and other methods, with particular attention to the Hilbert transform. Out of these, the spectrograms have been used to find the breakers in records of underwater acoustic noise (Bass & Hey, 1997; Babanin *et al.*, 2001, Section 3.5) and the phase-time method based on the Hilbert transform has been shown to be successful in detecting deep-water laboratory breaking under a variety of conditions (Zimmermann & Seymour, 2002).

In addition to these, the phase-averaging method has been widely employed for conditional sampling of one signal (for example, air pressure over a breaking wave) with respect to the phase of a reference oscillation (for example, a surface wave that induces this pressure as in Babanin *et al.*, 2007b, see also Section 8.3). This method first needs to obtain time series of phases of the reference oscillations and either the Hilbert transform or the wavelet method can be used for this purpose (see e.g. Hristov *et al.*, 1998; Donelan *et al.*, 2006, about the phase-averaging technique and its applications).

The wavelet method has also been broadly applied and proved successful in detecting wave breaking and quantifying the breaking statistics in surface-elevation time series (e.g. Liu, 1993; Mori & Yasuda, 1994; Liu, 2000; Liu & Babanin, 2004). Here, we will follow Liu & Babanin (2004) when describing the wavelet applications.

Over the past three decades, wavelet data analysis has emerged as an alternative to the Fourier transform. One of its distinguishing features is the capability to analyse time-varying signals with respect to both time and scale, which provides a link to capturing rapid changes of dynamic properties of the wave surface and associating them with the breaking processes. This kind of extension of Fourier analysis is particularly effective when using a continuous wavelet transform with the complex-valued Morlet wavelet (Farge, 1992; Liu, 2000) which practically provides a local energy spectrum for every data point of the time series. One of the earlier successful applications of this kind of wavelet towards studies of wave spectral properties was done by Donelan *et al.* (1996) where a method of non-stationary analysis of directional spectra was developed and shown to be able to obtain instantaneous wave-propagation directions, amplitude and phase of a spectral frequency component, as well as wavenumber-related time-dependent information.

With respect to wave breaking, Mori & Yasuda (1994) considered that there is a sudden surface jump and interpreted this as a shock wave and then defined a shock-wavelet spectrum, using a discrete wavelet transform and the Meyer (1989) wavelet to detect occurrences of such surface jumps. They verified their method on laboratory mechanically generated waves which were breaking randomly, and found a good detection rate. The use of discrete scales and the ratio of two adjacent wavelet coefficients as a criterion for breaking detection, however, needs to have a clearer physical meaning.

Liu (1993) used a continuous wavelet transform and the complex-valued Morlet wavelet to obtain a time-frequency wavelet spectrum that effectively provided a localised frequency energy spectrum for each data point in a given time series. This spectrum can then be used to define an average wave frequency ω, and thus combined with the local wave amplitude a, to obtain a local surface acceleration $a\omega^2$ which would be compared to a given limiting fraction γ of gravitational acceleration g to define the breaking event (2.60).

The classical concept of studying the wave-breaking process led to various usages of a limiting value of wave steepness beyond which a continuous surface cannot be sustained (e.g. Longuet-Higgins, 1969a, see also Sections 2.9, 3.2, 3.3, 3.8). Thus, for example, the wave surface will break when its downward acceleration exceeds the limiting fraction γ (2.60, see also (2.50)–(2.59)).

As mentioned in Section 2.9, the difficulties of applying the acceleration criterion to real sea waves are the uncertain value of γ and the fact that the natural wave fields are multi-scaled. Because of the latter, the quantity on the left-hand side of (2.60) cannot be readily calculated from a time series of wave data. If it were possible to estimate this quantity, then such a simple familiar notion could readily be used to identify breaking waves in the time series. It is, however, not immediately clear how to pertinently resolve the local amplitude a and the instantaneous frequency ω from the measured time series $\eta(t)$ of real multi-scaled seas.

Liu (1993) suggested use of the time-frequency wavelet spectrum, $W_\eta(\omega, t)$, to obtain instantaneous values of effective wave amplitude a and frequency ω from time series of surface elevations $\eta(t)$ as

$$W_\eta(\omega, t) = \frac{1}{C_\psi}\left|\int_{-\infty}^{\infty} \eta(\tau)|\omega|^{1/2}\psi^*[\omega(\tau - t)]d\tau\right|^2 \tag{3.32}$$

where $C_\psi < \infty$ is the admissibility condition and function ψ is the mother wavelet. Here, we use the Morlet wavelet given by

$$\psi(t) = \frac{1}{\pi^{1/4}}(e^{-imt} - e^{-m^2/2})e^{-t^2/2}, \tag{3.33}$$

with $m = \pi\sqrt{2/\ln 2}$ chosen to fit the wavelet shape.

Once a localised frequency spectrum at each time moment t_i is known:

$$\Phi_i(\omega) = \left[W_\eta(\omega, t)\right]_{t=t_i}, \tag{3.34}$$

then for each $t = t_i$ we can define, based on that spectrum, a characteristic wave amplitude and frequency at the measurement point. In other words, it replaces the localised spectrum by an equivalent characteristic monochromatic wave.

As the characteristic frequency, average frequency σ (Rice, 1954) was chosen:

$$\sigma_i = \left[\frac{\int_{\lambda_L\omega_p}^{\omega_n} \omega^2\Phi_i(\omega)\,d\omega}{\int_{\lambda_L\omega_p}^{\omega_n} \Phi_i(\omega)d\omega}\right]^{1/2} \tag{3.35}$$

where ω_p is the localised frequency at the local energy peak, ω_n is the cutoff frequency, and λ_L is a number that Liu (1993) introduced to denote the start of the frequency range covering the wave-breaking process. We generally carry cut-off frequency up to 2.5 times the peak frequency. The value of λ_L usually lies between 0 and 2. $\lambda_L = 1$, for example, means that we expect waves of peak frequency and higher to be breaking and therefore disregard the contribution of those below ω_p in the determination of the characteristic wave.

For the local instantaneous amplitude a, Liu (1993) used $a_i = \eta(t_i) - \overline{\eta}$ in an initial application. Another approach was employed in Liu & Babanin (2004) which was based on considering the case of a simple monochromatic wave that has an acceleration $A\sigma^2 \cos(\omega t + \varphi)$, in order to infer that an appropriate characteristic amplitude at local instantaneous time t_i should be given as

$$a_i = A_i \cos(p_i). \tag{3.36}$$

Here, the local amplitude A_i is obtained from the analytic envelope signal of $\eta(t_i)$ by means of a Hilbert transform:

$$A_i = |\text{Hilbert}(\eta_i)|, \tag{3.37}$$

and the local phase p_i can be obtained from the wavelet spectrum $W_\eta(\omega, t)$.

In order to get the phase information of the time series, the mother wavelet to be used should necessarily be a complex one such as the Morlet wavelet shown above. So the calculation of the phase is given as

$$p(\omega, t) = \tan^{-1}\left\{\frac{\Im\left[W_\eta(\omega, t)\right]}{\Re\left[W_\eta(\omega, t)\right]}\right\} \tag{3.38}$$

and thus the local phase p_i can be obtained from averaging the local wavelet phase spectrum at each $t = t_i$ over the same range between $\lambda_L\omega_p$ and ω_n.

Sample results of the average frequency and local amplitude as obtained from (3.35) and (3.36) respectively, using $\lambda_L = 1$ (that is, waves of the peak frequency and above are expected to break (Babanin *et al.*, 2001)), are shown in the middle and bottom panels of Figure 3.13 for the Lake George waves. There remains to be determined the limiting fraction γ as the threshold for wave breaking that can be rendered through assimilation with the measured data.

Liu & Babanin (2004) tested the breaking-detection approach suggested by Liu (1993) and determined factor γ on the basis of field data. The data were obtained under a variety of wind–wave conditions in deep water in the Black Sea and in a finite-depth environment in Lake George, Australia. The two data sets included time series of surface elevations, with breaking waves marked. Both had been extensively used to study breaking statistics, and detailed descriptions of the breaking detection procedure as well as of relevant environmental conditions for the Black Sea experiment are given in Babanin (1995), Babanin & Soloviev (1998a), Banner *et al.* (2000) and Babanin *et al.* (2001). The Lake George field experiment was described in detail in Section 3.5 (see also Young *et al.*,

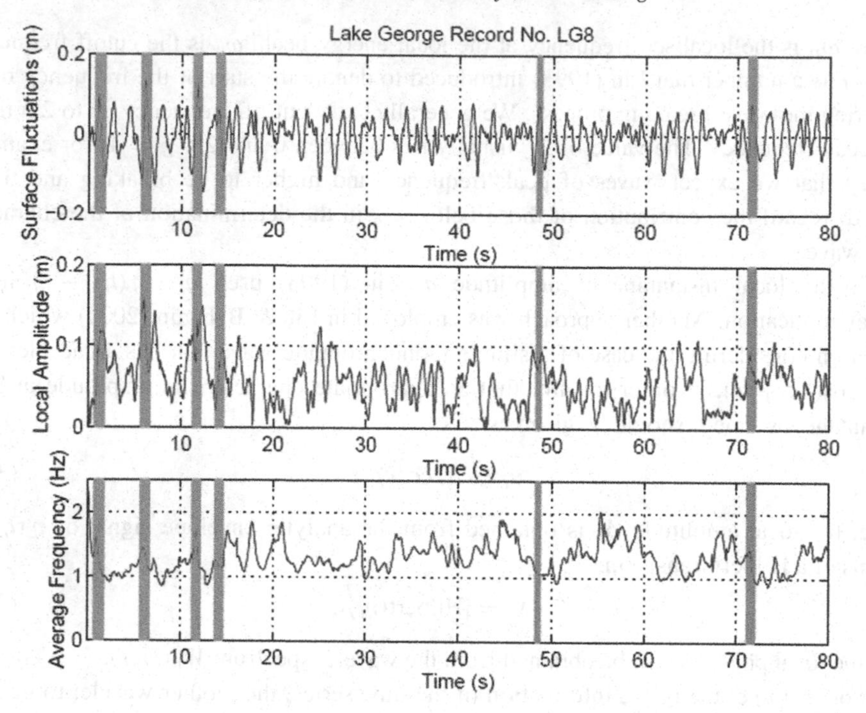

Figure 3.13 Lake George sample results. Vertical bars indicate breaking events. (top) Surface elevations. (middle) Instantaneous amplitude a (2.60). (bottom) Instantaneous spectrum-average frequency ω (2.60). Figure is reproduced from Liu & Babanin (2004) (copyright of Copernicus Publications on behalf of the European Geosciences Union)

2005), and here we will only summarise relevant details of the Black Sea data set and observations.

The Black Sea is a large enclosed water body, extending some 1200 km west–east and more than 400 km north–south (Figure 3.14). Most of the sea is over 1000 m deep, and therefore wind–waves develop in ocean-like conditions. A major difference, compared to the ocean, is the rarity of swell in the Black Sea because of its enclosed location, and absence of strong surface currents. This makes the Black Sea a convenient site for field observations of deep-water waves in their relatively pure state, not perturbed by wave–swell and wave–current interactions.

Four wind–wave records analysed by Liu & Babanin (2004) with the purpose of developing a wavelet-based breaking-detection procedure were taken from an oil rig situated on the 30 m-deep sea shelf in the northwest region of the Black Sea (Figure 3.14). Fetch and depth environment constituted ideal deep-water development conditions, with peak frequencies $f_p = 0.16$–0.27 Hz (wavelengths $\lambda_p \approx 20$–60 m), significant wave heights $H_s \sim 1$ m, wind speeds $U_{10} = 8.7$–10.7 m/s and mature-wave development stages of $U_{10}/c_p = 1$–1.7. A brief summary of relevant wind–wave properties is provided in the last four records of Table 5.1.

Figure 3.14 Location of research platforms in the Black Sea (pluses). Figure is reproduced from Babanin & Soloviev (1998a) © American Meteorological Society. Reprinted with permission

The waves were recorded by an array of high-precision wire wave gauges, deployed beyond the zone perturbed by the platform legs. The breaking events were labelled electronically by an observer (see also Section 3.2). The observer located 16 m above the wave probe array monitored one of the wave probes and triggered a signal whenever a whitecap of any size occurred at the probe. The signal was recorded synchronously with the wave data. An example of a wave record with breaking waves marked as vertical bars was shown in Figure 2.2. The waves propagate at a speed of 8 m/s, and thus it is difficult for the observer to place the marks precisely. So the marks, even though they are short, only indicate the presence of breaking waves and not an exact position of whitecapping over the wave phase.

Another set of wave breaking was recorded from a research vessel in the Black Sea and the Mediterranean Sea by means of an accelerometer buoy. Wave measurements by this buoy are described in Babanin *et al.* (1993) and environmental conditions during the measurements are given in Table 5.1 (see also Section 5.2). Again, the observer watched the buoy from the vessel and triggered the signal to register the passage of a whitecap over the buoy. Both the ship and the buoy were drifting, the buoy being deployed far beyond the zone perturbed by the ship and connected to the ship by a long loose cable. The observer was about 10 m above sea level, allowing a clear view of whitecaps with scales down to the size of the buoy (less than a metre). This set of data was used to obtain the histograms of maximal acceleration in the breaking wave (Figure 3.15) and to support the breaking criterion (2.60) in terms of the surface accelerations employed by the wavelet breaking-detection procedure.

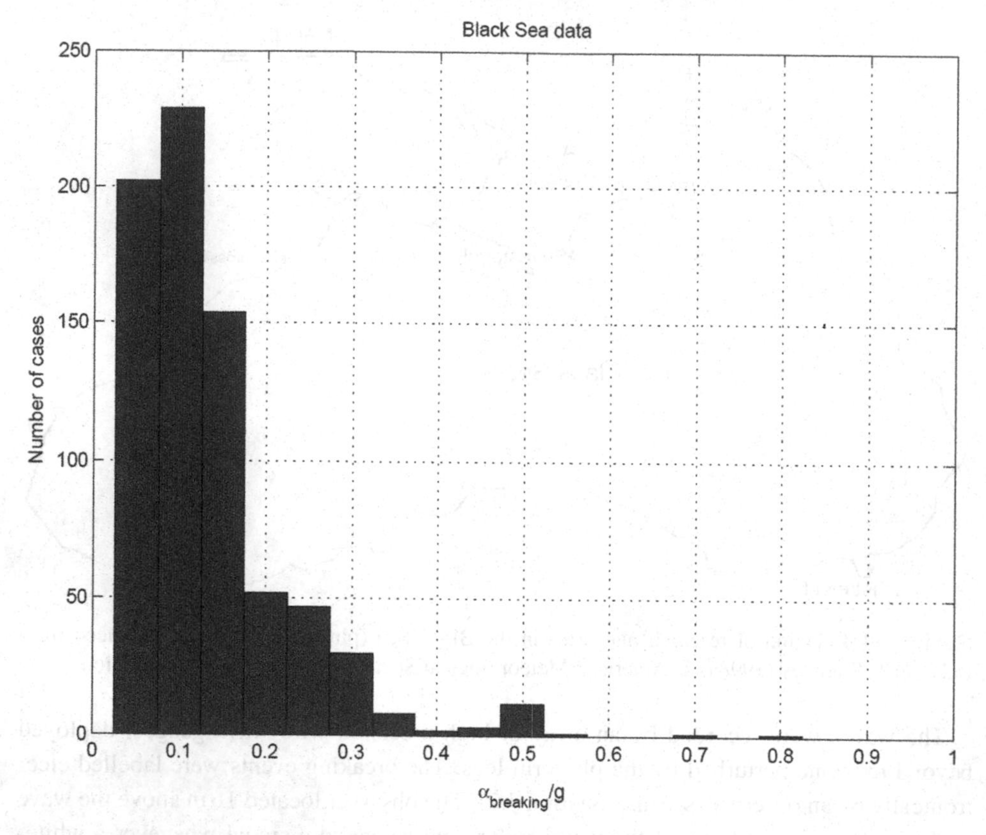

Figure 3.15 The Black Sea data. Histogram of $\alpha_{\text{breaking}}/g$ where α_{breaking} is the a_{downward} acceleration in waves detected as those breaking ((see (2.50), (2.59), (2.60)). Figure is reproduced from Liu & Babanin (2004) (copyright of Copernicus Publications on behalf of the European Geosciences Union)

In finite depths, the waves become steeper, and for the same frequency ω, the amplitude a of waves in (2.60) will change. Therefore, to verify the consistency of criterion γ obtained for the wavelet procedure on the basis of deep-water measurements, a finite-depth data set was also employed by Liu & Babanin (2004). These were data recorded during the Lake George experiment described in Section 3.5 above.

To measure the wave-breaking and associated effects reliably, a number of independent but integrated techniques were employed in Lake George (see also Young *et al.*, 2005). The data used by Liu & Babanin (2004) were wave recordings taken by wave resistance probes, and breaking detections were performed by means of bottom-pressure probes (Donelan *et al.*, 2005). The breaking waves generate enhanced pressure at high frequencies, and boosts of this pressure were detected by the pressure probes mounted at the base of the resistance probes. Successful detection of the breaking events by the probes was verified with synchronised video records. The pressure signal was smoothed by applying a running-average filter (see also Babanin *et al.*, 2007b, and Section 8.3).

Table 3.1 *Summary of the Lake George data used.* U_{10} *is the wind speed at 10 m height,* k_p *is the peak wavenumber,* $a = H_s/4$ *is the standard deviation wave amplitude,* d *is the water depth.*

Record no.	U_{10}, m/s	k_p, rad/m	a, m	d, m
4	6.6	7.2	0.013	0.31
8	11.9	2.3	0.039	0.32
9	12.0	2.1	0.034	0.29
10	8.1	3.1	0.019	0.33
11	10.6	2.2	0.020	0.32
14	7.1	5.6	0.015	0.27
15	7.3	2.5	0.016	0.28

An example of such a wave record with the bottom-pressure-indicated breaking events marked as vertical bars is shown in the upper panel of Figure 3.13. Data records used by Liu & Babanin (2004) were taken during the last stage of the Lake George study when the lake became very shallow (see Table 3.1). Therefore, these records, if compared to the deep Black Sea, represent the other end of wave development – bottom-limited and strongly wind-forced waves, under winds of $U_{10}/c_p = 4.7$–7.5, with peak frequencies $f_p = 0.53$–1.32 Hz (wavelengths $\lambda_p = 0.9$–3 m), and significant wave height H_s of about 0.1 m.

In Liu & Babanin (2004), the determination of the limiting fraction γ in deep water was based both on direct measurements of surface acceleration in the breaking waves (Figure 3.15) and on comparisons of the ability of the measurements and the wavelet breaking-detection technique to provide the same breaking statistics. The feasibility of the approach was then verified by means of predicting the shallow water γ on the basis of knowledge of the deep-water limiting fraction. A reality check was also discussed.

It has always been obvious from general reasoning that once the wave collapse is in progress, downward acceleration of the water particles on the breaking crest will be determined by a ratio γ in connection with the centrifugal acceleration of the particle and gravitational acceleration g. The exact measure of γ, however, remained elusive. As was mentioned earlier in this section, in studies which may already be regarded as classical (i.e. Longuet-Higgins *et al.*, 1963; Snyder *et al.*, 1983) it has been enacted generally as $\gamma = 0.5$, although indirect inference of the acceleration based on laboratory (Hwang *et al.*, 1989) and field measurements (Holthuijsen & Herbers, 1986; Srokosz, 1986) indicates that the value of γ should be 0.4 or even lower.

In the Black Sea, some of the breaking waves were measured by an accelerometer, and therefore direct estimates of maximal acceleration α_{breaking}, which is the a_{downward} acceleration in waves detected as those breaking ((see (2.50), (2.59), (2.60)), for such a wave are available. A histogram of distribution of these estimates for 742 breaking events is shown in Figure 3.15.

To define the limiting fraction γ, we are interested in the maximal detected value of acceleration α_{breaking}. While being on a breaking wave, the accelerometer was not necessarily located at a point of maximal acceleration, i.e. if it was riding a breaker in progress (see Section 2.2), and therefore not all the events depicted in Figure 3.15 are indicative of the limiting acceleration required to find γ.

As can be seen from the histogram in Figure 3.15, α_{breaking} values of up to 0.8 g were detected, although the number of these events is very low and their statistics is poor. Occasional high values of acceleration could have been registered if the accelerometer was shaken by, for example, a direct hit by a plunging breaker jet or by a jerk of the cable connecting the buoy to the mother ship. The histogram shows that continuous distribution with reliable statistics of α_{breaking} led to an $\alpha_{\text{breaking}}/g \approx 0.3$, and we shall use this as a reference value for γ in deep water. The limiting fraction γ in shallow water is expected to be different for reasons to be discussed below.

Following the supposition inferred in Liu (1993), Liu & Babanin (2004) persisted in trying to find the pertinent values for λ_L and γ used to detect wave-breaking events based on a wavelet approach. Amidst the redundant choices and combinations of the values of λ_L and γ, they settled on a plausible and practical approach: to match, as closely as possible, the number of breaking waves estimated from the wavelet approach for a time series – with the number of measured breaking waves. In so doing for each available time-series data set, they were able to resolve a rational γ value from a given λ_L value. A physically sound value of λ_L, in turn, can be chosen on the basis of knowledge of the lower bound of the frequency scale of waves that are expected to break. This knowledge is available for both deep-water breakers (e.g. Banner *et al.*, 2000) and finite-depth breakers (Babanin *et al.*, 2001).

The top panels of Figures 3.16 and 3.17 present the results of the applications of this approach to the same sample cases in the Black Sea and Lake George as shown in Figures 2.2 and 3.13, respectively. The results were obtained for $\lambda_L = 1$. With measured breaking waves marked by vertical bars, breaking events detected by the wavelet approach are shown by the dots. For reference, the figures also include plots of corresponding local steepness ak in the middle panel with resolved γ value indicated as a horizontal line, and the corresponding contours of the wavelet spectrum displayed in the lower panel.

Examination of the results shows that the wavelet approach along with measurements are generally successful in capturing breaking-wave events on many occasions, although on some other occasions one of them fails to detect a breaker while the other indicates that the breaking has occurred. Since measurement is also basically an approach to capturing breaking events, the latter is not unexpected as both the measurement and the wavelet approaches should be anticipated as relevant to different phases of wave breaking (see discussion in Sections 2.2–2.4).

Overall, it was expected that the breaking measurements and the wavelet approach would detect the same breaking events only at the developing stage of the breaking phases. The

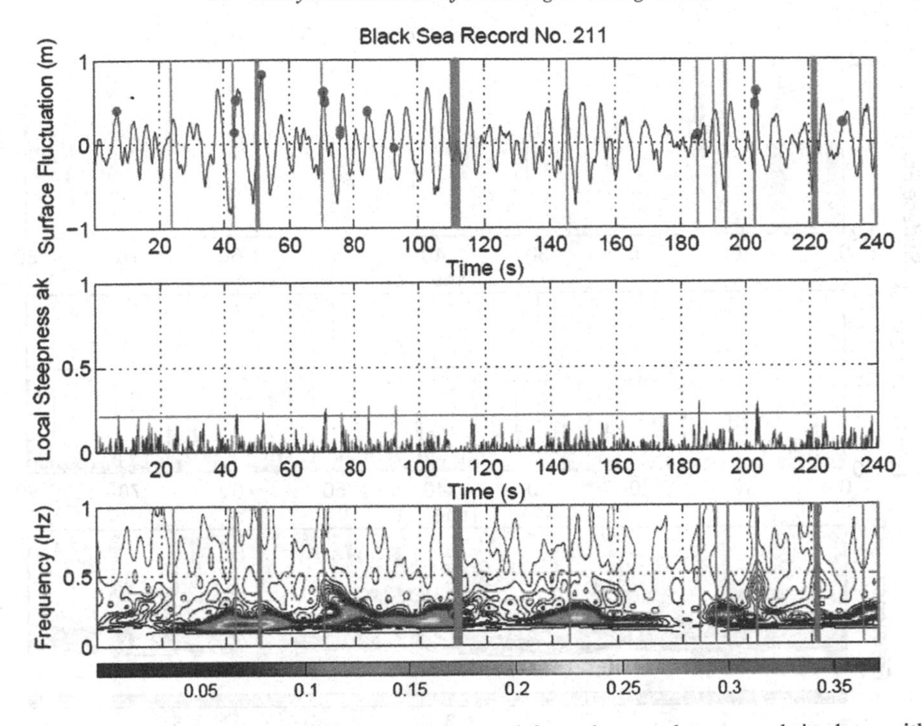

Figure 3.16 (top) Comparing breaking events detected from the wavelet approach in dots, with a Black Sea measurement (see last four records in Table 5.1) in vertical bars. (middle) Local steepness ak. The straight line shows $\gamma = 0.2$, the chosen threshold value. (bottom) Instantaneous wave spectra in relative units (shading scale below). The horizontal axis is the time axis, as in the top two panels. Figure is reproduced from Liu & Babanin (2004) (copyright of Copernicus Publications on behalf of the European Geosciences Union)

incipient breakers will be detected by the wavelet method and will not be detected by the measurements, and the subsiding breakers, on the other hand, will be detected by the measurements whereas the wavelet method will fail to pick them up. The relative duration of the different breaking phases was discussed in Section 2.4.

As one might have expected in general, there are breaking cases that the wavelet approach captured, there are breaking cases that the wavelet approach did not capture, and there are cases when the approach anticipated breaking but it was not confirmed by the observed whitecaps. These are all attributes of wave-breaking phases. Therefore, for individual breakings, a perfect match between the measurement and the detection from the wavelet procedure cannot be expected. Matching of total breaking percentages was used. The results of γ value assessments for different λ_Ls based on applying the matching-breaking percentage approach to the Black Sea and Lake George data are given in Tables 3.2 and 3.3, respectively.

Another interesting insight provided by the wavelet method pertains to the relation between wave breaking and wave grouping. Donelan *et al.* (1972), followed by a number

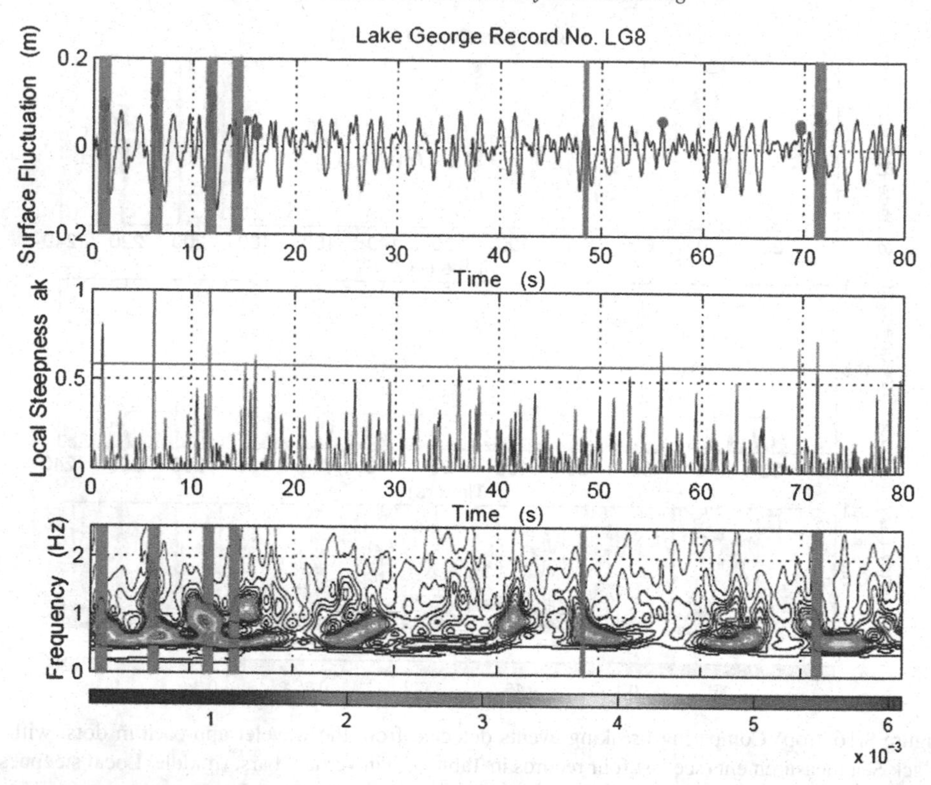

Figure 3.17 (top) Comparing breaking events detected from the wavelet approach in dots with a Lake George measurement (see Table 3.1) in vertical bars. (middle) Local steepness ak. The straight line shows $\gamma = 0.6$, which is the chosen threshold value. (bottom) Instantaneous wave spectra in relative units (shading scale below). The horizontal axis is the time axis, as in the top two panels. Figure is reproduced from Liu & Babanin (2004) (copyright of Copernicus Publications on behalf of the European Geosciences Union)

of experimental and analytical studies (e.g. Holthuijsen & Herbers, 1986; Babanin, 1995; Babanin *et al.*, 2007a, 2009a, 2010a, among others), reported that the majority of breaking events take place within the group structure close to the peak of the group envelope, although field observations indicated that an essential proportion of the breakings also occurred outside the distinct wave groups (Holthuijsen & Herbers, 1986; Babanin, 1995). Figures 3.16–3.17 show that breaking does happen beyond the obvious groups of dominant waves and the instantaneous spectra plotted in these figures indicate why this may be the case.

Once the limiting acceleration criterion (2.60) is applied to the instantaneous wave (3.35)–(3.38) which is characteristic of the instantaneous wave spectrum shown in the bottom panels of Figures 3.16–3.17, occurrence of the breaking event will depend on the product of the characteristic amplitude a and the characteristic average frequency σ squared. The peak of the envelope of dominant waves will give rise to the amplitude a,

Table 3.2 *Adapted γ values from Black Sea data.*

Rec. no.	$\lambda_L = 0.8$	$\lambda_L = 0.9$	$\lambda_L = 1.0$	$\lambda_L = 1.1$	$\lambda_L = 1.2$	$\lambda_L = 1.3$	$\lambda_L = 1.4$
211	0.1800	0.1940	0.2150	0.2630	0.2950	0.4100	0.6200
238	0.2050	0.2230	0.2550	0.3300	0.4350	0.6400	0.8000
242	0.2878	0.3070	0.3650	0.4750	0.6280	0.8750	1.1000
244	0.2690	0.2950	0.3500	0.4750	0.6780	0.9200	1.2000
Mean	0.2354	0.2547	0.2963	0.3857	0.5090	0.7113	0.9300
Std. dev	±0.0512	±0.0549	±0.0728	±0.1066	±0.1770	±0.2354	±0.2676

Table 3.3 *Adapted γ values from Lake George data.*

Record no.	$\lambda_L = 0.8$	$\lambda_L = 0.9$	$\lambda_L = 1.0$	$\lambda_L = 1.1$
4	0.3480	0.3825	0.4800	0.8170
8	0.4540	0.4900	0.5820	1.0500
9	0.4600	0.5150	0.6700	1.2355
10	0.2900	0.3280	0.4350	0.7680
11	0.3180	0.3500	0.4250	0.6690
14	0.2855	0.3180	0.4050	0.7070
15	0.4370	0.4820	0.5820	0.9730
Mean	0.3704	0.4094	0.5113	0.8885
Std. deviation	±0.0779	±0.0838	±0.1006	±0.2059

but not to the frequency σ. If the amplitude rise results in overshooting the threshold value (2.60), the wave will break at the wave group crest. If, however, deformation of the instantaneous spectrum leads to a rise in the average frequency σ, this will indicate wave breaking far from the envelope crest and in fact may happen outside a visible wave group. Such events can be seen around the 70th and the 205th seconds of the Black Sea record in Figure 3.16 where the second higher-frequency peak appears in the instantaneous spectrum and shifts up the average frequency: respective breakers are detected by both measurements and the wavelet method.

The secondary peak is a rather permanent feature of shallow-water spectra, and the Lake George spectra in particular (Young & Babanin, 2006b). Therefore, one would expect wave breaking to occur more frequently beyond obvious group structure at finite depths, and to occur more frequently in general. The first expectation is supported by Figure 3.17 where many breakers are associated with rising instantaneous secondary peaks, and the second expectation relates to the known fact that breaking rates are much higher in finite depths compared to deep water (e.g. Babanin *et al.*, 2001).

Mean values of γ shown in Tables 3.2 and 3.3 plotted versus the corresponding λ_L yield two distinct and smooth $\gamma - \lambda_L$ curves for finite-depth (Lake George) and deep (Black Sea) water conditions. The curves are shown in Figure 3.18.

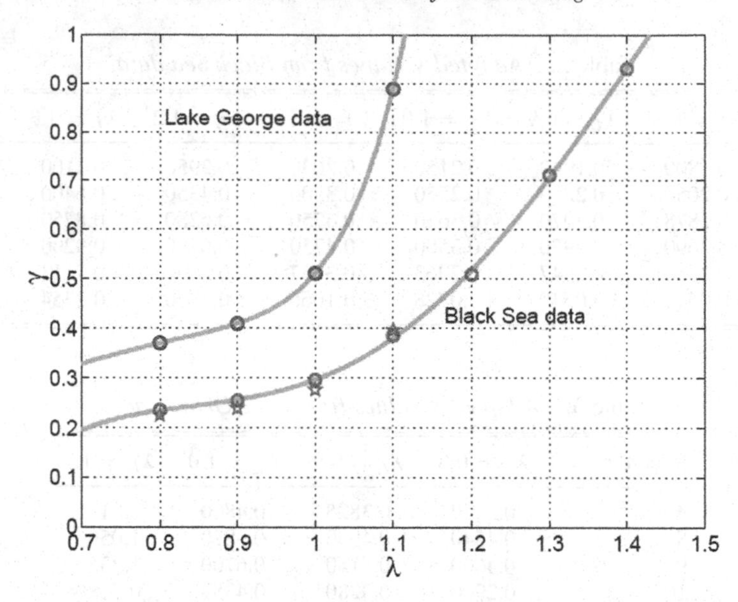

Figure 3.18 The λ_L–γ graph. Circles indicate data from Tables 3.2 and 3.3. Stars show conversion of the finite-depth γ according to (3.39)–(3.42). Figure is reproduced from Liu & Babanin (2004) (copyright of Copernicus Publications on behalf of the European Geosciences Union)

While the clear distinction between finite-depth and deep-water conditions is not unexpected, it presents a tangible challenge for an analytical and physically sound interpretation of these results. Previously, we postulated that the limiting downward acceleration is determined by the balance of the gravitational and centrifugal forces. Therefore, the difference between the two curves should be possible to explain if we take into account the difference in centrifugal accelerations for deep water (a_{deep}) and finite depth (a_{shallow}) waves of the same frequency ω. Conspicuously, this can be obtained if we consider what happens at the wave crest.

We expect that the wave surface breaks once it can no longer sustain itself for some reason and we surmise that, whenever it happens, the downward acceleration at the surface must have exceeded some threshold level we wish to resolve. Effective downward acceleration of the real physical particles at the crest is the difference between gravitational acceleration and centrifugal acceleration caused by the motion of the particle along its orbit. The latter is given by $\omega^2 R$, where R is the radius of curvature of the motion of a particle on the breaking crest:

$$\frac{a_{\text{deep}}}{a_{\text{shallow}}} = \frac{\omega^2 R_{\text{deep}}}{\omega^2 R_{\text{shallow}}} = \frac{R_{\text{deep}}}{R_{\text{shallow}}}. \tag{3.39}$$

Here, R_{deep} and R_{shallow} are now radii of curvature at the highest points of wave orbits configured as circles in deep water and ellipsoids in finite depth. The curvature of the two-dimensional curve defined by $y = y(x)$ is

$$\frac{1}{R} = \frac{\frac{d^2y}{dx^2}}{\sqrt{\left[1 + \left(\frac{dy}{dx}\right)^2\right]^3}}. \tag{3.40}$$

Now, in deep water we have

$$R_{\text{deep}} = a \tag{3.41}$$

where a is the local wave amplitude. For shallow water with water depth d it can be shown that

$$R_{\text{shallow}} = a \left(\frac{\cosh^2 [k(d+a)]}{\sinh [k(d+a)] \sinh(kd)} \right). \tag{3.42}$$

In essence, the orbits in shallow water are extended horizontally so that the centrifugal accelerations are larger for the same values of amplitude a and frequency ω, and thus the effective downward acceleration is smaller which in turn leads to larger γ for the shallow water cases. Clearly, a correction can be obtained from the ratio $R_{\text{shallow}}/R_{\text{deep}}$. Applying this formulation to the four γ values of Lake George data at λ_L values of 0.8, 0.9, 1.0 and 1.1 in Figure 3.18 successfully adjusted them back to the deep-water cases as shown by the plots of the four open stars.

The significance of this result extends further than merely resolving the differences between the two curves. Our ability to bring together two distinctly different results, obtained from explicitly different wave environments, by applying the principles of physics, shows that the wavelet breaking-detection approach based on the limiting-acceleration concept has a clear physical meaning rather than being just a technical measure. It allows us to apply a single approach to both deep and shallow waters and thus to translate derived parameters of γ and λ_L from one environment to another.

It is also of interest to note that based on the Black Sea acceleration measurement, as shown in Figure 3.15, we have picked a reference value of $\gamma = 0.3$. Now in Figure 3.18, the deep water $\gamma - \lambda_L$ curve based on the Black Sea breaking-wave measurements leads γ to be 0.3 at $\lambda_L = 1$. Thus, $\lambda_L = 1$ would be another appropriate reference value to use, particularly as in Figure 3.18 this value also corresponds to transition of the translated finite depth γs (stars) from underestimating to overestimating their deep-water counterparts. It is certainly consistent with the deep-water observations based on the same Black Sea data (Banner *et al.*, 2000) and with the finite-depth study based on the Lake George data (Babanin *et al.*, 2001) which both concluded that the peak waves do break. A more refined analysis may require us to introduce a wave-age dependence for λ_L as, according to Banner *et al.* (2000), the peak waves do not break if the waves are old enough and thus $\lambda_L > 1$ may be expected in such circumstances.

Now that a feasible rational approach had been devised, a reality check, i.e. final count of the breaking waves, was performed. To make an assessment of the performance of the approach, the γ values for $\lambda_L = 1$ were taken (Tables 3.2 and 3.3) and applied to the two available data sets in order to test the resulting breaking-event matchings between wavelet-approach detection and the measurements. The results of these final counts of the breaking cases are shown in Figure 2.4 and discussed in Section 2.4.

Thus, the wavelet transform provides an opportunity to look at each individual wave crest in a time series of wave data and assess whether or not it might be a breaking wave. In Liu & Babanin (2004), the approach was shown to be capable of producing the same breaking statistics as field measurements of wave-breaking conditions based on detection of whitecaps at a fixed point of observations.

The Liu & Babanin (2004) approach uses the classical limiting downward-acceleration concept, developed primarily for monochromatic waves (e.g. Longuet-Higgins, 1969a). With wavelet transform as a time-frequency analysis method, this concept can now be extended to spectral dominant waves when an instantaneous wave spectrum is replaced by an instantaneous characteristic wave (3.35)–(3.38), and applied to actual sea-wave measurements. The results can be interpreted through basic wave physics and a limiting value of the acceleration has been obtained from available field measurements. The approach is applicable to both deep-water and finite-depth environments.

3.8 Statistical methods for quantifying breaking probability and dissipation

The statistical methods for quantifying breaking probability and whitecapping dissipation, described in this section, are not actually based on direct or even indirect detection and measuring the breaking events as such. Based on some assumptions and, as is usual in wave-breaking studies, on some theoretical/empirical criteria, they try to interpret statistical properties of the wave fields in order to figure out a contribution of breaking waves to this statistics. It is argued that these properties prior to the breaking (or rather in the absence of breaking) are known theoretically, and therefore the differences observed are due to the breaking. In this regard, such statistical methods do not deal with either detection or the measurement of wave breaking, which is the topic of this chapter, but we thought it would be logical to place a brief description in the current chapter because these methods do appeal to measurements of the waves and, based on these measurements, infer breaking probabilities and severity, as with the other methods described in this chapter.

The first analytical approach of this kind was a probability model suggested by Longuet-Higgins (1969a) and further developed by Yuan *et al.* (1986), Hua & Yuan (1992) and Yuan *et al.* (2008, 2009). All of these studies used the Gaussian distribution of surface elevations to predict the appearance of wave heights exceeding the limiting steepness of the Stokes wave (2.47), or its limiting orbital velocity (2.49), or its limiting acceleration at the crest (2.50). In this regard, the models are based on sound physical principles.

Statistical approaches to surface elevation, or wave-height, or wave-crest probability distributions are now a textbook subject and we refer the reader to such books for details (e.g. Young, 1999; Holthuijsen, 2007). The surface elevation is often treated as having a normal/Gaussian probability distribution, and although obvious deviations from such distributions are reported (e.g. Babanin & Polnikov, 1995), the Gaussian probability density function is a reasonable approximation across a broad range of statistical wave applications.

For a Gaussian process with a narrow-banded spectrum, which is what the spectrum of surface waves is, the probability density function of wave heights $p(H)$ has a Rayleigh distribution:

$$p(H) = \frac{H}{4m_0} \exp\left(-\frac{H^2}{8m_0}\right). \tag{3.43}$$

Thus, the probability that wave height H is greater than, for example, some limiting height H_{lim} is

$$P(H > H_{\text{lim}}) = \int_{H_{\text{lim}}}^{\infty} p(H)dH = \exp\left(-\frac{H_{\text{lim}}^2}{H_{\text{rms}}^2}\right) \tag{3.44}$$

where

$$H_{\text{rms}}^2 = 8m_0. \tag{3.45}$$

Now, for any given wave field with significant wave height H_s (2.39), the probability of occurrence of waves exceeding, for example, the Stokes limiting wave steepness (2.47) can be estimated. In a similar fashion, in terms of the wave amplitude $a = H/2$, the probabilities of occurrence of waves exceeding a threshold acceleration (2.60) or limiting orbital velocity (2.49) can be estimated.

Longuet-Higgins (1969a) investigated the probability of waves having an acceleration greater than the limiting downward acceleration for Stokes waves (2.50). Such waves were assumed to break until the wave height is reduced back to the limiting value, and the difference was attributed to dissipation. Deductions were made for a narrow spectrum in order to obtain the dissipation of wave energy as a function of the spectrum, which is effectively an essential component of this probability model if assumptions are made with respect to the breaking severity.

Yuan *et al.* (1986) extended the theory of Longuet-Higgins (1969a) by removing the restriction of a narrow-banded spectrum. Hua & Yuan (1992) further argued that breaking does not stop once the wave steepness reaches down to the Stokes limit (2.47), and applied a similar probability model to investigate wave-breaking dissipation by assuming that the lower limit of wave height in the course of breaking is determined by the mean value at a particular frequency derived from the Phillips (1958) equilibrium spectrum. In all cases, the dissipation was found to be a linear function of the wave spectrum.

More recently, as was discussed in Sections 3.2 and 3.7, it was argued that breaking waves do not necessarily have the (2.50) acceleration (Holthuijsen & Herbers, 1986; Hwang *et al.*, 1989; Liu & Babanin, 2004). In addition, once they are breaking they do not stop at the Stokes limiting steepness but may keep losing energy until their steepness is well below the Stokes limit and even below the wave mean steepness (Liu & Babanin, 2004; Babanin *et al.*, 2009a, 2010a; Figure 2.3) Therefore, even though conceptually attractive, the probability models, as they have been derived, are so far not well justified quantitatively.

Yuan *et al.* (2008) continued the probability-function limiting-value approach by using a modification of the kinematic criterion (2.49), that is by assuming that the waves at the tail of the probability distribution cannot exist if their surface orbital velocity exceeds the wave phase speed, which is the limiting velocity of the Stokes wave. An essential new ingredient of their model is the surface wind-drift velocity, which modifies (increases) the speed of water particles at the wave crest and thus promotes breaking.

Yuan *et al.* (2009) combined this probability model with further empirical and analytical argument and applied it to investigating oceanic properties related to breaking, such as energy loss, both potential and kinematic, whitecap coverage (see also Section 3.1) and breaking entrainment. They continued on and compared expressions obtained with field observations, to find satisfactory agreements.

A different class of probability models was developed over the years to target the probability of breaking occurrence, rather than breaking strength as in the studies described above (e.g. Longuet-Higgins, 1975b, 1983; Houmb & Overvik, 1976; Nath & Ramsey, 1976; Tayfun, 1981; Huang *et al.*, 1983, 1984; Ochi & Tsai, 1983; Snyder & Kennedy, 1983; Papadimitrakis & Huang, 1988; Papadimitrakis, 2005). They usually employ comparison of probability functions of some properties of the wave systems, sometimes *ad hoc* properties, and refer to empirical criteria for such comparisons, rather than to physics explicitly. Since the wave measurements are most often conducted as time series of surface elevations, the majority of these probability models employed joint probability distributions of wave height H, and wave period T rather than wavelength or wavenumber.

As an example of such a model, which does not have a narrow-band spectrum limitation, we will describe the probability model of Ochi & Tsai (1983). They used T^2 which can be converted into wavelength or wavenumber by means of a linear dispersion relationship (see 2.17, 2.55), and therefore the probability of occurrence of waves exceeding the Stokes limit (2.47), (2.55) or other limiting steepness can be estimated. Ochi & Tsai (1983) argued that, based on their laboratory observations, the Stokes limiting-steepness criterion (2.55) is too high and the waves will actually break if

$$H \geq 0.020gT^2. \tag{3.46}$$

They concluded that (2.55) is applicable to what they called regular waves, and the irregular (or spectral waves) have to obey the limit (3.46). From what we know about wave breaking now, this difference can perhaps be explained in physical terms.

If the waves break due to modulational instability, they do not follow the simple regular/irregular notion because shortly before breaking onset their steepness reaches the Stokes limit (2.47) whereas their wavelength shrinks and the period is reduced accordingly. The reduction of the period as observed by Babanin *et al.* (2007a, 2009a, 2010a) is some 10–15%, which is in perfect accord with the difference between (2.55) and (3.46) observed by Ochi & Tsai (1983).

Ochi & Tsai (1983) continued on to derive the prediction formula (3.46) by using a joint probability distribution of wave excursion and the associated time interval of a non-narrow-band random process. They subdivided the wave breakings into two types, those whose excursion crosses the zero line (i.e. effectively dominant breaking) and those whose excursion is above the zero line (i.e. short waves breaking near the crest of the dominant waves, see e.g. Longuet-Higgins & Stewart, 1960; Phillips, 1963; Donelan *et al.*, 2010). Ochi & Tsai (1983) found that the second type of breaking is approximately 27% of the first type.

The significance of this study is hard to overestimate. Because the model was intended for a non-narrow-band process, it can be applied both at the spectrum peak which has a characteristic bandwidth and at the spectral tail which does not. This is a very essential advantage as the present understanding of wave breaking indicates two-phase physics of the phenomenon: inherent breaking of dominant waves, determined by their narrow-banded nature, and induced breaking of short waves (see Chapters 5, 6 and Section 7.3.4). The Ochi & Tsai (1983) results point to a general steepness threshold for both phases of wave breaking, which is consistent with the present understanding that the waves break because they reach a limiting steepness regardless of the nature of physical processes that led to this steepness being achieved (see comments in Section 2.9). The relative fraction of the short-scale breakers of 27%, however, appears very low and in this regard the Ochi & Tsai (1983) probability model would perhaps need further revision.

Unrealistic also is the conclusion that no breaking is expected unless the significant wave height is greater than 4 m. Mathematically, the most important result of the Ochi & Tsai (1983) model is the outcome that breaking probability depends on the shape of the wave spectrum and that the fourth moment of the spectrum is the main parameter in this regard (see also Srokosz (1986)).

A probability model of a different kind was proposed by Snyder & Kennedy (1983). They introduced an artificial 'breaking variable' and then considered the statistics of this variable and its relation with the wave directional spectrum. The variable was set equal to one inside a whitecap and zero outside, the whitecap having both horizontal and vertical extents. Defining the whitecapping volume was done in terms of another variable which had a dynamic threshold such that it would indicate a breaking. A number of further assumptions were made with respect to the dynamic variable, one of which was that it has a maximum at the free surface.

To clarify this reasoning in physical terms, we should simply say that the acceleration was eventually used as the dynamic variable. If it was over some threshold limit, the point of the wave body in space and time was regarded as breaking and the breaking variable was set to one. Then, mapping of the whitecapping surface was done in terms of geometric moments of this variable. Further on, the dynamic variable (acceleration) was related to the wave spectrum, and thus the statistics of the breaking variable in terms of its moments was connected to this spectrum.

A number of very insightful conclusions regarding breaking probability were obtained with this model. It was found that this probability, and whitecap coverage, are a function of the ratio between the rms of the vertical acceleration $\sqrt{m_4}$ (see eq. 2.10) and its critical level γ in (2.60). The probability appeared as a simple inverse function of the wave fetch, and interestingly the cross-wave scale of whitecaps was concluded to be greater than the down-wave scale. The presence of breaking water at some point had certain positive and negative correlations with some phases of other waves in the wave train: for example, if a wave crest was breaking, there was a tendency for successive crests to also break.

Some of the outcomes, however, indicate the limitations of the method. Thus, it was concluded that the whitecaps should travel with a speed appropriate to their scale. Such a finding is of course perfectly consistent with the expected behaviour of waves breaking due to inherent reasons, but not of induced breaking, i.e. breaking of short waves forced near crests of dominant waves. Obviously, one cannot blame the model for not reproducing what it was never designed to reproduce, but this fact outlines important constraints of such models. Since the induced breaking is expected to dominate at small scales (see Chapter 5), this means that in practical terms the Snyder & Kennedy (1983) approach is restricted to some range of frequencies around and above the spectral peak. Snyder & Kennedy (1983) in fact had a self-imposed cutoff of some $5f_p$–$10f_p$, due to divergence of the integral of the acceleration spectrum which they relied on. The lower bound of this cutoff is actually in reasonable agreement with the limits that indicate dominance of the induced breaking (Babanin & Young, 2005; Babanin *et al.*, 2007c).

To bypass the integration of probability densities for the acceleration, the technique was further tested by means of Monte Carlo simulations of the vertical accelerations (Kennedy & Snyder, 1983). An additional interesting finding of this statistical study was that

> "the propagation velocity of the whitecap was typically 45% of the phase velocity associated with the frequency of peak energy"

(see also references to Smith *et al.* (1996) and Stevens *et al.* (1999) in Section 3.6). Kennedy & Snyder (1983) further conjectured that

> "while this velocity is close to the group velocity, the similarity between the two velocities is probably coincidental, as there appears to be no reason to believe that group velocity is a pertinent parameter. The low velocity of the whitecaps probably reflects the importance of higher frequency wave components to the breaking process".

Both the conclusion and the conjecture now find experimental support (e.g. Gemmrich *et al.*, 2008).

Snyder *et al.* (1983), the final paper in the series of Snyder & Kennedy (1983) and Kennedy & Snyder (1983), was intended to provide experimental support to the probabilistic model and its rich set of interesting and important conclusions, and to estimate experimentally the key parameter of the model, i.e. threshold value of the downward acceleration. A field experiment was conducted in order to measure the statistical geometric properties of whitecaps, by means of synchronised photographing of the breaking waves and recording them with an array of wave probes. Many theoretical findings of the statistical model were confirmed, with the most important conclusion being that the threshold acceleration should correspond to the theoretical limit for monochromatic Stokes waves (2.50), i.e. 0.5 g. The authors cautiously warned that

> "this conclusion is less than definite because our analysis is limited to wave components with frequencies less than twice the frequency of the spectral peak".

Since, as was discussed above, from a practical point of view the model is basically designed for dominant waves anyway, the uncertainty is not too essential and the limit perhaps reflects well the critical level suitable for this kind of model.

Srokosz (1986) followed the approach of Snyder & Kennedy (1983), but offered an alternative estimate of the breaking probability which is closer to the definitions used here (i.e. Section 2.5). The statistics of Snyder & Kennedy (1983)

"is essentially a spatial quantity representing the fraction of the area of the sea surface over which breaking occurs, while B represents the proportion of crests that break at a given point".

A very simple expression for B was proposed:

$$B = \exp\left(-\frac{\gamma^2 g^2}{2m_4}\right). \tag{3.47}$$

In technical terms, Srokosz (1986) relied on the limiting acceleration at the wave crest only, as opposed to Snyder & Kennedy (1983) who considered the entire wave surface (in this regard, see Liu & Babanin, 2004, and discussion in Section 3.7 above). Srokosz concluded that the limiting fraction of the gravitational acceleration in (2.60) should be 0.4 g which was also consistent with his interpretation of the experimental data of Ochi & Tsai (1983).

The relevance of inclusion/exclusion of high-frequency breaking, already discussed above, is again highlighted in Srokosz (1986). The author points out that his estimate of the acceleration variance m_4 depends on the choice of high-frequency cutoff. If, for the Pierson–Moscowitz spectrum this cutoff is chosen as $2f_p$, then $B = 10^{-8}$, and the $6f_p$-cutoff leads to $B = 0.002$, i.e. five orders of magnitude difference in predicted breaking rates when small-scale breaking is included.

To summarise the overview of probability methods and models, we have to conclude that they of course do not deal with individual breaking events, like the other approaches described in this chapter, but appear to be extremely capable in quantifying statistical characteristics of breaking waves in the overall surface-wave field. This should not come as a surprise as, unlike many or even most of the empirical breaking-detection techniques, they refer to limiting surface properties based on fundamental physical grounds. In most cases those are limiting steepness, orbital velocity or downward acceleration, or their derivatives (see Section 2.9), which signify definite conditions such that beyond these conditions the water surface cannot sustain itself and collapses. One can argue that the water surface may become unstable even before it reaches these limiting conditions, but ultimately the wave certainly cannot persist without breaking after. Therefore, although quantitative conclusions of the probability models perhaps need revision, particularly those that try to predict the dissipation as described above, such statistical models are very sound and promising in the physical and theoretical sense.

4

Fully nonlinear analytical theories for surface waves and numerical simulations of wave breaking

The previous chapter was dedicated to experimental methods of detecting wave breaking, quantifying the breaking probability and severity, and measuring effects related to the breaking, including the wave energy dissipation. In the next chapter, it is logical to describe theoretical methods of describing wave-breaking physics or phenomena leading to the breaking.

While experimental oceanography has produced an abundant variety of techniques and approaches to detect and measure breaking, the theories capable of dealing with wave breaking are few. These should not be confused with analytical methods intended to detect the breaking events in surface-wave records (Section 3.7) or with the statistical methods of quantifying the breaking probability and strength (Section 3.8). Both such groups of analytical techniques are placed into experimental Chapter 3 for a good reason – they principally rely on empirical criteria.

Another significant group of analytical approaches, dealing with the dissipation due to breaking, rather than with the breaking as such is also not included in this chapter. Some of these models are based on assumptions intended to interpret pre-breaking or post-breaking properties of the waves, rather than on working with the physics leading to breaking or driving the breaking and its consequences. Others attempt to deduce the dissipation from differences between wave-evolution predictions done by means of kinetic and dynamic equations. In any case, these are indirect techniques that do not depict the wave-breaking event explicitly. They will be described in Section 7.1.

In this regard, here we will consider as a wave-breaking theory an analytical method that is able to describe the evolution of nonlinear waves to the point/moment of breaking onset, or even beyond, without relying on empirical criteria, or assumptions yet to be proved, or some interpretations of wave properties that supposedly allow us to reveal wave-breaking impacts provided that those would take place or have already taken place. In this section, we will refer to wave theories based on first principles.

The first analytical theory that produced limiting steepness for two-dimensional steady monochromatic water waves (2.46) was that by Stokes (1880). It is based on perturbation expansion terms added to the solution of Eulerian linear wave theory, and can be extended to three-dimensional irregular waves (Laing, 1986). As a result the wave profile appears much more realistic, it does exhibit such nonlinear features of wave shape as skewness S_k (1.2), but as was argued in Section 1.2, this is not how breaking waves appear (see Figure 1.2).

In this regard, at least as far as the visual appearance of surface waves is concerned, the Lagrangian approach is more convincing (see also Section 4.2). Even at the second-order expansion, it is capable of reproducing both skewness S_k and asymmetry A_s of waves with respect to the vertical axis (1.3) in two dimensions (Fouques *et al.*, 2006) and horse-shoe patterns in three dimensions (Fougues & Stansberg, 2009).

Analytical and numerical means of describing the evolution of nonlinear wave trains can be broadly subdivided into five major groups. The nonlinear Schrödinger equation (NLSE, Benney & Newell, 1967; Zakharov, 1967, 1968) and its extensions (i.e. Dysthe, 1979; Stiassnie, 1984; Shemer & Dorfman, 2008) work in the physical space and model the complex wave envelope. The Zakharov equation (Zakharov, 1968) and its modifications (Gramstad *et al.*, 2010) are for the amplitude spectrum. The Alber equation (Alber, 1978; Stiassnie *et al.*, 2008) is a stochastic counterpart of the deterministic dynamic equations where the evolving variable is the correlation function. The most frequently used stochastic model is the kinetic equation for wave spectra (Hasselmann, 1962; Zakharov, 1968; Krasitskii, 1994; Janssen, 2003; Annenkov & Shrira, 2006). The latter, as mentioned above, gives the interesting possibility of estimating the dissipation indirectly if compared to evolution simulated by means of dynamic equations (Zakharov *et al.*, 2007 and see 7.1.2).

In this chapter, however, we will only be interested in the fifth group, fully nonlinear analytical theories for potential surface waves, as opposed to the above theories which are based on first principles, but then involve further assumptions, like a 'small parameter' in perturbation theories, usually wave steepness (1.1) in Eulerian approaches, or the ratio of the particle displacement to wavelength in Lagrangian derivations, or narrow-banded spectrum in NLSE, and so on. Theories that rely on a small steepness, or even on a finite steepness are indeed also nonlinear, and they do depict some nonlinear effects as mentioned above, including some very important nonlinear characteristics of wave shape, wave fields and wave dynamics, but they can hardly be expected to adequately attend to the problem of wave breaking where the steepness of the individual wave is by definition ultimately extreme. Some techniques use a combination of solutions of fully nonlinear equations and one of the approaches outlined above, and such a combination is another potentially very promising direction which does not depend on limiting assumptions and at the same time allows us to scrutinise and better understand the physics underlying specific analytical methods (e.g. Irisov & Voronovich, 2011).

Analytical solutions of fully nonlinear equations and boundary conditions have not so far proved feasible, therefore the existing fully nonlinear approaches are numerical models. In this regard, we can mention modelling by Watson & West (1975), Longuet-Higgins & Cokelet (1976), Dold & Peregrine (1986), West *et al.* (1987), Dold (1992), Craig & Sulem (1993), Tulin *et al.* (1994), Dyachenko *et al.* (1996), Chalikov & Sheinin (1998, 2005), Landrini *et al.* (1998) and Dyachenko & Zakharov (2005). The model by Landrini *et al.* (1998) is Lagrangian and will be described in Section 4.2. The Eulerian approach will be considered next in Section 4.1.

Non-potential phase-resolvent models which have to treat the turbulence and deal with the air explicitly, that is with bubbles in the water and spray above the surface, will also not be considered in this chapter. They allow us to describe wave-breaking dissipation directly and will be reviewed in Section 7.2 together with other dissipation theories and applications.

The Chalikov–Sheinin (CS) model (Chalikov & Sheinin, 1998, 2005, see page 131) has been mentioned a number of times throughout the book as a model based on a fully nonlinear approach. In very simple terms, this model employs solution of the Euler equation, the fundamental equation of hydrodynamics and Newton's law applied to fluids in the absence of friction force. The model does not involve any initial assumptions on wave-steepness magnitude or any other physical/spectral conditions. As mentioned previously, there are several models based on fully nonlinear equations, and it is their accuracy, stability and ability to integrate the evolution equations in space and time without accumulating numerical errors, that differentiate and distinguish them from one another. We will give a brief overview of the fully nonlinear models, based primarily on the CS example, largely following Babanin *et al.* (2007a, 2009a, 2010a).

4.1 Free surface at the wave breaking

Numerical computations of nonlinear surface waves have previously been undertaken based on solutions of the potential flow equations (e.g. Watson & West, 1975; Longuet-Higgins & Cokelet, 1976; West *et al.*, 1987) and with a Cauchy-type integral algorithm (Dold & Peregrine, 1986; Dold, 1992). Both schemes have no limitation in terms of wave steepness, and both are capable of simulating the initial phase of wave breaking (the later stages are rotational and remain extremely difficult to simulate directly). More recently, a method based on a Taylor expansion of the Dirichlet–Neumann operator was developed by Craig & Sulem (1993). The capabilities of this method were illustrated by computing the evolution of modulated wave packets and a low-order approximation of a Stokes wave for relatively short periods of time. We should point out that this appears to be a principal limitation of all the above schemes: for a steep wave field, they have only been used for simulations of relatively short time/space evolution. These approaches could not be applied to longer periods of time because none of them appear to provide conservation of integral invariants (mass, energy, horizontal momentum).

A numerical scheme for direct hydrodynamical modelling of two-dimensional nonlinear gravity and gravity-capillary waves was developed by Chalikov & Sheinin (1998) (see also Chalikov & Sheinin, 2005; Chalikov, 2005, 2007). This approach is based on a non-stationary conformal mapping, which allows the equations of potential flow with the inclusion of a free surface to be written in a surface-following coordinate system. This transformation does not impose any restriction on the shape of the surface, except that it has to be possible to represent this surface in terms of a Fourier series. An analogous approach was developed by Dyachenko *et al.* (1996) and Dyachenko & Zakharov (2005).

Let us consider periodic two-dimensional deep-water waves whose dynamics is described by principal potential equations. Because of the periodicity condition, the conformal mapping for infinite depth can be represented by the Fourier series (see details in Chalikov & Sheinin, 1998; Chalikov & Sheinin, 2005):

$$x = \xi + \sum_{-M \le k < M, k \ne 0} \eta_{-k}(\tau) \exp(k\zeta) \vartheta_k(\xi), \tag{4.1}$$

$$z = \zeta + \sum_{-M \le k < M, k \ne 0} \eta_k(\tau) \exp(k\zeta) \vartheta_k(\xi); \tag{4.2}$$

where x and z are Cartesian coordinates, ξ and ζ conformal surface-following coordinates, τ is time, η_k are coefficients of the Fourier expansion of a free surface $\eta(\zeta, \tau)$ with respect to the new horizontal coordinate ζ:

$$\eta(\zeta, \tau) = h(x(\zeta, \xi = 0, \tau), t = \tau) = \sum_{-M \le k \le M} \eta_k(\tau) \vartheta_k(\zeta), \tag{4.3}$$

ϑ_k denotes the functions

$$\vartheta_k(\xi) = \begin{cases} \cos k\xi & \text{for } k \ge 0, \\ \sin k\xi & \text{for } k < 0 \end{cases} \tag{4.4}$$

and M is the truncation number.

Non-traditional presentation of the Fourier transform with definition (4.4) is, in fact, more convenient for computations with real numbers, such as $(\vartheta_k)_\xi = k\vartheta_{-k}$ and $\sum (A_k \vartheta_k)_\xi = -\sum k A_{-k} \vartheta_k$. So, the Fourier coefficients A_k form a real array $A(-M : M)$, thus making possible a compact programming in Fortran90. Such a presentation can be generalised for the three-dimensional case.

Note that the definition of both coordinates ξ and ζ is based on Fourier coefficients for surface elevation. It then follows from (4.1) and (4.2) that time derivatives z_τ and x_τ for Fourier components are connected by a simple relation:

$$(x_\tau)_k = \begin{cases} -(z_\tau)_{-k} & \text{for } k > 0, \\ (z_\tau)_k & \text{for } k < 0. \end{cases} \tag{4.5}$$

As a result of conformity, the Laplace equation retains its form in (ξ, ζ) coordinates. It is shown in Chalikov & Sheinin (1998) and Chalikov & Sheinin (2005) that the potential wave equations can be represented in the new coordinates as follows:

$$\Phi_{\xi\xi} + \Phi_{\zeta\zeta} = 0, \tag{4.6}$$

$$z_\tau = x_\xi G + z_\zeta F, \tag{4.7}$$

$$\Phi_\tau = F\Phi_\xi - \frac{1}{2} J^{-1} \left(\Phi_\xi^2 - \Phi_\zeta^2 \right) - z, \tag{4.8}$$

where (4.7) and (4.8) are written for the surface $\zeta = 0$ (so that $z = \eta$, i.e. the surface elevation), J is the Jacobian of the transformation:

$$J = x_\xi^2 + z_\xi^2 = x_\zeta^2 + z_\zeta^2, \tag{4.9}$$

G is an auxuliary function:

$$G = (J^{-1}\Phi_\zeta)_{\zeta=0}, \tag{4.10}$$

and F is a generalisation of the Hilbert transform of G, which for $k \neq 0$ may be defined in Fourier space as

$$G_k = \begin{cases} -F_{-k} & \text{for } k > 0, \\ F_k & \text{for } k < 0, \end{cases} \tag{4.11}$$

actually following from (4.5). Above, Φ is the velocity potential (and Φ_ζ is the derivative of the potential with respect to the 'vertical' coordinate ζ at the surface) and z represents the shape of the surface.

Equations (4.6)–(4.11) are written in non-dimensional form with the following scales: length L, where $2\pi/L$ is the (dimensional) horizontal wavenumber, time $L^{1/2}g^{-1/2}$ and the velocity potential $L^{3/2}g^{-1/2}$. Capillary effects and external pressure were not taken into account in this formulation. Note that the adiabatic equations for surface waves outside the capillary interval are self-similar: they are invariant over length scale L, which makes the numerical approach very effective and allows for broad interpretations.

The boundary condition assumes vanishing vertical velocity at infinite depth

$$\Phi_\zeta(\xi, \zeta \to -\infty, \tau) = 0. \tag{4.12}$$

Solution of the Laplace equation (4.6) with boundary condition (4.12) yields a Fourier expansion which reduces the system (4.6)–(4.8) to a one-dimensional problem:

$$\Phi = \sum_{-M \le k \le M} \phi_k(\tau) \exp(k\zeta)\vartheta_k(\xi), \tag{4.13}$$

where ϕ_k are Fourier coefficients of the surface potential $\Phi(\xi, \zeta = 0, \tau)$. Equations (4.6)–(4.8), (4.10) and (4.11) constitute a closed system of prognostic equations for the surface functions $z(\xi, \zeta = 0, \tau) = \eta(\xi, \tau)$ and the surface velocity potential $\Phi(\xi, \zeta = 0, \tau)$. For more detailed descriptions of the analytical and numerical model, we refer to Chalikov & Sheinin (1998) and Chalikov & Sheinin (2005).

Remarkably, this new formulation is also simpler than the original set of equations since the nonlinear conformal coordinate transformation removes a number of nonlinear terms. For the stationary case, this method coincides with the classical complex variable method (e.g. Crapper, 1957); an efficient numerical scheme (CS) for this was developed by Chalikov & Sheinin (1998). Note that this scheme is more precise than the popular surface integral scheme (Dold, 1992). As mentioned above, the Dold scheme as well as all schemes based on truncated equations (like those of West *et al.* (1987) and Craig & Sulem (1993)) do not conserve energy without some artificial means. Thus, they are only acceptable for short-duration processes (as indicated by Dold (1992) and Chalikov (2011)). In addition, compared to the CS scheme, the surface integral method is cumbersome: a complete set of its equations occupy several pages. For the CS scheme, the equations take three lines and the core of the numeric scheme takes 11 lines in Fortran90.

The accuracy of this scheme was demonstrated by a long-term simulation of very steep Stokes waves ($ak = 0.42$). The stability of Stokes waves has been the subject of much speculation. The reality is quite simple: Stokes waves are always unstable to any disturbances, but the rate of development of the instability depends on the amplitudes of the perturbations and their phases. In general terms, a Stokes wave is always unstable if it has any perturbation from the pure Stokes form. In the CS case, 11 decimal places of precision and a fourth order Runge–Kutta scheme were sufficient to simulate the propagation of a virtually undisturbed Stokes wave for up to one thousand periods (e.g. Chalikov, 2007).

The conformal mapping even made it possible to reproduce the initial stages of the breaking process where the surface ceases to be a single-valued function. It should be mentioned that the Dold (1992) scheme is also capable of achieving this, but with special smoothing and regularisation (for capabilities of Lagrangian models in this regard, see Tulin & Waseda (1999), Tulin & Landrini (2001) and Section 4.2). The CS model, however, has a number of important advantages: (1) comparison with an exact solution showed that the scheme has extremely high accuracy; (2) it preserves integral invariants; (3) it is very efficient: its computation time scales as $M \cdot \log(M)$ where M is the number of modes, whereas the Dold scheme scales as M^2; (4) the scheme demonstrates stability over millions of time steps (thousands of periods of the dominant wave). This scheme is able to reproduce a nonlinear concentration of energy in physical space resulting in wave breaking and potentially in the appearance of freak waves.

In the CS model, the wave model is also coupled with an atmospheric boundary-layer model (see Chalikov & Rainchik, 2011):

$$\Phi_\tau = F\Phi_\xi - \frac{1}{2}J^{-1}\left(\Phi_\xi^2 - \Phi_\zeta^2\right) - z - p \tag{4.14}$$

where p is surface pressure, which describes the exchange of momentum and energy between the air and water. In the wind-influence investigation described later in this section, in order to speed up the computations, the coupling was conducted by means of a β-function which parameterised the connection of the surface pressure and the surface shape on the basis of an exhaustive set of numerical simulations by means of the coupled model. Real and imaginary Fourier amplitudes of pressure p_r and p_i are calculated as linear functions of amplitudes of water elevation η_r and η_i:

$$p_r + ip_i = (\beta_r + i\beta_i)(\eta_r + i\eta_i). \tag{4.15}$$

The real and imaginary parts of this β-function are functions of non-dimensional frequency $\Omega_n = u(l_k/2)\omega$ where $u(l_k/2)$ is wind velocity at height equal to half of the wave length l_k, $\omega = |k|^{1/2}$ (here, both ω and k are nondimensional variables).

4.1.1 Simulating the evolution of nonlinear waves to breaking

For the purposes of studying wave breaking, the model's ability to reproduce wave evolution without limitations in terms of steepness or duration of propagation is crucial. For this reason, the CS model was chosen in Babanin *et al.* (2007a, 2009a, 2010a) for detailed numerical simulations of physical characteristics of strongly nonlinear waves leading to the onset of breaking. Before, the CS model was extensively verified and tested (Chalikov & Sheinin, 1998; Chalikov & Sheinin, 2005) and used in a number of strongly nonlinear applications (e.g. Chalikov, 2005, 2007). It was then additionally checked in terms of its capacity to model nonlinear wave features associated with wave breaking.

As mentioned above, one of the essential checks for a wave-breaking model is its ability to describe wave asymmetry with respect to the vertical axis. Definitions of the asymmetry A_s (1.3) and skewness S_k (which is asymmetry with respect to the horizontal axis (1.2)) are given in Figure 1.2 and Section 1.2. The capacity of the CS model in this regard is demonstrated in Figure 2.1 where a transient steep wave dynamically develops very large asymmetry and skewness.

In numerical simulations of the fully nonlinear evolution of steep two-dimensional waves to the point of breaking, we will concentrate on three physical properties featuring nonlinearity, i.e. wave steepness, skewness and asymmetry, and their inter-relationships. We will then try to reproduce and investigate these properties in a laboratory experiment with two-dimensional waves (Section 5.1.1). If these properties are indeed linked to wave breaking, but the percentage of breaking waves is small, as it usually is (e.g. Babanin *et al.*, 2001), then examination of average steepness, skewness or asymmetry is likely to be of little use. Therefore, the numerical analysis here will be dedicated to nonlinear properties of individual waves.

In Figure 4.1, development of an unforced wave (no wind) to the point of breaking is shown. The wave shown had the initial monochromatic steepness IMS = 0.16 and is regarded as moderately steep in terms of the modulational instability. This moderate steepness will allow a reasonably long evolution before breaking occurs. Therefore, the simulation will produce a general, rather than detailed picture to begin with (the time scale is expressed in wave periods, i.e. the wave breaks after 82 periods).

As seen in the figure, the steepness of individual waves stays reasonably constant for a significant number of periods (~30), before it starts oscillating noticeably. The magnitude of the oscillation increases significantly beyond the 60th period, and from this point grows rapidly until the simulation ceases (wave breaks) after the 80th period mark.

Similar behaviour is exhibited by the skewness and asymmetry. The simulation starts from a Stokes wave of $S_k = 0.18$ and $A_s \approx 0$. It is informative to note that prior to breaking, the magnitude of the skewness oscillation is so large that at times the wave even becomes negatively skewed (i.e. the trough is deeper than the crest). At the termination of the simulation, however, $S_k \approx 1$, that is the crest is twice as high as the trough (1.2). In Section 5.1.1, it is shown that two-dimensional laboratory waves asymptote to

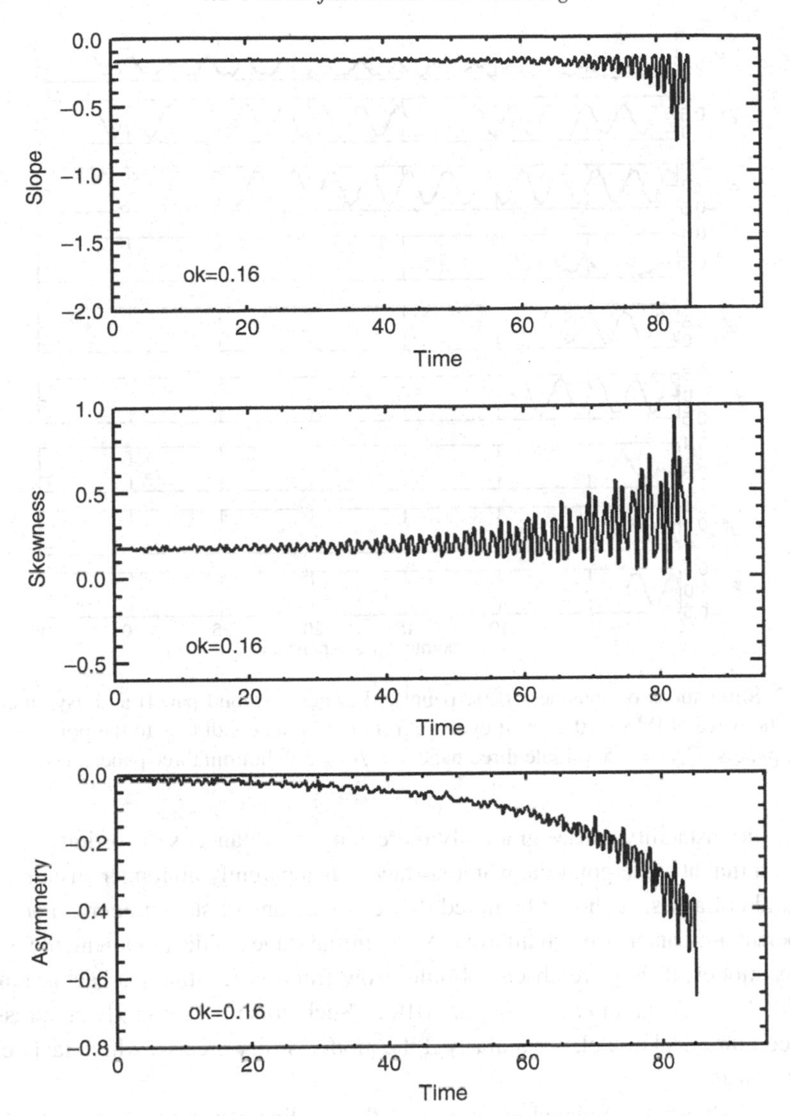

Figure 4.1 Numerical simulation of evolution of a wave with IMS $= 0.16$ to the point of breaking (no wind). Time scale is in wave periods. (top) Wave slope (steepness where minus sign signifies the forward slope); (middle) skewness; (bottom) asymmetry

this value of skewness at the onset of breaking. The asymmetry also oscillates through the simulation and reaches the experimentally observed breaking magnitude of $A_s \sim 0.5$ (i.e. Caulliez, 2002).

Therefore, the inherent instability of nonlinear waves leads to a breaking even in the absence of wind forcing. In such a scenario, the wave cannot gain energy to grow on average.

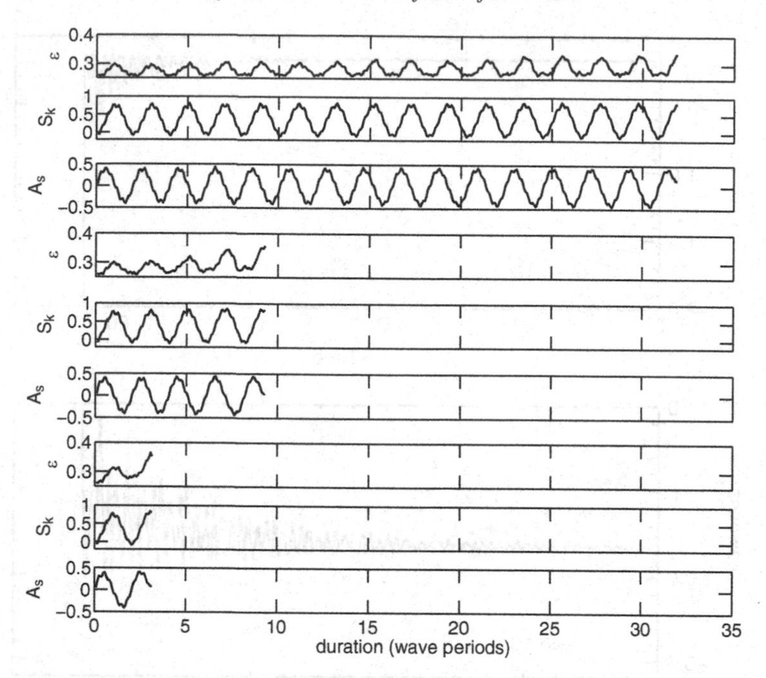

Figure 4.2 Simulations of steepness (first panel), skewness (second panel) and asymmetry (third panel) of the wave of IMS = 0.26 as it evolves from the initial conditions to the point of breaking. Top three panels: $U/c = 2.5$; middle three panels: $U/c = 5.0$; bottom three panels: $U/c = 10.0$

However, the instability causes gradually increasing instantaneous distortions of the wave shape, such that at some point the water surface can apparently no longer sustain the wave profile, and collapses. It should be noted that comparisons of such numerical simulations with experiments can only be qualitative. At the initial stages of development, the necessary instability modes, if they are absent, should grow from the continuous background noise (e.g. Reid, 1992; Babanin *et al.*, 2007a, 2010a). Such noise is essentially suppressed in a discretised numerical model, particularly if the model is very precise, which fact delays the instability onset.

Figure 4.2 shows a simulated evolution of the nonlinear wave properties to the point of breaking in the presence of wind forcing. As above, in each set of three panels the top panel shows the evolution of individual wave steepness, middle panel – wave skewness, and bottom panel – asymmetry. Three sets of subplots correspond to three wind-forcing conditions: $U/c = 2.5$ (moderate forcing), $U/c = 5.0$ (strong forcing) and $U/c = 10.0$ (very strong forcing) where U is a characteristic wind speed at a characteristic half-wavelength height (the model is non-dimensional and therefore there is no standard 10 m height and respective wind U_{10}). The initial steepness chosen is IMS = 0.26, which should lead to a faster evolution to breaking onset. No initial modulations were imposed. Note that the minimum value plotted on the steepness scale is $\epsilon = 0.25$ and not zero, and that the simulation starts from a harmonic wave with $S_k = A_s = 0$.

The most obvious feature of the simulations is the oscillations of the values of steepness, asymmetry and skewness. These are not the waves that are shown in the figure, these are the characteristics of wave nonlinearity evolving in physical space.

The top three panels correspond to a moderate wind-forcing condition of $U/c = 2.5$. Under such a wind, it takes approximately 32 wave periods to reach the point of breaking. Oscillations of wave steepness begin immediately and the wave reaches a steepness of $\epsilon = 0.3$ (first maximum) within one period which then relaxes back to $\epsilon = 0.26$ (first minimum) within the next wave period. The period of the modulation is equal to twice the wave period, which is consistent with the theoretical expectations for Benjamin–Feir instability (Longuet-Higgins & Cokelet, 1978). During each oscillation, the steepness relaxes back to almost $\epsilon = 0.26$ (magnitude of the last steepness trough before breaking is $\epsilon = 0.269$). With the wind energy input imposed, however, the maxima of instantaneous steepness keep growing and reach a value of $\epsilon = 0.34$ at the point that is interpreted as incipient breaking by the model (2.1).

The skewness and asymmetry oscillate with the same double-wave period, but without a noticeable increase in magnitude of the oscillation. For example, the value of $S_k = 0.84$ of the skewness at breaking is repeatedly reached by the wave in its progress without breaking. Therefore it appears that the local steepness, if anything, defines the breaking. Visually, skewness is in phase with the steepness oscillations, and it relaxes back to zero when the steepness is minimal.

The oscillations of asymmetry are apparently shifted in phase with respect to steepness and skewness. The asymmetry oscillates about zero in the range ± 0.45 which means that the waves are periodically tilted backward and forward. When steepness (skewness) is maximal, asymmetry is zero, i.e. the wave is symmetric with respect to the vertical. If the point of maximum steepness (skewness) is passed without breaking, the asymmetry becomes negative. That is, the wave begins to lean forward. If this point signifies the breaking onset, the wave is apparently still continuing to tilt forward, and this explains why all the breaking waves exhibit negative asymmetry. The negative asymmetry thus is not an indication of breaking but is rather an indication of the modulation phase at which breaking in progress may or may not occur.

The second set of three panels correspond to wind forcing of $U/c = 5.0$. Whilst such forcing is quite strong, and therefore the steepness growth rate is much faster than above, apart from the steepness growth almost all the other breaking and non-breaking properties of nonlinear evolution remain similar to the previous test. The wave steepness, skewness and asymmetry oscillate with the same period and their phase-shifting pattern is qualitatively the same, steepness ($\epsilon = 0.36$) and skewness ($S_k = 0.82$) values at breaking are close to those of the above test, and asymmetry at breaking approaches zero.

It is interesting to note that, according to known results on wave amplification by wind, the wave growth increment at non-extreme conditions should be approximately a quadratic function of the wind (e.g. Donelan *et al.*, 2006). If indeed there is some critical steepness signifying breaking onset, then doubling the wind speed in numerical tests should lead to this limiting value being reached four times as fast. This conjecture produces a result close to that simulated: the duration of the evolution to breaking, when the wind forcing was doubled from $U/c = 2.5$ to $U/c = 5.0$, was reduced from 32 to 9 wave periods (almost four times).

A further doubling of the wind input, as shown in the bottom set of three panels, led to another reduction of the evolution duration – from 9 to 3 periods. This is again consistent with Donelan *et al.* (2006) who showed that at very strong winds the relative wave growth slows down. The other patterns of nonlinear wave evolution appear unaltered. The whole picture again points to the critical local steepness as the parameter responsible for the onset of the water-surface collapse. The maximum values of steepness $\epsilon = 0.36$ and skewness $S_k = 0.83$ are almost the same as previously. These values also demonstrate that the instantaneous effect of the wind on breaking onset is negligible. The wind forcing of $U/c = 10.0$ is now very strong, but this wind is still not capable of pushing the wave over and reducing, even marginally, the critical steepness at breaking.

Let us summarise observations made with this instructive Figure 4.2. Values of steepness, skewness and asymmetry oscillate at a frequency half that of the carrier wave. While the simulation begins with both skewness and zero asymmetry (sinusoidal wave), the sinusoidal wave immediately turns into a Stokes wave. It is, however, only conditionally a Stokes wave. The shape of this wave oscillates, and it is only at a particular phase of these oscillations that the wave shape is clearly that of the Stokes wave again: at the point of maximal skewness, when the wave is symmetric. As the skewness is decreasing or growing, the asymmetry is also changing, that is the wave is tilting forward (negative asymmetry) or backward (positive asymmetry). Values of S_k and A_s oscillate between their maximum and minimum levels, but remain bounded, their maximum and minimum do not increase in magnitude if the initial steepness is already large enough. In contrast, the oscillations in steepness progressively grow in amplitude until a point is reached where breaking takes place. It is therefore evident from Figure 4.2 that it is the steepness which is the limiting parameter for breaking to occur.

The coherence and phase relationships of steepness, skewness and asymmetry, outlined qualitatively above, are analysed in Figures 4.3 and 4.4. Figure 4.3 compares spectra of running instantaneous steepness ϵ (1.1, top subplot) and skewness S_k (1.2, second subplot), their coherence (third subplot) and phase (bottom subplot). Since the time scale of the simulations is dimensionless (i.e. presented in terms of wave periods), the frequency scale is expressed in inverse wave periods. Therefore, as expected from visual examination of Figure 4.2, the peak of the steepness/skewness modulation occurs at twice the wave period (0.5 of the inverse wave period). This frequency dominates the spectrum, in agreement with theory (Longuet-Higgins & Cokelet, 1978). The spectral density decreases very rapidly away from the peak.

The peak is rather broad and covers a range of frequencies of 0.4–0.6 of inverse wave periods. The coherence of the steepness and skewness oscillations in this range is 100%, as could have been expected for numerical simulations of the theory with a model of such high precision. The phase shift between the dominant oscillations of steepness and skewness is

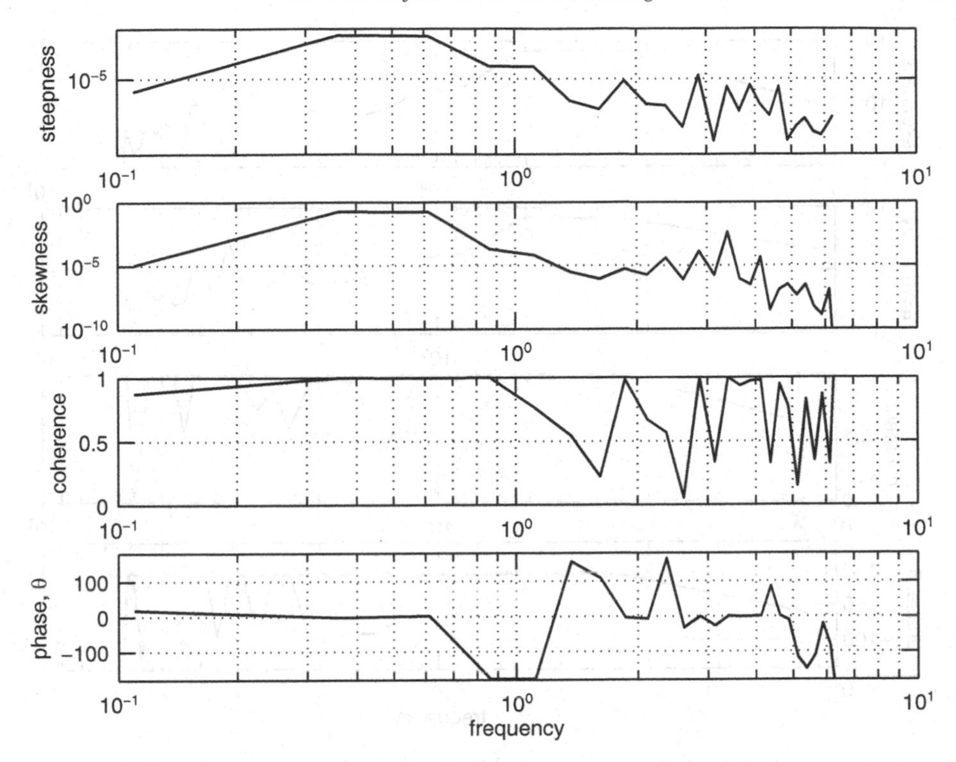

Figure 4.3 Numerical simulations (see Figure 4.2). Dimensionless wave period is 1. Co-spectra of running steepness and skewness for waves of IMS = 0.26, $U/c = 2.5$. (top panel) Steepness spectrum. (second top panel) Skewness spectrum. (second bottom panel) Coherence spectrum. (bottom panel) Phase spectrum (in degrees)

zero, as was observed visually. Thus, the steepness and skewness are in phase, and the maximum steepness is achieved at the same instant as the maximum skewness.

Similarly, Figure 4.4 compares the co-spectra of running instantaneous skewness S_k (1.2) and asymmetry A_s (1.3). Again, the broad peak of the asymmetry spectrum falls in the 0.4–0.6 range of inverse wave periods. Spectra of skewness and asymmetry are almost perfectly coherent, with a phase shift of about 90° (asymmetry is leading). The latter means that the asymmetry reaches its positive maximum (i.e. wave is tilted backwards) when skewness is approximately zero (wave crest and trough are of equal magnitude) and the local steepness is half-way through rising from its minimum to the maximum value in an oscillation. From the positive maximum, the asymmetry begins to decrease and reaches zero half a wave period later (quarter of the period of the oscillation) – i.e. the wave becomes symmetric with respect to the vertical. At this point, steepness and skewness are at their maximum, and it is at this point that the wave may break. Whether the wave breaks or not, the asymmetry will keep decreasing into negative values (wave is tilting forwards), while the steepness/skewness start subsiding in quadrature with the asymmetry. It is interesting to look at this moment from the point of view of an observer who encounters the

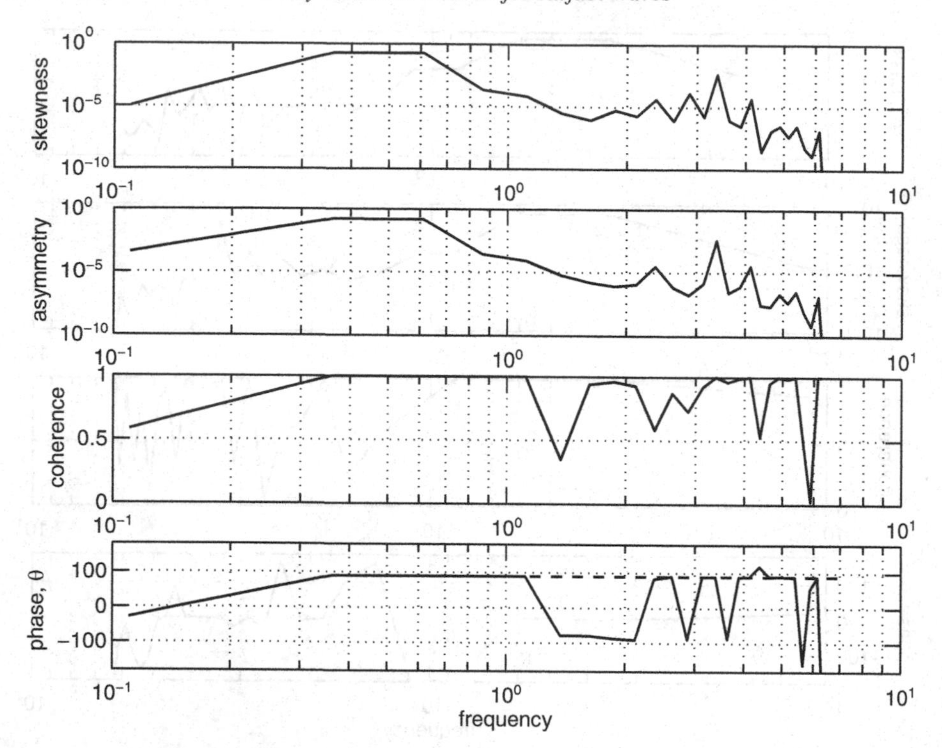

Figure 4.4 Numerical simulations (see Figure 4.2). Dimensionless wave period is 1. Co-spectra of running skewness and asymmetry for waves of IMS $= 0.26, U/c = 2.5$. (top panel) Skewness spectrum. (second top panel) Asymmetry spectrum. (second bottom panel) Coherence spectrum. (bottom panel) Phase spectrum (in degrees), positive phase means asymmetry is leading. Dashed line shows 90° phase shift

breaker: he sees a very tall crest which begins to break with the water mass falling down from the top and at the same time the front face is growing steeper as the wave is leaning forward – obviously a very dangerous situation.

To briefly summarise the intermediate conclusions: we can speculate that a two-dimensional nonlinear wave will break when, due to inherent modulations of its height, it reaches some limiting steepness. The skewness and asymmetry also oscillate, in phase and in quadrature with the steepness, respectively. In the simulations, however, they do not appear to exhibit some specific limiting value at the point of breaking.

We shall now conduct a similar set of numerical simulations for a wave with initially half the steepness IMS $= 0.13$ (Figures 4.5 and 4.6). A very strong forcing of $U/c = 10.0$ is applied in order to achieve breaking in a reasonably short period of time.

The three panels in Figure 4.5 show the time evolution of individual wave steepness (top), skewness (middle) and asymmetry (bottom) as in Figure 4.2. The wave steepness in the top subplot grows under the $U/c = 10.0$ wind forcing so rapidly that its oscillations are barely visible over the strong mean trend. The value of steepness $\epsilon = 0.33$ at the point of

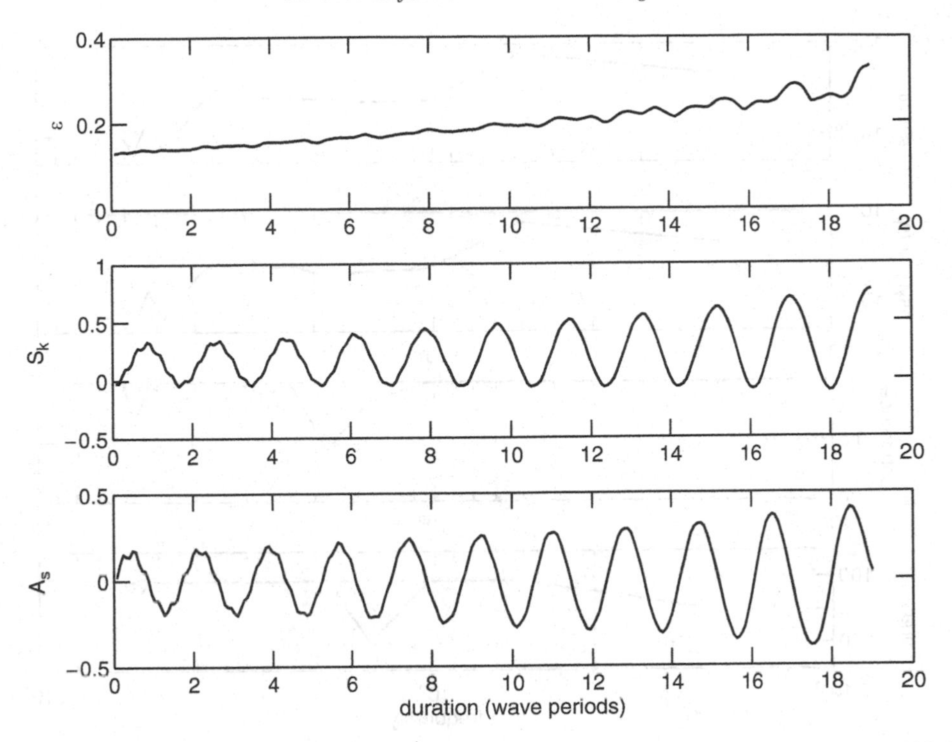

Figure 4.5 Simulations of steepness (first panel), skewness (second panel) and asymmetry (third panel) of the wave of IMS = 0.13, $U/c = 10.0$ as it evolves to breaking

breaking, similar to the previous tests, supports the concept of a limiting breaking-onset value. Compared with the same wind forcing in Figure 4.2, it now takes much longer (19 wave periods versus 3) to achieve this steepness, but as soon as it is reached, the wave breaks.

It is instructive to observe the behaviour of the skewness and asymmetry of these much less steep waves (bottom two panels). Since nonlinearity is now obviously weaker, it could have been expected that the oscillations would start from much smaller magnitudes of $S_k = 0.33$ and $A_s = 0.18$. They do eventually grow to maxima of $S_k = 0.77$ and $A_s = 0.41$ similar to values observed previously. These may be indicative of limiting values of the wave skewness and asymmetry, but they are obviously not a breaking criterion, as in the previous test they did not lead to breaking unless the limiting steepness was also reached.

Co-spectra of running skewness and asymmetry of the IMS = 0.13 wave are shown in Figure 4.6. These values are quite similar to those for the case of IMS = 0.26 in Figure 4.4 above, but it is noticeable that the peak at half the inverse wave period is now narrower and the coherence is stronger across almost the entire frequency band. Apparently, the stronger nonlinearity tends to somewhat randomise the wave-shape oscillations in Figure 4.4. The obvious decrease in the coherence between one and two inverse wave periods still requires an explanation.

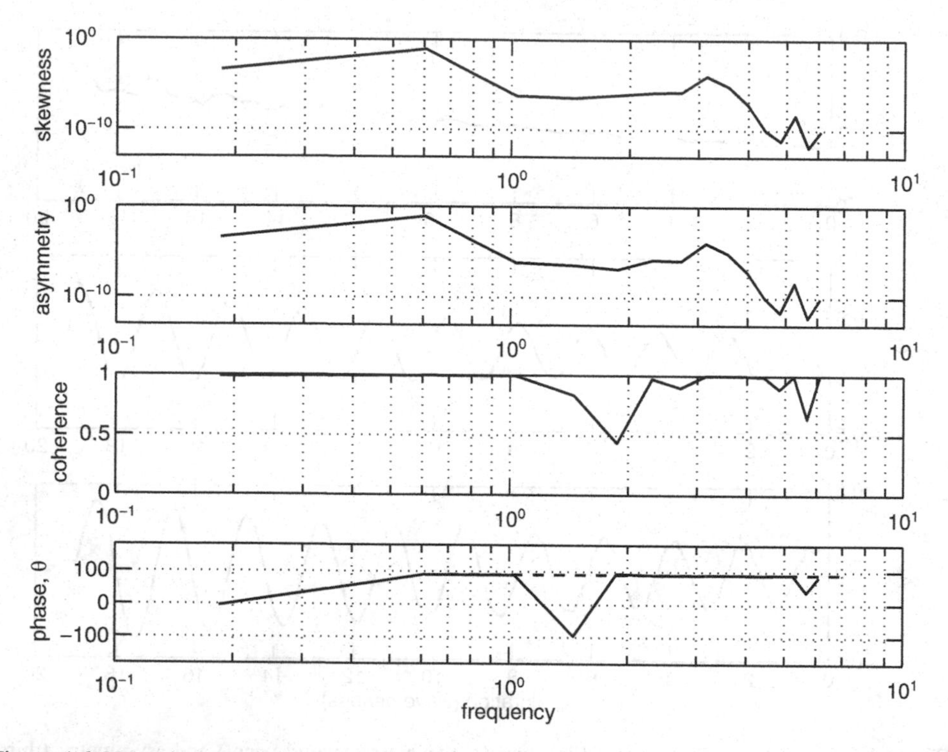

Figure 4.6 Numerical simulations (see Figure 4.5). Dimensionless wave period is 1. Co-spectra of running skewness and asymmetry for waves of IMS = 0.13, $U/c = 10.0$, 19 wave periods to breaking. (top panel) Skewness spectrum. (second top panel) Asymmetry spectrum. (second bottom panel) Coherence spectrum. (bottom panel) Phase spectrum (in degrees), positive phase means asymmetry is leading. Dashed line shows 90° phase shift

Figure 4.7 shows a composite set of fetch-versus-steepness dependences for different values of wind forcing $U/c = 1$–11. The fetch is expressed in dimensionless terms of number of wavelengths to breaking at a particular IMS $= ak$.

As shown in Babanin *et al.* (2007a), a wave with no superimposed wind forcing and IMS < 0.1 will never break, even though it will exhibit oscillations of steepness, asymmetry and skewness similar to those shown in Figures 4.1, 4.2 and 4.5. The evolution of such waves is not plotted in the figure, and in fact we do not plot IMS < 0.17 because such development to breaking is too slow for the purpose of demonstration. The upper limit of steepness included is IMS $= 0.28$ as waves with IMS > 0.3 will break immediately, within one wavelength/period. Between these two limits, the dimensionless distance to breaking decreases with increasing IMS.

Figure 4.7 allows the estimation, based on numerical simulations with the CS model, of when a two-dimensional wave breaks. For example, it will take a wave of IMS $= 0.24$ six wavelengths to reach the point of breaking under $U/c = 8.5$, and 11 wavelengths under $U/c = 8.0$, and 11 wavelengths under $U/c = 7.5$. If wind forcing is reduced

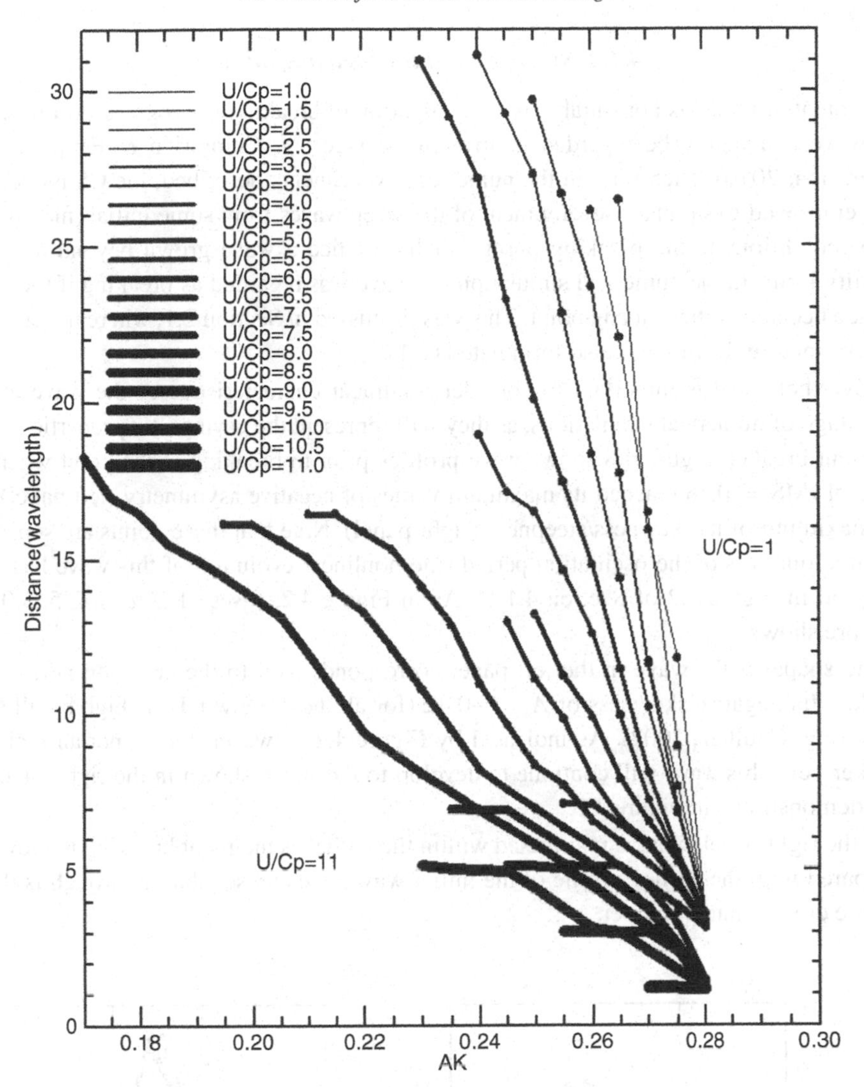

Figure 4.7 Numerical simulations, number of wave lengths to breaking versus IMS = ak, different wind forcing U/c. Chalikov (2007, personal communication)

significantly, i.e. $U/c = 3.0$, such a wave will only break after 31 wavelengths. Under the same wind forcing, the wave will break faster if its IMS is greater. That is, at a forcing of $U/c = 3.0$, but IMS = 0.26, it will take 15 wavelengths to reach the point of breaking, and at IMS = 0.27 only nine. Quantitative application of these strictly two-dimensional numerical results may be limited, but as will be shown in Section 5.1.1, qualitatively this picture agrees well with the experiment.

4.1.2 Simulation of the breaking onset

The potential model is not suitable for investigation of breaking in progress. At this stage, the wave can hardly be regarded as irrotational (see e.g. Gemmrich & Farmer, 2004; Gemmrich, 2010). Therefore, in the numerical experiments described, the CS model was only employed to simulate development of the steep waves from some initial (mostly uniform) conditions to the breaking onset which signified a wave grown beyond a certain stability limit. In the numerical simulations, a wave was regarded as breaking if the water surface became vertical at any point. This was discussed in Section 2.1, where the criterion for terminating the run was also formulated (2.1).

Nevertheless, it is important to consider nonlinear characteristics of the wave at this final stage of numerical simulations, as they will represent the asymptotic properties of the incipient breaker. Figure 4.8 shows wave profiles prior to breaking at the point when the wave of IMS = 0.26 reached its maximum values of negative asymmetry (left panel) and the maximum of its skewness/steepness (right panel). Note that these points are separated by three-quarters of the oscillation period (the nonlinear evolution of this wave has been analysed in Figure 4.2 of Section 4.1.1). As in Figure 4.2, cases of $U/c = 2.5, 5.0$ and 10.0 are shown.

The shape of the wave in the left panel corresponds well to the common notion of a breaker. Its negative skewness of $A_s = -0.42$ (for all the three winds) is high by all standards (e.g. Caulliez, 2002). As indicated by Figure 4.2, however, this is not an incipient breaker yet. This wave will continue to develop to the shape shown in the right panel, as also demonstrated in Figure 4.2.

In the right panel, what is interpreted within the model as incipient breaking is shown. If compared with the dashed profile of the initial wave, one can see that the trough is flatter and the crest is much sharper.

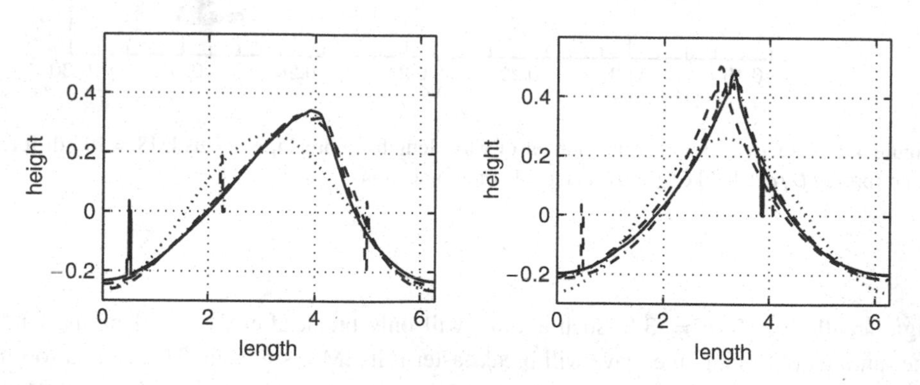

Figure 4.8 Numerically simulated wave shape prior to breaking. Waves propagate from left to right. (left) minimal (maximal negative) asymmetry A_s (1.3); (right) maximal skewness/steepness S_k/ϵ (1.2)/(1.1). IMS = 0.26, U/c = 2.5 (solid line), 5.0 (dashed line), 10.0 (dash-dotted line). Initial wave is shown by dotted line

In both the left and right panels, wave profiles have evolved significantly from the initial harmonic wave shown by the dashed line. In spite of the fact that the evolution occurred under very different wind-forcing conditions, and took very different times to reach the breaking point, the magnitudes of the asymmetry ($A_s = -0.42$), skewness ($S_k = 0.84, 0.82, 0.83$) and steepness ($\epsilon = 0.34, 0.36, 0.36$), as well as the profiles of the three waves, are virtually identical. This highlights again the important role of the hydrodynamic mechanism in redistributing the wave energy and forming the nonlinear wave profile, whereas the wind here appears to serve merely as the source of energy to the wave system.

The statistical properties obtained by means of the CS model for such incipient breaking are shown in Figure 4.9. Note, again, that steepness/skewness and asymmetry are measured at the different phases of the last prior-to-breaking oscillation. The estimates shown in Figure 4.9 were obtained from a comprehensive set of numerical runs covering a range of initial steepness IMS = 0.10–0.30 and wind forcing $U/c = 0$–11. The steepness is shown in terms of $kH = 2\epsilon$.

Since we have identified the steepness as a possible reason for wave collapse, it is most instructive to investigate the limiting values of kH. The data points cluster (top subplots),

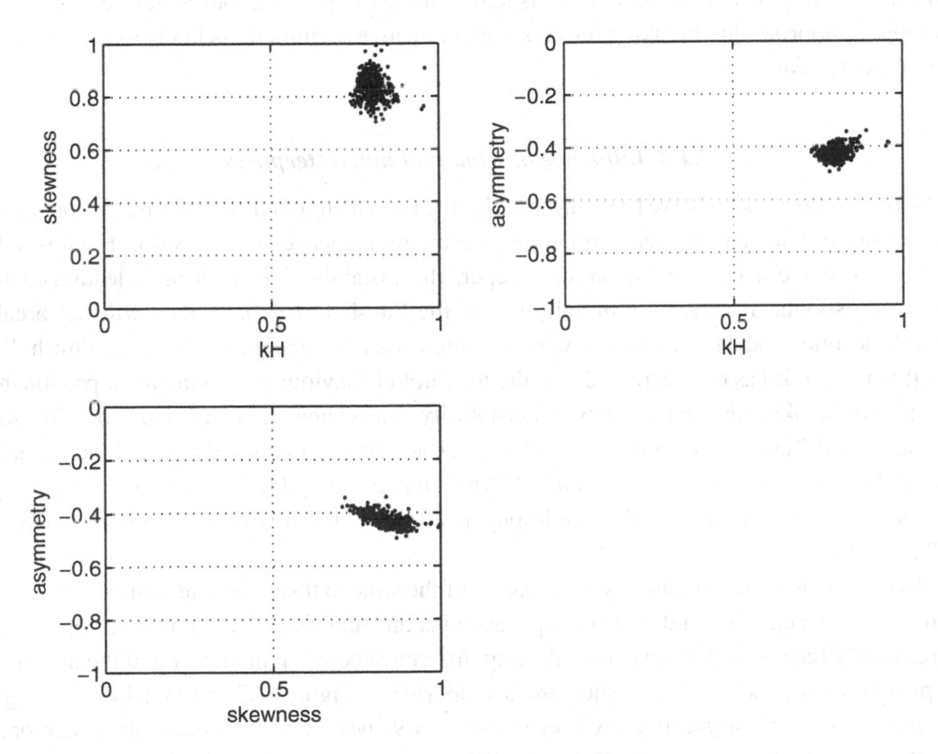

Figure 4.9 Numerically simulated incipient breaking. (top left) skewness versus steepness; (top right) asymmetry versus steepness; (bottom) asymmetry versus skewness

although the scatter is noticeable: $kH = 0.75$–0.85 with some outliers reaching values above 0.9. The mean value of the limiting steepness is $\epsilon \sim 0.4$, which is very high, particularly considering that the value of the local slope near the crest is even higher than this mean value. Dyachenko & Zakharov (2005) specifically investigated the shape of the incipient breaker with a high-resolution version of a fully nonlinear model similar to that of CS (Dyachenko *et al.*, 1996). They found a remarkable agreement of the limiting steepness with the Stokes limiting-steepness criterion (2.47). The shape of the wave, however, is different to that of the Stokes wave with the crest pointed at a 120° angle (Stokes, 1880). What is most encouraging is that such an outcome of numerical simulations finds full experimental support (see Section 5.1). At such magnitudes of water surface slope, the surface may simply collapse because of gravity, depending on the velocity field in the water. If so, the role of the instability which led to the occurrence of these waves is not in generating wave breaking as such, but rather in simply producing a very steep wave.

As indicated previously, skewness and asymmetry are not breaking criteria, but they do have limiting values (see Figure 4.2). In Figure 4.9, the incipient-breaking skewness is scattered in the range of $S_k = 0.7$–1 (top left) and the asymmetry in the range from $A_s = -0.35$ to -0.5 (bottom). Within the scatter, there is no dependence of one property on the other, except a possible negative correlation between the skewness and asymmetry in the bottom panel. The latter result is supported by Figure 4.2, but is not necessarily a feature of approaching breaking onset: a larger negative asymmetry is likely to be followed by a greater skewness.

4.1.3 Influence of wind and initial steepness

The role of the wind in wave breaking has already been mentioned several times throughout the book. It is apparently very important in growing the wave steepness (i.e. Figures 4.2, 4.5, 4.7). Once a wave is becoming steeper, the instability mechanism is leading it to breaking sooner. If waves are initially below the threshold $\epsilon = 0.10$, they will not break at all, despite modulations, unless wind forcing raises the steepness above the threshold. In this regard, it has to be stressed that the threshold behaviour is also the most prominent feature of breaking due to modulational instability in mechanically generated waves in two-dimensional flumes (Babanin *et al.*, 2007a, 2010a) and in directional (three-dimensional) wind-forced field waves (Banner *et al.*, 2000; Babanin *et al.*, 2001), which fact most likely links all the observations to the single physical course, the modulational instability (see Chapter 5).

In this section, we will discuss the capacity of the wind to instantaneously affect breaking onset. That is: can the wind push a steep wave over and thus reduce the limiting steepness at breaking? Because of the very large density difference between the water and the air, such a possibility seems low. This conjecture is supported by Figures 4.2, 4.5 and 4.8. The large scatter of the limiting-steepness values in Figure 4.9 above, however, needs investigation.

In Figure 4.10, the nonlinear features of the incipient breaker are shown as a function of IMS for a variety of wind-forcing conditions. Note that the simulation was run within a

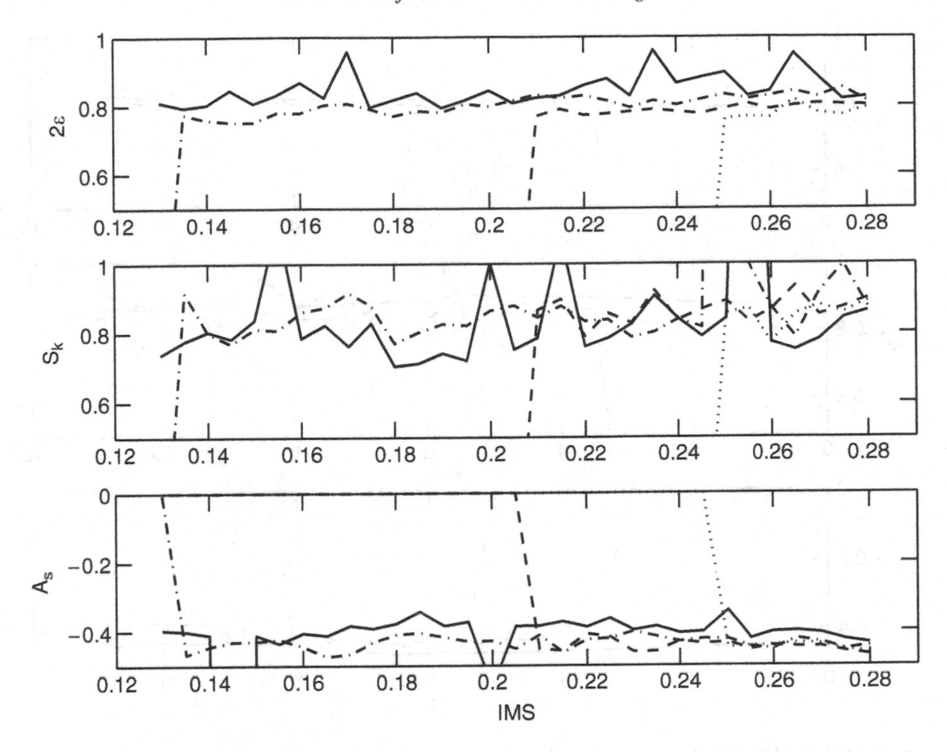

Figure 4.10 Simulations of incipient breaking. (top panel) Steepness 2ϵ (1.1); (middle panel) skewness S_k (1.2); (bottom panel) asymmetry A_s (1.3) – versus IMS for $U/c = 3$ (dotted line), 5 (dashed line), 8 (dash-dotted line) and 11 (solid line)

limited number of wave periods (i.e. simulation was stopped after some 400 wave periods regardless of whether the wave had reached breaking onset or not). Thus, in these numerical experiments, for different wind-forcing situations waves stop breaking at different IMS (e.g. for $U/c = 3$, the waves do not have enough time to break if IMS < 0.25).

The limiting steepness at breaking onset 2ϵ (1.1) in the top subplot is plotted versus IMS for $U/c = 3$ (dotted line), 5 (dashed line), 8 (dash-dotted line) and 11 (solid line). The incipient-breaking steepness grows both for higher IMS and stronger wind forcing. The latter is particularly counter-intuitive. Even though the growth is marginal, the four lines clearly separate and therefore instantaneous steepness at breaking appears to be somewhat larger at stronger winds.

The skewness and asymmetry of the incipient breakers (middle and bottom panels) do not exhibit a dependence on IMS or wind forcing except at extreme forcing of $U/c = 11$. At such winds, the skewness is somewhat lower (the wind flattens the wave crests) and the wave is slightly less tilted. Thus, it is only at extreme conditions that the wind is capable of influencing the wave shape at breaking and even then the effect is only marginal.

The same properties of the incipient breaker are plotted versus U/c for three different IMS values of 0.20, 0.24 and 0.28 in Figure 4.11. The value of IMS $= 0.28$ is extreme and

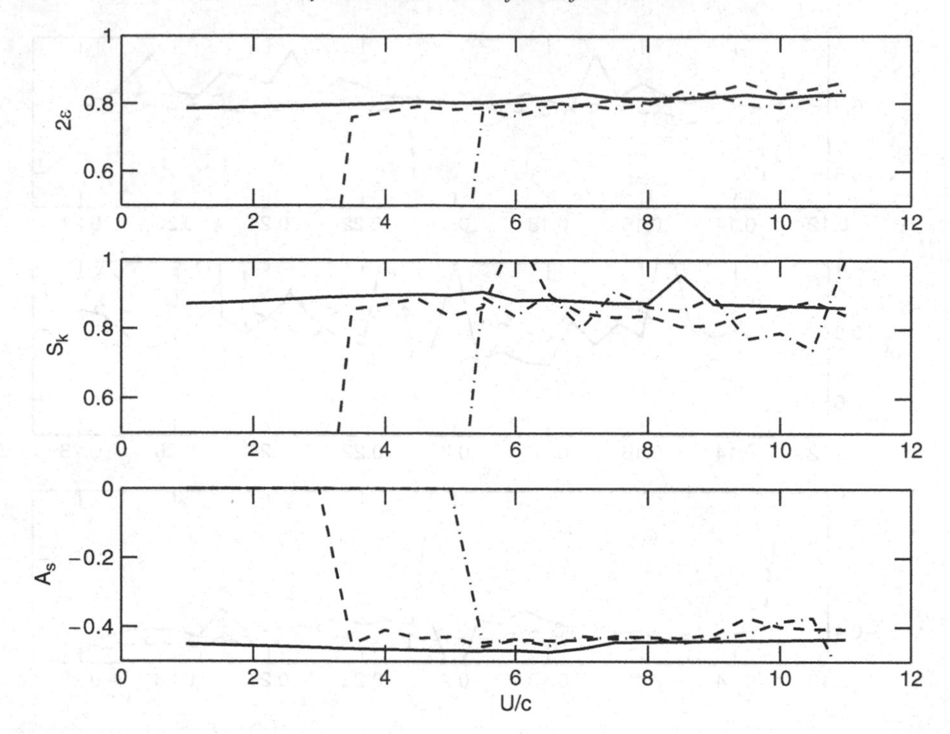

Figure 4.11 Simulations of incipient breaking. (top panel) Steepness 2ϵ (1.1); (middle panel) skewness S_k (1.2); (bottom panel) asymmetry A_s (1.3) – versus U/c for IMS = 0.20 (dash-dotted line), 0.24 (dashed line) and 0.28 (solid line)

the wave grows to the limiting steepness almost instantaneously – within 1–3 periods (see Figure 4.7). Note that the simulation was again run for a limited number of wave periods. As a result, in the case of IMS = 0.20, for example, the waves do not have enough time to break if $U/c < 5$.

Apart from the relatively weak growth of the limiting steepness as a function of wind in the top panel (as previously noted), the skewness (middle) and asymmetry (bottom) panels exhibit another marginal feature. For the critical initial steepness of IMS = 0.28, both skewness and asymmetry magnitudes at breaking are, for all wind forcing cases, greater than the respective values at less steep initial conditions.

Thus, numerical simulations of the breaking onset reveal some marginal effects which wind forcing and initial-steepness conditions have on the onset of breaking. Some of these effects are only noticeable at extreme winds and critical initial steepness.

Therefore, the wind plays a dual role in this process. Firstly, it accelerates the growth of individual wave steepness. In the simulations shown in Figure 4.2, doubling the wind speed resulted in the wave growing to its critical height almost four times faster. Secondly, the wind can push the wave over and thus reduce the critical steepness, but this reduction was found to be small and only relevant at very strong wind forcing $U/c > 10$ (see also Babanin *et al.*, 2010a).

To finalise this section, based on the numerical simulations of initially monochromatic steep two-dimensional irrotational waves by a fully nonlinear model, it can thus be summarised that there are breaking-onset criteria in terms of free surface. First, there is a critical initial monochromatic steepness for the wave train. If IMS is greater than this value,

$$\text{IMS} \approx 0.1, \tag{4.16}$$

then according to simulations based on the CS model a breaking will always occur. Even if the wave is initially sinusoidal and linear, the nonlinear evolution of the wave will ultimately lead to breaking. The distance to breaking will be a function of this initial steepness (Babanin *et al.*, 2007a, see also Chapter 5). The second criterion is the steepness of an individual wave at the breaking onset. As seen in Figures 4.2 and 4.9, there appears to be a limiting value of such a steepness of

$$2\epsilon \sim 0.8. \tag{4.17}$$

Nonlinear characteristics of the wave shape, i.e. steepness (1.1), skewness (1.2) and asymmetry (1.3) all oscillate, that is the profile of a nonlinear wave does not remain self-similar and, strictly speaking, is not even that of a Stokes wave: asymmetry is present and oscillates in quadrature to steepness/skewness. Oscillations, as well as the critical steepness, are only marginally affected by the wind unless the wind forcing is very strong.

4.2 Lagrangian nonlinear models

Within Lagrangian methodology, the motion of individual particles, rather than fluid dynamics with respect to a fixed coordinate system, is modelled. One can argue that for surface waves Lagrangian governing equations are more complicated, but on the other hand the boundary conditions are simpler. One way or another, however, the Lagrangian approach in fluid mechanics in general and in wave modelling in particular has enjoyed much less attention compared to Eulerian methods.

A few available Lagrangian nonlinear models, however, demonstrate the impressive capacity of such approaches. Fouques *et al.* (2006) and Fougues & Stansberg (2009) showed that even first- and second-order Lagrangian simulations are capable of reproducing many two-dimensional and three-dimensional features of surface waves, which are such complicated and delicate issues in Eulerian models as discussed in Section 4.1. Fully nonlinear Lagrangian models can possibly go much further than that, although their potential difficulties are due to problems with conservation of fundamental physical properties. These need special attention and treatment. For example, in order to satisfy the conservation of mass, Fougues & Stansberg (2009) used a second-order residual in the continuity equation to shift the fluid particles vertically.

A few Lagrangian models have been suggested and applied to wave-breaking studies by Landrini *et al.* (1998), Tulin & Landrini (2001), Dalrymple & Rogers (2006) and

Dao *et al.* (2010). Based on the smoothed particle hydrodynamics method (SPH, Lucy, 1977; Monaghan, 1977), the water volume is treated as a set of large numbers of particles, each having its own mass and characterised by a kernel of its interactions with neighbouring particles.

In Tulin & Landrini (2001), spline kernels of the third and fifth orders were used, which accommodated 30 to 50 particles within interaction range of each individual particle, with the total number of particles being up to the order of 10^5. As the authors put it:

"Each particle moves in the force field generated by the whole particle system and the physical quantities evolve according to suitable evolution laws following from the original differential field equations".

As usual, caution should be exercised in identifying the reality with the model, see also a general discussion of potential limitations of models in Section 5.1. With that in mind, however, such a Lagrangian model was capable of simulating the nonlinear behaviour of wave trains and going far beyond the point where wave-breaking simulations by Eulerian means would be stopped.

That is, Tulin & Landrini (2001) were able to model the breaking progress after breaking onset, including splashing, mixing and air entrainment, i.e. effectively to simulate a multi-phase medium. The motion of individual particles is tracked, and thus the kinematics and dynamics of the primary and secondary plunging jets is investigated, formation of the underwater vortical structures is analysed, and the collapse of the air cavity, series of splash-up cycles and other breaking and post-breaking features are described. The authors stressed that, technically speaking their method can also deal with weak compressibility effects and thus potentially handle even acoustic-noise generation in the course of breaking.

This potential capacity of the method was achieved by a more recent SPH model of Dalrymple & Rogers (2006). The authors essentially extended earlier applications of the SPH methods to free-surface fluid flows by improving and implementing the treatment of water density, viscosity and turbulence.

By introducing the compressibility of the fluid, the Dalrymple & Rogers (2006) model was able to produce sound associated with the breaking. This sound is one of the principal proxies for wave-breaking events and associated passive-acoustic instrumentation is among the most promising means of remote-sensing studies of wave breaking and dissipation (see Section 3.5). Examples shown in Dalrymple & Rogers (2006) for waves breaking at a beach clearly and realistically exhibit the acoustic impact of breaking waves similar to that in Bass & Hey (1997) and Babanin *et al.* (2001) (see also Figures 3.1 and 3.4 in this book).

Modelling two-dimensional breaking at the beach realistically reproduced a number of apparent and less apparent known features of shallow-water breaking, such as the plunging jet, formation of vorticity, the splash-ups, including the reverse breaking occurrence and associated downbursting all the way to interacting with the bottom. The latter two less obvious features have independent observational support (Li & Raichlen, 2003; Kubo & Sunamura, 2001).

Dalrymple & Rogers (2006) went further and simulated three-dimensional breaking at the beach. This revealed new spatial features such as the occurrence of counter-rotating vortices which remained after the breaking wave had passed and continued descending obliquely. Again, this feature finds independent experimental support (Nadaoka *et al.*, 1989). Demonstration of this capacity of the method is particularly important as breaking is an essentially three-dimensional phenomenon (see e.g. Phillips *et al.*, 2001).

Therefore, the fully nonlinear models described in Sections 4.1 and 4.2 are sufficiently self-consistent to approach the evolution of the steep wave train from initial conditions which have no indication of any singularity in the system, all the way to incipient breaking. They also provide quantitative characteristics both for the wave train, in which the breaking will occur, and for the individual wave, which will break. Lagrangian models are potentially promising in further simulating the breaking-in-progress and post-breaking effects in water, but are less developed and common at this stage. Further discussion on models capable of reproducing the breaking in progress will be given in Section 7.2 dedicated to simulations of wave-energy dissipation in two-phase models (Abadie *et al.*, 1998; Zhao & Tanimoto, 1998; Chen *et al.*, 1999; Watanabe & Saeki, 1999; Mutsuda & Yasuda, 2000; Christensen & Deigaard, 2001; Grilli *et al.*, 2001; Guignard *et al.*, 2001; Tulin & Landrini, 2001; Hieu *et al.*, 2004; Song & Sirviente, 2004; Zhao *et al.*, 2004; Iafrati & Campana, 2005; Dalrymple & Rogers, 2006; Lubin *et al.*, 2006; Liovic & Lakehal, 2007; Iafrati, 2009; Dao *et al.*, 2010; Janssen & Krafczyk, 2010; Lakehal & Liovic, 2011, among others).

Thus, fully nonlinear numerical modelling allows us to move from mathematical abstractions to physical reasoning and verify the theoretical conclusions by experimental means. Such verification will be discussed in Chapter 5.

5
Wave-breaking probability

As already mentioned in Chapter 2 dedicated to definitions of wave properties and phenomena related to wave breaking, the breaking probability, or as it is also often called breaking rate or frequency of breaking occurrence is one of the most important statistical characteristics of wave fields that contain the breaking events. Technical definitions for the breaking probability are given in Section 2.5.

Together with the breaking severity (Section 2.7, Chapter 6), the probability defines the wave-energy dissipation due to wave breaking. Knowledge of such dissipation is required across a broad range of wave-related applications, with the wave forecast being the most frequent and obvious, and therefore the breaking occurrence has enjoyed key attention within wave-breaking studies.

Experimental and statistical techniques of breaking-probability studies have been discussed in detail in Chapter 3, and theoretical approaches in Chapter 4. As described in these chapters, in the past parameterisations of the breaking rates in terms of environmental characteristics have usually relied on wind speed or its derivatives. In this book, we have argued that, although the wind is of course essentially responsible for the formation of fields of wind-generated waves, its capacity to directly trigger or even affect a breaking event is only marginal, except perhaps for very strong wind forcing. Breaking mainly happens due to hydrodynamic phenomena, that is due to processes in the wave train itself. Therefore, as far as the wave-breaking probability is concerned, the wind influence is indirect and parameterisations have to be done in terms of the properties of wave fields. In this chapter, we will concentrate on the latter approach and will describe the hydrodynamics which controls wave-breaking occurrence and rates.

In Section 5.1, we start from breaking events that develop within trains of initially monochromatic waves due to the modulational instability of such wave trains. While the frequency of breaking occurrence depends on the initial monochromatic steepness, there is a steepness threshold below which the breaking does not happen. Such a threshold is of key importance, both for monochromatic wave trains and directional wave fields, and points to the course of breaking in real oceanic conditions. It will be scrutinised in Section 5.2. Breaking in wave fields with a continuous spectral distribution of wave scales will be discussed in Section 5.3, which is subdivided into further subsections. The wave spectrum usually has a narrow and sharp peak, and the physics of breaking of the dominant waves, which correspond to the spectral peak and exhibit pronounced group-modulated structure (Section 5.3.1), appears essentially different to that of waves which are relatively shorter compared to the dominant waves and whose breaking may be induced by larger and longer waves (Section 5.3.2). In directional wave fields, the very existence of modulational instability is an issue. Since the real waves are directional, with the exception of swell which has low steepness and does not break anyway, this issue is of significant importance for wave breaking and will be discussed in Section 5.3.3. Finally, wind-forcing effects will be described in Section 5.3.4.

5.1 Initially monochromatic waves

In Section 4.1, numerical simulations were described of the breaking development within initially monochromatic wave trains, conducted by means of CS models. Results of the simulations were further used in a laboratory experiment to verify the model's ability to predict the breaking onset and to parameterise the breaking probability for such waves (Babanin *et al.*, 2007a, 2009a, 2010a).

Following the logic of Section 4.1, the current section is subdivided into a number of subsections on laboratory measurements of the wave evolution to breaking, breaking onset, wind influence, and breaking probability. Each subsection has a brief introduction on the main and side topics discussed. Before these subsections, we will outline the issue of comparing numerical simulations and laboratory experiments, provide necessary definitions for the modulational instability, and describe the ASIST wave tank and laboratory experiments conducted in this tank. Examples of time series and modulational instability will be shown, and a definition of the depth of the modulation, important for the breaking-severity issue and used throughout the rest of the book, will be given.

Although it has already been mentioned a number of times, it should be emphasised again that comparisons of numerical simulations of nonlinear wave evolution with laboratory experiments can only be qualitative. Firstly, no matter how sophisticated the model is, it is still a simplification of the physical environment and disregards or possibly suppresses some natural features. One such feature is the three-dimensionality of wave motion. Even in the quasi-two-dimensional environment of the wave tank, some directional features may play an essential role. For example, Melville (1982) showed that for steepness $\epsilon > 0.3$, the wave crests develop a crescent-shaped perturbation and this three-dimensional instability manifests itself in a more complicated way compared to the strictly two-dimensional case. This has a significant consequence for numerical simulations. The two-dimensional CS model predicts immediate breaking onset for $\epsilon > 0.29$, whereas in the laboratory experiments of Melville (1982) such waves become short-crested but can persist without breaking for some time.

Another significant difference between the laboratory and the model is the continuous nature of modes in the experimental environment, even if those modes are only minor background noise, and the discrete nature of numerical modes. It is important to understand that at the initial stages of development, the necessary modulational modes should grow from continuous noise. If the modes are not imposed, some sidebands naturally appear from the background and are expected to be defined by the ratio of characteristic wave steepness $\epsilon = ak_0$ to spectral bandwidth $\Delta\omega/\omega_0$, where k_0 and ω_0 are some characteristic wavenumber and angular frequency, respectively, and a is the mean amplitude at this wavenumber:

$$M_I = \frac{\epsilon}{\Delta\omega/\omega_0}. \tag{5.1}$$

This ratio was shown to be important in the original studies of instabilities of weakly modulated trains of monochromatic carrier waves of small amplitude (e.g. Yuen & Lake, 1982). Here, we denote this ratio as M_I (modulational index), and in Section 5.3.3 we will introduce directional modulational index M_{Id} by analogy.

The evolution of wave trains described in this section mainly deals with slowly modulated two-dimensional monochromatic waves. Such wave trains are subject to modulational instability which is commonly termed Benjamin–Feir instability after the work of Benjamin & Feir (1967). Many authors point out, however, that it was first discovered by Lighthill (1965) who established the growth rate for this instability in the limit of very long modulation. The growth rate was proportional to the wavenumber of the modulational perturbation, a result which has obvious physical limitations if applied to short waves (large wavenumbers). The general description for the behaviour of the growth rate was found by Zakharov (1966) before Benjamin & Feir (1967), but in English-language literature the papers were published independently in the same year (Zakharov, 1967). Feir (1967) was the first to observe modulational instability in the experiment and Zakharov (1968) further developed and summarised its theory.

The Benjamin–Feir instability was developed for nearly-linear waves, and in this book, dedicated to wave breaking, the waves even initially are not of small amplitude. Therefore, the analogy of the observed empirical modulational interplay with the small-amplitude near-monochromatic theoretical phenomenon should be treated with caution and we will avoid the term of Benjamin–Feir instability. Here, M_I signifies the fact that the wave steepness and length of wave modulation (or number N of waves in the modulation), where

$$1/N \sim \Delta\omega/\omega_0, \tag{5.2}$$

are not independent quantities, i.e. steeper waves will correspond to fewer waves in a modulation (similarly, as far as wave breaking is concerned, wave steepness and directional spread are not independent quantities in the directional modulational index (5.57)). Thus, if nonlinear waves are allowed to evolve naturally, they will form groups where N is not a free parameter, but will be defined by the initial steepness ϵ (1.1).

Therefore, as mentioned above, in the experiment we expect the necessary resonant modes to develop naturally from the background turbulent noise. These modes, however, can be suppressed or even prohibited in a discretised numerical model. In such circumstance, the waves, even if they are steep Stokes waves, will propagate for an indefinitely long period without breaking.

If the model is constructed so as to allow multiple modes, another numerical feature still distinguishes the model from nature. The background noise, the source of the necessary modes dictated by M_I, cannot be completely reduced to zero in nature. In a digitised medium it can, however, be made very small. For example, in the CS numeric scheme, the 11th-order decimal accuracy is employed. Such accuracy is essential for precise simulations, but since it is the only source of noise in the system, it can obviously slow down the development of the initial modes. As is sometimes done in numerical simulations (e.g. Dold & Peregrine, 1986; Banner & Tian, 1998; Song & Banner, 2002), the modes can be deliberately introduced as initial conditions. Such an approach was not, however, employed in the simulation described in Section 4.1, since in this scenario M_I of the system is pre-defined rather than formed naturally, and wave development to breaking may be altered. Other implications of numerical modelling, essential for the present discussion, have already been considered in Sections 2.1 and 4.1 above.

The laboratory experiment described in Babanin *et al.* (2007a, 2009a, 2010a) was conducted at the ASIST wind–wave facility at RSMAS, University of Miami (http://peas.rsmas.miami.edu/groups/asist). The tank is of stainless-steel construction with a working section of 15 m × 1 m × 1 m. Its programmable fan is capable of generating centreline wind speeds in the range 0 to 30 m/s. Immediately downstream of the fan, extensive flow-straightening devices are installed to condition the air flow and introduce appropriately scaled turbulence. Further values of wind speed used here will be those of U_{10}, i.e. extrapolated to 10 m height.

The ASIST facility includes a fully programmable wave maker able to produce both monochromatic waves and waves with a predefined spectral form. These waves are dissipated at the opposite end of the facility by a minimum-reflection beach. The ASIST beach design had been the subject of a special research project. A gently sloping (10 degrees) grid of 2.5 cm-diameter acrylic rods is used. A perforated acrylic plate is placed beneath the rods to split wave orbital velocities into multiple turbulent jets to increase viscous dissipation. The energy of the reflected component is approximately 5–10% of the incident energy depending on the initial wavelength.

In the experiment described, monochromatic deep-water two-dimensional wave trains were generated by the wave paddle. The water depth was held at 0.4 m, thus providing deep-water conditions for the wave frequencies involved. With a tank length of 13.24 m, surface elevations were recorded at 4.55 m, 10.53 m, 11.59 m and 12.56 m from the paddle. For each record, the initial monochromatic steepness (IMS) was varied in such a way that the waves would consistently break just after one of the wave probes. In this way, the dimensional distance to breaking (and therefore the breaking probability), wave train properties immediately prior to breaking and detailed properties of the incipient breaker could be measured. Note that this breaker is the result of natural nonlinear wave evolution, rather than being an outcome of an imposed modulation, being forced or simulated by means of, for example, coalescing linear wave packets. The fact that breaking could be predicted and controlled by manipulating IMS only is a strong corroboration of the numerical model.

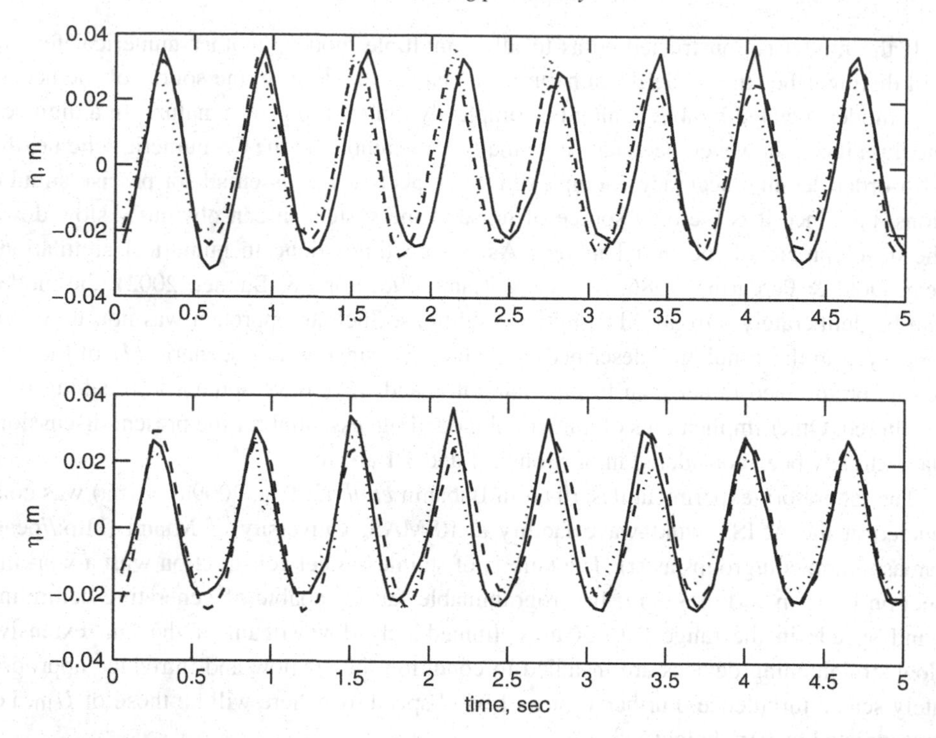

Figure 5.1 Time series of surface elevations η measured at the first wave probe. (top panel) Waves of $U/c = 0$ and IMF $= 1.6$ Hz for different IMS: 0.31 (solid line), 0.25 (dashed line), 0.23 (dotted line). (bottom panel) Waves of IMS $= 0.23$ and IMF $= 1.6$ Hz under different wind forcing: $U/c = 0$ (solid line), $U/c = 1.4$ (dashed line), $U/c = 11$ (dotted line)

In Figures 5.1 and 5.2, time series of surface elevations η at the first and the second wave probes are shown. All the waves in these time series are generated with the same initial monochromatic frequency IMF $= 1.6$ Hz, but with different initial steepness IMS and wind forcing U/c, as indicated.

At the first probe (Figure 5.1), 4.55 m from the wavemaker, the waves are still near-monochromatic, with only marginal modulation due to developing instabilities and perhaps some parasitic modes present in the tank (i.e. non-potential parts of the oscillations generated by the paddle, seiches etc.). The latter, if present, are part of the background noise from which the necessary modulational modes will grow in Figure 5.2, as dictated by M_I (5.1) (see e.g. Reid, 1992; Babanin *et al.*, 2007a, 2010a).

The top subplot of Figure 5.1 has zero wind forcing. Waves with three different IMS $= 0.31$ (solid line), 0.25 (dashed line) and 0.23 (dotted line) are shown. Differences other than those due to the initial wave height are hardly distinguishable.

In the bottom subplot, waves of IMS $= 0.23$ are plotted with no wind forcing (solid line, for cross-reference with the top panel), $U/c = 1.4$ (dashed line) and a very strong wind of $U/c = 11$ (dotted line). The effect of the wind on the profile of the mechanically

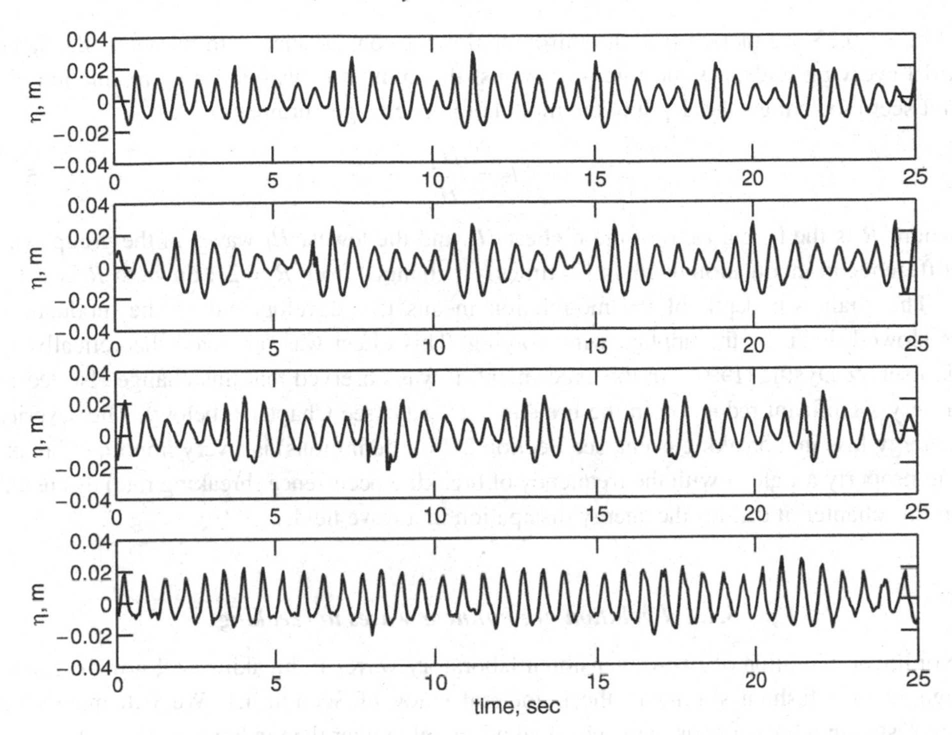

Figure 5.2 Time series of surface elevations η measured at the second wave probe, IMF = 1.6 Hz. (top panel) IMS = 0.31, $U/c = 0$. (second top panel) IMS = 0.25, $U/c = 0$. (second bottom panel) IMS = 0.23, $U/c = 0$. (bottom panel) IMS = 0.23, $U/c = 11$

generated wave is not noticeable at this first probe, except for the extreme forcing case, where wind-generated ripples are visible in the time series.

The wave profiles look very different at the second probe, 10.53 m from the paddle, some 10 wavelengths downstream (Figure 5.2). In all cases, breaking has still not occurred. Waves in the top three panels evolve without wind forcing, and in the bottom subplot waves are shown strongly forced ($U/c = 11$).

The top subplot shows initially very steep waves of IMS = 0.31. By the time they reach probe 2, they have developed into a very strongly modulated group of six waves. Less initially steep waves (IMS = 0.25, second subplot) evolve into a more elongated modulated group of some seven waves. Even less steep waves (IMS = 0.23, third subplot) evolved into a subsequently weaker modulated group of approximately 7.5 waves (15 waves in two modulations). Note that no initial modulation was introduced (see Figure 5.1). This interesting observation is, however, not unexpected and is in full qualitative agreement with the discussion above: i.e. if M_I for the system does not change, a larger initial steepness should lead to fewer waves in the modulation which results from the Benjamin–Feir-like instability.

The effect that the wind forcing has on the modulation is demonstrated in the bottom subplot of Figure 5.2. Here, very strongly wind-forced mechanically generated waves of

IMS $= 0.23$ are plotted (see the third subplot for comparisons with waves of the same IMS evolving without wind forcing). Whilst the number of waves in the modulation did not seem to change, the depth of the modulation R changed dramatically:

$$R = \frac{H_h}{H_l} \tag{5.3}$$

where R is the height ratio of the highest H_h and the lowest H_l waves in the group. The difference in modulation depth is 1.6 times – it changed from $R = 2.1$ down to $R = 1.3$.

The shallower depth of the modulation means that development of the modulation is slowed down by the applied wind forcing. This effect was predicted theoretically by Trulsen & Dysthe (1992). In the experiment, it was observed that this change also led to a very significant reduction in the breaking severity (see Chapter 6 below). The severity (energy loss in a breaking event, see Section 2.7 for definitions) is a very important breaking property as, along with the frequency of breaking occurrence (breaking rate) discussed in this chapter, it defines the energy dissipation in a wave field.

5.1.1 Evolution of nonlinear waves to breaking

Nonlinear evolution of two-dimensional laboratory waves to breaking will now be investigated in a fashion similar to the numerical study of Section 4.1. We will mainly be interested in what happens in the physical wave field rather than in Fourier space. Also, we will deal with individual waves, rather than with average nonlinear properties of the wave ensemble. The nonlinear characteristics of interest (i.e. individual wave steepness, skewness and asymmetry), will be obtained by means of zero-crossing selection and analysis of individual waves. In addition to these three characteristics, we will scrutinise the behaviour of the period (frequency) of individual nonlinear waves. This feature was not examined in dimensionless numerical simulations, but in the laboratory it appears to be quite variable, even in the train of waves of initially uniform frequency. The effect of such local-frequency variation is significant for the breaking-onset study since, when wave-height growth is accompanied by a synchronous reduction of wave period, this has a combined impact on the local wave steepness. Like the background-steepness threshold mentioned above, this modulational-instability feature provides a further hint on this instability as a cause of wave breaking in the ocean. Multiple indications of such a prior-to-breaking wavelength decrease are scattered throughout the book. For example, if the wavelength shrinks, they have to slow down before breaking which is what the remote-sensing methods observe (see e.g. Section 3.6).

Figure 5.3 shows a wave record with an initial monochromatic frequency IMF $= 1.8$ Hz and an IMS $= 0.30$, with no wind forcing. It should be noted that there is a conceptual change in the frame of reference compared to the numerical-model results. In the case of the model, a single wave was followed as it approached the point of breaking. Here, observations are made at a single point as a succession of waves passes. One can move approximately from the fixed frame of reference in Figure 5.3 to the moving frame by

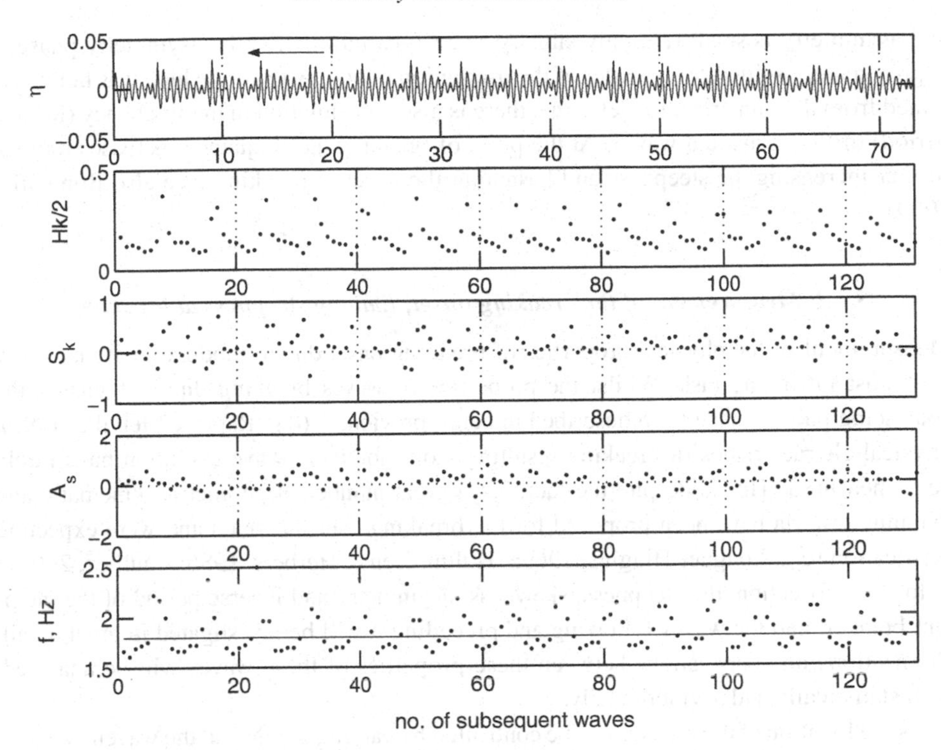

Figure 5.3 Segment of a time series with IMF = 1.8 Hz, IMS = 0.30, $U/c = 0$. a) Surface elevation η. b) Rear-face steepness ϵ (1.1). c) Skewness S_k (1.2, rear trough depth is used). d) Asymmetry A_s (1.3). e) Frequency. IMF = 1.8 Hz is shown by the solid line. The waves propagate from right to left. Figure is reproduced from Babanin *et al.* (2007a) by permission of American Geophysical Union

considering the waves shown propagating from right to left, as indicated by the arrow in the figure.

The top panel in Figure 5.3 shows the measured water surface elevation η as a function of time (horizontal axis). Interpreting this as a wave moving from right to left shows that, within each wave group, the maximum value of the water-surface elevation gradually decreases and then suddenly increases until a point, where breaking occurs. The saw-tooth shape of the wave envelope once again indicates modulational instability as the mechanism behind this behaviour of the nonlinear wave train (e.g. Shemer & Dorfman, 2008). The point of breaking was located immediately after the probe at a distance of 10.73 ± 0.1 m from the wave maker. Each successive wave passing the wave gauge was analysed to determine its steepness, skewness, asymmetry and frequency, which are shown in the four panels of Figure 5.3.

The major features seen in the numerical model are confirmed by the laboratory data. The incipient breaking waves are the steepest waves in the wave train, with the steepness oscillating in a periodic fashion. Skewness and asymmetry also oscillate, but behave in a less ordered fashion. However, at the point of breaking skewness is positive (i.e. peaked up) and asymmetry is small (i.e. only slightly tilted forward), that is the asymmetry phase is approximately shifted by a quarter of the oscillation period. A feature which was not determined from the numerical model is that there is also a modulation in the frequency (inverse period) of the individual waves. At the point of breaking the frequency increases rapidly, further increasing the steepness and hastening the onset of breaking (see also Bonmarin, 1989).

5.1.2 Measurement of the breaking onset; limiting steepness at breaking

Measurement of the physical properties of breaking onset due to modulational instability is a most intriguing task. Whilst the properties of waves breaking due to focusing the coalescing packets have been described in detail previously (e.g. Rapp & Melville, 1990), physical characteristics of breaking resulting from nonlinear wave evolution have rarely been measured. This is despite the fact that a great number of geometric, kinematic and dynamic criteria have been proposed for the breaking over the years and were expecting verification (e.g. Longuet-Higgins, 1969a; Holthuijsen & Herbers, 1986; Caulliez, 2002).

In this subsection, the steepness, skewness, asymmetry and inverse period of the incipient breaker, and the waves following and preceding it will be investigated in great detail. Connections and correlations between these properties of these waves will be analysed, both statistically and asymptotically.

As the location of the breaker can be controlled by varying the IMS at the wavemaker, the waves were made to break immediately after a wave probe and thus the properties of incipient breakers could have been measured directly. Figure 5.4 shows the four steepest incipient breakers. A slight variation of the recorded height of the breakers is visible here and was also the case in Figure 5.3. There are two reasons for this variation in height. Firstly, there is some low-frequency modulation of the wave-group amplitude which can be either natural or an artifact of the tank. Secondly, there was a ± 10 cm variation of the observed breaking location. This variation in the breaking position could have been due to a non-integer number of waves in the modulation. These reasons were not investigated in detail as they did not influence our main results. In any scenario, transition of the wave to the incipient-breaking stage is very rapid and difficult to capture precisely. Therefore, we chose the highest breakers for examination, as those are apparently the closest to breaking onset.

Typical features of waves just prior to breaking can now be analysed. The nearly breaking wave is the highest and most skewed, but is almost symmetric. The two waves immediately preceding and following the breaker are asymmetric: the preceding wave is tilted backwards (positive asymmetry) and the following wave is tilted forwards (negative asymmetry). The preceding wave is smaller than the following wave, and, at least in these observations, the preceding trough is shallower. This may be a key feature to distinguish linear-focusing breaking onset from modulational-instability breaking, as the former is not expected to exhibit uneven front and rear troughs. Thus, the boat that climbed the front face of the breaker, would then fall into the very deep trough followed by an even steeper wave front.

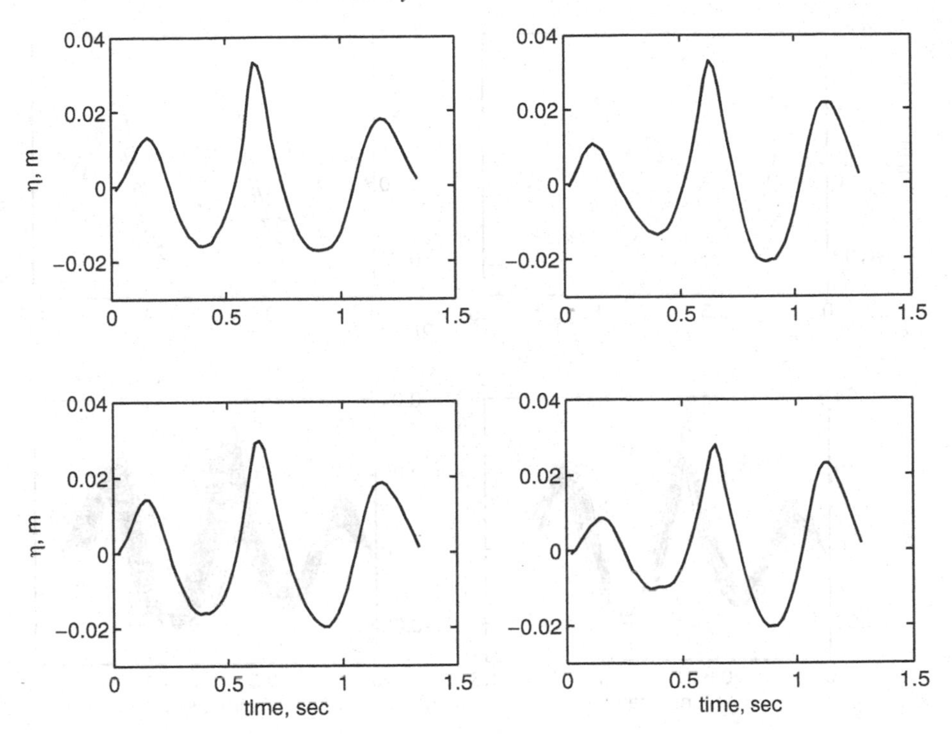

Figure 5.4 IMF = 1.8 Hz, IMS = 0.30, $U/c = 0$. Four steepest incipient breakers (from top left to bottom right)

Figure 5.5 shows one, five, twenty and fifty steepest breakers in consecutive panels. Note that no normalisation was applied and the waves are plotted as they appear in physical space, from the front zero-crossing of the wave preceding the breaking to the rear zero-crossing of the following wave. Despite this, the pictures are remarkably similar even in the subplot with 50 waves and this prompts a universality of such incipient-breaker shapes. The quantitative characteristics of these waves are further scrutinised in Figures 5.6–5.12.

Figure 5.6 shows data for the five steepest breakers. The analysis is at first limited to these steepest cases as wave quantities close to the breaking point change rapidly, as shown in Figure 5.3. These steepest cases are considered to be on the point of breaking. According to the numerical simulations, the steepness seems to be the single robust criterion for breaking. For the 20 steepest breakers (see further Figure 5.7), steepness was confined to the narrow range $Hk/2 = 0.37$ to 0.44, whilst skewness was scattered over the wide range $S_k = 0$ to 1 and asymmetry $A_s = 0.8$ to -0.4. Considering only those waves at the point of breaking, however, as in Figure 5.6, shows a clearer trend. The steepness appears to approach an asymptotic limit of

$$\frac{Hk}{2} \approx 0.44, \tag{5.4}$$

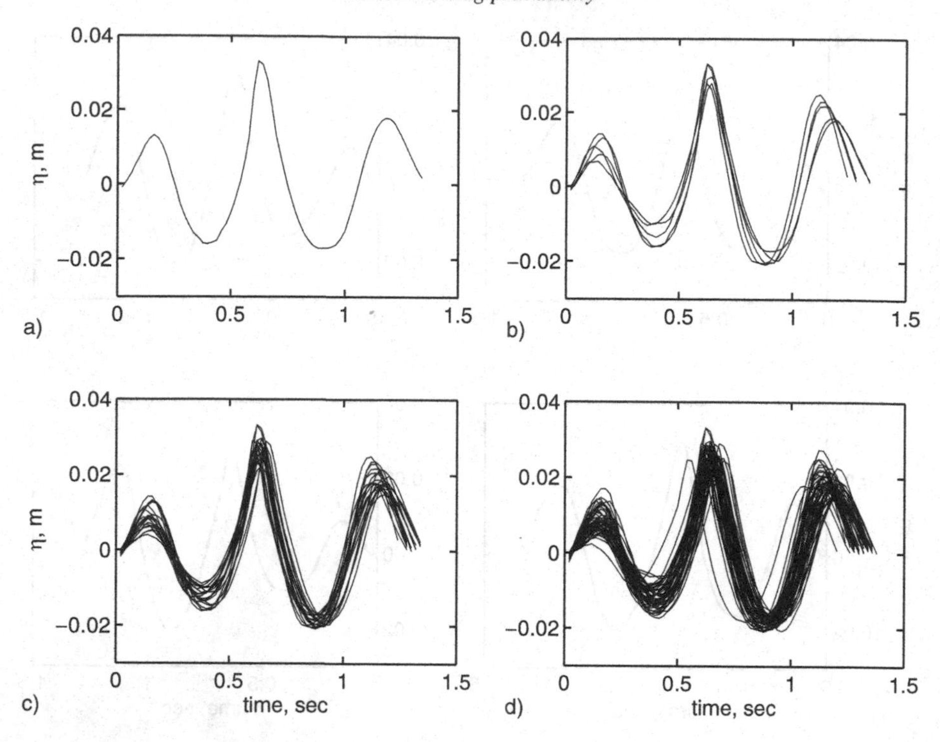

Figure 5.5 IMF = 1.8 Hz, IMS = 0.30, $U/c = 0$. (top left panel) The steepest incipient breaker. (top right panel) Five steepest incipient breakers. (bottom left panel) Twenty steepest incipient breakers. (bottom right panel) Fifty steepest incipient breakers

which may represent an absolute steepness limit for two-dimensional or quasi-two-dimensional waves (see also Toffoli *et al.*, 2010a). We should point out that this limit is remarkably close to the theoretical steady limiting steepness of $ak = 0.443$ (2.47), i.e. the Stokes limit $H/\lambda = 1/7$ (2.46). It is also in excellent agreement with the dedicated fully nonlinear simulations of the breaking onset (see Section 4.1). The skewness at the breaking point asymptotes the value of

$$S_k \approx 1 \tag{5.5}$$

which means that the crest is twice as high as the trough (1.2). Values of asymmetry (1.3) in the imminent breaker are transitional rather than definite.

Such an observation is very important because it signifies that the waves break once they achieve the well-established state beyond which the water surface cannot sustain its stability. It may be postulated that the other geometric, kinematic and dynamic criteria of breaking, explored in the literature (see Section 2.9), are indicative of a wave approaching this state, but are not a reason or a cause for the breaking. As this limit is approached, the skewness increases very rapidly and immediately after the limit is reached the asymmetry becomes negative (i.e. the wave starts tilting forward at the point of breaking).

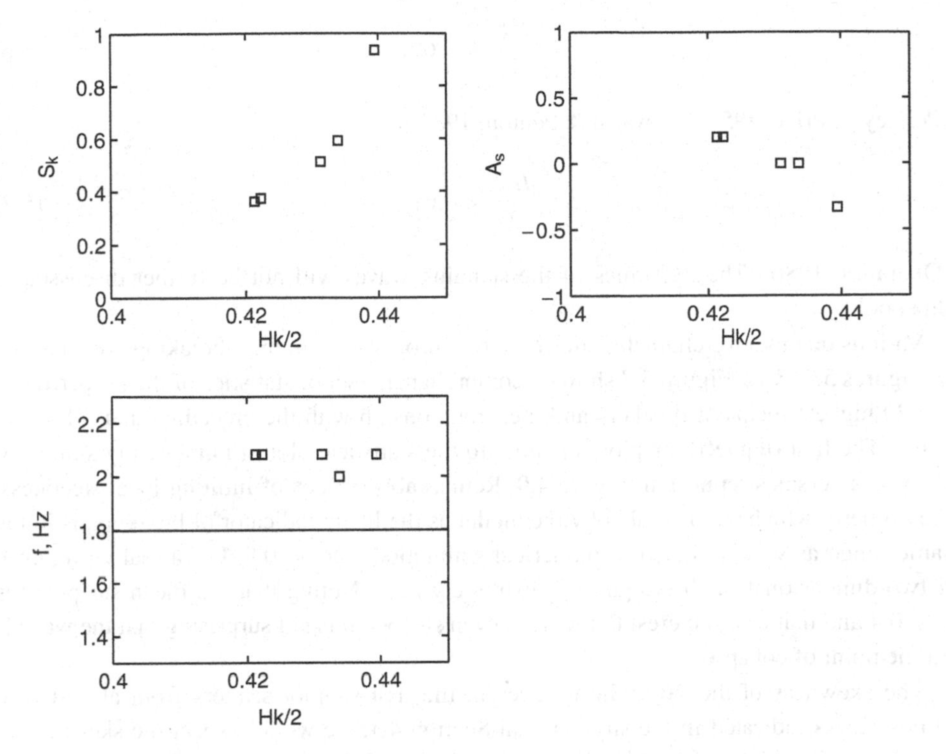

Figure 5.6 Laboratory statistics for the incipient breakers (five steepest waves). IMF = 1.8 Hz, IMS = 0.30, $U/C = 0$. (top left panel) Skewness versus steepness. (top right panel) Asymmetry versus steepness. (bottom panel) Frequency (inverse period) versus steepness. IMF = 1.8 Hz is shown with the solid line

In this regard, it is necessary to highlight that, although the limit (5.4) replicates the Stokes limiting steepness (2.46)–(2.47), the dynamical nonlinear waves we observe here should not be unambiguously identified as Stokes waves. They do exhibit positive skewness (1.2) on average, but this skewness goes through oscillations to the extent of waves periodically becoming almost sinusoidal (Figures 4.2, 5.3). And although their average vertical asymmetry (1.3) is zero, it oscillates too, and for most of the time the wave is actually asymmetric. Also, the crest geometry is far from the steady pointy 120° corner flow (2.48), and is likely to be driven by crest instabilities characteristic of very steep waves (Longuet-Higgins & Dommermuth, 1997; Longuet-Higgins & Tanaka, 1997). A good illustration of the differences between the stationary Stokes wave and a transient wave at the breaking onset is Figure 1.2. In the measured breaking-onset Figures 5.4–5.5, the crest instability is actually visible (the tip of the crest is leaning forward, see also the simulated maximal-height wave picture in Dyachenko & Zakharov (2005)).

It should be mentioned, even if in passing, that the limiting steepness of standing waves at breaking is different to that of the propagating waves. In a series of analytical studies it was identified as

$$\frac{Hk}{2} \approx 0.69, \tag{5.6}$$

(Penney & Price, 1952; Schwartz & Fenton, 1982) or

$$\frac{Hk}{2} \approx 0.61, \tag{5.7}$$

(Okamura, 1986). The dynamics of the standing waves will not be further discussed in this book.

Various quantitative characteristics of waves propagating to their breaking are analysed in Figures 5.7–5.12. Figure 5.7 shows a comprehensive set of statistics of the properties of the 20 highest incipient breakers and their relationship with the preceding and following wave. The first (top left) subplot is similar to the statistical plot of numerically simulated skewness versus steepness in Figure 4.9. Remarkably, values of limiting local steepness, the property which was revealed by the model as the likely indicator of breaking, is in the same range as was predicted in numerical simulations: $2\epsilon \approx 0.8$. For a real wave, even if two-dimensional, such steepness is extremely high. Noting that the mean steepness is $\epsilon \approx 0.4$ and that near the crest the wave is even steeper, it is not surprising that the wave is on the point of collapse.

The skewness of the 20 highest waves in the first subplot scatters from almost 0 to almost 1. As indicated in the simulational Section 4.1, we would expect the skewness to also have a limiting value. Clearly, however, such a limit is not a very robust breaking characteristic, although for the five waves closest to the breaking point in Figure 5.6 the skewness does exhibit the limiting value. Also in the top row of the subplots, asymmetry is scattered from $A_s = -0.33$ to $A_s = 0.75$ (second panel), with a possible dependence of S_k on A_s in the fourth panel, consistent with the numerical simulations (Figure 4.9).

A robust property of the breaking, in the third panel, is the wave frequency. The scatter of this property is small, with all the values falling into a range $f = 2{-}2.08$ Hz, that is the ratio is

$$\frac{f}{\text{IMF}} = 1.11\text{–}1.16. \tag{5.8}$$

It should be pointed out that, for $ak \approx 0.44$, a frequency increase of about 10% is simply expected from the second-order term of the basic dispersion relationship (2.14). Thus, the wave clearly reduces in length prior to collapse. We should mention that the measured steepness $\epsilon = kH/2$ in the figure is the physical rear-face steepness, and therefore the effect of period contraction has already been accounted for.

In the second row of plots, the skewness of the wave following the incipient breaker (first panel) and its asymmetry (second panel) are much less scattered than the skewness and asymmetry of the breaker itself: $S_k = 0.32$–0.70, asymmetry changes from $A_s = -0.29$ to 0.33. We have already discussed the double-breaking in observational Chapter 3, which means that this following wave will break soon after the incipient breaker. Thus, its

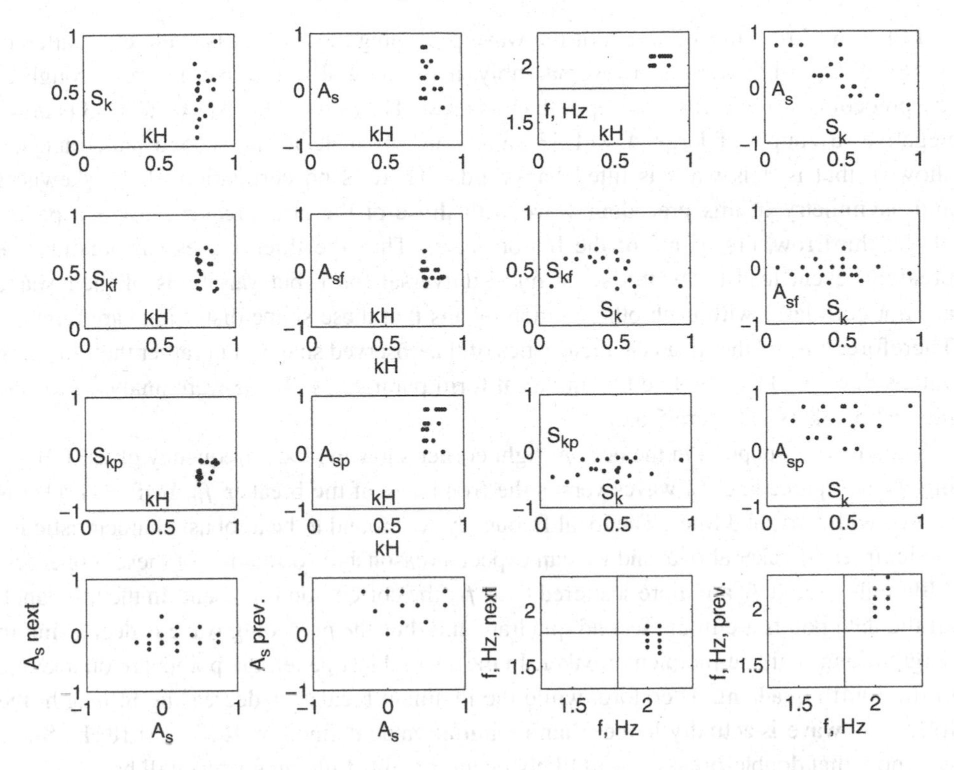

Figure 5.7 Laboratory statistics for the incipient breaker (20 steepest breakers). IMF = 1.8 Hz, IMS = 0.30, U/c = 0. (top left panel) Skewness versus steepness. (top line, second left panel) Asymmetry versus steepness. (top line, second right panel) Frequency (inverse period) versus steepness. IMF = 1.8 Hz is shown with solid line. (top line, right panel) Asymmetry versus skewness. (second top line, left panel) Skewness of the following wave versus steepness of the breaker. (second top line, second left panel) Asymmetry of the following wave versus steepness of the breaker. (second top line, second right panel) Skewness of the following wave versus skewness of the breaker. (second top line, right panel) Asymmetry of the following wave versus skewness of the breaker. (second bottom line, left panel) Skewness of the preceding wave versus steepness of the breaker. (second bottom line, second left panel) Asymmetry of the preceding wave versus steepness of the breaker. (second bottom line, second right panel) Skewness of the preceding wave versus skewness of the breaker. (second bottom line, right panel) Asymmetry of the preceding wave versus skewness of the breaker. (bottom line, left panel) Asymmetry of the following wave versus asymmetry of the breaker. (bottom line, second left panel) Asymmetry of the preceding wave versus asymmetry of the breaker. (bottom line, second right panel) Frequency of the following wave versus frequency of the breaker. IMF = 1.8 Hz is shown with solid lines. (bottom right panel) Frequency of the preceding wave versus frequency of the breaker. IMF = 1.8 Hz is shown with solid lines

physical shape is not random and should exhibit some characteristic properties leading to breaking. The skewness and asymmetry of the following wave, however, do not correlate with the skewness and asymmetry of the breaker (third and fourth panels of the second row, first panel of the bottom row).

In the third row, the skewness of the wave preceding the breaker is even less scattered: $S_k = -0.55$–0.12 (first panel). Remarkably, it is essentially negative, i.e. rear trough of the preceding wave is always deeper than its crest. The asymmetry $A_s = 0$ to 1.33 is never negative (a couple of large $A_s = 1.33$ values are off scale in the second panel and not shown), that is, this wave is tilted backwards. There is no correlation of the skewness and asymmetry of this preceding wave with those of the near-breaker (last two panels of the third row, first panel of the bottom row). Thus the three waves surrounding the breaking event tend to exhibit some quasi-universal form, but variations of their shape are not correlated with each other, which means that these shape distortions are random. Therefore, it is not the mean characteristics of the observed shapes, but rather their limiting values that should asymptote the universal form parameters. These were analysed for the highest breakers in Figure 5.6.

The last two subplots in the bottom right corner show the local frequency of the following f_f and preceding f_p waves versus the frequency of the breaker f_b. IMF $= 1.8$ Hz is shown with two solid lines. The local frequency was found to be a robust characteristic for the incipient breaker above, and we can expect a reasonable correlation of these properties. Although f_f and f_p are more scattered than f_b, the correlation is present. In the last panel, all the data points are in the second quadrant and thus the preceding wave is decreasing in length along with the incipient breaker. In the second last panel, the points are on average in the fourth quadrant. Therefore, while the incipient breaker is decreasing in length, the following wave is actually longer than its initial value defined by IMF $= 0.18$ Hz. Since we know that double-breaking will likely occur, i.e. this following wave will break shortly after the current breaker, then it should now be shrinking rapidly. Both the preceding and the following waves are quite steep, but the preceding wave is less steep than the following wave (see also Figures 5.8, 5.9), and therefore the observed opposite deviations of their frequency with respect to IMF $= 0.18$ Hz are more complicated than just second-order effects in the dispersion relationship (2.14). Thus, some very active physics must be involved in the short-time-scale evolution of this set of very nonlinear waves.

Figure 5.8 provides further similar detailed analysis of the shape of the wave following the incipient breaker. Note that this wave will break shortly after the current incipient breaker and therefore, whatever its properties are now, they are progressing towards breaking. In the figure, kH, S_k, A_s, f are properties of the currently analysed wave.

The steepness $2\epsilon = 0.28$–0.57 is on average about half the steepness of the incipient breaker (first subplot). Since this wave is still the second largest in the modulated group, this fact highlights how much higher is the incipient breaker. The ranges of skewness $S_k = 0.32$–0.70 and asymmetry $A_s = -0.29$ to 0.33 have been mentioned above; here they do not appear to correlate with the steepness (first and second plots) or with each other (fourth plot).

There is a 94% correlation between the steepness and local frequency (third subplot), and the frequency grows as steepness increases. Whilst this following wave is on average longer than IMF, it crosses the IMF values at approximately $2\epsilon = 0.5$ and continues to decrease:

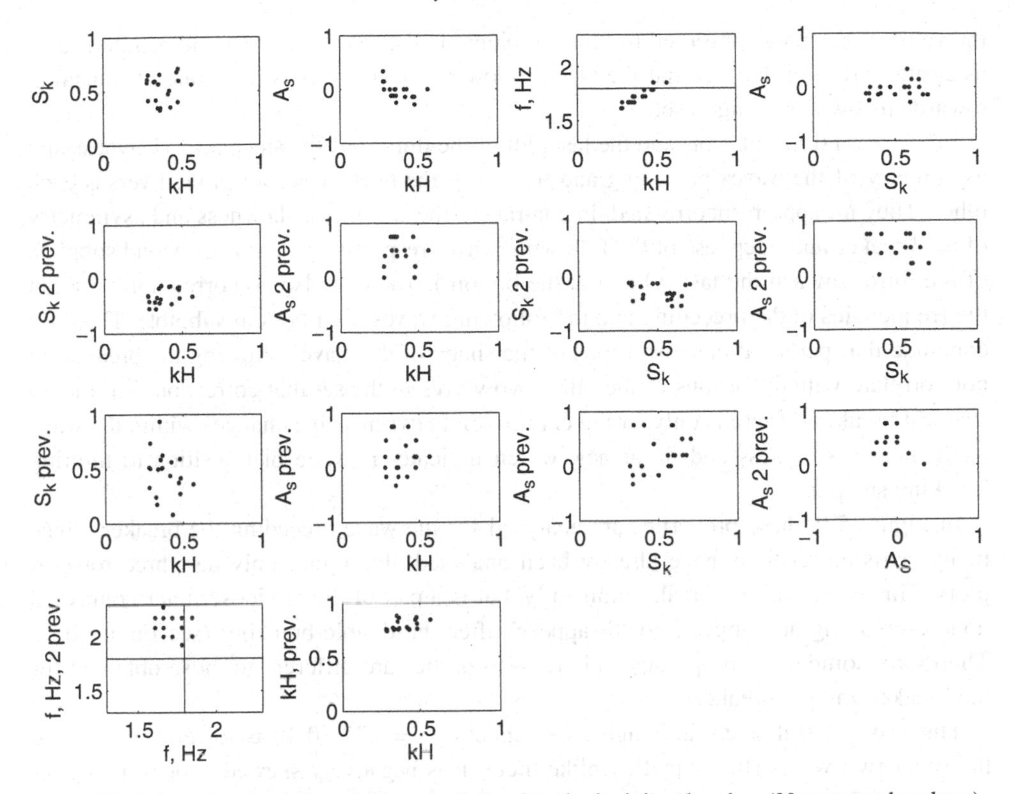

Figure 5.8 Laboratory statistics of the wave following the incipient breaker (20 steepest breakers). IMF = 1.8 Hz, IMS = 0.30, $U/c = 0$. (top left panel) Skewness versus steepness. (top line, second left panel) Asymmetry versus steepness. (top line, second right panel) Frequency (inverse period) versus steepness. IMF = 1.8 Hz is shown with solid line. (top line, right panel) Asymmetry versus skewness. (second top line, left panel) Skewness of the wave preceding the breaker versus steepness of the wave following the breaker. (second top line, second left panel) Asymmetry of the wave preceding the breaker versus steepness of the wave following the breaker. (second top line, second right panel) Skewness of the wave preceding the breaker versus skewness of the wave following the breaker. (second top line, right panel) Asymmetry of the wave preceding the breaker versus skewness of the wave following the breaker. (second bottom line, left panel) Skewness of the breaker versus steepness of the wave following the breaker. (second bottom line, second left panel) Asymmetry of the breaker versus steepness of the wave following the breaker. (second bottom line, second right panel) Asymmetry of the breaker versus skewness of the wave following the breaker. (second bottom line, right panel) Asymmetry of the wave preceding the breaker versus asymmetry of the wave following breaker. (bottom line, left panel) Frequency of the wave preceding the breaker versus frequency of the wave following the breaker. IMF = 1.8 Hz is shown with solid lines. (bottom line, right panel) Steepness of the breaker versus steepness of the wave following the breaker

$$f = 1.7\epsilon + 1.39. \tag{5.9}$$

If this expression is extrapolated to the ultimate frequency value of $f = 2.08$ of the breaker in the third subplot of Figure 5.7, it yields $2\epsilon = 0.81$, which is approximately the value of steepness observed for this incipient breaker. This result should be interpreted as confirmation of the fact that the wave following our breaker is in a state of transition towards its own breaking onset.

In the second row of plots and the last plot of the third row, the steepness, skewness and asymmetry of the waves preceding and following the breaker are all plotted versus each other. They all appear uncorrelated, in contrast to the steepness, skewness and asymmetry of the breaker and steepness of the following wave (respectively, first and second subplots of the third row and the last subplot at the bottom). There is also no correlation between the frequencies of the preceding and the following waves (first bottom subplot). Thus, we conclude that perturbations/distortions of the shape of the wave following the breaker do not correlate with distortions of the other two waves in the set that corresponds to the 20 steepest breakers. There is only strong correlation between shape changes within the wave itself, i.e. its steepness and frequency, which indicates its state of transition to limiting breaking steepness.

In Figure 5.9, these properties are analysed for the wave preceding the breaker. Since many cross-correlations have already been analysed, this figure only has three rows of plots. This wave will not break imminently, but is a part of the obviously inter-connected double-breaking and appears to 'disappear' after the double-breaking (see Figure 6.2). Therefore, some of its properties and cross-properties are different to those obtained for the breaker and pre-breaker.

The wave is still steep, although its steepness $2\epsilon = 0.27-0.49$ is lower compared to the other two waves (first panel). Unlike them, it is negatively skewed (trough is deeper than crest, first plot) and tilted backwards (positive asymmetry, second plot). It is very short, even if compared with the incipient breaker, but its local frequency is more scattered: $f_P = 1.92-2.27$. There is no noticeable correlation between these properties of the wave itself, or between its steepness, skewness and asymmetry and those of the breaker (bottom three plots) and the wave following the breaker (first three plots in the second row). It is only in the last subplot of the middle row that a marginal correlation (68%) may be identified between the steepness of this wave and the skewness of the incipient breaker. As discussed above, this skewness is transient and is expected to asymptote to some value at the point of breaking (see Figure 5.6). This correlation may be an indication of some interaction between the breaker and the wave preceding it which will very soon lead to dissipation of the preceding wave (Figure 6.2). The issue of the interaction between the breaker and preceding wave, however, may be of significance and may even hold a key to the downshifting observed in Figures 6.1–6.2. Babanin *et al.* (2011a) describe their visual observation of instability breaking in a directional wave tank as a breaking wave, whose length had apparently shrunk prior to the breaking onset, rapidly accelerated in the course of breaking as its wavelength was bouncing back, increasing to the point of catching up and merging with the preceding individual wave. This issue needs further investigation.

In Figures 5.10–5.12 asymptotic, rather than statistical properties of the incipient breaker are considered. Three subplots in the top row of Figure 5.10 were zoomed in and scrutinised

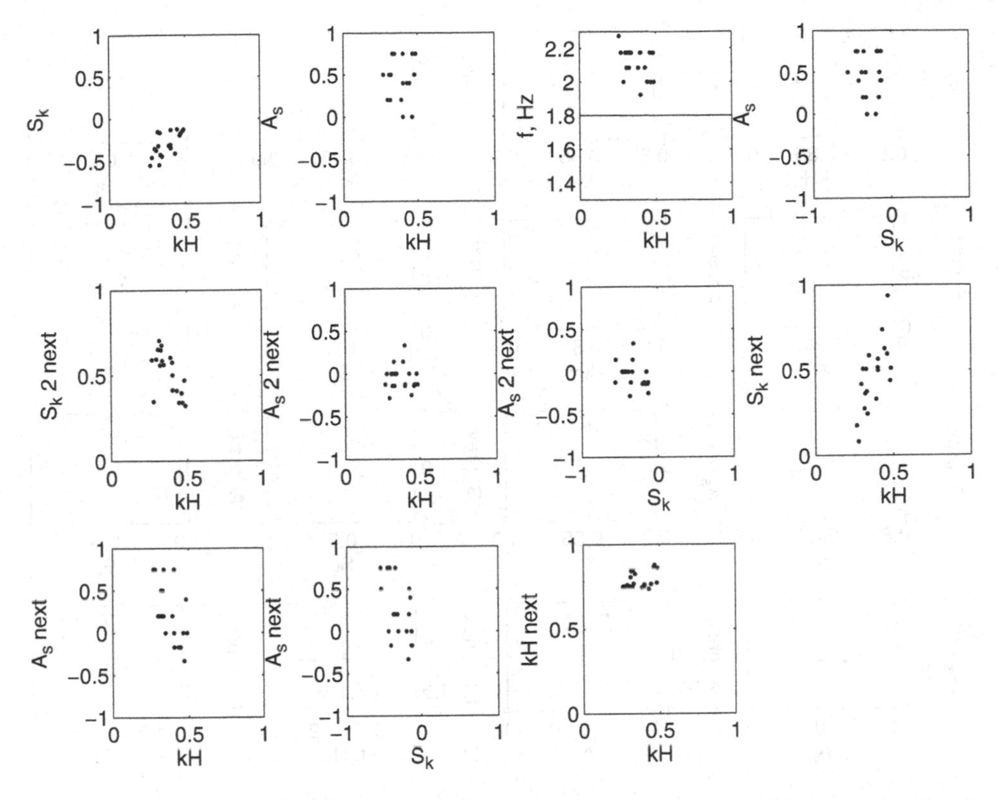

Figure 5.9 Laboratory statistics of the wave preceding the incipient breaker (20 steepest breakers). IMF = 1.8 Hz, IMS = 0.30, $U/c = 0$. (top left panel) Skewness versus steepness. (top line, second left panel) Asymmetry versus steepness. (top line, second right panel) Frequency (inverse period) versus steepness. IMF = 1.8 Hz is shown with solid line. (top line, right panel) Asymmetry versus skewness. (middle line, left panel) Skewness of the wave following the breaker versus steepness of the wave preceding the breaker. (middle line, second left panel) Asymmetry of the wave following the breaker versus steepness of the wave preceding the breaker. (middle line, second right panel) Asymmetry of the wave following the breaker versus skewness of the wave preceding the breaker. (middle line, right panel) Skewness of the breaker versus steepness of the wave preceding the breaker. (bottom line, left panel) Asymmetry of the breaker versus steepness of the wave preceding the breaker. (bottom line, second left panel) Asymmetry of the breaker versus skewness of the wave preceding the breaker. (bottom line, right panel) Steepness of the breaker versus steepness of the wave preceding the breaker

in Figure 5.6 and are repeated here for consistency of comparisons with other figures and analysis of other wave-shape characteristics. In these three figures, the characteristics of the five steepest waves are plotted. As discussed above, transition to breaking happens very rapidly, and breaking onset and its location may be somewhat modulated due to, for example, uneven numbers of waves in the nonlinear oscillations. Thus, we would expect that it is the highest waves measured that would be closest to the ultimate limiting characteristics of the incipient breaker and its following and preceding counterparts.

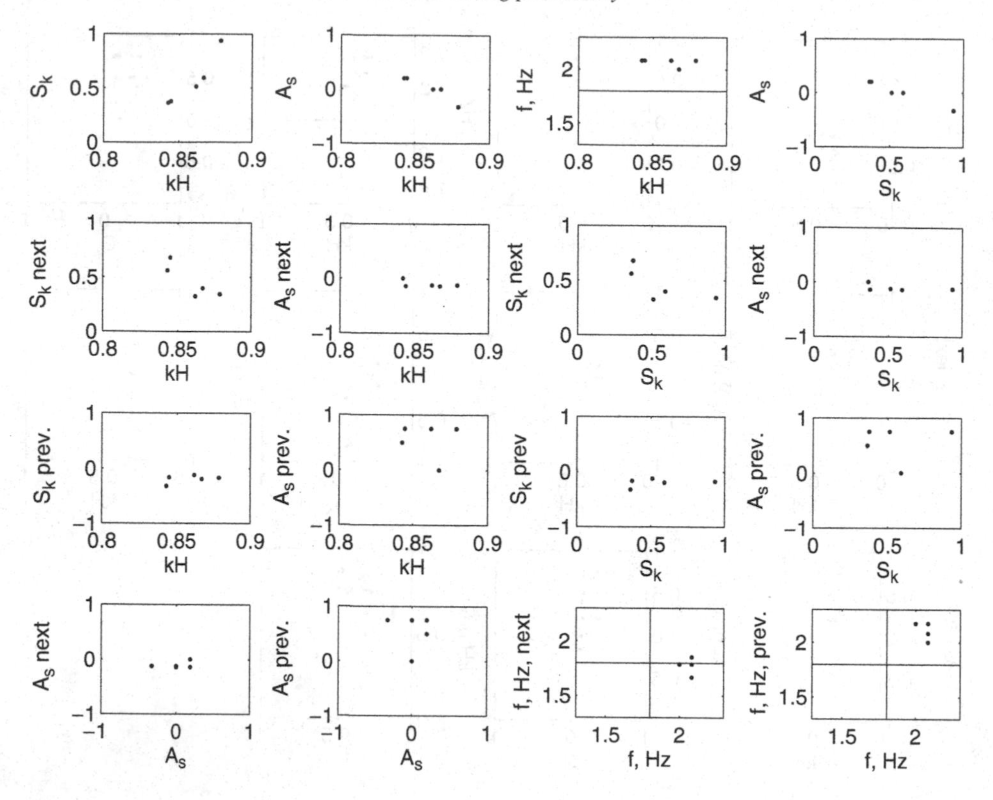

Figure 5.10 As in Figure 5.7, for the five steepest breakers

In Figure 5.10, as the steepness limit is approached, the skewness increases very rapidly (first subplot) and the asymmetry starts to decrease and becomes negative (subplots 2 and 4), i.e. the wave starts tilting forward at the point of breaking. The latter conclusion slightly differs from what was simulated numerically with the CS model (see Section 4.1) where the wave started tilting forward after the point of maximum steepness was passed. Thus, the model simulates the very late stages of the breaking onset with some deviation, which is not unexpected as discussed above.

We will not discuss the scatter, correlations and cross-correlations of the five steepest breakers in Figures 5.10, 5.11 and 5.12 in great detail as this largely repeats relationships observed in Figures 5.7, 5.8 and 5.9. We will only highlight correlations which were marginal within the 20-wave statistical plots and only become visible in the asymptotic plots.

In Figure 5.11, for the waves following the five steepest breakers, some dependence of skewness on steepness can be identified (first subplot). This could be expected since this wave is developing towards breaking. Its asymmetry is almost zero (second and fourth subplots). This fact indicates some interaction between the breaker and this wave. Apparently, as the breaker asymptotes its point of collapse, the shape of the following wave does not appear random, but is almost perfectly symmetric and therefore is somehow locked with

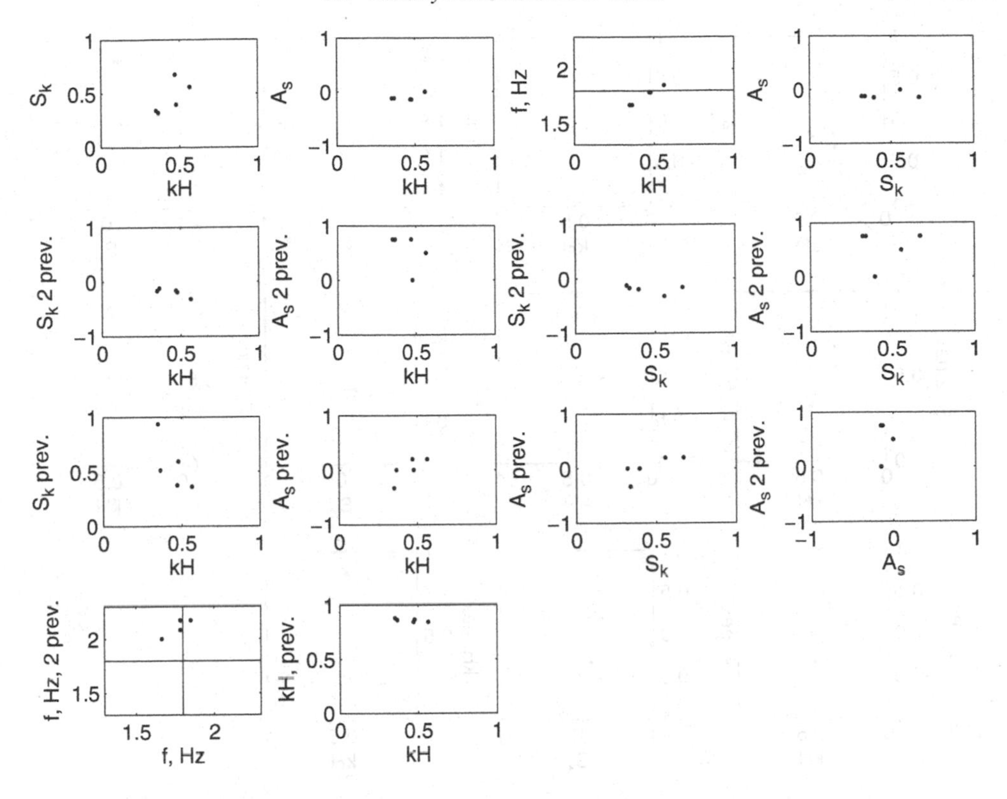

Figure 5.11 As in Figure 5.8, for the five steepest breakers

the shape of the incipient breaker. The dependence of the local frequency on steepness, which was noticed previously, is now very clear (third subplot).

Some marginal dependences between the skewness and asymmetry of the breaker and the steepness and skewness of the following wave (first three subplots of the third row) are visible. The frequencies of this wave and the wave preceding the breaker are also correlated, and tend to increase together (bottom left subplot). Some correlations are visible between the skewness of this wave and the steepness of the wave preceding the breaker (first subplot in the middle row of Figure 5.12), and the asymmetry of the breaker and the steepness of the preceding wave (first subplot in the bottom row of Figure 5.12). Together with the correlation of the skewness of the breaker and the steepness of the preceding wave noticed before (last subplot of the middle row), these properties likely indicate interactions between the three waves, one of which is now at the point of breaking.

5.1.3 *Laboratory investigation of wind influence*

Following the same approach as in the numerical simulations in Section 4.1, we now investigate the influence of wind on nonlinear wave evolution and breaking onset. As in the

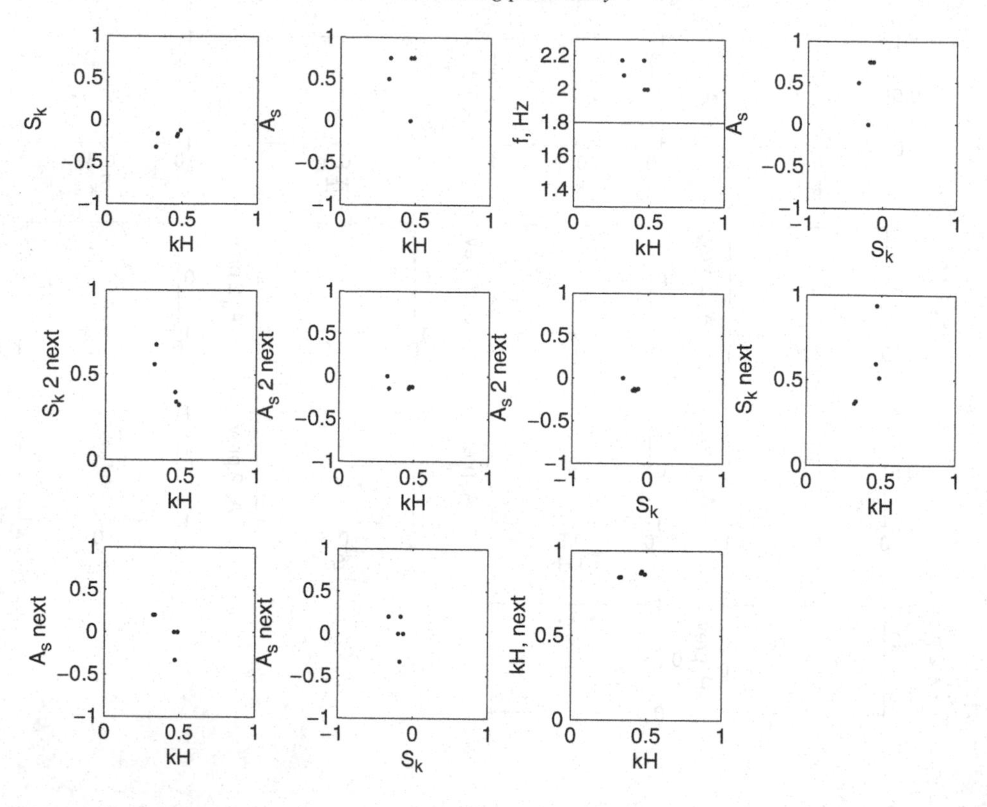

Figure 5.12 As in Figure 5.9, for the five steepest breakers

subsections above, the influence of the wind is analysed with respect to its effects on the modulation of the surface elevations, and on statistical and asymptotic properties of the individual waves at the breaking onset.

The wind has been shown to play multiple roles in wave evolution to breaking. At larger scales, the wind generates waves and pumps energy into the waves, increasing their steepness, and thus leads to more frequent breaking. At moderate wind conditions, doubling the wind speed resulted in the breaking occurring four times faster (e.g. Figure 4.2). Instantaneously, at the point of breaking onset, the capacity of the wind to push the wave over and thus affect the onset is small, even if the wind is very strong and the wave is very steep (Figures 5.14–5.16). At medium scales, however, the wind was observed to change the depth R (5.3) of the modulation of already existing waves, that is the wind affects the instability growth and in this way alters the severity of the breaking (Figures 6.2–6.3, see also Galchenko *et al.*, 2010).

Figure 5.13 is similar to Figure 5.3, but moderately strong wind forcing of $U/c = 3.9$ is now applied to the mechanically generated waves. Note that IMF $= 1.5\,\text{Hz}$ is different to IMF $= 1.8$ in Figure 5.3. This is done in order to have incipient breakers, as before, at

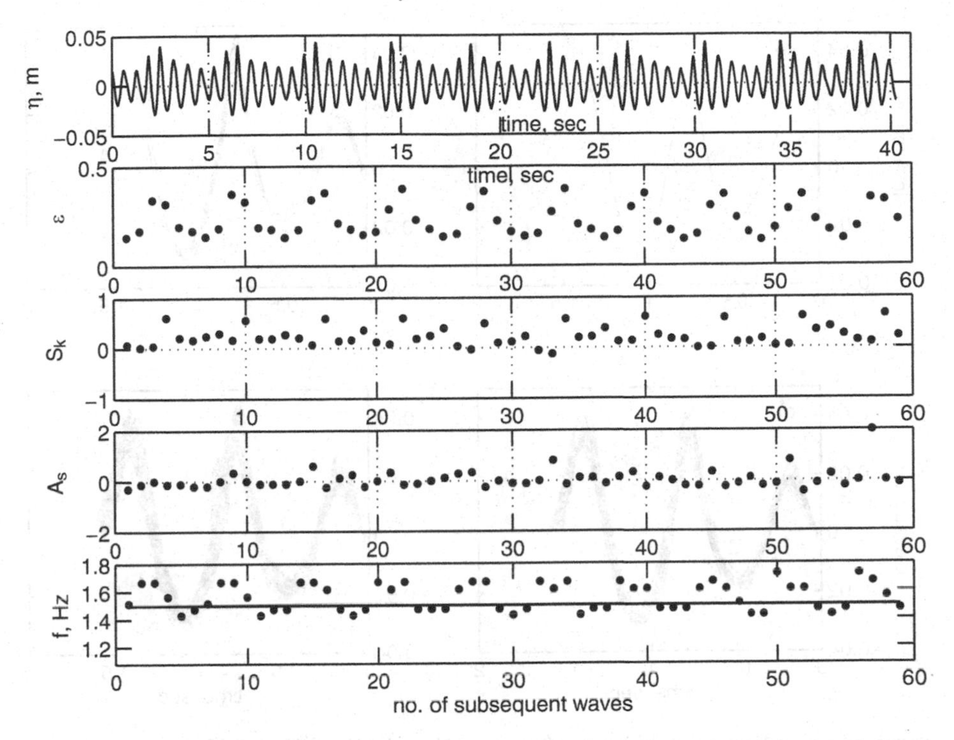

Figure 5.13 As in Figure 5.3, with wind forcing. Segment of time series with IMF = 1.5 Hz, IMS = 0.30, $U/c = 3.9$. (top panel) Surface elevation η. (second top panel) Rear-face steepness ϵ. (middle panel) Skewness S_k (rear trough depth is used). (second bottom panel) Asymmetry A_s. (bottom panel). Frequency (inverse period). IMF = 1.5 is shown with solid line

probe 2 where data are recorded. With the wind superimposed, the waves of IMF = 1.8 Hz, IMS = 0.30 would break before this probe (see discussion in Section 4.1 and Figure 4.2).

In Figure 5.3, the modulation depth (5.3) is $R = 4$ whereas in the current Figure 5.13 it is $R = 2.9$, i.e. the difference in the modulation depth is 1.4 times. Thus, we observe the expected feature of smearing of the modulation by the wind (see Trulsen & Dysthe, 1992).

This smearing is reflected in all other nonlinear properties shown in the figure. Steepness (second panel), skewness (third panel) and asymmetry (fourth panel) are intentionally plotted at the same vertical scale as those in Figure 5.3 even though their range of oscillations is now noticeably reduced. Because of the IMF change, the scale of the frequency plot (bottom) could not be left the same, but scale limits were kept proportional to those in Figure 5.3. Reduction of the local frequency oscillations, moderated by the wind, is also apparent.

The major features of breaking onset in the presence of wind forcing are shown in Figures 5.14, 5.15 and 5.16. Visually, the breakers in Figure 5.14 did not qualitatively change compared to those in Figure 5.5 with no wind. Quantitative properties, however, were altered by the wind.

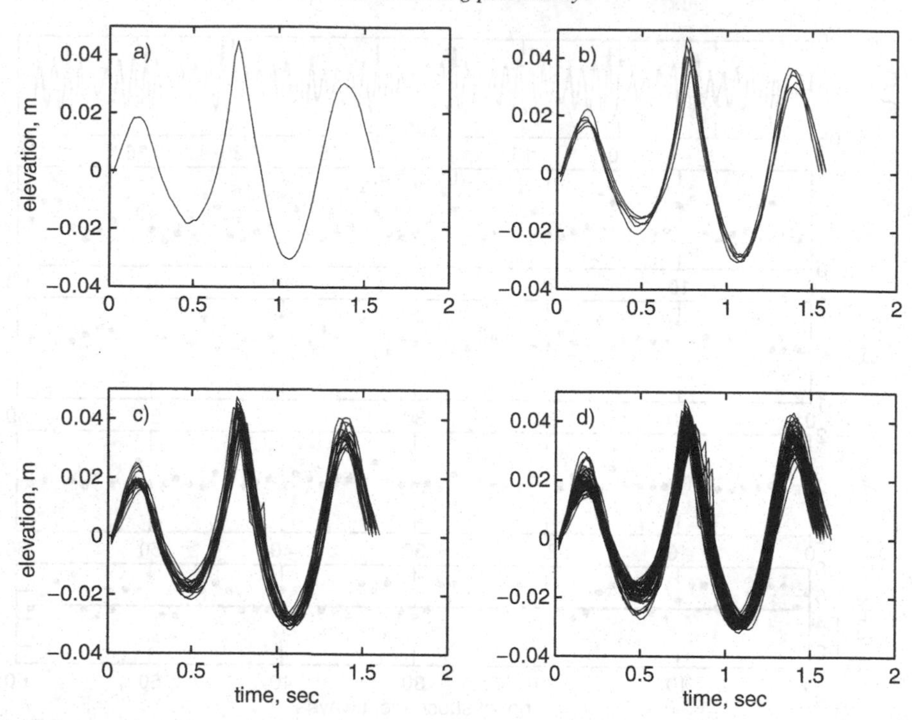

Figure 5.14 As in Figure 5.5, with wind forcing. IMF = 1.5 Hz, IMS = 0.30, $U/c = 3.9$. (top left panel) The steepest incipient breaker. (top right panel) The five steepest incipient breakers. (bottom left panel) The 20 steepest incipient breakers. (bottom right panel) The 50 steepest incipient breakers

In Figure 5.15, analogous to the no-wind Figure 5.7, the statistics of a comprehensive set of properties for the 20 highest incipient breakers and their links to the preceding and following wave are shown. Quantitatively, for the 20 waves approaching breaking, the wind influence generally brought more order to their shapes, as the scatter of almost all of the properties is reduced and the marginal dependences became clearer. Qualitatively, the wind changed the shape of the preceding wave which is now not skewed negatively on average (first subplot in the third row) and increased the steepness of the following wave from $\epsilon = 0.19$ to $\epsilon = 0.27$ on average (not shown).

With regard to the asymptotic shape of the breaker, the wind in Figure 5.16 has a scattering rather than stabilising influence. In this figure, analogous to Figure 5.10, characteristics of the five steepest waves are plotted in the presence of wind. Apparently, at these very last pre-breaking stages, the wind is capable of modifying the wave, which is about to lose its stability, and to somewhat randomise its characteristics.

In the first subplot of the top row, the limiting skewness is plotted versus limiting steepness. Skewness no longer approaches 1, but steepness extends beyond the $2\epsilon = 0.88$ limit and reaches $2\epsilon = 0.97$ (see also Toffoli *et al.*, 2010a). The asymmetry is no longer

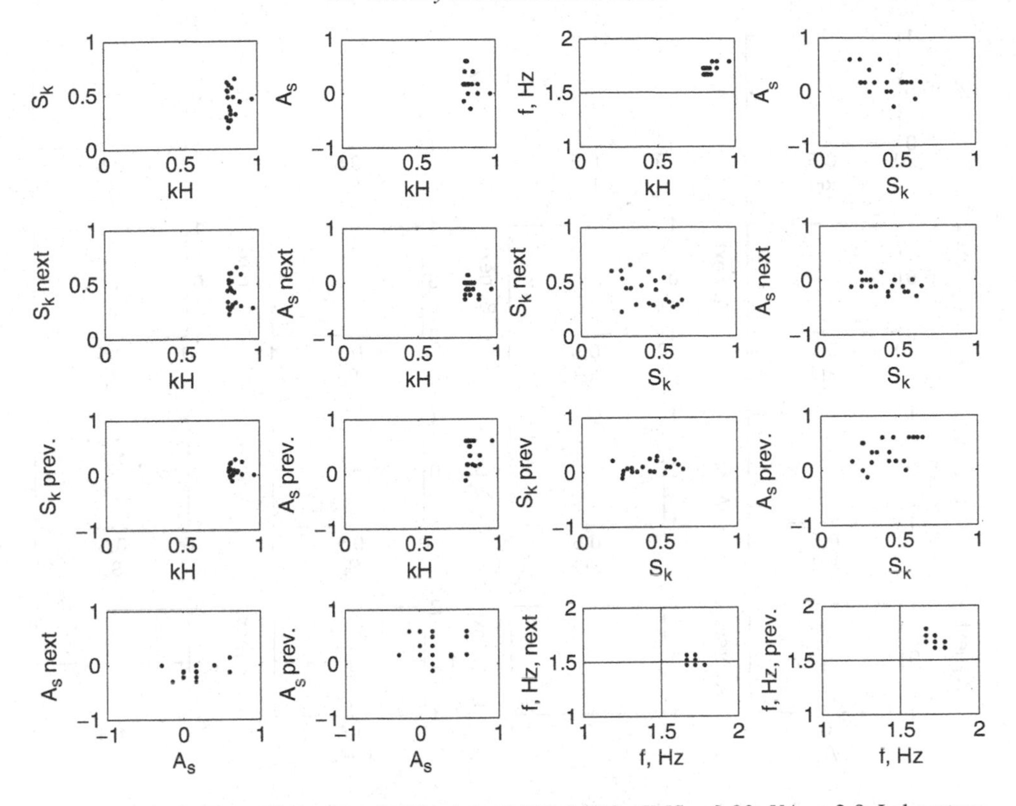

Figure 5.15 As in Figure 5.7, with wind forcing. IMF = 1.5 Hz, IMS = 0.30, $U/c = 3.9$. Laboratory statistics for the 20 steepest incipient breakers

negative, that is, the breakers do not tilt forward. Frequency remains a robust property and stays in almost the same range of $f = 1.11–1.19$ IMF.

Therefore, while breaking is mainly a hydrodynamic process, wind, if present, does influence incipient breaking. This is apparently achieved both through slowing down the modulational-instability growth and directly. As was noticed in the numerical simulations, this influence is small, but it is noticeable and diverse – from the shape-stabilising effect when approaching breaking onset to the shape-randomising effect at the point of breaking.

5.1.4 Distance to the breaking

In this subsection, the probability of breaking of the monochromatic waves is analysed in terms of the spatial extent of their evolution to the breaking occurrence. In this context, differences between predicting the breaking within monochromatic wave trains and within full-spectrum wave fields are discussed, as well as the issue of comparison of laboratory breaking rates, which are based on deterministic detection of breaking onset, and statistical field estimates for the progressive, rather than incipient breakers.

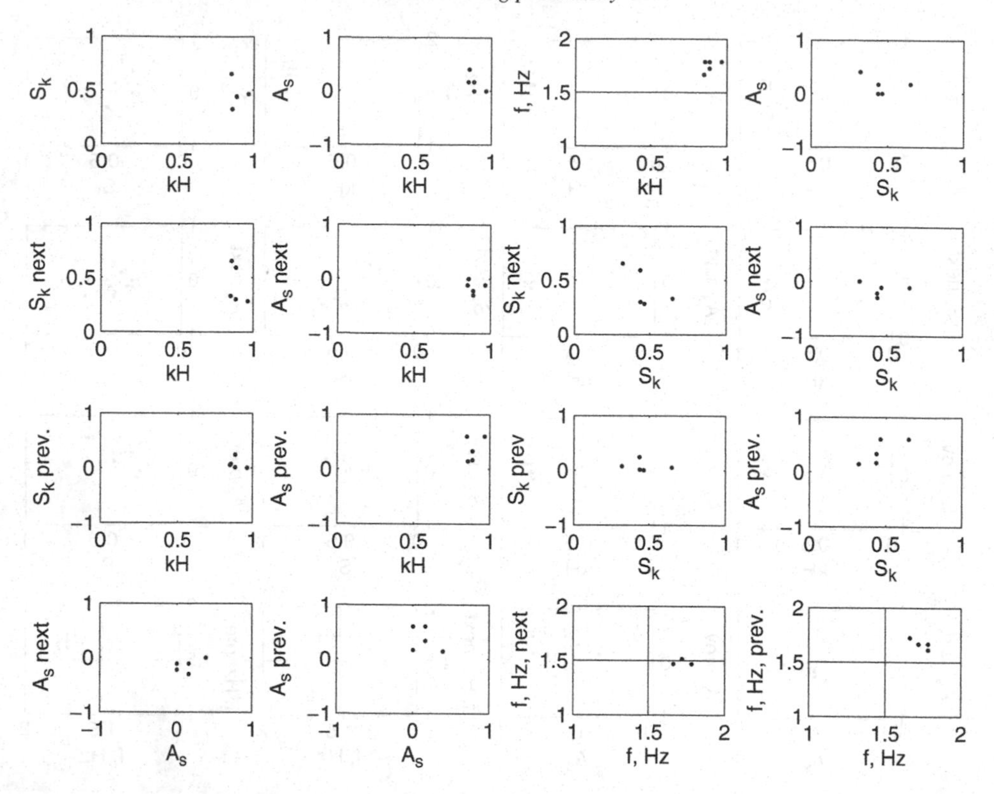

Figure 5.16 As in Figure 5.10, with wind forcing. IMF = 1.5 Hz, IMS = 0.30, $U/c = 3.9$. Laboratory statistics for the five steepest incipient breakers

As mentioned earlier in this section, distance to breaking in a train of steep initially monochromatic waves can be controlled by varying IMS and therefore can be predicted. Based on the laboratory measurements, such predictions are summarised in Figure 5.17 (top), which shows the non-dimensional distance to breaking $N = x_b/\lambda$ as a function of IMS, where x_b is the dimensional distance to breaking. A range of values of IMS are shown, along with cases with and without wind forcing. As expected, the addition of wind forcing reduces the non-dimensional distance to breaking. However, this reduction is not so great that the data points would deviate markedly from the functional relationship between N and IMS, and the nonlinear effect obviously dominates over the wind forcing.

In accordance with the numerical simulations, for each wave length an increase of its initial steepness resulted in the breaking occurring closer to the wavemaker. In dimensionless terms, this dependence was parameterised as follows:

$$N = -11 \operatorname{atanh}(5.5(\epsilon - 0.26)) + 23, \quad \text{for } 0.08 \le \epsilon \le 0.44. \tag{5.10}$$

Consistent with the model results, the formula imposes two threshold values of IMS. For $\epsilon > 0.44$, the wave breaks immediately (compared to $\epsilon = 0.3$ for the two-dimensional

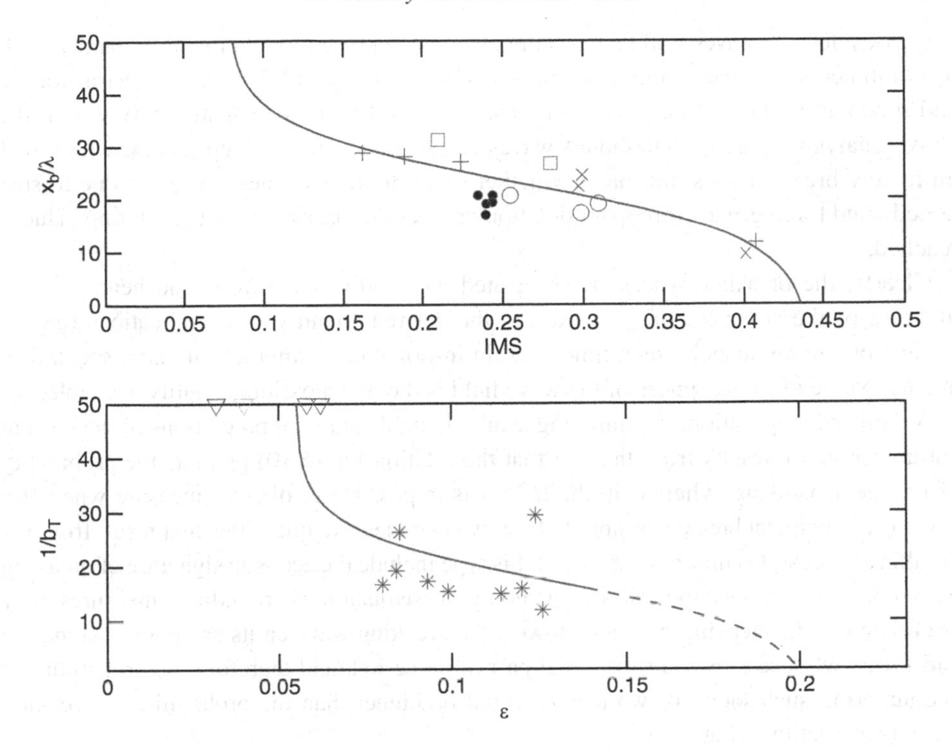

Figure 5.17 Parameterisation of the breaking probability. (top panel) Laboratory data. Number of wavelengths N to breaking versus IMS. No wind forcing: o – IMF = 1.6 Hz; x – IMF = 1.8 Hz; + – IMF = 2.0 Hz. Filled circles represent IMF = 2.0 Hz, with wind forcing applied. Squares are data points derived from Melville (1982). The parameterisation (5.10) is shown with a solid line. (bottom panel) Field data. Inverse breaking probability b_T, measured by visually detected whitecaps, versus the peak spectral steepness ϵ_{peak}. Triangles signify measured $b_T = 0$. The line identifies the approximation (5.27) (the dotted part is the extrapolation based on parameterisation of Babanin *et al.* (2001)). Figure is reproduced from Babanin *et al.* (2007a) by permission of American Geophysical Union

model – see discussion of modelling limitations at the beginning of Section 5.1) and if $\epsilon < 0.08$ the wave, in the absence of wind forcing, will never break (compared to $\epsilon = 0.1$ for the model).

In Figure 5.17 (top) two points (squares) are shown which were derived from Figs. 1 and 2 of Melville (1982) for comparison. The two measurements in Melville (1982) were conducted for initially uniform wave trains, their initial steepness and approximate dimensionless distance to breaking being known. Although recorded under different conditions, for much longer waves in a different wave flume, these points agree very well with the above parameterisation and provide strong support for laboratory results presented here.

The relationship (5.10) potentially provides a means of predicting the onset of breaking in the open ocean, although some further modification is required for application to such a case. In a field situation, the notion of an initial monochromatic steepness does not exist. Besides, the waves will be three-dimensional, and the modulational mechanism will be combined with wind forcing, current shear, superposition of dispersive spectral waves, and modulation due to linear wave groups, among other relevant features. However, the above analysis suggests that should waves reach some critical steepness then they will ultimately break. It does not matter whether this limiting steepness occurred due to sustained wind forcing, wave-group modulation or other means, as long as the limiting value is reached.

Clearly, the breaking process is associated with individual waves, and hence a local measure of the steepness of each wave is the desired quantity. For applications (e.g. in a wave-prediction model), such time-domain information is impractical and a spectral or average value of the steepness of the wave field is the only possible quantity available.

A further complication in comparing available field data with predictions of the current parameterisation results from the fact that the relationship (5.10) predicts the probability of incipient breaking, whereas in the field it is impossible to directly measure whether a wave is an incipient breaker or not. At best, we can measure quantities that result from the breaking process. Common measures of this type include the acoustic signature of breaking waves or surface whitecap coverage. Although these quantities are indirect measures, they are related to the breaking process. However, a breaking wave emits an acoustic signature and forms whitecaps over a substantial part of its period, and therefore the probability of encountering such sound or whitecaps would be higher than the probability of breaking onset (see Section 2.4 above).

Given the uncertainties, comparison of the present parameterisation of the breaking probability (5.10) with field data can only be qualitative at this stage, as the quantities being compared are not identical. This is done in Figure 5.17 (bottom), but will be discussed in detail in Section 5.3 after the probability dependences for spectral waves are introduced in Section 5.2.

5.2 Wave-breaking threshold

Before discussing the breaking probability in spectral environments (Section 5.3), here we would like to formulate and quantify the wave-breaking threshold, below which breaking does not happen in a wave train or a wave field, in terms of wave steepness. While applicable in any wave-breaking scenario, the threshold has to be introduced differently in monochromatic wave trains, where it is the steepness of individual waves, and in the spectral wave field where the information on the steepness of individual waves is usually not available. Essentially, the meaning of this threshold will be discussed and its difference from the breaking criterion that signifies breaking onset (5.4) will be explained. Examples from the deep-water (Black Sea) and finite-depth (Lake George) field sites will be used and respective data will be described in detail (see also Sections 3.7 and 3.5 for general descriptions of the two field experiments, correspondingly).

In the numerical simulations of Section 4.1 and the laboratory experiments of Section 5.1, breaking threshold was established in terms of the initial monochromatic steepness of an initially uniform train of steep waves (Babanin *et al.*, 2007a, 2010a). According to the numerical simulations, if IMS is less than

$$\mathrm{IMS}_{\mathrm{threshold}} = 0.1 \qquad (5.11)$$

the wave train will exhibit all the nonlinear behaviour described in Section 4.1, including oscillations of the steepness, skewness and asymmetry, but will never break. This is because in the course of the modulational evolution of nonlinear groups no wave will grow to reach the limiting steepness (2.47). Therefore, the threshold (5.11) does not signify any new kind of nonlinear behaviour of the wave system, but merely the magnitude of the modulation. In this regard, the modulational instability may lead either to a breaking wave, if limit (2.47) is achieved, or to a high non-breaking wave. The latter may even qualify as a freak (rogue) wave if, by definition, its height satisfies the criterion

$$H_{\mathrm{freak}} \geq 2.2 H_s, \qquad (5.12)$$

but its steepness is still below the limiting steepness.

An interesting confirmation of this conjecture can be found in Hwung *et al.* (2005). In this report, a variety of wave behaviours due to modulational instability were investigated in the supertank of the National Cheng Kung University, Taiwan (see also Hwung *et al.*, 2007). In their Figure 2.2.6.13, the authors plotted the ratio of the maximal crest height $a_{\max}$ in the unstable wave train to initial wave amplitude a in this train, as a function of the initial mean background steepness $\mathrm{IMS} = \epsilon = ak$ (1.1) of such a nonlinear wave train. Up to steepnesses of

$$\epsilon \approx 0.12\text{–}0.13, \qquad (5.13)$$

this ratio was growing and then it started to drop.

In view of what we would expect from modulational-instability behaviour, the wave trains below the mean-background-steepness threshold (5.11) (which is apparently slightly greater in the measurements of Hwung *et al.* (2005), i.e. (5.13)) would experience the instability and produce a high wave, but this wave would not break. Therefore, the higher the IMS, the higher is the individual tallest wave. This, however, can only be true until these individual waves start reaching the limiting steepness (2.47) and start breaking. From this background steepness on, the highest wave will be determined by this limiting-breaking steepness of the individual wave (2.47), and its height will no longer be increasing if IMS is kept tuned up. On the contrary, the ratio of the highest crest and the initial amplitude will start decreasing, since regardless of the increase of this initial amplitude, the nonlinear evolution will always be stopped at exactly the same highest crest by the breaking onset.

In the experiments of Hwung *et al.* (2005), the maximal ratio at steepness (5.13) was

$$\frac{a_{\max}}{a} \approx 3.5. \qquad (5.14)$$

Since this is achieved at the breaking point where skewness is 1 (5.5), then the crest is twice as high as its trough and the maximal ratio of the largest individual-wave height $H_{\max}$ to the height of initial monochromatic waves is

$$\frac{H_{\max}}{2a} \approx 2.6. \tag{5.15}$$

This ratio satisfies definition (5.12) of a freak wave and is apparently the highest wave that can be achieved due to modulational instability if measurement (5.14) describes the breaking onset precisely enough. Estimates of the individual-wave steepness in this measurement, with account for the frequency increase (5.8) and corresponding wavelength shrinkage, give $Hk/2 \approx 0.43$, very close to the breaking-onset limit (5.4), and therefore the estimate (5.14) can be regarded as a good approximation of what happens at the breaking point. Such an observation therefore means that freak waves due to this mechanism can only be produced in wave fields/trains with background steepness close to (5.13), not much lower or higher (see also Babanin *et al.*, 2011b).

In Banner *et al.* (2000) and Babanin *et al.* (2001), a similar mean-background-steepness threshold was established for dominant waves in a continuous wave spectrum. This was done on the basis of two deep-water and a finite-depth data set. The deep-water sets included the Black Sea data (Babanin (1995), see Section 3.7) and Lake Washington observations (Katsaros & Atakturk, 1992), and the finite-depth measurements were those conducted at Lake George (Young *et al.* (2005), see Section 3.5). The overall range of the dominant frequencies involved as a result was very broad: $f_p = 0.2$–0.4 Hz for the Black Sea, $f_p > 0.5$ Hz for Lake Washington, and $f_p = 0.3$–0.5 Hz for Lake George. Also, two deep-water data points from the Southern Ocean were included with $f_p < 0.1$ Hz. The ten-metre reference wind speed U_{10} ranged from 3 to 20 m/s, and the dimensionless depth of Lake George – from deep water down to values of $k_p d = 0.7$ (see Young & Babanin, 2006b).

The Black Sea data are tabulated in Table 5.1. The experimental arrangement involved visual surveillance of waves passing over a wave probe, with collocated breaking events labelled electronically by an observer. With this data, it was possible to investigate the distribution of breaking probability with respect to the distance from the spectral peak. The first 12 records (Table 5.1) were obtained in 1993 from the research vessel "Professor Kolesnikov" (henceforth PK) operated by the Marine Hydrophysical Institute (MHI) in Sebastopol.

The PK wave data were recorded using an accelerometer buoy, as described by Babanin *et al.* (1993). Briefly, the buoy diameter was 0.58 m and its operational bandwidth was 0.08–1.0 Hz, which easily covered the wave frequencies of interest. It was deployed around 100 m from the ship to avoid any interference between the ship and the wind and wave fields. Record lengths of 34 min to 68 min were acquired using a sampling frequency of 4 Hz. An observer watched the buoy from the vessel and triggered an electronic signal to register the passage of a whitecap over the buoy. This signal was recorded synchronously with the buoy data. The observer varied the duration of the electronic label according to the geometrical size of the whitecap, providing an approximate indication of individual breaker dimensions for future analysis. The observer was about 10 m above sea level, allowing a clear view of whitecaps with scales down to the size of the buoy. The environmental data collected simultaneously included 10-minute averages of the 10 m wind speed and

Table 5.1 *Summary of Black Sea data used for the breaking-probability analysis. Here,* f_p *is the spectral peak frequency,* H_s *is significant wave height,* U_{10} *is wind speed at* 10 m, ϵ_{peak} *is the significant steepness of the spectral peak,* $\Delta_{shear} = u_s/u_o$ *is the shear parameter defined by the ratio of the surface drift* u_s *to wave orbit* u_o *velocity,* $\tilde{f}_p = \frac{U_{10} f_p}{g} = \frac{1}{2\pi}\frac{U_{10}}{c_p}$ *is the non-dimensional peak frequency or the parameter of the inverse wave age, and* b_T *is the spectral peak breaking probability (2.3).*

Run	f_p, Hz	H_s, m	U_{10}, m/s	ϵ_{peak}	Δ_{shear}	$\tilde{f}_p$	b_T
6(1)	0.36	0.39	11.7	0.085	0.318	0.430	0.038
7(2)	0.34	0.49	12.7	0.099	0.280	0.441	0.065
9(3)	0.30	0.53	14.0	0.080	0.336	0.428	0.060
10(4)	0.31	0.54	14.4	0.084	0.341	0.455	0.052
11(5)	0.44	0.38	15.0	0.120	0.344	0.657	0.063
13(6)	0.39	0.45	14.6	0.114	0.320	0.581	0.067
14(7)	0.41	0.45	13.7	0.126	0.285	0.571	0.084
16	0.17	1.19	8.4	0.062	0.148	0.146	0
18	0.16	1.32	11.2	0.058	0.198	0.183	0
211	0.16	0.83	9.5	0.032	0.303	0.154	0
238	0.17	0.89	10.7	0.040	0.293	0.186	0
242(8)	0.27	0.99	10.0	0.124	0.139	0.274	0.034
244(9)	0.25	0.88	8.7	0.093	0.152	0.224	0.058

direction, measured by a standard Russian M63-MP anemometer on the ship's bow, together with mean water- and air-temperature data.

The first ten records were obtained with the ship anchored about 3.2 to 6.4 km offshore on the relatively flat west coast of the Crimea. Relatively young, strongly forced, fetch-limited waves with $U_{10}/c_p = 2.7$–4.1 were observed. Their spectral peak frequencies f_p ranged from 0.3 to 0.44 Hz. The wind speed U_{10} ranged from 11.7 to 15.0 m/s. Records 16 and 18 were gathered from the drifting vessel during the same PK cruise in the eastern part of the Mediterranean Sea, during which older waves were observed approaching full development, with $U_{10}/c_p = 0.9$ and 1.15. Their rather low peak frequencies of 0.16 Hz and 0.17 Hz corresponded to 8.4 m/s and 11.2 m/s wind speeds respectively.

The last four wave records were recorded from an oil rig situated 60 km offshore in the north-west region of the Black Sea in horizontally uniform water of 30 m depth. These are the deep-water records which have been used extensively in wavelet analysis of wave breaking and were described in Section 3.7 above. The complete environmental data set included one-minute averaged wind speeds, five-minute averaged surface-current speed and direction measurements at three-hour intervals as well as water and air temperatures recorded every three hours (see Babanin, 1988). To avoid possible air-flow distortion by the platform, an M63-MP anemometer was deployed on a tower 42 m above sea level, well above the deck buildings. The drag-coefficient dependence of Large & Pond (1981) was used to estimate the 10 m wind speed from the observed data.

The surface-elevation time series with labelled individual breaking events were analysed to determine individual breaker properties and breaking statistics. The accelerometer data series from the PK cruise were segmented to reduce the effects of low-frequency trends and integrated twice prior to analysis. With shorter waves riding on longer ones, occasional confusion could occur as to which wave scale was actually breaking, but since dimensional information on the individual whitecaps was available from the label length, it is believed that uncertainties of this type were not essential.

The Lake Washington data set described by Katsaros & Atakturk (1992) represents a short-wave extreme of natural deep-water wind–waves. These short fetch-limited waves were generated by light winds. For this data set, U_{10} ranged from 3.4 m/s to 6.8 m/s, with peak wave frequencies $f_p = 0.55$ Hz to 0.75 Hz and wave-development stages of $U_{10}/c_p = 1.5$ to 2.5. The waves were measured by means of a wire wave gauge, with breaking events recorded by a video camera observing the wave gauge. A detailed analysis provided the number of plunging, spilling and micro-scale breakers for each of the sixty-six 17-minute-long records. For the breaking-probability analysis, available plunging and spilling breaker statistics were combined to quantify the breaking waves in the spectral peak band.

The finite-depth wave-breaking data were obtained at the experimental site at Lake George near Canberra in south-eastern Australia (Figure 3.6) during October–December 1997. These data were described in detail in Section 3.5 (see also Table 5.2). The two Southern Ocean data points were processed from the videos taken from a low-flying plane, i.e. in the spatial domain, and were added to extend the probability study to very long waves, see Banner *et al.* (2000).

As mentioned above, Banner *et al.* (2000) and Babanin *et al.* (2001) concentrated on the breaking statistics of dominant waves. These statistics were investigated and the dominant-breaking probability was parameterised in terms of average spectral-peak steepness. The significant wave steepness given by

$$\epsilon_{\text{significant}} = \frac{H_s k_p}{2} \tag{5.16}$$

and contains some contribution from the higher-frequency components, which is irrelevant from the point of view of the evolution of dominant-wave groups to the breaking. Therefore, the significant steepness of the spectral peak was introduced:

$$\epsilon_{\text{peak}} = \frac{H_p k_p}{2}, \tag{5.17}$$

where

$$H_p = 4\left[\int_{0.7f_p}^{1.3f_p} F(f)df\right]^{1/2}, \tag{5.18}$$

i.e. the dominant waves were assumed to have frequencies within ±30% vicinity of the spectral peak (see (2.6) and discussion in Section 2.5). Intrinsically, ϵ_{peak} is an appropriate parameter as it provides a direct measure of the nonlinearity of the dominant waves.

Table 5.2 *Summary of Lake George data used for the breaking-probability analysis. Here,* f_p *is the spectral peak frequency,* H_s *is significant wave height,* U_{10} *is wind speed at* 10 m, ϵ_{peak} *is the significant steepness of the spectral peak,* $\Delta_{shear} = u_s/u_o$ *is the shear parameter defined by the ratio of the surface drift* u_s *to wave orbit* u_o *velocity,* $\tilde{f}_p = \frac{U_{10} f_p}{g} = \frac{1}{2\pi}\frac{U_{10}}{c_p}$ *is the non-dimensional peak frequency or the parameter of the inverse wave age,* d *is water depth and* b_T *is the spectral peak breaking probability (2.3).*

	Run	f_p, Hz	H_s, m	U_{10}, m/s	ϵ_{peak}	Δ_{shear}	$\tilde{f}_p$	H_s/d	b_T
1	311501.oc7	0.48	0.21	11.0	0.078	0.509	0.672	0.222	0.016
2	311615.oc7	0.48	0.17	8.5	0.059	0.474	0.517	0.178	0.007
3	311638.oc7	0.49	0.17	9.4	0.062	0.475	0.581	0.183	0.005
4	311757.oc7	0.42	0.35	17.1	0.114	0.496	0.977	0.337	0.375
5	311823.oc7	0.36	0.45	19.8	0.120	0.513	1.035	0.410	0.600
6	311845.oc7	0.33	0.40	15.0	0.096	0.538	0.797	0.372	0.388
7	311908.oc7	0.35	0.37	12.9	0.101	0.500	0.721	0.380	0.279
8	311930.oc7	0.38	0.34	12.8	0.107	0.436	0.723	0.347	0.265
9	311958.oc7	0.39	0.33	11.5	0.100	0.457	0.646	0.324	0.210
10	312021.oc7	0.40	0.39	13.7	0.132	0.397	0.761	0.365	0.303
11	312048.oc7	0.37	0.37	13.1	0.112	0.390	0.718	0.348	0.182
12	312111.oc7	0.40	0.25	9.3	0.074	0.434	0.536	0.290	0.087
13	312207.oc7	0.48	0.20	8.5	0.083	0.399	0.532	0.227	0.047
14	312232.oc7	0.50	0.22	9.0	0.100	0.399	0.568	0.239	0.077
15	312254.oc7	0.49	0.22	9.1	0.092	0.380	0.565	0.240	0.058
16	312316.oc7	0.49	0.21	8.6	0.084	0.384	0.532	0.228	0.086
17	312339.oc7	0.50	0.21	8.6	0.090	0.400	0.541	0.230	0.060
18	010004.no7	0.52	0.22	9.8	0.101	0.396	0.629	0.240	0.113
19	010030.no7	0.48	0.24	10.7	0.106	0.412	0.665	0.266	0.119
20	010055.no7	0.46	0.26	11.8	0.105	0.414	0.723	0.285	0.165
21	010140.no7	0.43	0.28	12.6	0.096	0.461	0.759	0.303	0.157
22	010204.no7	0.40	0.31	13.3	0.105	0.497	0.782	0.332	0.192
23	010226.no7	0.40	0.35	13.9	0.122	0.461	0.810	0.370	0.257
24	010248.no7	0.39	0.35	14.8	0.115	0.479	0.850	0.364	0.271
25	151238.de7	0.48	0.19	11.1	0.067	0.573	0.716	0.227	0.009
26	151301.de7	0.45	0.21	11.8	0.071	0.524	0.737	0.250	0.021

Huang (1986) previously used a similar parameter, the significant wave slope, in his modelling of breaking influence on upper-ocean dynamics. Values of ϵ_{peak} for the data records of this book are given in Tables 5.1 and 5.2.

Figure 5.18 shows the number of breakers per wave period as a function of the dominant wave steepness ϵ_{peak} (5.17) which is denoted as ϵ in the figure, for the Black Sea data (left panel) and Lake Washington data (right panel). Two conclusions may be drawn from these results. Firstly, these figures show that steeper dominant waves have higher breaking probabilities and these are well-correlated with the spectral peak steepness parameter ϵ_{peak}. Secondly, there is strong visual evidence of a threshold value for ϵ_{peak} around 0.05–0.06 which determines whether dominant wave breaking occurs or not. This level is consistent with the threshold level of $\epsilon = \text{IMS} \sim 0.1$ (5.10)–(5.11) reported by Babanin *et al.* (2007a)

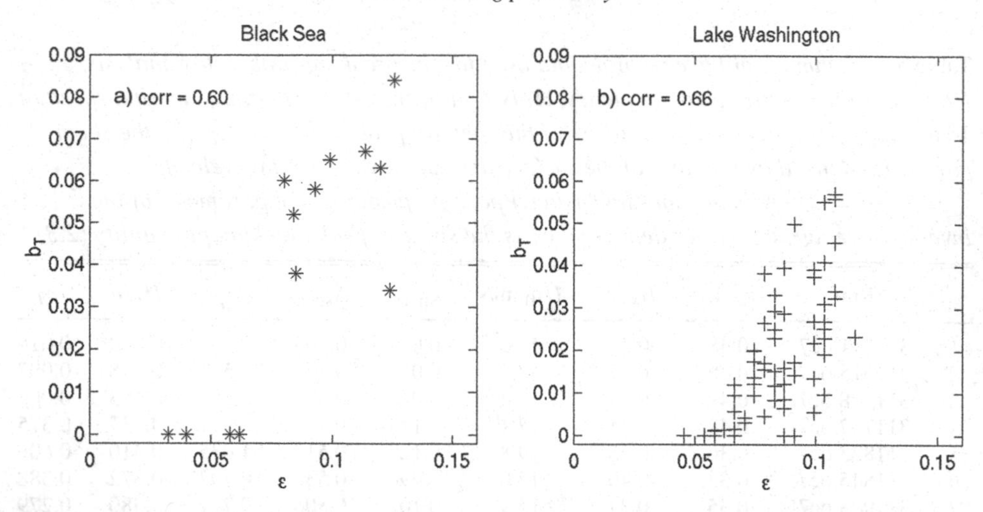

Figure 5.18 Observed dominant wave-breaking probability b_T (2.3) versus the dominant wave steepness ϵ (5.17) for two diverse field sites (a) Black Sea data (*), (b) Lake Washington data (+). The legend shows the correlation coefficient based on a linear fit. Figure is reproduced from Banner *et al.* (2000) © American Meteorological Society. Reprinted with permission

on the basis of numerical simulations and laboratory experiments, once this is converted to an equivalent root-mean-square measure. Therefore, a link between the modulational-instability mechanism of two-dimensional wave trains and breaking in three-dimensional wave fields can be expected.

The finite-depth Lake George breaking probability is tabulated in Table 5.2 and plotted in Figure 5.22a. As can be seen in this figure, the Lake George data are consistent with the ϵ_{peak} threshold observed for deep-water data in Figure 5.18. Therefore, the value of

$$\epsilon_{\text{peak}_{\text{threshold}}} = 0.055 \tag{5.19}$$

can be adopted as a universal wave-breaking threshold for dominant spectral waves, applicable both in deep water and in finite depths.

Thus, the presence of a wave-breaking threshold is established both for monochromatic wave trains and spectral waves. The threshold represents some characteristic mean steepness value. Whether this is a uniform two-dimensional wave train or a spectral three-dimensional directional wave field, this characteristic value signifies whether evolution of the train will result in a breaking event or events, or not. In the spectral sense, the threshold as described here is established for dominant waves.

For shorter waves of the spectral tail, such a threshold is impossible to define unambiguously, in terms of a characteristic steepness similar to (5.17)–(5.18), because there is no characteristic bandwidth at the broad-banded spectral tail (see also discussion in Section 2.5). A breaking threshold, however, does exist at the smaller scales too, as the wave-breaking dissipation clearly exhibits thresholded behaviour in terms of the spectral density of wave steepness (see Babanin & Young, 2005; Babanin *et al.*, 2007c). This will be discussed in Section 5.3.2 in the context of breaking probabilities for spectral waves and its value is given by (5.36).

Therefore, a characteristic breaking threshold always exists and bears a principal significance in wave-breaking dynamics and statistics. Breaking probability will depend on the characteristic steepness or its equivalent, but only if it exceeds the threshold value, and the probability appears to depend on the excess above the threshold steepness rather than on the steepness itself (see Sections 5.1.4 and 5.3.1).

From the point of view of nonlinear hydrodynamics of wave trains/fields, however, this threshold does not signify any fundamental transition from one type of behaviour to another. As mentioned at the beginning of this section with respect to two-dimensional trains, the non-breaking trains, with their characteristic steepness being below the threshold (5.11), will exhibit all the same nonlinear dynamic features as the breaking trains (except for the breaking itself of course). The breaking/non-breaking segregation is not because of a transition to a different underlying physics of wave evolution, but due to the ability/inability of the nonlinear wave system to produce individual waves that exceed the limiting steepness (2.47).

Differences between the breaking threshold (5.11) and (5.19) and breaking criterion (2.47) should be reiterated, emphasised and stressed at this stage. The threshold is a mean steepness of the wave train/field, whereas the criterion is a limiting steepness for individual waves in this train/field. That is if the mean steepness is above (5.11) and (5.19), there will appear individual waves reaching up to the steepness of (2.47). Therefore, individual waves can be steeper than (5.11) and (5.19), but that does not warrant an eventual breaking unless the mean steepness exceeds this level. Once these individual waves break, the mean steepness may fall below the threshold and the breaking will stop, unless the steepness is pumped up, for example, by the wind, by an adverse current, by the bottom proximity etc. High individual waves will still be produced, but they will not be reaching the limit and will not be breaking unless there is an input of energy into the wave system (e.g. wind input in the field or mechanical generation of waves in the laboratory).

5.3 Spectral waves

Now that the meaning of the breaking threshold is understood (Section 5.2 above), the breaking probability of spectral waves, i.e. waves with continuous distribution of energy along different wave periods/lengths and different directions can be analysed. While it is tempting to extend an analogy of the distance to breaking (which, with some caution, can be interpreted as breaking probability) of the monochromatic trains, parameterised in Section 5.1.4, to the spectral waves, this is not a straightforward exercise because the amplitudes and therefore steepness of waves with a specific wavenumber are not defined in the continuous-spectrum environment. An even less straightforward notion is at the spectral tail where there is not even a characteristic bandwidth that can be employed to produce a definition of a characteristic steepness similar to that for the spectral peak (5.17)–(5.18).

Therefore, in Section 5.3.1, we start from parameterising the spectral-peak breaking probability first, following Banner *et al.* (2000) and Babanin *et al.* (2001), and will then attempt the analogy with monochromatic wave trains (Figure 5.17) for these dominant waves of continuous-spectrum wave fields. In Section 5.3.2, the breaking probability of relatively short waves (i.e. short with respect to the waves at the spectral peak) will be discussed.

5.3.1 Breaking probability of dominant waves

As defined in Section 2.5, the breaking probability b_T (2.3) for spectral dominant waves will be considered as the mean passage rate past a fixed point of dominant wave-breaking events per dominant wave period. The dominant waves are taken within the spectral band of $0.7f_p$ to $1.3f_p$ (2.6), which contains spectral components determining the group structure of the dominant wave field. Measurements of b_T require averaging over a large number of wave groups since the breaking process is characterised by long-period intermittencies (e.g. Donelan *et al.*, 1972; Holthuijsen & Herbers, 1986; Babanin, 1995; Babanin *et al.*, 2010a).

As discussed in Chapter 3, wave breaking properties such as breaking probability, white-cap area coverage etc. had been assumed to have a primary dependence on the wind speed U, the dependence ranging from linear up to the fourth power according to different authors. In Figure 5.19, the breaking probabilities for two deep-water data sets are plotted against the wind speed U_{10}. These are the Black Sea data (Table 5.1, first panel) and Lake Washington data (Katsaros & Atakturk, 1992, second panel) (see also Section 5.2 for more details on these data). It is seen in these panels of Figure 5.19a and b, that in isolation the Black Sea data and the Lake Washington data do correlate rather well with U_{10}. When these two data sets are plotted together in Figure 5.19c, however, they have distinctly different offsets. The two additional data points obtained from analysis of the Southern Ocean video records show yet another dependence on U_{10}. Therefore, it is an obvious conclusion that it is not possible to establish a common dependence of the dominant wave-breaking probability on U_{10}. While for individual data sets such a dependence can be a good fit to the data, there is no universal direct dependence of the wave-breaking probability on the wind speed.

In Figure 5.20, the breaking probability is plotted against another plausible wind-forcing parameter, the inverse wave age U_{10}/c_p denoted as $2\pi\gamma$ in this figure:

$$\gamma = \tilde{f}_p = \frac{U_{10} f_p}{g} = \frac{1}{2\pi}\frac{U_{10}}{c_p}. \tag{5.20}$$

Here, it is seen that the individual data sets have similar offsets, but exhibit quite different rates of change of breaking probability as a function of wave age. Therefore, wind forcing does not appear to be a universal breaking-probability property either, although it provides a relevant secondary parameter as will be seen below.

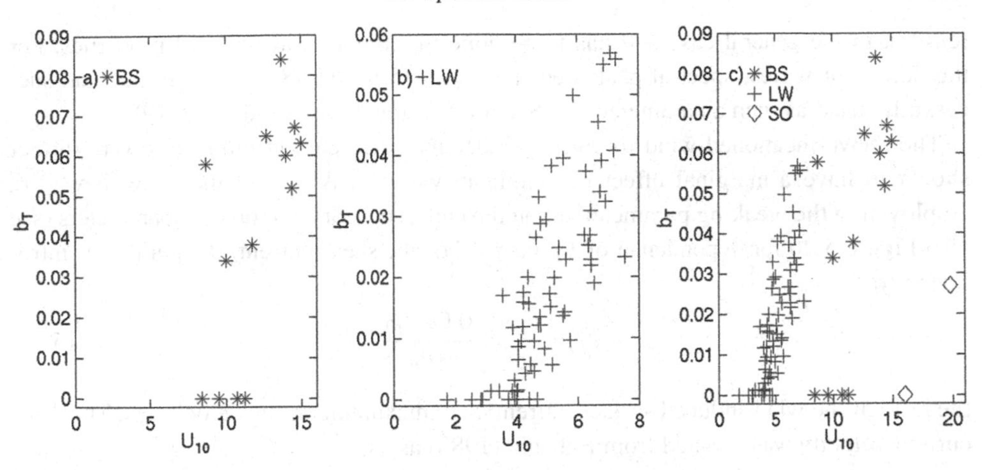

Figure 5.19 Observed dominant wave-breaking probability b_T versus U_{10} wind speed for three diverse field sites (a) Black Sea data (*), (b) Lake Washington data (+), (c) composite of the Southern Ocean data (diamonds) with (a) and (b). Figure is reproduced from Banner *et al.* (2000) © American Meteorological Society. Reprinted with permission

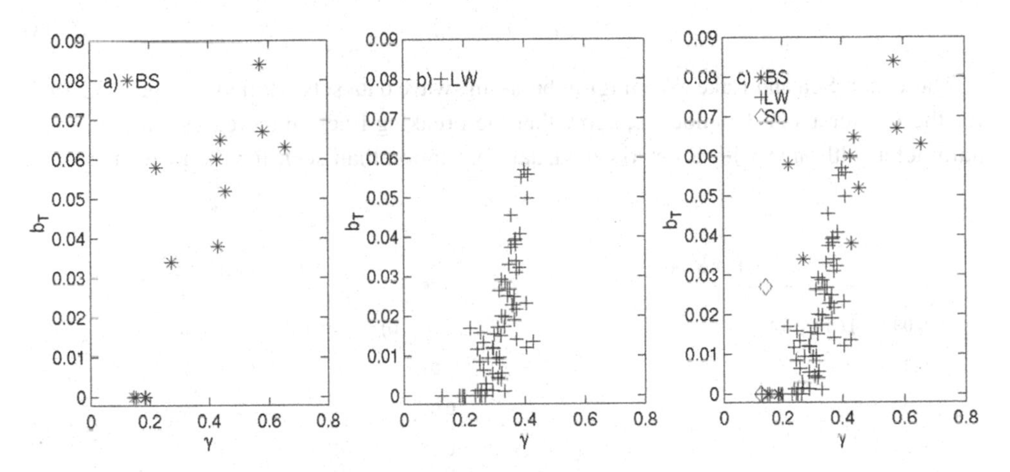

Figure 5.20 Observed dominant wave-breaking probability b_T versus wind-forcing parameter U_{10}/c_p, denoted as $2\pi\gamma$ in this figure, for three diverse field sites (a) Black Sea data (*), (b) Lake Washington data (+) (c) composite of the Southern Ocean data points (diamonds) with (a) and (b). Figure is reproduced from Banner *et al.* (2000) © American Meteorological Society. Reprinted with permission

Thus, Figures 5.19–5.20 support the argument made throughout this book that the wind's influence on wave breaking is indirect and, unless the wind forcing is very strong, breaking mainly comes through the slow increase of wave steepness which is then linked to faster nonlinear hydrodynamic processes which may lead on to the breaking. Correspondingly, parameterising the breaking probability by means of wind-speed characteristics is not feasible in the general case and has to be done in terms of wave-related properties. For the dominant waves, spectral-peak steepness ϵ_{peak} (5.17)–(5.18) was suggested and successfully used as such a parameter (see Section 5.2, Figure 5.18 and eq. (5.19)).

The above-mentioned wind forcing γ (5.20) and shear-current influences were further shown to have a marginal effect on dominant-wave breaking and therefore they were employed in the breaking parameterisation through secondary-importance parameters (see also Figure 5.20 for dependence of b_T on γ). For the shear current, the parameter introduced was

$$\Delta = \frac{u_s}{u_0} = \frac{0.01U_{10}}{\epsilon c_p}, \tag{5.21}$$

the ratio of the wind-induced surface current u_s to maximum orbital velocity u_0. The drift-current velocity was adopted from Babanin (1988) as

$$u_s \sim 0.01U_{10}, \tag{5.22}$$

and the maximal orbital velocity used was that of a linear surface gravity wave with height equal to the peak wave height (5.18)

$$u_0 = \epsilon_{peak} c_p. \tag{5.23}$$

The Black Sea and Lake Washington breaking-wave data sets, plotted in Figure 5.21a, b for the b_T-versus-Δ dependence, show that the breaking fraction increases with the shear parameter, although with a far lower visual correlation than seen for the peak steepness.

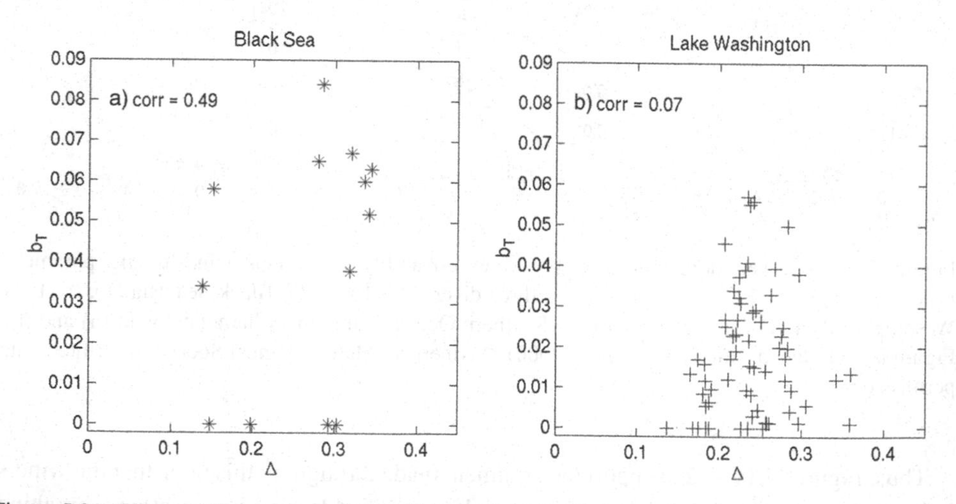

Figure 5.21 Observed dominant wave-breaking probability b_T (2.3) versus surface-shear parameter Δ (5.21) for two diverse field sites (a) Black Sea data (*), (b) Lake Washington data (+). The legend shows the correlation coefficient based on a linear fit. Figure is reproduced from Banner *et al.* (2000) © American Meteorological Society. Reprinted with permission

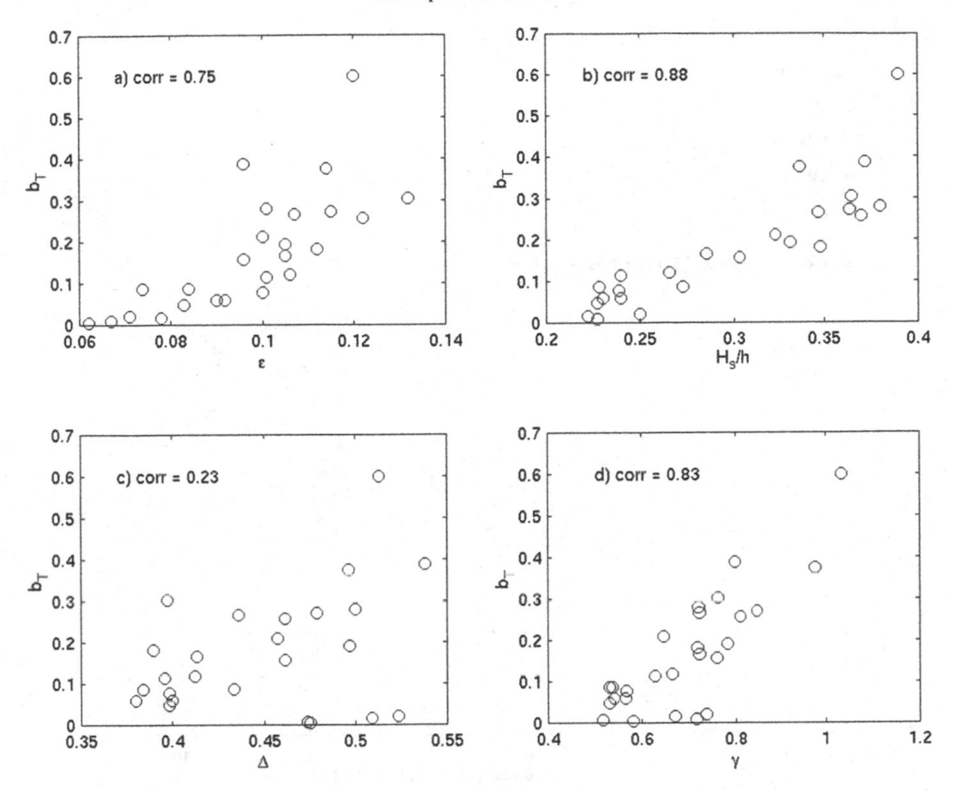

Figure 5.22 Lake George dominant wave-breaking probability b_T versus (a) steepness ϵ_{peak} (5.17), (b) bottom-interaction parameter H_s/d, (c) shear-stress parameter Δ (5.21), and (d) non-dimensional peak frequency γ (5.20). The legend shows the correlation coefficient based on a linear best fit. Figure is reproduced from Babanin *et al.* (2001) by permission of American Geophysical Union

Also, the existence of a shear threshold is not as evident as with the steepness. These indicate the secondary role of vertical shear as outlined above.

For the finite-depth Lake George data set, these marginal influences, as well as the primary dependence on ϵ_{peak} (5.17), are demonstrated in Figure 5.22a, c and d. Since waves observed at Lake George were affected by bottom proximity, ratio H_s/d was also used as a parameter to characterise finite-depth effects on the wave-breaking statistics (denoted as H/h in Figure 5.22b). In the final parameterisation (5.24), the secondary properties were introduced as perturbation terms of the form $1+\gamma$, $1+\Delta$ and $1+H_s/d$, which makes their effect negligible when the parameters are small. For example, H_s/d reduces to zero when the water becomes deep or when the waves vanish, and in such circumstances the parameter will have no effect on the overall dependence of breaking probability b_T on steepness ϵ_{peak}. We note that extrapolations of the expression (5.24) into conditions when γ, Δ or H_s/d are very large have to be done with caution as such extrapolations would take the dependence beyond the range of actual experimental data used to obtain the parameterisation.

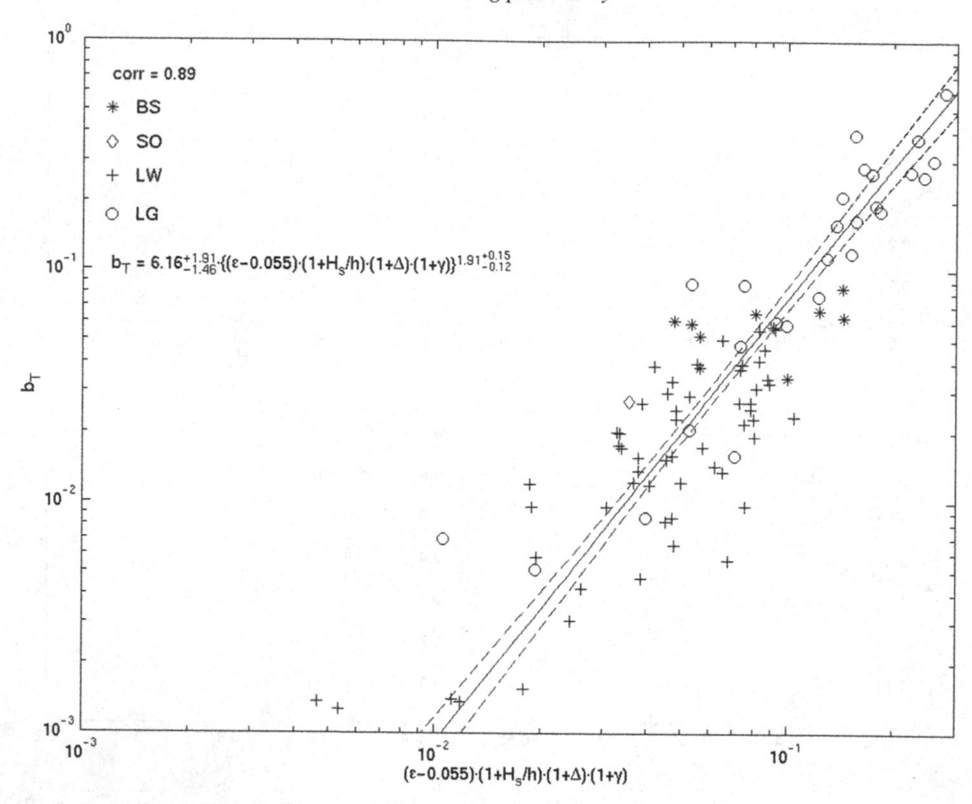

Figure 5.23 Combined log–log plot of the observed dominant-wave breaking probability b_T versus the modified significant peak steepness adjusted for the water-depth, wind-forcing and shear-current influences. Four diverse field sites are shown: Black Sea (BS), Southern Ocean (SO), Lake Washington (LW) (all deep water), and Lake George (LG) (finite depth). The legend shows the correlation coefficient based on a linear best fit in the log–log domain together with the parameterisation (5.24) and the ±90% confidence limits. Figure is reproduced from Babanin *et al.* (2001) by permission of American Geophysical Union

Values of b_T, Δ, γ, and H_s/d are shown in Table 5.1 for the Black Sea and Table 5.2 for Lake George, respectively.

For the data sets, combined from the Black Sea, Lake Washington and Southern Ocean data points (deep water) and Lake George measurements (finite depth), dependence of the spectral peak breaking probability b_T on the composite parameter is shown in Figure 5.23. The data sets agree very well, and the correlation coefficient of 0.89 is very high, particularly if we take into account the diversity of the data used. The exponent is approximately quadratic in the final proposed parameterisation of

$$b_T = 6.16^{+1.91}_{-1.46}[(\epsilon_{\text{peak}} - 0.055)(1 + H_s/d)(1 + \Delta)(1 + \gamma)]^{1.94^{+0.15}_{-0.12}}, \tag{5.24}$$

where 90% confidence limits are shown.

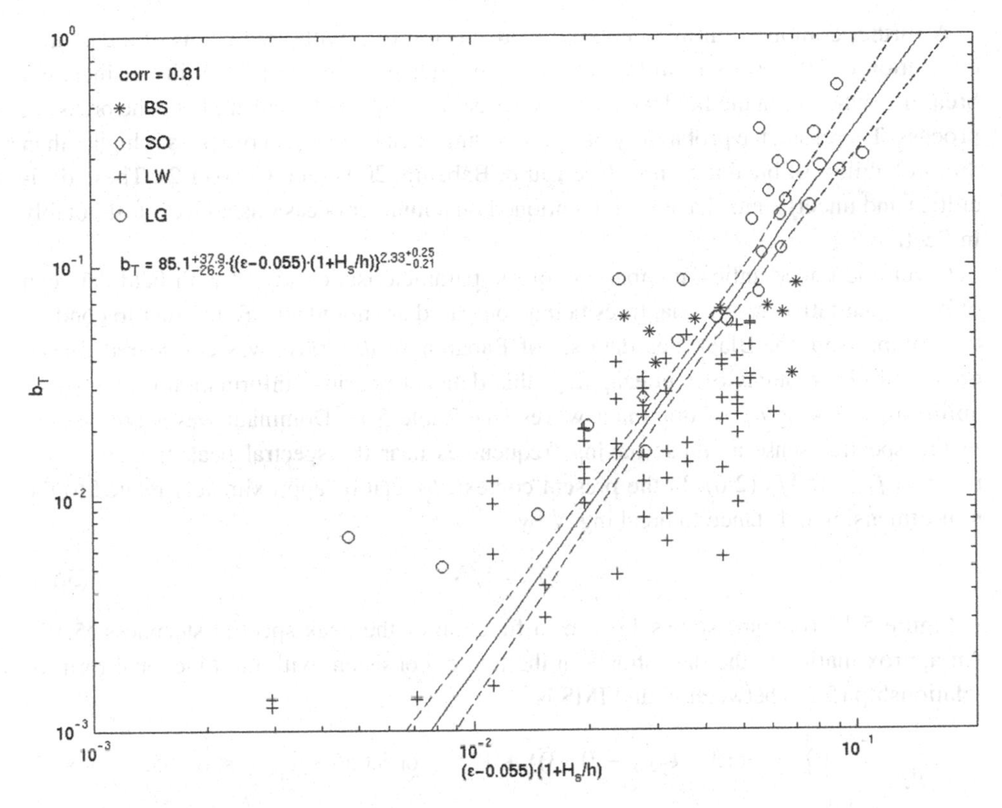

Figure 5.24 Combined log–log plot of the observed dominant wave breaking probability b_T versus the modified significant peak steepness adjusted for the water depth $(\epsilon - 0.055)(1 + H_s/h)$ (5.25). Four diverse field sites are shown: Black Sea (BS), Southern Ocean (SO), Lake Washington (LW) (all deep water), and Lake George (LG) (finite depth). The legend shows the correlation coefficient based on a linear best fit in the log–log domain together with the parameterisation and the ±90% confidence limits. Figure is reproduced from Babanin *et al.* (2001) by permission of American Geophysical Union

For practical purposes, when information on the wind forcing and shear is not available, a simplified parameterisation can be used, in terms of wave and water-depth characteristics only:

$$b_T = 85.1^{+37.9}_{-26.2}[(\epsilon_{\text{peak}} - 0.055)(1 + H_s/d)]^{2.33^{+0.25}_{-0.21}}. \tag{5.25}$$

This dependence is shown in Figure 5.24 and provides a reasonable linear correlation of 0.81 within the range of the data available.

Now, the comparison of this dominant-wave spectral breaking probability with the distance to breaking of the monochromatic wave trains (5.10) can be conducted. Establishing a link between parameterisations of the breaking probability in terms of individual-wave steepness and in terms of spectral steepness, even if approximate, has a significant practical value.

A further complication in comparing field spectral data with predictions of the parameterisation (5.10) results from the fact that the latter predicts the probability of incipient breaking, whereas in the field we can only measure quantities that result from the breaking process. Therefore, the probability of encountering such breaking in progress is higher than the probability of breaking onset (see Liu & Babanin, 2004, and Chapter 2). These difficulties and uncertainties have been mentioned on a number occasions above, most notably in Section 5.1.4.

Given the uncertainties, comparison of the parameterisation (5.10) with field data can only be qualitative, as the quantities being compared are not identical. In order to conduct the comparison, the Black Sea data set of Babanin *et al.* (2001) was considered. Based on visual observations of whitecapping, this data set provides information on the probability of breaking b_T of dominant waves (see Table 5.1). Dominant waves are defined in the spectral sense as those having frequencies near the spectral peak frequency f_p, i.e. $f = f_p \pm 0.3 f_p$ (2.6). In the present context, b_T can be approximately related to the non-dimensional distance to breaking N by

$$b_T \approx 1/N. \tag{5.26}$$

Figure 5.17 (bottom) shows $1/b_T$ as a function of the peak spectral steepness (5.17). An approximation to the data shown in the figure, consistent with the functional form of relationship (5.24) between N and IMS is

$$\frac{1}{b_T} = -10\,\mathrm{atanh}(13.3(\epsilon_{\mathrm{peak}} - 0.13)) + 17, \quad \text{for } 0.055 \le \epsilon_{\mathrm{peak}} \le 0.205. \tag{5.27}$$

The lower limit (no breaking if $\epsilon_{\mathrm{peak}} < 0.055$) is obtained from the experimental data (Banner *et al.*, 2000; Babanin *et al.*, 2001) and the upper limit ($\epsilon_{\mathrm{peak}} = 0.205$) is obtained by extrapolating the parameterisation developed in Babanin *et al.* (2001) to the 100% breaking condition.

Thus, we conclude that the distance before breaking occurs, which is related to the breaking probability, is a function of the background mean wave steepness in the wave train/field. In the latter case, this concept/analogy can only be applied to the dominant waves. Another potential method of estimating the breaking rates of dominant waves, based on measurements of ensemble-average asymmetry (1.3) in the wave trains, without having to actually detect the breaking events, will be discussed in Section 7.3.3 (see parameterisation (7.19)). Breaking of relatively short waves, i.e. waves of scales smaller than the dominant waves, is a separate topic in many regards and will be discussed in Section 5.3.3.

5.3.2 Breaking probability of small-scale waves

Breaking of waves shorter than those at the spectral peak is altered in a number of ways and at smaller scales is perhaps driven by physics different to that of dominant-wave breaking. Therefore, this subsection of the spectral-breaking section is important and is necessarily large, and needs a summary overview to begin with.

The subsection begins from outlining the physics that makes small-scale breaking different to the dominant breaking and from formulating the cumulative-effect concept. Experimental estimates of the breaking probability are then described, and the problem of the count of the small waves riding larger waves is discussed. The important spectral wave-breaking property, the dimensionless and dimensional threshold which identifies spectral levels below which breaking does not occur, is introduced and quantified. Formulations are suggested for breaking probability and breaking dissipation which account for the threshold and for the cumulative effect. The transition frequency between the inherent- and induced-breaking domination in the spectrum is identified quantitatively. Finally, an alternative method of approaching breaking probability in terms of breaking-wave-height distributions is described.

Lately, the frequency distribution of the breaking probability $b_T(f)$ (2.4) has been a sought after function (Ding & Farmer, 1994; Phillips *et al.*, 2001; Banner *et al.*, 2000, 2002; Melville & Matusov, 2002; Babanin & Young, 2005; Gemmrich, 2006; Manasseh *et al.*, 2006; Babanin *et al.*, 2007c; Filipot *et al.*, 2008, 2010; Babanin, 2009). There is a reasonable expectation in the wave-modelling community that, once some universal function for $b_T(f)$ is obtained, such a parameterisation will provide a major step forward towards an experimental, rather than speculative dissipation function.

Breaking of small-scale waves, however, apart from inherent reasons such as modulational instability or linear focusing, can be affected by longer waves (e.g. Longuet-Higgins & Stewart, 1960; Phillips, 1963; Donelan *et al.*, 2010). As a result, parameterisation of the breaking rates at short wavelength is problematic, if not impossible in terms of the steepness of the short waves or spectral density at the respective wavelength. Here, by short/small-scale we mean waves with temporal and spatial relative scales smaller than those of the dominant waves in a wave field.

There are a number of ways in which the long waves can affect the breaking of shorter waves. One of them is modulation of the train of the short waves riding the underlying large-scale waves. The latter compress the short wavelengths at their front face and extend those at the rear face. As a result, the front-face small waves become steeper and frequently break (Donelan, 2001). In the absence of underlying long waves, the breaking, due to the regular reasons only, would be much less frequent. This mechanism must be most essential in deep water and finite depths outside the surf zone, as for every sequence of shorter waves such a modulation inevitably occurs over every period of any underlying longer wave.

Another effect is due to breaking of the large waves (Banner *et al.*, 1989; Tulin & Landrini, 2001; Manasseh *et al.*, 2006; Young & Babanin, 2006a). This is how Tulin & Landrini (2001) describe the short waves in such circumstances:

> "As we have observed in our large wind–wave tank, the growth of these waves is much effected by the existence of breaking energetic waves, which not only modulate the microbreakers, but virtually eliminate them as the microbreaker train is overcome by the energetic breaker at the peak of the wave group. It would seem, observing this striking phenomenon, that it is the action of the energetic breakers which causes the microbreakers to disappear, and that they begin growing again from very short waves. Their eventual length is thus determined by the effective fetch between energetic breakers."

The fetch, apparently, becomes shorter as the breaking rates of larger waves go up which would be under conditions of strong wind forcing, finite depths and, most essentially, the surf zone where this second effect should dominate over the above-mentioned modulation of short waves.

The quoted observational description is helpful in many regards. First of all, while it is mentioned in definite terms that the short waves are eliminated, it is not said explicitly that these microwaves actually break too. In principle, they can be dissipated without breaking, because of interaction, for example, with the turbulent wake of the large breaker (e.g. Banner *et al.*, 1989). In this regard, it is difficult if not impossible to separate small-scale breaking and small-scale wave-energy dissipation.

In the context of such dissipation, parasitic capillary waves should also be mentioned (Crapper, 1970; Perlin *et al.*, 1993; Longuet-Higgins, 1995; Fedorov *et al.*, 1998). These high-frequency waves are generated by steep waves and actively dissipate wave energy at the respective scales, or may in fact make the large waves break (i.e. Crapper, 1970), but the dynamics of this dissipation is determined, again, by their interaction with the longer waves rather than by their local steepness alone.

The quote above also points out that, in the presence of frequent large breaking, for the short waves the induced breaking/dissipation dominates. Indeed, the small-scale waves only exist between the large breakers and their breaking/dissipation is determined by their effective fetch between such breakers, regardless of the physics which would inherently drive them to breaking. This is the so-called cumulative effect, very important for the spectral dissipation of wind–waves (see also Section 7.3.4). This effect signifies the fact that the breaking/dissipation of short waves of a certain small scale above the peak in the wave spectrum is determined by the integral of the wave spectrum below this scale, rather than by the value of the spectral density at this particular frequency/wavenumber (Babanin & Young, 2005; Manasseh *et al.*, 2006; Young & Babanin, 2006a; Babanin *et al.*, 2007c).

Quantitative observations, and particularly parameterisation of the breaking probability of short spectral waves are not very many. This is primarily due to difficulties of detecting small breaking and of separating the dominant breaking and small breaking, particularly as the latter is often correlated or even linked to the large breaking as mentioned above.

One of the first clear experimental evidences of the spectral distribution of breaking events was provided by Gemmrich & Farmer (1999) in terms of rate of occurrence of breaking events with different phase speeds. Banner *et al.* (2000) showed results on the frequency distribution of the breaking probability based on information gathered in the Black Sea data set. A total of 2121 individual breakers of the 13 records listed in Table 5.1 were analysed. The sampling frequency of 4 Hz allowed us to compile frequency distributions for the breakers as histograms over the range from f_p to twice the peak frequency $2f_p$. Binning of breaking-event probabilities was carried out for $\pm 15\%$ constant-percentage wavenumber bands centred on k_p, $1.35k_p$, $1.83k_p$, $2.48k_p$ and $3.35k_p$, thereby covering the wavenumber range k_p to $4k_p$ or, equivalently, the frequency range f_p to $2f_p$.

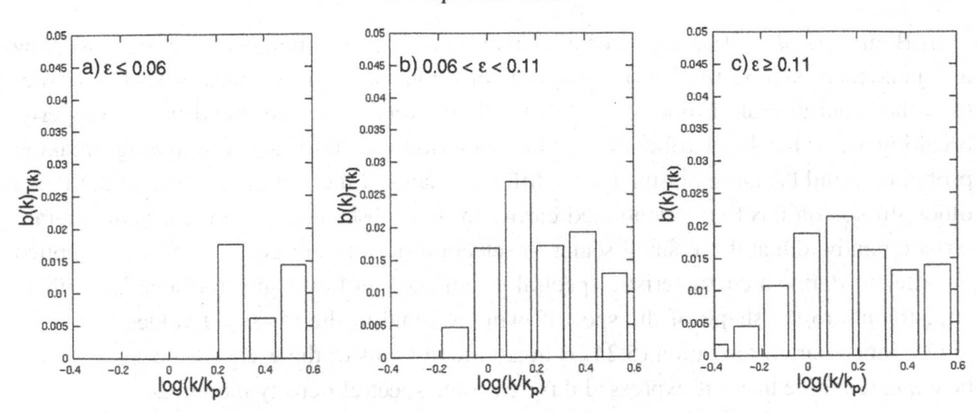

Figure 5.25 The shift towards the spectral peak of the histogram of the breaking probability as a function of the mean steepness in each ±15% relative wavenumber bin. (a) for $\epsilon_{peak} < 0.06$ (corresponding to well-developed seas) breaking occurs well above the spectral peak. For the steeper seas in (b) $0.06 < \epsilon_{peak} < 0.11$ and (c) $\epsilon_{peak} > 0.11$, the histogram peak moves towards the spectral peak. Figure is reproduced from Banner *et al.* (2000) © American Meteorological Society. Reprinted with permission

Figure 5.25 shows how the peak of the breaking probability distribution shifts from well above the spectral peak to close to the spectral peak as the dominant wave slope ϵ_{peak} increases. This is consistent with the findings of Ding & Farmer (1994) who found from their relatively old wind–sea conditions that the mean breaking-event speed is considerably smaller than the phase speed of the dominant wind seas. It also follows the trend reported by Gemmrich & Farmer (1999), who observed that the peak of the normalised breaking-frequency distribution was located well above the spectral peak frequency for typical mature wind seas.

While negligible breaking is observed at the spectral peak for the wave fields with $\epsilon_{peak} < 0.06$, the wave breaking occurs at higher wavenumbers (Figure 5.25a). If the peak steepness is higher ($0.06 < \epsilon_{peak} < 0.11$ in Figure 5.25b), the wave breaking histogram moves closer to the spectral peak, and even below the peak at $\epsilon_{peak} > 0.11$ (Figure 5.25c).

For a standard unimodal spectral shape, which occurs under simple wave-development conditions, e.g. JONSWAP spectrum (2.7), the peak steepness ϵ_{peak} can be related to the wind forcing or inverse wave age U_{10}/c_p. That is, for young waves, frequent breaking of the dominant waves is expected, whereas old dominant waves will not be breaking at all. As mentioned above, this is consistent with other observations (i.e. Ding & Farmer, 1994; Gemmrich & Farmer, 1999), but it has to be kept in mind that generally speaking the breaking parameter ϵ_{peak} is not unambiguously related to the wave age. Waves of the same nominal development stage U_{10}/c_p may have quite different values of ϵ_{peak} due to a variety of circumstances, for example due to the presence of swell which is quite typical for the open oceans. These could lead to different breaking probabilities at the spectral peak and different breaking-frequency distributions for the same nominal wave age.

In Banner *et al.* (2000) and later in Banner *et al.* (2002), attempts were made to draw a dependence, similar to (5.24) through a selection of frequency bins at scales smaller than the spectral peak. Filipot *et al.* (2008, 2010) were interested in a different property, breaking wave height distribution, for the same ultimate purpose of assigning breaking probability and breaking dissipation to different scales of the wave spectrum. In these and other studies on this topic mentioned earlier in this subsection, since there is no characteristic bandwidth at these small scales, a selection of different Δf in (2.5) were applied in order to define a characteristic spectral steepness ϵ in those spectral bins. Even then, adjustments to the shape of the spectral windows and to the threshold values had to be made. Since parameterisation (5.24) is quadratic in terms of the wave steepness, it should be expected to be linear if expressed through some spectral-density measure.

In this regard, two issues must be clarified. The absence of the characteristic bandwidth away from the spectral peak is not merely a technical question, this is a physical problem. As mentioned above, the modulational-instability mechanism cannot be active in the broad-banded process which the small-scale waves appear to be. Even if the induced-by-long-waves breaking of short waves is disregarded, this means that the physics of the breaking at the shorter scales is altered compared to the dominant waves. The induced breaking, however, cannot be disregarded as it is already significant close to the spectral peak and apparently becomes dominant at higher frequencies/wavenumbers.

Another possibility may be that the short waves, as well as the dominant waves, are also represented by coherent wave groups, information about which is averaged out in such a space-time mean characteristic as the wave spectrum. Within such trains, the usual instability mechanisms would be working, but such a concept would be mere speculation at the present stage of our knowledge.

The second issue is the actual count of waves falling within a spectral band, whatever width for such a band is found to be relevant. In Manasseh *et al.* (2006), the bubble-detection method described in Section 3.5 was applied to the Lake George wave-breaking data in order to estimate the breaking probability at wave frequencies beyond the spectral peak, and to obtain the distribution of breaking probability $b_T(f)$ (2.4) with wave frequency. To do this, the number of waves at each frequency $N(f)$ was redefined. As discussed in Section 2.5, if the waves are counted by the zero-crossings, the resulting count $N_c(f)$ will be less than the nominal reference count $N(f)$ given by (2.13), because in real seas, waves of periods different to $1/f$ will occupy some part of the record. The breaking probability b_T used by Manasseh *et al.* (2006) was defined as

$$b_T(f) = \frac{n(f)}{N_c(f)}, \tag{5.28}$$

where $N_c(f)$ is the number of waves actually counted by the zero-crossing analysis within the bandwidth

$$f = f_c \pm 0.1 f_p, \tag{5.29}$$

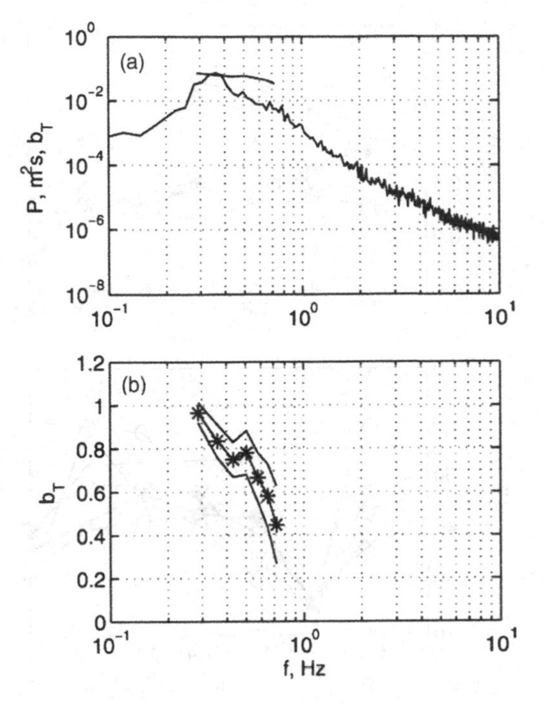

Figure 5.26 Wave power spectrum and breaking probability versus wave frequency f. (a) Wave power spectrum $P(f)$; (b) Wave-breaking probability $b_T(f)$ (5.28). In (a), probability (b) is shown, normalised by the spectrum-peak value. Figure is reproduced from Manasseh *et al.* (2006) © American Meteorological Society. Reprinted with permission

with the set of central frequencies being

$$\frac{f_c}{f_p} = 0.8, 1.0, 1.2, 1.4, 1.6, 1.8, 2.0. \tag{5.30}$$

Shown in Figure 5.26 is the wave power spectrum created from the surface-elevation data, and the breaking probability, b_T as a function of wave frequency for a 19.8 m/s wind speed (rec. 5 of Table 5.2). The breaking distribution in the top panel of Figure 5.26 was normalised so that it matches the spectral density at the peak frequency. This was done purely to make comparison of the two curves easier. The $b_T(f)$ curve in the bottom panel is bracketed by two lines which are the calculated 95% confidence intervals on $b_T(f)$ (Walpole & Meyers, 1978). Although the b_T curve only covers a fraction of the frequency spectrum, it is clear that the downward trend in breaking probability with wave frequency is statistically significant.

Figure 5.27 presents more derived analysis made possible by the passive acoustic method. Even though this figure was only intended by Manasseh *et al.* (2006) to demonstrate the potential of the passive acoustic method, it is worth pointing out some apparent features of

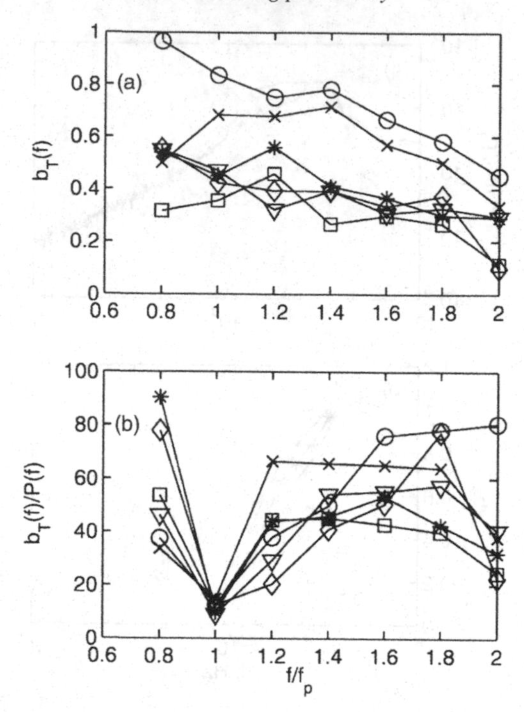

Figure 5.27 Breaking probabilities versus wave frequency f normalised by the peak frequency f_p. (a) $b_T(f)$ (5.28); (b) b_T normalised by the spectral density $P(f)$. Squares: 12.8 m/s; *: 12.9 m/s; ∇: 13.2 m/s; diamonds: 13.7 m/s; $\times$: 15.0 m/s; circles: 19.8 m/s. The records are from Table 5.2. Figure is reproduced from Manasseh *et al.* (2006) © American Meteorological Society. Reprinted with permission

the breaking-probability frequency distributions which had hardly been examined before experimentally.

In this figure, distributions of $b_T(f)$ (5.28) are plotted versus relative frequency f/f_p for records 5 through 11 of Table 5.2, which correspond to different wave spectra developed under different wind speeds. In Lake George's bottom-limited environment, well-developed and even fully developed waves can still be strongly forced (Young & Babanin, 2006b) and therefore are expected to break at the spectral peak.

Out of the six wave records analysed, only the first one (19.8 m/s mean wind speed), corresponds to full development for the given water depth (Young & Babanin, 2006b). Therefore, although the waves are strongly forced, the wave spectrum will not develop further: the total wave energy will not grow and the spectral peak will not shift to lower frequencies. Since both the wind and the nonlinear interactions keep pumping energy into these lower frequencies, it must be rapidly dissipated at these scales due to interaction of the longer waves with the bottom and subsequent breaking. This is seen in the upper curve of the top panel of Figure 5.27: nearly 100% breaking is measured for frequencies below the spectral peak.

Breaking rates $b_T(f)$ normalised by their respective spectral densities $P(f)$ are shown in Figure 5.27, bottom panel. At the spectral peak, these normalised breaking rates merge together very clearly, but they stay separated both above and below the peak. Thus, if there is a linear or quasi-linear dependence of b_T on $P(f)$, it would only be applicable at the spectral peak. Away from the peak, other influences make the dependence of b_T on $P(f)$ nonlinear or affect this dependence in another way.

This uncertainty was investigated, based on the Lake George measurements, by Babanin & Young (2005) and Babanin *et al.* (2007c). They attempted to draw an analogy with the findings of Banner *et al.* (2000) and Babanin *et al.* (2001) for the spectral peak and sought a dependence of the breaking probability at different frequencies as a function of the so-called saturation density at that frequency, a spectral analogue of the squared wave steepness introduced by Phillips (1958, 1984):

$$\sigma_{\text{Phillips}}(f) = \frac{(2\pi)^4 f^5 F(f)}{2g^2}. \tag{5.31}$$

In Babanin & Young (2005) and Babanin *et al.* (2007c), the saturation $\sigma(f)$ was normalised by a directional spreading parameter:

$$\sigma(f) = \sigma_{\text{Phillips}}(f)A(f). \tag{5.32}$$

Here, $A(f)$ is the integral characteristic of the inverse directional spectral width introduced by Babanin & Soloviev (1987, 1998b):

$$A(f)^{-1} = \int_{-\pi}^{\pi} K(f,\theta)d\theta \tag{5.33}$$

where θ is the wave direction, and $K(f,\theta)$ is the normalised-by-its-maximum-value directional spectrum:

$$K(f,\theta_{\text{maximum}}) = 1. \tag{5.34}$$

Normalisation by directional spreading was brought in by Banner *et al.* (2002), who also investigated breaking probability across the frequency as a function of the saturation spectrum. Banner *et al.* (2002) needed an additional normalisation in order to explain why the wave-breaking threshold that they observed is not universal in terms of the saturation-density values. As will be shown below, the results of Banner *et al.* (2002) must have been influenced by the cumulative effect and the directional normalisation is in fact not necessary. In Babanin & Young (2005) and Babanin *et al.* (2007c), values of A were

$$A(f) \approx 1, \tag{5.35}$$

and therefore the normalisation did not impact the value of the universal threshold significantly. When revisited, the threshold saturation level (5.36) did not change.

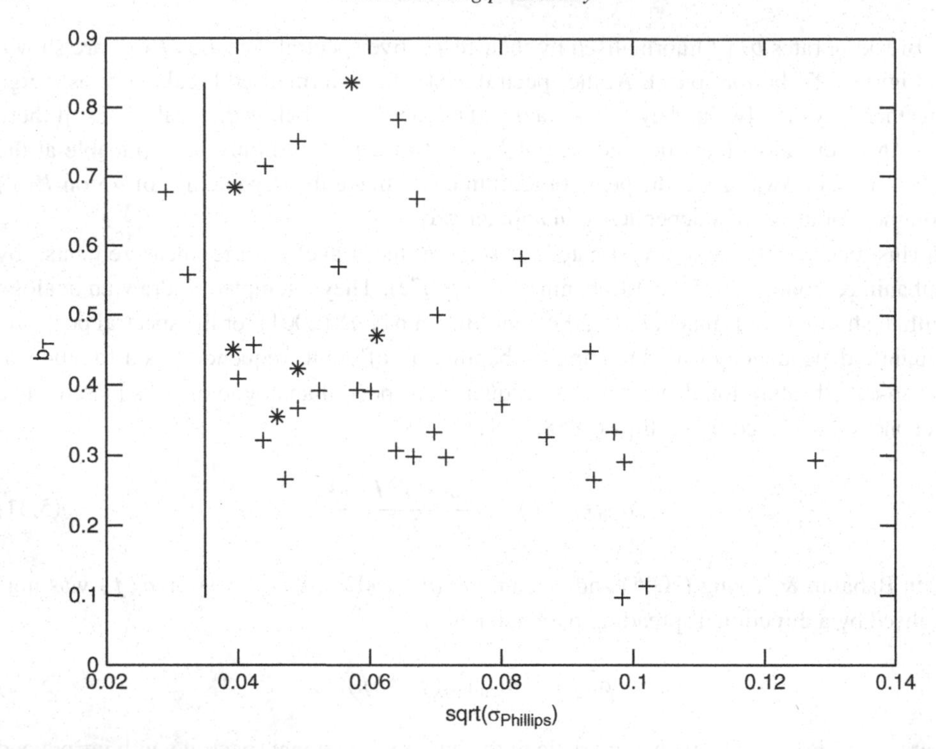

Figure 5.28 Breaking probability $b_T(f)$ (2.4) versus saturation parameter $\sqrt{\sigma(f)}$. Asterisks denote spectral-peak points. Threshold $\sigma = 0.035$ is shown with the solid line

In Figure 5.28, the revised plot of the breaking probability $b_T(f)$ (2.4) versus $\sqrt{\sigma(f)_{\text{Phillips}}}$ instead of $\sqrt{\sigma(f)}$ is shown; $\sqrt{\sigma_{\text{Phillips}}}$ rather than σ_{Phillips} is used to provide a qualitative analogy with Figures 5.22–5.25 where spectral steepness ϵ was employed. Asterisks denote the spectral-peak values, crosses are all the other data points from the frequency range (5.30). Based on such a figure, Babanin & Young (2005) and Babanin *et al.* (2007c) concluded that the saturation (5.31)–(5.32) is not the most suitable parameter for wave-breaking dependences. It is not possible to draw a general dependence through the data cloud in the figure with any degree of certainty. The saturation is the fifth moment of the spectrum, and any variations of the spectral shape, particularly at higher frequencies, cause large scatter of this characteristic.

As a threshold property, however, the saturation produced quite a robust value. When verified in the spectral model (Babanin *et al.*, 2007d, 2010c), the threshold was chosen as

$$\sqrt{\sigma(f)_{\text{threshold}}} = 0.035, \tag{5.36}$$

with only a few outliers below this value, which is also suitable as the revised threshold $\sqrt{\sigma(f)_{\text{Phillips}_{\text{threshold}}}}$ in Figure 5.28.

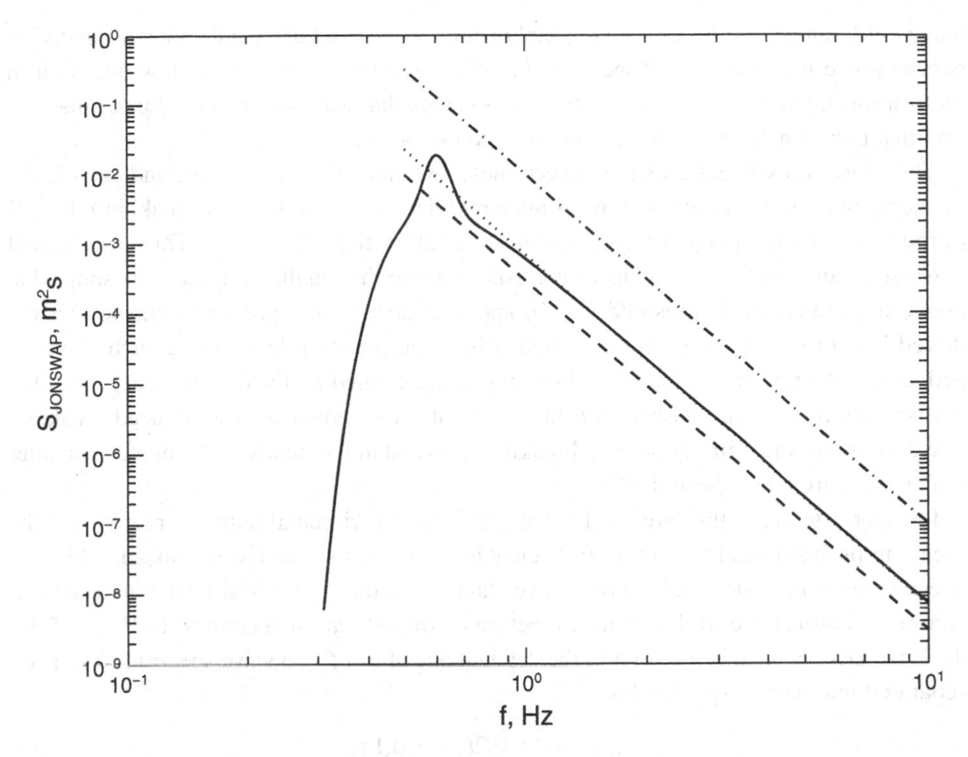

Figure 5.29 Illustration. JONSWAP spectrum (2.7) is shown with solid line (peak enhancement of $\gamma = 7$ is chosen for clarity), zero-breaking saturation level with dashed line, 5% breaking level with dotted line and 100% breaking level with dash-dotted line. The breaking limits are drawn qualitatively for illustration purposes only and must not be used for estimates. Figure is reproduced from Babanin & van der Westhuysen (2008) © American Meteorological Society. Reprinted with permission

Since a universal dimensionless saturation-threshold value $\sigma_{\text{threshold}}$ can be established, the dimensional threshold can then be obtained at every frequency:

$$F(f)_{\text{threshold}} = \frac{2g^2}{(2\pi)^4} \frac{\sigma(f)_{\text{threshold}}}{f^5}. \tag{5.37}$$

The meaning of the dimensional threshold is illustrated in Figure 5.29.

In this figure, JONSWAP-spectrum parameterisation (2.7, solid line) and arbitrary breaking limits are used for illustration purposes, but the real spectra and breaking threshold exhibit qualitatively similar behaviour (Babanin & Young, 2005; Babanin *et al.*, 2007c; Babanin & van der Westhuysen, 2008). The dashed line indicates the dimensional threshold (5.37), and the dash-dotted line is the imagined ultimate spectral limit in saturation terms – i.e. the spectral density cannot physically reach over this limit because the steepness of waves will be such that all the waves will be breaking (Babanin *et al.*, 2007c; Stiassnie, 2010). If the spectral density drops below the dashed level, there will be no breaking in the wave field, but unless the waves are swell or forced by very light winds, they normally do exhibit some breaking (e.g. Rogers *et al.*, 2003). The spectrum drops below this limit in its front face which signifies the obvious knowledge that waves, which are longer than the dominant ones in the wind–wave spectrum, do not break.

Therefore, wave spectra exist between these two lines. The dotted line indicates a 5% breaking rate. As is known, at typical moderate deep-water conditions breaking rates are of the order of a few percent (e.g. Babanin *et al.*, 2001, their Figure 13). Thus, in a typical wave spectrum, the spectral density corresponding to the small/short waves is somewhat above the dimensional threshold (5.37), approximately at the position indicated by the dotted line in Figure 5.29. This means that inherent breaking in trains of such waves is active, and the rate of such inherent breaking is determined by the degree of excess of the real spectrum above the dashed threshold level at each frequency. The induced breaking, however, alters such simply derived breaking rates and in fact tends to dominate at smaller scales away from the spectral peak.

Let us try to apply the threshold value (5.37) to experimental data in order to test the breaking probabilities in different frequency bins in dimensional terms. Indeed, while the overall scatter in Figure 5.28 is prohibitive, data points for the spectral-peak bin (asterisks) exhibit a reasonable correlation as a function of the saturation spectrum. In Figure 5.30, the Lake George data, in a search of the dependence of $b_T(f)$ on wave spectrum $F(f)$, are separated into narrow spectral bins

$$f_i = f_p + 0.2 i f_p \pm 0.1 f_p \tag{5.38}$$

with $i = 0, 1, \ldots 4$, i.e. starting from the spectral peak. Only records with breaking rates in excess of 2% across all the frequencies were chosen to avoid bias due to zero-breaking contributions when the rates are low. A riding wave removal (RWR) procedure (Schulz, 2009) was used to identify the periods of the breaking waves. The standard zero-crossing analysis becomes naturally noisier towards higher frequencies when the riding shorter waves may not necessarily cross the mean level. The RWR technique works, once the bubble detection signals a breaking, by finding the shortest riding waves first, and then removing all of them from the signal before re-processing the signal to look for the next scale of largest riding waves.

At the spectral peak (first panel), consistent with the two-phase behaviour of the breaking/dissipation discussed above, dependence in terms of the wave spectrum is linear:

$$b_T \approx 2(F - F_{\text{threshold}}). \tag{5.39}$$

If, however, this dependence, as shown with solid lines in subsequent subplots, is applied to the breaking rates at higher frequencies, it exhibits a progressively larger underestimation. Such a result is consistent with our expectations that follow from the documented cumulative behaviour. Inherent (linear) dependence of the wave-breaking rates on the spectrum excess should be present at each frequency. However, at every next frequency away from the spectrum peak, the contribution of the induced breaking (due to waves breaking at lower frequencies) has to become progressively larger.

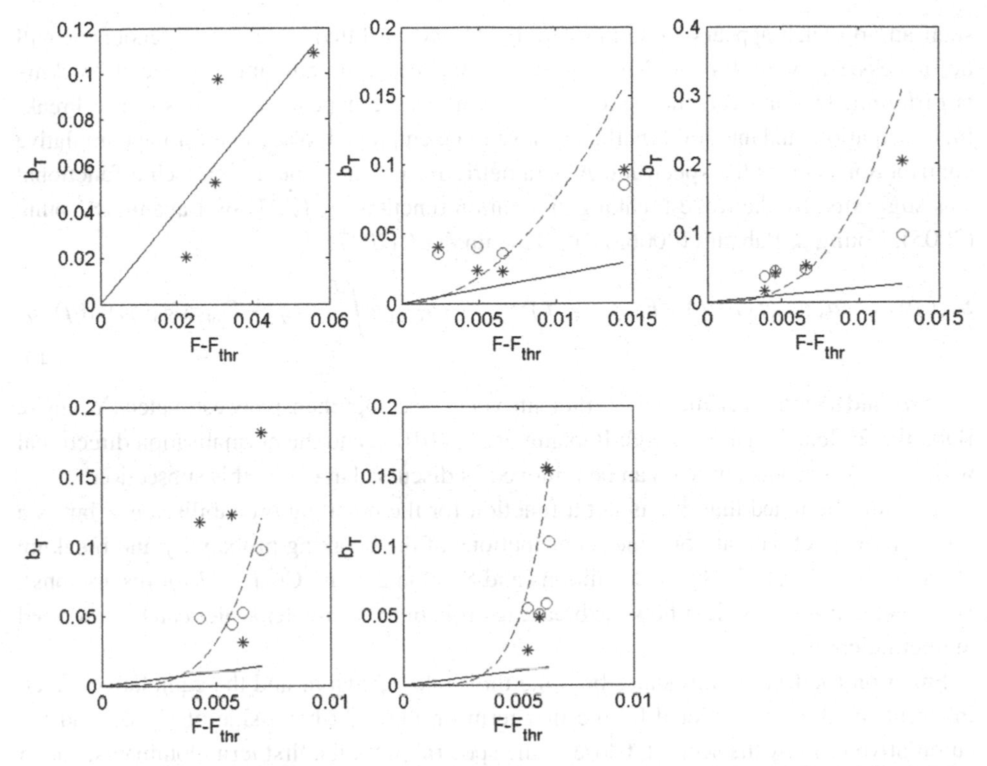

Figure 5.30 Breaking probabilities (from left to right) for frequencies of f_p, $1.2f_p$, $1.4f_p$, $1.6f_p$, and $1.8f_p$ in the $\pm 0.1f_p$ frequency range. Asterisks are experimental points. The solid line in all plots identifies the linear dependence obtained in the first panel. Dashed lines, from left to right, are $b_T \sim (F - F_{\text{threshold}})^2$, $b_T \sim (F - F_{\text{threshold}})^3$, $b_T \sim (F - F_{\text{threshold}})^4$, and $b_T \sim (F - F_{\text{threshold}})^5$. Circles in the higher-frequency bins are estimates of the breaking probability based on accounting the cumulative effect (5.41). Figure is reproduced from Babanin *et al.* (2007c) (public domain site http://www.waveworkshop.org/ sponsored by Environment Canada, the U.S. Army Engineer Research and Development Center's Coastal and Hydraulics Laboratory, and the WMO/IOC Joint Technical Commission for Oceanography and Marine Meteorology)

What happens if the cumulative effect is disregarded, as it is now in most breaking/dissipation parameterisations? As seen in Figure 5.30, it would still be possible to draw a linear dependence in each frequency bin, but at higher frequencies such dependence will become steeper and the intercept will move further from the origin (i.e. the threshold value will be growing). This is similar to what was done, for example, in Banner *et al.* (2002).

In the case of Figure 5.30, the universal threshold value has already been subtracted at the bottom scale, and therefore all the dependences have to go through zero. If we now try to fit a best power function at each frequency, this will result in a quadratic function at $f = 1.2f_p$, a cubic function at $f = 1.4f_p$, a fourth power at $f = 1.6f_p$, and a fifth power at $f = 1.8f_p$ as shown in the figure.

Thus, fitting some functions expressed in terms of local spectral density at each frequency can be done across the spectrum as a matter of tuning, but as a matter of physics such an approach appears to be misleading. Moreover, fitted once, the dependences will not necessarily work for another wave spectrum, where the amount of induced breaking is different. In our view, there are no simple algebraic dependences for spectral breaking/dissipation, and integral functionals have to be employed to account for the cumulative contributions across the spectrum. A parametric form that accounts for such a functional was suggested for the wave-breaking dissipation function S_{ds} (2.21) by Babanin & Young (2005), Young & Babanin (2006a) and Babanin *et al.* (2007c,d):

$$S_{ds}(f) = -a_1 \rho_w g f (F(f) - F_{\text{threshold}}(f)) A(f) - a_2 \rho_w g \int_{f_p}^{f} (F(q) - F_{\text{threshold}}(q)) A(q) dq. \tag{5.40}$$

Here a_1 and a_2 are experimental coefficients yet to be comprehensively estimated (Young & Babanin, 2006a; Tsagareli, 2009; Babanin *et al.*, 2010c), and the normalisation directional width $A(f)$ is redundant and can be removed as discussed above in this subsection.

It should be noted that S_{ds} is not a function for the breaking probability only, but is a dissipation function that combines contributions of the breaking probability and breaking severity (see (2.20)–(2.21) for definitions and Section 2.7 and Chapter 7 for discussions). With that in mind, it is clear how the breaking probability at smaller scales can be described mathematically.

Function (5.40) accommodates both the threshold behaviour and the cumulative effect. Inherent breaking is depicted by the first term on the right-hand side of (5.40), and the cumulative term by the second. Close to the spectral peak, the first term dominates, and at the spectral peak the inherent breaking is the only breaking mechanism as the integral in the second term is zero. Away from the peak, the integral grows and, while the inherent breaking may persist throughout, the cumulative term will rapidly increase and become the dominant cause of the breaking at higher frequencies. These higher-frequency breaking rates are in fact influenced by the induced breaking already as close to the peak as $1.2 f_p$; this can be seen in Figures 5.27 and 5.30.

Now, to prove that it is the cumulative effect that is working in Figure 5.30, at each higher-frequency bin of (5.38), the cumulation was estimated based on the spectral-peak wave-breaking approximation (5.39), i.e.

$$b_T(f_i) = 2 \sum_{i=0}^{N} (F(f_i) - F(f_i)_{\text{threshold}}). \tag{5.41}$$

Expression (5.41) implicitly assumes two things with respect to the higher frequency breaking: that is, an induced breaking at a high frequency occurs every time that a lower-frequency breaking takes place, and the inherent high-frequency breaking is driven by the same dependence as the spectral-peak breaking (5.39). The first assumption is true for the wave-breaking dissipation (Young & Babanin, 2006a), but as has been discussed immediately above such a dissipation is not exactly the same property as the breaking probability. For the breaking probability, the two assumptions need to be further verified, but we will use them here for the cumulation-feasibility check.

The (5.41)-estimated data points are shown in Figure 5.30 with circles. With the amount of uncertainty involved, they agree remarkably well with the measurements and with the local-in-frequency-space dependences which fit these measurements, and therefore provide a very strong corroboration of the cumulative effect, which in fact was estimated here in a quite simple way, defined by (5.41).

Other mathematical and physical expressions for the cumulative terms are possible of course (e.g. Donelan, 2001; Ardhuin *et al.*, 2010), but one way or another these breaking-probability functions/parameterisations have to include integral functionals, rather than be a function of spectral density or other distributed property local in frequency/wavenumber space. The cumulative term has a principal importance, particularly as the contribution of the inherent breaking becomes so small at shorter scales that it renders little connection between $b_T(f)$ and wave spectrum $F(f)$, to an extent that the inherent breaking can in fact be neglected.

The principal question in this regard is what is the relative frequency (with respect to the spectrum peak) at which the inherent-breaking and induced-breaking terms are equal? Then, even though both terms can exhibit their influence, below this frequency the inherent term will dominate, and above, the induced term will prevail. Once again we should remember that the question about breaking has not necessarily the same answer as the question about inherent dissipation and induced dissipation in (5.40) being equal.

There is not much direct quantitative guidance on this topic, apart from an experimental investigation by Filipot (2010, personal communication). He demonstrated that the breaking of short waves with frequencies in excess of $3f_p$ is locked in phase with the dominant-wave crests. Qualitatively, such an effect should theoretically be expected for wave breaking due to the modulated effects of the underlying long waves which make the short waves steeper close to the dominant crests (e.g. Longuet-Higgins & Stewart, 1960; Phillips, 1963; Donelan *et al.*, 2010).

Indirectly, discussion of this transition can be reduced to the question about transition from the f^{-4} to f^{-5} tail of the wave spectrum (see Section 8.2). Tsagareli (2009), Tsagareli *et al.* (2010) and Babanin *et al.* (2010c), based on the physical constraint of the total wind-to-wave momentum flux being equal to the integral of the wind-input spectral term $S_{in}(f)$ in (2.61), argued that such a transition must necessarily exist. The f^{-4} behaviour of the equilibrium interval is consistent both with observations (e.g. Donelan *et al.*, 1985) and theory (e.g. Pushkarev *et al.*, 2003). If, however, extended to high frequencies, Tsagareli *et al.* (2010) found that such a spectrum cannot satisfy the above-mentioned principal constraint.

Thus, the Phillips (1958) f^{-5} equilibrium interval has to be invoked at higher frequencies. The existence of such an interval has also been confirmed by experiments (i.e. JONSWAP, Hasselmann *et al.*, 1973), as well as the presence of both the subintervals in the single spectrum and the transition between them (e.g. Forristall, 1981; Evans & Kibblewhite, 1990; Kahma & Calkoen, 1992; Babanin & Soloviev, 1998b; Resio *et al.*, 2004). Because of the cumulative breaking behaviour, significant wave breaking is predicted at these frequencies and thus the original Phillips concept can be applied.

According to this concept, if wave breaking dominates the dynamics and thus defines the spectral shape at certain scales, f^{-5} behaviour of the spectrum should be expected.

In this regard, the transition from f^{-4} to f^{-5} can be interpreted as transition from inherent-breaking (nonlinear fluxes form the spectrum) to induced-breaking (breaking forms the spectrum) domination. In relative terms, typically such a transition would occur at

$$f_t \approx 3 f_p. \tag{5.42}$$

At the end of this subsection, we should mention that expressions/parameterisations of breaking rates $b_T(f)$ (2.4) or (5.28) are not the only way to characterise the breaking probability across the spectrum. Filipot *et al.* (2010), for example, provided a parameterisation of breaking wave height distributions (BWHD) within selected frequency (wavenumber) bins.

The target of Filipot *et al.* (2010) was the dissipation function $S_{ds}(f)$ in (2.61). Therefore, like everybody before them, they had to deal with the choice of the spectral bandwidth (2.5) for converting the physical wave count into spectral values. For convenience of comparisons with the experimental outcomes available for the spectral peak, they chose $\Delta f = 0.3$ as in (2.6) of Banner *et al.* (2000) and Babanin *et al.* (2001), smoothed by a Hann window.

This windowed band was then applied across the spectrum. It is quite broad, and, for example, in the parts of the spectra characterised by the f^{-4} or f^{-5} behaviour the wave height attributed to such bands will be biased towards wave heights at a lower-frequency end of a respective spectral bin. For Filipot *et al.* (2010), such a bias was undesirable since they expected the dissipation rates to be a nonlinear function of wave height. This is why the BWHD, rather than the bin-averaged breaking rates $b_T(f)$ (2.4) or (5.28), were chosen as a characteristic for the breaking probability.

Following Thornton & Guza (1983), they introduced a formulation for BWHD based on a Rayleigh distribution $p(H)$ (3.43) of non-breaking wave heights:

$$p_B(H) = p(H)W(H), \tag{5.43}$$

where the weighting function $W(H)$ is subject to choice, research and calibration. Filipot *et al.* (2010) targeted a universal deep-to-shallow-water function, which if converted into deep-water conditions, was chosen as

$$W(H, f_c) = a\left(\frac{k_r(f_c)H_r(f_c)}{\tilde{\beta}}\right)^2 \left(1 - \exp\left(-\left(\frac{kH}{\tilde{\beta}}\right)^p\right)\right). \tag{5.44}$$

Here, $a = 1.5$ and $p = 4$ are tuned parameters, $\tilde{\beta} = 0.25$–0.3 was found empirically, and k_r and $H_r = H_{\mathrm{rms}}$ are average wavenumber and wave height for the bins centered at the frequency f_c within the Hann window, respectively. In finite depths, the function is somewhat more complicated:

$$W(H, f_c) = a\left(\frac{\beta_r}{\tilde{\beta}}\right)^2 \left(1 - \exp\left(-\left(\frac{\beta}{\tilde{\beta}}\right)^p\right)\right). \tag{5.45}$$

Here, the depth-adjusted steepness property β is used such that its f_c-centered value is

$$\beta_r = \frac{k_r(f_c)H_r(f_c)}{\tanh(k_r(f_c)d)}. \tag{5.46}$$

Tuning, calibration and verification of the suggested BWHD was done on the basis of comparisons with data available in the literature and through direct validation by means of wave-breaking data from the Black Sea and Lake George field sites (see Sections 3.5, 3.7 and 5.2 about these field experiments and respective data). Both integral properties of BWHDs and the distributions themselves were used for this detailed analysis.

The integral of BWHD at different frequencies should be consistent with estimates of the breaking rates at those frequencies:

$$b_T(f_c) = \int_0^\infty p_B(H, f_c)dH. \tag{5.47}$$

In Figure 5.31, such available breaking rates denoted as $Q_{\text{obs}}(f)$ are plotted versus the modelled integral of (5.47).

In the figure, Lake George data processed by Filipot *et al.* (2010) and those adopted from Banner *et al.* (2000) and Babanin *et al.* (2001) are shown. The processed data cover records 5–8 and 10–11 of Table 5.2 and are subdivided into bins centered at $0.55 f_p$ (asterisks), f_p

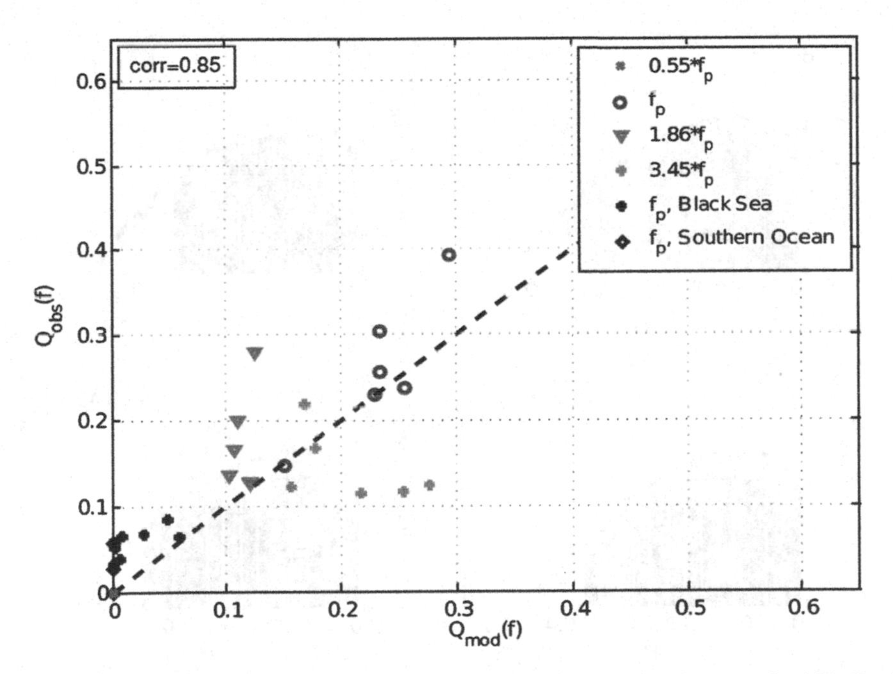

Figure 5.31 Observed versus modelled (according to (5.43)–(5.47)) breaking rates $b_T(f_c)$. Dashed line signifies perfect fit. The top four symbols in the legend denote results from the indicated wave scales for the records 5–8 and 10–11 of Table 5.2. The other two symbols represent the Black Sea and Southern Ocean data (see Banner *et al.*, 2000, and Section 5.2). Figure is reproduced from Filipot *et al.* (2010) by permission of American Geophysical Union

(circles), $1.86f_p$ (triangles) and $3.45f_p$ (pluses). Filled circles present the Black Sea data points, and diamonds Southern Ocean points – these points correspond to the spectral peak.

Overall agreement between the model and the data is very good, given the variety of data and differences between the breaking mechanisms responsible for the observed breaking probabilities. Within the individual frequency bins, however, some further sub-trends are visible. These sub-trends were not picked up by the BWHD model, and they may be due to different physical mechanisms in different frequency bands.

Indeed, below $3f_p$, regardless of how diverse the breaking rates are, for example, at the peak frequency for the Lake George and the deep-water records, the integrated BWHDs tend to underestimate the observed breaking rates, and at $3.45f_p$ these rates are overestimated. This transition corresponds to the transition from breaking being due to inherent reasons to the induced breaking as discussed above in this subsection.

Direct comparison of the (5.43)–(5.46) model with the data of record 5 (Table 5.2) is shown in Figure 5.32. In this figure, histograms of the wave height distributions (light colour) and the breaking wave height distributions (dark colour) are plotted and compared with the Rayleigh distribution and with the model, respectively.

For the model, two options based on two different sets of the tunable parameters are shown: the preferred option ($a = 1.5$ and $p = 4$) and one of the rejected options ($a = 1$ and

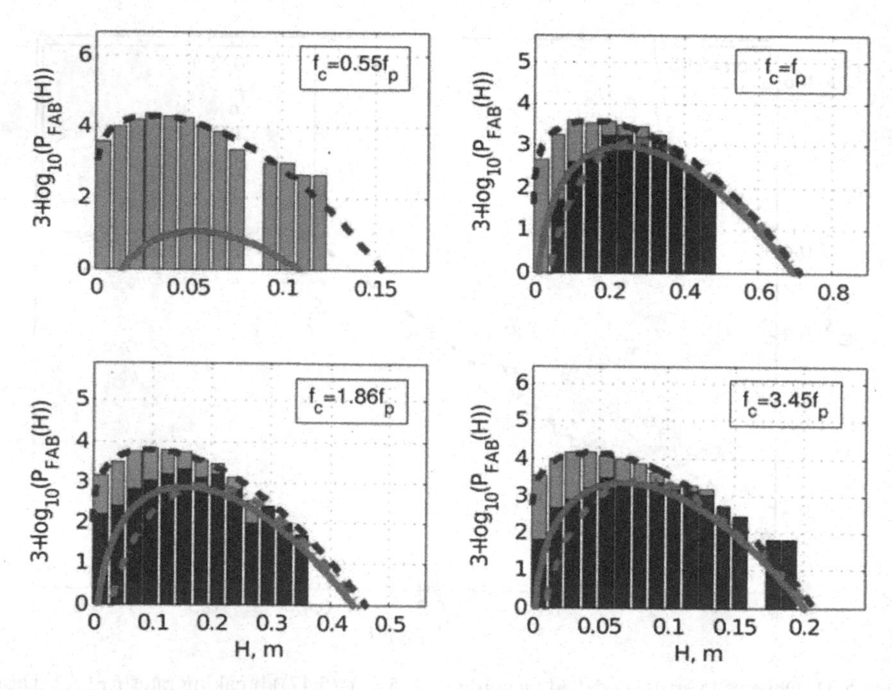

Figure 5.32 Wave height H distributions (light colour) and BWHD (dark colour) in the designated frequency bins for record 5 of Table 5.2. The distribution values are denoted as P_{FAB} and plotted in logarithmic scale. Rayleigh distribution is shown by dark dashed line and BWHD (5.43)–(5.46) with light dashed line. The solid line indicates modelled BWHD with tunable parameters $a = 1$ and $p = 2$. Figure is reproduced from Filipot *et al.* (2010) by permission of American Geophysical Union

$p = 2$). While both the options reproduce the observations well, when there is a substantial amount of breaking, the second option clearly fails the no-breaking case of $f_c = 0.55 f_p$ in the top left panel.

Thus, Filipot *et al.* (2010) introduced an alternative method for quantifying breaking probabilities across the spectrum, which does not rely on the empirical breaking-threshold value, but employs other empirical parameters. The method was shown to work both for deep-water and finite-depth waves and to approximate the average probability distribution of breaking-wave height reasonably well both for inherent-breaking and induced-breaking scales in the wave spectrum. The latter gives it an advantage compared to the probability formulations which have to rely on the cumulative integral as in (5.40)–(5.41), but there are indications that, with respect to the average distribution at each scale, BWHD may still vary in a different way depending on whether the nature of the breaking is spontaneous or induced (i.e. Figure 5.31).

5.3.3 Breaking in directional wave fields

In this subsection, the effects of directionality on breaking occurrence and probability are discussed. The main question to answer is whether the modulational instability, which is often regarded as a two-dimensional phenomenon, is still active in directional wave fields. The other question is whether it is more frequent than the linear-focusing, including directional-focusing breaking. The possibility of modulational instability is discussed in terms of the directional spectrum and the directional modulational index defined in this subsection. Lower limits for instability existence in these terms are identified and parameterisation of the separation of the two breaking types is proposed.

Waves on the ocean surface are directional, i.e. three-dimensional, and this is their principal feature. Apart from the case of pure swell, which appears unidirectional and even monochromatic but due to its low steepness has little relevance as far as wave breaking is concerned, the oceanic wind-generated and wind-forced waves are characterised not only by continuous distribution of their energy along temporal/spatial (or frequency/wavenumber) scales, but also along directions θ (5.33)–(5.34). Investigations of the directional wave fields present apparent technical difficulties both experimentally (see Mitsuyasu *et al.*, 1975; Hasselmann *et al.*, 1980; Donelan *et al.*, 1985; Babanin & Soloviev, 1987, 1998b) and numerically (e.g. Ducrozet *et al.*, 2010).

This is obviously true with respect to all other wave-related scale-distributed properties and characteristics, including the wave-breaking probability, severity and the dissipation term. These functions are spectra, i.e. they describe the distribution of the respective properties along frequencies/wavenumbers, and should in principle describe their distribution along directions too. That is, the breaking probability in (2.4), technically speaking, should be:

$$b_T(f,\theta) = \frac{n(f,\theta)T}{t} = \frac{n(f,\theta)}{N(f,\theta)} \tag{5.48}$$

and the dissipation in (2.20)

$$D(f, \theta) = b_T(f, \theta)E_s(f, \theta). \tag{5.49}$$

Here, $n(f, \theta)$, for example, is the number of breaking crests in the frequency bin $f \pm \Delta f$ and directional bin of $\theta \pm \Delta\theta$.

Little is known about these breaking-related directional distributions, however, to such an extent that the dissipation function $S_{ds}(f, \theta)$ (2.25) in wave forecast models is routinely treated as an isotropic function. It is hard to imagine an isotropic energy sink; however, where all the other acting spectral sources have quite defined directional distributions, there is experimental evidence that the dissipation term is also anisotropic and perhaps even bimodal (see Section 7.3.6 below).

There are two issues as far as wave breaking and directional properties of wave fields are concerned. The first one relates to the modulational instability in the three-dimensional wave fields, and the second one to linear directional focusing.

Indeed, it is a known experimental and theoretical fact that the modulational-instability mechanism experiences limitations in broadband, and particularly in three-dimensional fields (e.g. Brown & Jensen, 2001; Onorato *et al.*, 2002, 2009a,b; Cherneva *et al.*, 2009; Waseda *et al.*, 2009a,b). Since field waves are spectral and directional, this issue has to be considered seriously when talking about wave-breaking issues with respect to the real oceanic waves.

Brown & Jensen (2001) studied the focusing of unidirectional waves and found that the Benjamin–Feir instability is impaired in focusing (i.e. spectral) wave trains. Such a study needs to be extended into spectra typical of field waves. There is a reasonable expectation, however, that the modulational mechanism may work for reasonably steep waves with a narrow spectrum (e.g. Waseda *et al.*, 2009a). For unidirectional spectral waves, Alber (1978) derived a requirement that can be expressed as

$$M_I > 1 \tag{5.50}$$

(see (5.1) for the definition of M_I), and this condition can be satisfied for spectra of young wind–waves (e.g. Onorato *et al.*, 2001).

There is no analogue of M_I and condition (5.50) available for three-dimensional characteristics of the modulational-instability mechanism. In Onorato *et al.* (2002), directional effects were investigated and quantitative criterion β was obtained in terms of the width of directional spectrum $D(\theta)$ where θ is angle:

$$D(\theta) = \cos^2\left(\frac{\pi}{2\beta}\theta\right) \tag{5.51}$$

– i.e. if the directional width is greater than $\beta = 15$, then the modulational instability appears to be suppressed. There was a clerical error in Onorato *et al.* (2002), and the value of β has to be actually multiplied by $\pi/180$, that is the criterion is

$$\beta \approx 0.26 \tag{5.52}$$

(Onorato, 2007, personal communication). Since the Onorato *et al.* (2002) model is weakly nonlinear rather than fully nonlinear, the criterion should only be regarded as an approximation, but we will use it as a reference point here.

To compare the width of the (5.52) spectrum with observations, integral value A (5.33) was estimated:

$$A^{-1} = \int_{-\beta}^{\beta} D(\theta) d\theta \tag{5.53}$$

which was used in the field study of Babanin & Soloviev (1998b) to measure directional distributions (the higher A, the narrower is the spectrum). For $\beta = 0.26$,

$$A = 3.8 \tag{5.54}$$

which is well above the experimentally observed values of typical directional spectra of wind-generated waves (Babanin & Soloviev, 1987, 1998b). It should be mentioned that this theoretical estimate is in excellent agreement with the laboratory measurement of Waseda *et al.* (2009a) who concluded that such a critical directional spread for modulational instability in a wave system with typical JONSWAP spectrum (2.7) is

$$A \approx 4. \tag{5.55}$$

When measuring directional spectra in various field conditions, Babanin & Soloviev (1987, 1998b) demonstrated that, typically in field conditions $A \sim 1$, and for dominant waves it can reach up to values of $A \approx 1.8$. Both Babanin & Soloviev (1987, 1998b) and Waseda *et al.* (2009a) measured directional spectra by means of wave arrays, but different methods of data processing were used: wavelet directional method (WDM, Donelan *et al.*, 1985) in Waseda *et al.* (2009a) and maximum likelihood method (MLM, Capon, 1969) in Babanin & Soloviev (1987, 1998b). Therefore, before conclusions are drawn, comparisons of the two methods need to be made.

This is conducted in Figure 5.33. In this figure, data from a new experiment in the directional ocean engineering tank of the University of Tokyo are shown (Babanin *et al.*, 2011a). Apart from recording the waves by a directional array, the main observational task was to distinguish the breaking between those caused by linear directional focusing and by modulational instability. This was done visually, and visual observations were verified and confirmed by other direct and indirect means, including measuring kurtosis (7.16) of the surface elevations and growth of the instability sidebands. In the figure, circles correspond to the modulational instability and diamonds to the linear focusing. As one could expect, there were records when both linear focusing and modulational instability were present. These would be either co-existing, or replacing each other's patterns. Such records were labelled transitional and in the figure they are designated with stars.

For the ocean engineering tank directional records, values of A were obtained both by means of WDM (A_{WDM}) and MLM (A_{MLM}), and as seen in the figure the former is on average three times larger. This means that, based on the same directional-array input data,

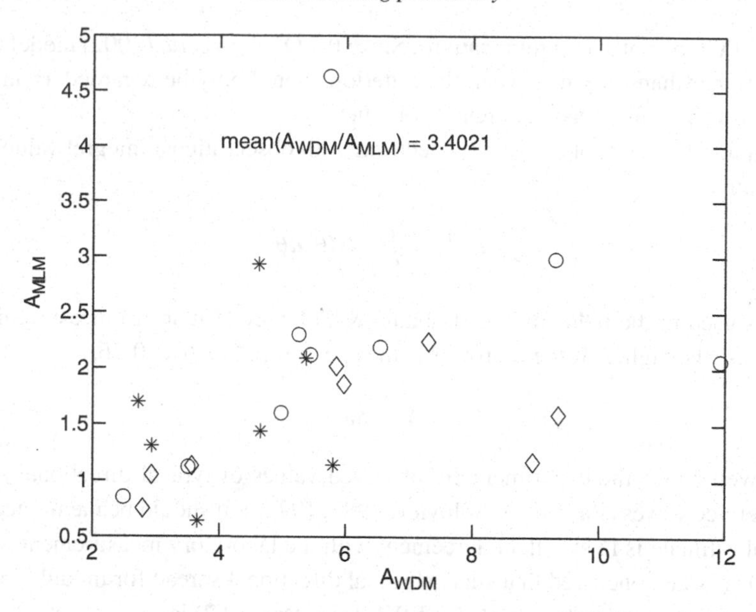

Figure 5.33 Comparison of A_{MLM} and A_{WDM} for the ocean engineering tank records. Circles correspond to modulational instability, diamonds to linear focusing, stars to transitional cases. Figure is reproduced from Babanin *et al.* (2011a)

WDM indicates much narrower spectra. While the WDM estimates should be more accurate and have other advantages (e.g. Waseda *et al.*, 2009a; Young, 2010), here we will use the MLM estimates in order to be able to compare new results with the field-observed directional spectra of Babanin & Soloviev (1987, 1998b) which were done by means of MLM.

An important implication of the difference observed in Figure 5.33 is immediately obvious if we boldly divide criterion (5.55) of Waseda *et al.* (2009a) obtained with the use of WDM, by three: then the transition from no visible modulational instability to detected modulational instability happens at $A \approx 1.3$ which, according to Babanin & Soloviev (1987, 1998b) falls right into the range of directional spreads typical for dominant waves.

In Babanin *et al.* (2011a), this limit was further investigated on the basis of the directional data of the experiment. It was shown that, for mean wave steepnesses typical of those observed in the field, the instability exists at as low values of A as

$$A \sim 0.8, \tag{5.56}$$

which are quite realistic directional widths in field conditions; in fact quite broad by any standards, certainly at the spectral peak where the modulational instability is expected to work.

Thus, modulational instability may still be found to be applicable, at least for the dominant waves if they are steep enough. It is not unreasonable to expect a directional condition analogous to (5.50) to be relevant. Parameter A (5.53) can be used for this purpose as it has the proper physical meaning of the inverse relative width of the directional spectrum whose peak is normalised to be 1.

At the spectral peak, a relative steepness (as the wave spectrum develops) is defined by $\sqrt{\gamma}$ where γ is the peak enhancement of the JONSWAP spectrum (2.7). That is, for the peak, we can define a directional analogue of M_I as

$$M_{Id} = A\sqrt{\gamma}. \tag{5.57}$$

Now, it is informative to look at how this index evolves over the wave development. From (eq. 19) of Babanin & Soloviev (1998b), at the spectral peak

$$A = 1.12\left(\frac{U_{10}}{c_p}\right)^{-0.50} + (2\pi)^{-1}, \tag{5.58}$$

and from (eq. 44) of Babanin & Soloviev (1998a)

$$\gamma = \frac{7.6}{2\pi}\frac{U_{10}}{c_p}, \tag{5.59}$$

that is

$$\sqrt{\gamma} = 1.10\left(\frac{U_{10}}{c_p}\right)^{0.50}. \tag{5.60}$$

Therefore,

$$M_{Id} = 1.23 + \frac{1}{2\pi}\left(\frac{U_{10}}{c_p}\right)^{0.50} \tag{5.61}$$

is a weak function of the wind forcing, and its value at the spectral peak varies from 1.40 to 1.79 for $\frac{U_{10}}{c_p}$ in the range from 0.89 to 10 where $\frac{U_{10}}{c_p} = 0.89$ signifies the limit of full development (Pierson & Moskowitz, 1964).

Now, if the M_{Id} assumption is valid and the critical value for this index is in the range of $M_{Id} = 1.4$–1.8, the 'de-focusing' effect of directionality can be overcome by a stronger nonlinearity if waves grow steeper. It is worth noting here that the directional spectra broaden towards frequencies above the peak (e.g. Babanin & Soloviev, 1998b). This means that, even if applicable at the peak, the directional modulational instability may not be working at higher frequencies and some other causes of breaking and dissipation will have to be found in that spectral band. In this regard, two-phase behaviour of breaking has indeed been observed in field experiments of Babanin & Young (2005), Manasseh *et al.* (2006) and Babanin *et al.* (2007c) (see Sections 5.3.2 and 7.3.4 for a detailed discussion) – i.e. the direct dependence of breaking on spectral density at the peak and an induced breaking/dissipation at higher frequencies.

If the steepness of waves is known, then the directional modulational index (5.57) can be formulated explicitly:

$$M_{Id} = A \cdot ak. \tag{5.62}$$

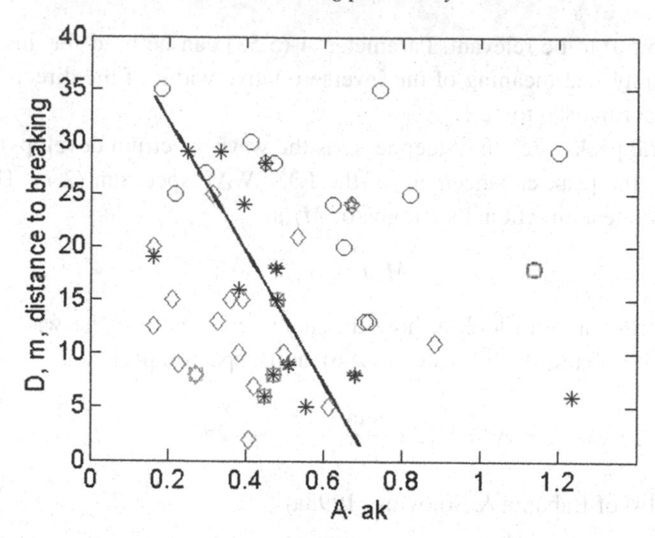

Figure 5.34 Distance to breaking D versus $M_{Id} = A \cdot ak$. Solid line is dependence (5.64). Symbols as in Figure 5.33, for seeded modulation with $M_I = 1$. Additionally, squares signify $M_I = 2$, pentagrams $M_I = 0.5$. Figure is reproduced from Babanin *et al.* (2011a)

Babanin *et al.* (2011a) found that the limit for existence of the modulational instability in terms of M_{Id} (5.62) is

$$A \cdot ak \approx 0.18 \tag{5.63}$$

which is again absolutely feasible. With $A \sim 1$, there should be $ak \sim 0.2$ which is possible in the field, and for $A \sim 1.8$, there should be $ak \sim 0.11$ which is a typical steepness of ocean waves.

Babanin *et al.* (2011a) further provided parametric means to separate directional-focusing breaking from modulational-instability breaking. Figure 5.34 verifies the dependence of distance to breaking D on $A \cdot ak$. In the figure, the instability breaking (circles) and linear-superposition breaking (diamonds) appear reasonably split. It is interesting that the dividing line, even visually, goes through the asterisks that were independently identified as transitional cases. The separation line was parameterised as the fit to asterisk data points:

$$D = -62A \cdot ak + 68. \tag{5.64}$$

In this figure the same symbols as in Figure 5.33 are used. In Babanin *et al.* (2011a), the modulation was seeded, typically with the modulational index $M_I = 1$. For comparison, a few cases of $M_I = 2$ (squares) and $M_I = 0.5$ (pentagrams) were also considered. These cases do not exhibit any essential trend, being basically in the middle of the data cloud. Marginally, perhaps, higher M_I cases (longer groups) lead to transitional points being under the curve (i.e. in the focusing-breaking area), and $M_I = 0.5$ to the transitional points being in the modulational-breaking area – but this needs further statistically significant verification.

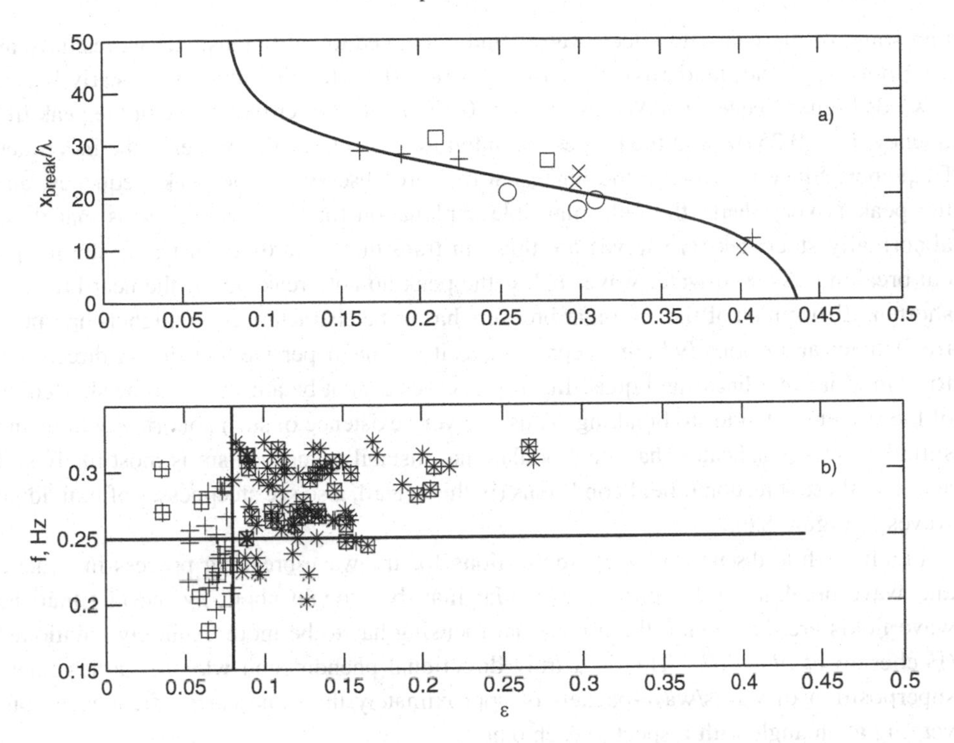

Figure 5.35 a) As in Figure 5.17 (the filled-circle wind-forced data points are omitted). b) Black Sea. As in Figure 2.3. Figure is reproduced from Babanin *et al.* (2010a) by permission

That is, with caution we will try to apply our modulational-instability results to the field data. Another problem, of the technical kind, still prevents direct comparisons of breaking rates obtained by means of (5.10) and field observations. Relationship (5.10) predicts the probability of incipient breaking, whereas in the field it is impossible to directly measure whether a wave is an incipient breaker or not. What is usually measured are quantities which result from the breaking process whose probability is significantly higher than the probability of breaking onset (e.g. Liu & Babanin, 2004, see also Sections 2.4 and 5.1.4).

Qualitative comparisons of laboratory and field breaking-probability dependences were made in Babanin *et al.* (2007a) and featured well. Here, the Babanin *et al.* (2007a) dependence, already shown in Figure 5.17, is reproduced in the top panel of Figure 5.35 for comparison with the bottom panel which is the reproduced Figure 2.3. In this bottom panel, we plot the frequency (inverse period) of individual dominant waves (from frequency range of $f = f_p \pm 0.3 f_p$) versus the steepness of these individual waves. This is done for a Black Sea record with $f_p = 0.25$ Hz (Rec. 244 of Table 5.1) used by Babanin *et al.* (2001) to obtain field breaking rates in the same frequency band.

If there were no shrinking of the wavelength prior to breaking, as described in Section 5.1.1, at each steepness the distribution of the frequencies around $f_p = 0.25$ Hz would be approximately even. This is so for waves of $\epsilon <\approx 0.12$. For steeper waves, and some of these detected deep-water wind-generated three-dimensional waves have an enormous (by field standards) steepness up to $\epsilon = 0.27$, the distribution is clearly biased towards higher frequencies. Waves with $\epsilon > 0.17$ are all shorter than those of the peak frequency $f_p = 0.25$ Hz, and the higher the individual steepness, the higher is the individual frequency. Since on average the highest waves are observed at the peak frequency, and the peak is very sharp, the only plausible explanation for this observation is that these abnormally steep, but rare waves are those in transition towards or just after the incipient breaking. As deep-water waves below the peak do not break, and if the near-breakers shorten, distribution of the incipient breakers has to be characterised by higher-than-peak frequencies at abnormally high steepnesses, as it is. The dispersive focusing or directional focusing, or other linear and quasi-linear processes cannot be attributed to the shortening of the wavelength prior to breaking. Thus, the very existence of such abnormally high and shrunken waves indicates that the modulational instability mechanism is most likely still active in these directional field conditions (in this regard, also see steepnesses of individual waves in Figure 8.6).

Finally, while discussing what implications for the wave-breaking process in general and wave-breaking probability in particular may be brought about by the fact that the wave fields are directional, the directional focusing has to be more explicitly mentioned (Fochesato *et al.*, 2007). This is a truly directional phenomenon which signifies linear superposition of waves/wave-packets of approximately the same carrier frequency converging at an angle with respect to each other.

If the carrier waves are steep enough, the superposition can lead to a very high crest (i.e. double the height of the waves involved in the superposition, in which case nonlinear effects may start playing a role (e.g. Brown & Jensen, 2001)), reaching the steepness limit (2.47) and ultimately breaking. Depending on the angle and the length of the converging crests, the breaking can have different severities and spatial/temporal extents.

How frequent is such breaking? In order to achieve the limit (2.47), there has to be a superposition of either multiple wave crests, which should be quite a rare event, or, for example, superposition of only two crests with half-the-limit steepness of $Hk/4 = \epsilon_{\text{limiting}}/2 = 0.22$; such crests are themselves quite rare events to begin with (see Figures 5.35 and 8.6).

In the experiments by Babanin *et al.* (2011a), dedicated specifically to separating modulational-instability and directional-focusing breaking, where such initially steep crests were intentionally mechanically produced, it was noticed that the linear focusing only exists at average steepness $ak > 0.24$. This signifies superposition of two waves leading to the breaking at limiting steepness (2.47). Thus, superposition of three waves is unlikely, at least it did not happen in the course of the records which encompassed some 500–700 waves in the experiment, whereas breaking did happen. This is an important observation, since in typical conditions, with much less steep waves and dispersive rather than directional focusing, i.e. focusing of waves of different lengths and therefore heights, superpositions of even greater numbers of waves would be required. Such an observation indicates that the probability of linear superposition, which would lead to the limiting steepness (2.47) due to directional convergence of crests, is low. At the other extreme, in the laboratory experiments of Babanin *et al.* (2011a), the directional focusing did not happen for very narrow (i.e. near unidirectional) spectra of $A > 2.25$ ($A > 9.4$ for WDM-estimated spectra).

Therefore, the breaking due to directional focusing and other types of focusing on that matter could not be expected as a frequent occurrence in realistic wave fields and thus linear focusing is hardly expected to be the main cause of wave breaking in directional fields. Indirectly, this conjecture is also supported by the wavelet directional method of Donelan *et al.* (1996). The main assumption of WDM is that at any given time there is only one wavelet of a particular frequency present at the measurement point. If not, the WDM reading fails at that particular instant. The level of the noise in the directional spectra produced by WDM would be an indicator of how often wavelets coming from different directions superpose. The answer is – not that often. Noise in the WDM-estimated field directional spectra is remarkably low (i.e. Donelan *et al.*, 1996; Young, 2010).

On the other hand, the other possible cause of wave breaking, modulational instability, is likely to be active in wave fields with typical directional-spread/wave-steepness properties as discussed above. It has to be emphasised again that this fact does not cancel the possibility of linear or amplitude focusing of course, but places instability breaking as a likely more-frequent cause of wave-breaking onset in a directional wave field.

5.3.4 Wind-forcing effects, and breaking threshold in terms of wind speed

If the wind forcing is superimposed, it can play multiple roles in affecting wave-breaking probability. We would like to start, however, not from these roles, but from another wind-related effect principal to wave breaking: breaking threshold in terms of the wind speed (see also Section 3.1). That is, similarly to the threshold of the breaking occurrence imposed by the wave-steepness/spectral-density (Section 5.2), there is a wind-speed threshold below which no breaking will happen in a wave field.

As has been discussed starting from Chapter 1, dissipation is an important and inevitable balance holder in the wave system, but in order to start breaking the waves have to grow beyond some average steepness. It is known that not every wind forcing can provide such conditions (e.g. Donelan, 1978). In Babanin *et al.* (2005), this threshold was investigated quantitatively for the breaking of dominant waves by simultaneously measuring the volumetric energy-dissipation rate in the water column and the energy flux from the wind to the waves in the air.

The volumetric rate of total turbulent kinetic energy dissipation ϵ_{dis} can be obtained from the Kolmogorov inertial subrange of the velocity spectrum in water (e.g. Terray *et al.*, 1996; Veron & Melville, 1999). If the velocity spectrum $V(f)$ exhibits a

$$V(f) \sim f^{-\frac{5}{3}} \tag{5.65}$$

Kolmogorov interval, the level of this interval depends on the dissipation ϵ_{dis}:

$$V(f) = \frac{7}{110} 2^{\frac{4}{3}} \Gamma\left(\frac{1}{3}\right) \left(\frac{8\epsilon_{dis}}{9\alpha} \frac{u_{rms}^{orb}}{2\pi}\right)^{2/3} f^{-\frac{5}{3}} \tag{5.66}$$

where $u_{\text{rms}}^{\text{orb}}$ is the rms orbital velocity and $\alpha \approx 0.4$ is Heisenberg's constant (Veron & Melville, 1999). The larger the dissipation rate ϵ_{dis}, the higher will be the Kolmogorov interval of the spectrum $V(f)$.

Term ϵ_{dis} is the dissipation rate per unit of volume, and to obtain the total dissipation in the water column per unit of area D_a, one needs to integrate $\epsilon_{\text{dis}}(z)$ over the water depth z from the surface $z = 0$ to the bottom $z = d$:

$$D_a = \int_0^d \epsilon_{\text{dis}}(z) dz. \tag{5.67}$$

To perform the integration, either continuous measurements of the $\epsilon_{\text{dis}}(z)$ profile or its parameterisation as a function of depth are required. Knowledge of the parameterisation is obviously preferable as it enables estimation of the total dissipation on the basis of a single-depth measurement of the spectrum (5.66).

There is, however, no general agreement on the parameterisation of the vertical dissipation distribution $\epsilon_{\text{dis}}(z)$. In the classical theory of the boundary layer over a solid wall, ϵ_{dis} is a simple inverse function of distance z to the wall:

$$\epsilon_{\text{dis}}(z) \sim z^{-1} \tag{5.68}$$

(see e.g. Landau & Lifshitz, 1987). Early measurements in the boundary layer beneath the wavy surface found the ϵ_{dis}-depth distribution to be consistent with this wall-layer theory (Arsenyev *et al.*, 1975; Dillon *et al.*, 1981; Oakey & Elliott, 1982; Jones, 1985; Soloviev *et al.*, 1988). More recently, however, both by direct and indirect means it was shown that, at least at strong wind forcing, the dissipation ϵ_{dis} close to the water surface may exceed the wall-layer values by up to two orders of magnitude (Agrawal *et al.*, 1992; Melville, 1994; Drennan *et al.*, 1996; Terray *et al.*, 1996). Terray *et al.* (1996) and Drennan *et al.* (1996) parameterised the vertical dissipation profile as

$$\epsilon_{\text{dis}}(z) = \begin{cases} \text{const} & z < H, \\ \sim z^{-2} & z \geq H. \end{cases} \tag{5.69}$$

Based on considerations of the expected total wind input which should match the total dissipation, it was found that H approximately scales with significant wave height H_s as

$$H = 0.6 H_s \tag{5.70}$$

(Terray *et al.*, 1996; Drennan *et al.*, 1996). Later refined studies of Soloviev & Lukas (2003) and Gemmrich & Farmer (2004) confirmed the existence of enhanced near-surface turbulence due to breaking, but pointed out that the scaling H of the constant-dissipation level is still an issue.

At Lake George (see Section 3.5 about the Lake George experiment), turbulence spectra $V(f)$ were measured by acoustic Doppler velocimeters (ADV) as described by Young *et al.* (2005) in greater detail. Under reasonably strong wind forcing, such spectra exhibited distinct Kolmogorov intervals as shown in Figure 5.36. A vertical profile of these turbulence spectra is plotted in the figure. The ADV was traversed down from the surface in

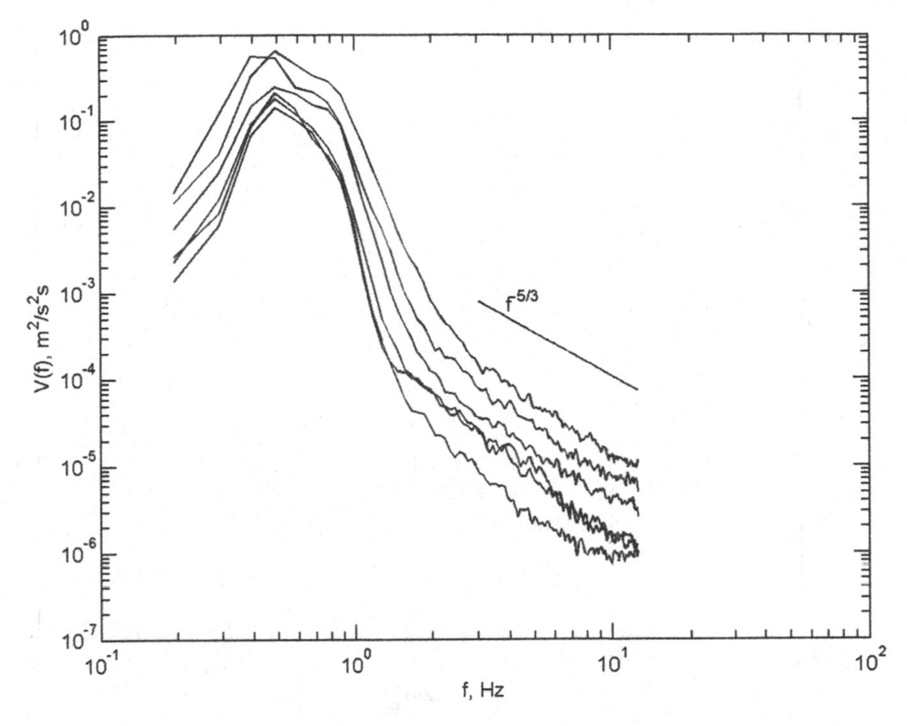

Figure 5.36 Velocity spectrum $V(f)$ measured at 10 cm, 20 cm, 30 cm, 40 cm, 50 cm and 60 cm distances from the surface for a 9.7 m/s mean wind speed (the more energetic spectra are closer to the surface). The Kolmogorov interval slope of $f^{-5/3}$ is shown with a straight line. Figure is reproduced from Young & Babanin (2006a) © American Meteorological Society. Reprinted with permission

10 cm increments and the six 20 min-averaged spectra shown in Figure 5.36 were recorded at 10 cm, 20 cm, 30 cm, 40 cm, 50 cm and 60 cm from the mean water level, respectively. The wind was steady in speed and direction over the two-hour time period of measuring the profile, $U_{10} = 9.7\,\text{m/s}$ on average, with a maximum of 10.9 m/s and a minimum of 8.3 m/s. The more energetic spectra shown in Figure 5.36 were recorded closer to the surface, with the energy level decaying with depth.

Dissipation rates ϵ_{dis} obtained on the basis of such spectra using (5.66) are shown in Figure 5.37. This figure plots ϵ_{dis} as a function of z in dimensionless form

$$\frac{\epsilon_{dis}\kappa z}{u_{*w}^3} = \text{function}\left(\frac{gz}{u_{*w}^2}\right) \tag{5.71}$$

where u_{*_w} is the friction velocity in the water, $\kappa \approx 0.4$ is the von Karman constant, and the mean water level is now treated as the wall. Wall-law scaling is applied to the dissipation ϵ_{dis} and height-of-fully-developed-waves scaling is applied to the distance to the surface z (see Agrawal *et al.*, 1992; Melville, 1994). For boundary layers over solid walls, such scaling of the dissipation would give values of

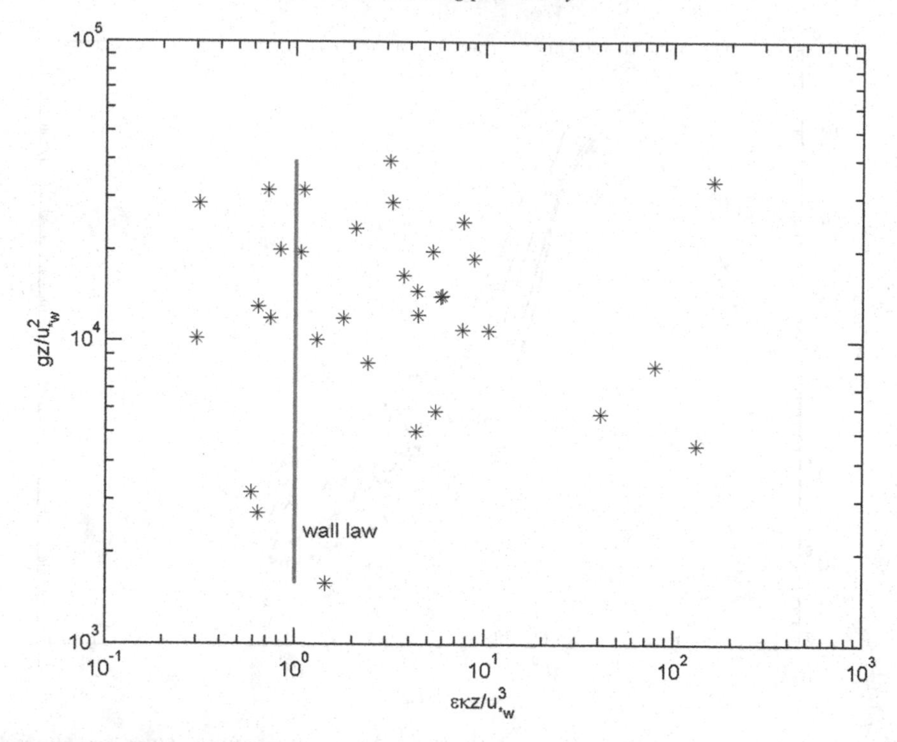

Figure 5.37 Dissipation ϵ_{dis} (denoted as ϵ in this figure) versus distance to the surface z in wall-layer coordinates. The vertical line represents the dissipation level in the boundary layer over a solid wall. Figure is reproduced from Young & Babanin (2006a) © American Meteorological Society. Reprinted with permission

$$\frac{\epsilon_{\text{dis}} \kappa z}{u_{*w}^3} = 1 \tag{5.72}$$

shown by the vertical line in the figure. The enhancement of the dissipation rates compared to the wall layer is obvious. Maximal values of the enhancement are up to 200 times greater than the wall-layer magnitude (even greater than those in Agrawal *et al.* (1992) where they were up to 70 times).

Dissipation of the energy of wind-generated waves in the finite-depth water column, that is in the Lake George conditions, consists of two parts: turbulent kinetic energy dissipation D_a (5.67), which would occur in the deep water as well as in finite depths, and dissipation per unit area due to bottom friction B_a. The total dissipation T_a

$$T_a = D_a + B_a \approx I_a \tag{5.73}$$

should match the wind energy input per unit area I_a for steady wind–wave conditions. Technically speaking, some of the wave energy dissipated from the waves does not necessarily convert into turbulence and may be spent on, for example, work against buoyancy forces when the bubbles are entrained into the water in the course of wave breaking. Such

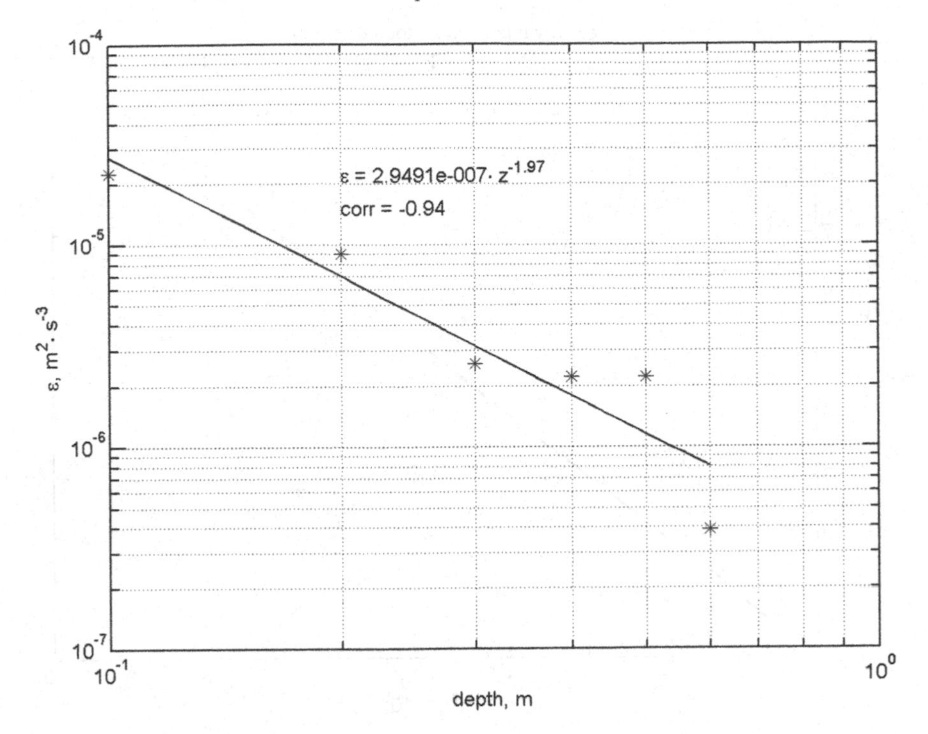

Figure 5.38 Dimensional profile of dissipation ϵ_{dis} (denoted as ϵ in this figure) versus distance to the surface z, obtained from the turbulence spectra shown in Figure 5.36. The line of best fit and its correlation coefficient are shown. Figure is reproduced from Young & Babanin (2006a) © American Meteorological Society. Reprinted with permission

an amount, however, is never greater than 50%, usually much less (Melville *et al.*, 1992), and therefore a reasonable match indicated by (5.73) has to be expected.

In order to verify which of the parameterisations (5.68) or (5.69) is to be used for the integration (5.67) in the general case, Babanin *et al.* (2005) estimated and compared B_a, D_a and I_a. For that, a set of records for which the wind input I_a was directly recorded (Donelan *et al.*, 2006) was chosen for the analysis. Bottom friction B_a was measured by means of physical modelling of the respective wave conditions in a laboratory flume with the bottom covered with Lake George mud (Babanin *et al.*, 2005), and the ADV-obtained ϵ_{dis} were integrated over the depth using both (5.68) and (5.69).

The vertical profile of ϵ_{dis} for the Figure 5.36 spectra is shown in Figure 5.38. The profile is very close to quadratic and therefore parameterisation (5.69) was first used to estimate the dissipation D_a.

The corresponding total dissipation T_a (5.73) is plotted versus the total wind input I_a in Figure 5.39. The data separate into two groups. For records with wind speed $U_{10} < 7.5\,\mathrm{m/s}$ (five points on the left), the total dissipation is significantly overestimated. It is, however, somewhat underestimated for the winds $U_{10} > 7.5\,\mathrm{m/s}$ (points on the right).

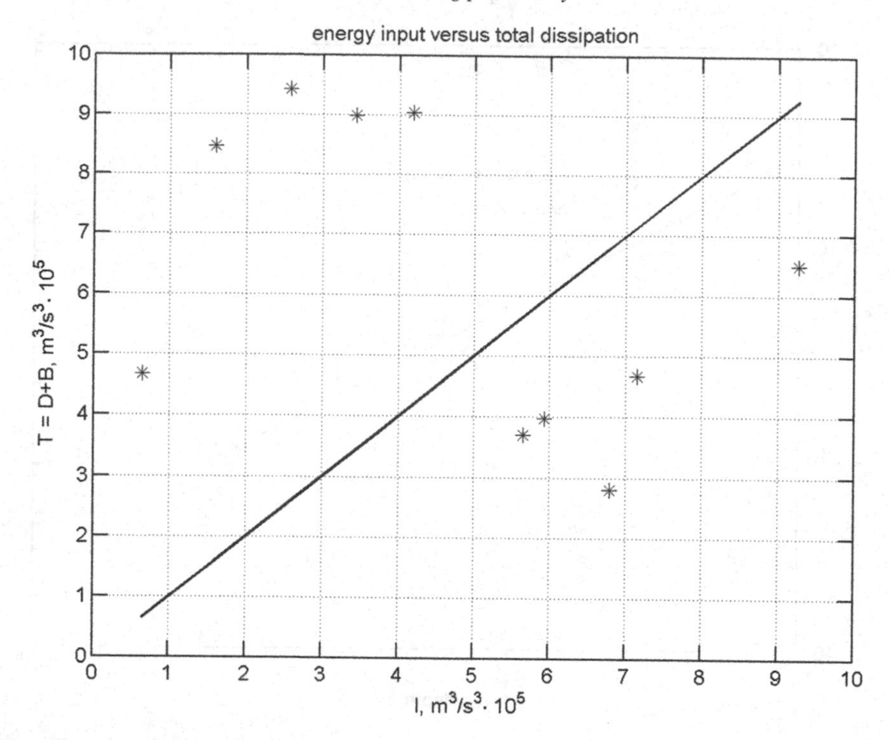

Figure 5.39 Total dissipation in the wave water column T_a (5.73) versus measured total wind input I_a (denoted as T and I, respectively). Parameterisation (5.69) is used for integrating D_a in (5.67). Figure is reproduced from Babanin *et al.* (2005)

Figure 5.40 shows the total dissipation T_a (5.73) plotted versus the total input I_a, while dissipation D_a (5.67) was estimated on the basis of the wall-layer distribution for ϵ_{dis} (5.68) as was suggested by a number of authors mentioned above. Dissipation for the light-wind points now matches the wind input quite well, whereas the dissipation at winds $U_{10} > 7.5$ m/s is greatly underestimated.

An obvious conclusion to be drawn is that the volumetric rate of total turbulent kinetic energy dissipation ϵ_{dis} is distributed according to the $\sim z^{-1}$ law (5.68) for waves generated by light winds and as $\sim z^{-2}$ (i.e. similar to predictions of (5.69)) for waves under stronger winds. Since the inverse-quadratic dissipation has always been associated with wave breaking, such a conclusion is consistent with observations that the breaking does not occur for waves forced by light winds of $U_{10} \leq 5$–7 m/s.

Finally, to provide better agreement between the dissipation and the energy input of the strong-wind points in the top panel, the scale for H in (5.69) had to be adjusted. To obtain the $H = 0.6\, H_s$ scale in (5.69), Terray *et al.* (1996) had to rely on an inferred wind-input rate. Babanin *et al.* (2005) used the total wind input I_a measured, and the comparisons led to a conclusion that the constant-dissipation layer does not reach below $H = 0.4\, H_s$.

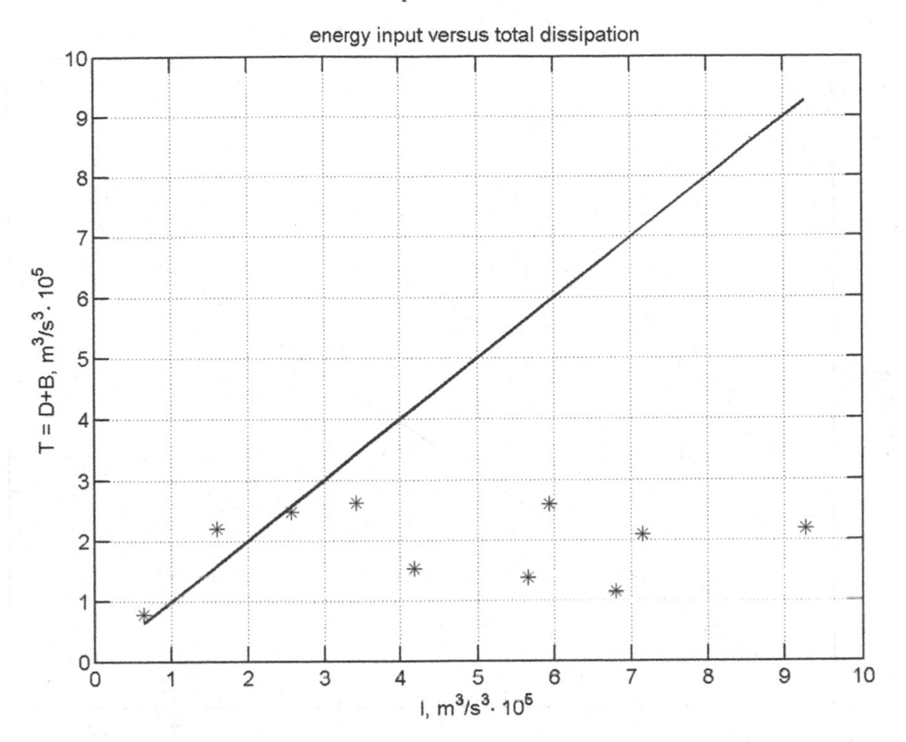

Figure 5.40 Total dissipation in the wave water column T_a (5.73) versus measured total wind input I_a (denoted as T and I, respectively). Parameterisation (5.68) is used for integrating D_a in (5.67) at $z > 0.6\,H_s$. Figure is reproduced from Babanin *et al.* (2005)

The outcome is shown in Figure 5.41 and the resulting parameterisation for the dissipation rate ϵ_{dis} is now as follows:

$$\epsilon_{\text{dis}}(z) = \begin{cases} \text{const} & z \leq 0.4\,H_s, \\ \sim z^{-1} & z > 0.4\,H_s,\; U_{10} < 7.5\,\text{m/s}, \\ \sim z^{-2} & z > 0.4\,H_s,\; U_{10} > 7.5\,\text{m/s}. \end{cases} \tag{5.74}$$

Thus, based on the Lake George measurements, the wind speed of $U_{10} \approx 7.5\,\text{m/s}$ is evident as a threshold for breaking of dominant waves starting to occur (see also Figure 9.3 and discussion in Section 9.1.1). The universality of this conclusion has to be further verified in larger water bodies (see also discussion in Section 3.1). Additionally, Bortkovskii (1997, 1998), for example, points out that this wave-breaking threshold wind speed depends on the water temperature. According to Bortkovskii (1998), the whitecaps start to appear at a wind speed of 7 m/s if the temperature is 30° C and at 8.5 m/s if it is 0° C.

Other roles of the wind can now be discussed. Wind action is important on longer scales in altering breaking statistics because of enhancing the wave steepness. At moderate winds, doubling the wind speed leads to achieving the limiting steepness and breaking four

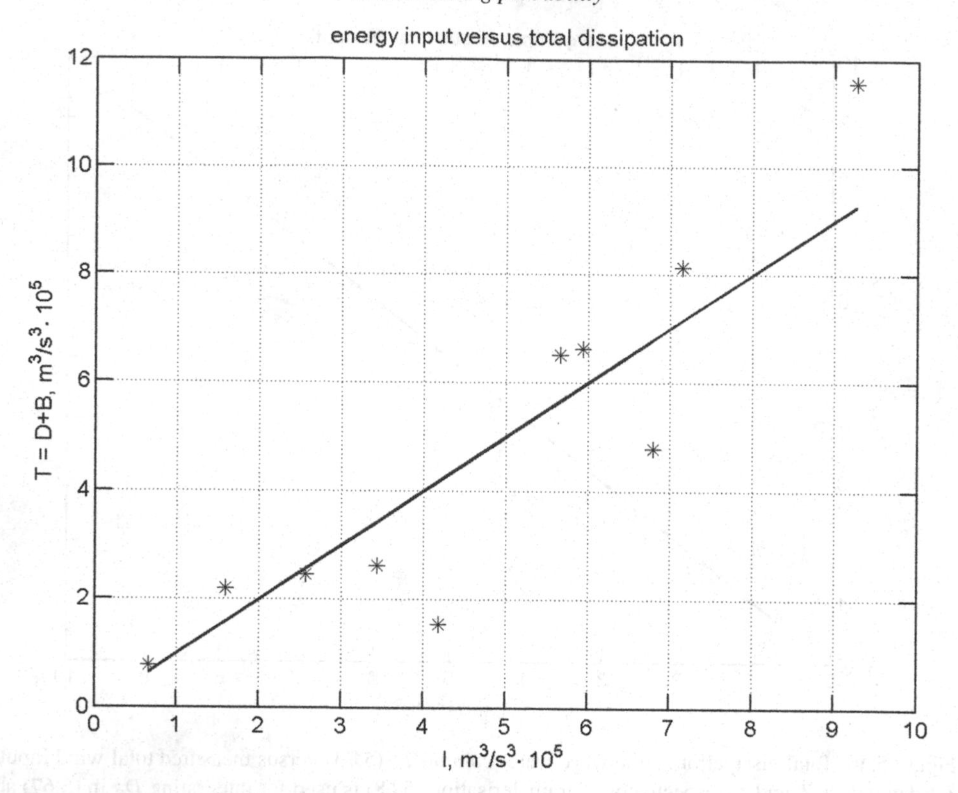

Figure 5.41 Total dissipation in the wave water column T_a (5.73) versus measured total wind input I_a (denoted as T and I, respectively). Parameterisation (5.74) is used for integrating D_a in (5.67). Figure is reproduced from Babanin *et al.* (2005)

times as fast. At stronger wind forcing, this effect slows down (see Donelan *et al.*, 2006, Section 4.1.1). Instantaneously, the wind capacity to effect breaking onset is marginal unless the wind forcing is very strong. See Section 4.1 and above sections of this chapter for more discussion on this topic.

These conclusions were obtained by analysis of quasi-monochromatic wave trains. The field waves are spectral and directional, and this fact adds an additional level of complexity to the topic of wind forcing.

In the spectral environments, waves of different scales are present at the same time and what is a low wind forcing for some components can inevitably be a strong or even very strong forcing for others. The phase speed $c(f)$ of waves across the spectrum depends on their frequency (2.15), and therefore so does the wind forcing $U/c(f)$.

If doubling wind forcing causes wave breaking four times faster, then doubling the wave frequency according to (2.18) hypothetically should have the same effect in terms of the wind forcing U/c. For example, if the wind speed is $U_{10} = 7.8\,\mathrm{m/s}$, then at frequency $f_1 = 0.2\,\mathrm{Hz}$ there will be $U_{10}/c(f_1) = 1$, whereas at frequency $f_2 = 2f_1 = 0.4\,\mathrm{Hz}$ it will be $U_{10}/c(f_2) = 2$. Therefore, at some small scales (low phase speeds, high relative winds) in the spectrum, the waves theoretically should be breaking very often.

The general pattern in real wave fields, however, is much more complicated, and this is due to a number of reasons. If the frequencies f_1 and f_2 above represent respective spectral peaks in the respective spectra (see Figure 5.29 for a typical view of the wave power spectrum), then the $U_{10}/c(f)$ wind forcing is applied to dominant waves in both cases and the values of the forcing, as far as the flux of momentum/energy is concerned, are unambiguous. If, however, the first frequency is the spectral peak $f_1 = f_p$ and the second frequency $f_2 = 2f_p$, then the translation of $U_{10}/c(f)$ into the wind forcing as such is not that straightforward (the mean wind force per unit area is the mean wind stress, i.e. equals the mean-over-wave-period momentum flux).

Indeed, as has been shown in a number of air–sea interaction studies (e.g. Donelan *et al.*, 2006; Babanin *et al.*, 2007b; Kudryavtsev & Makin, 2007), at strong wind forcing and in the presence of steep dominant waves, particularly if those are breaking, air-flow separation in the lee of these waves can occur. In this case, if the shorter waves riding the dominant ones are under the separated air bubble they experience much lower wind stresses compared to what would be expected from the respective values of $U_{10}/c(f)$ based on their phase speed $c(f)$.

At very strong wind forcing such a relative reduction of the wind stress/input is applicable to the dominant waves themselves. The separated-over-the-crest-of-a-dominant-wave air flow does not reattach to the surface until close to the crest of the next wave. As a result, the wind effectively skips the wave troughs and 'does not know' how high/deep the waves are. Consequently, the wave-induced pressure oscillations in the lower boundary layer are weakened, and so is the wind input/stress (Donelan *et al.*, 2006).

Absolute values of the stress always increase as the $U_{10}/c(f)$ goes up, but relative values of the wind forcing can go down, depending on a combination of wind-speed and wave-steepness magnitudes, or whether the waves with corresponding frequency f are the spectral-peak or spectrum-tail waves, whether these waves experience the flow separation, are in fact under the separated bubble, or the flow remains attached along the wave profile. The wave breaking occurrence/probability will respond accordingly. Figure 4.2 in Section 4.1.1 illustrates this effect: the first wind-speed doubling led to the breaking happening four times faster, whereas subsequent doubling the wind only reduced the distance to breaking three times.

Another complication of the general pattern in the spectral wave fields is dictated by the physics depicted in radiative transfer equations (2.61). There exists a competition between source terms responsible for the wave evolution in RTE. As the wind grows, some of the excessive energy/momentum flux provided to short waves by the wind will be transferred to lower frequencies (faster waves in the spectrum) through nonlinear interactions and will not contribute to the growth of the wave height/steepness of these short waves.

One way or another, however, but at some stage of the growing wind forcing, neither the air-flow separation nor the nonlinear interactions seem to be able to digest the amount of energy input by the wind, and wave breaking (dissipation) across the spectrum suddenly goes up. This was demonstrated by Babanin *et al.* (2007c) and this effect can be seen in Figure 5.27 of Section 5.3.2.

In the top panel of this figure, the peculiarity of the wave-breaking-probability behaviour at strong winds is demonstrated. According to (5.10), (5.24) and (5.40), the breaking probability and the dissipation function are expected to be determined by the wave spectrum, at least near the spectral peak. As concluded above, the wind influence on wave breaking and energy attenuation is indirect: the wind changes the wave spectrum first, and this change brings about alterations of the breaking as a consequence. In Figure 5.27 (top) the breaking distributions merge together for moderate winds and are clearly enhanced for the two stronger-wind cases, across the entire spectral band. Therefore, we could expect that if the wave spectra solely define the breaking/dissipation, the spectra for the last two cases should also be enhanced as a result of the stronger wind forcing.

This is, however, not the case. Figure 5.42 shows the full spectra in log–log scale in the left panel, and in the right panel these spectra are plotted in expanded linear scale in the frequency range of $f = 0.8\,f_p$–$3\,f_p$. The wave spectra do merge as expected for the moderate winds, but at strong winds of $U_{10} > 14$ m/s a further increase of the wind speed and the wind input does not cause noticeable changes of the wave spectrum either, except at the peak. The excessive wind input, or at least a significant part of it, appears to be dissipated locally through the enhanced breaking.

Alternatively, the excessive input could have been handled by the nonlinear interactions (S_{in} term in (2.61)) and converted into the growth of the spectral peak. To add an uncertainty, however, we must point out that in the scenarios studied here the peak waves were strongly forced, with $U_{10}/c_p = 2.5$ and 6.5, respectively (records 6 and 5 of Table 5.2), and thus extensively received the energy for their growth directly from the wind.

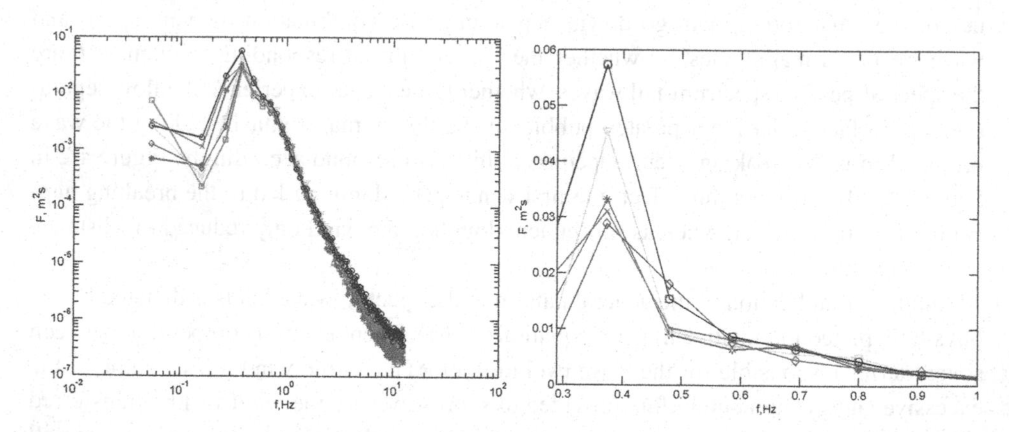

Figure 5.42 Wave spectra for the records shown in Figure 5.27. (left) Full spectra in log–log scale. (right) Spectra in $f = 0.8\,f_p{-}3\,f_p$ range in linear scale. Figure is reproduced from Babanin *et al.* (2007c) (public domain site http://www.waveworkshop.org/ sponsored by Environment Canada, the U.S. Army Engineer Research and Development Center's Coastal and Hydraulics Laboratory, and the WMO/IOC Joint Technical Commission for Oceanography and Marine Meteorology)

Therefore, we should conclude that wind influence on wave breaking in spectral environments accepts additional roles compared to its effects in the case of wind-forced uniform or modulated wave trains. The latter are obviously applicable to the trains of dominant waves and inherent breaking of shorter waves, but important new features of the wind-breaking connection are revealed at smaller scales in the spectrum.

At light-to-moderate winds, the spectrum responds to the growing wind forcing U_{10}/c_p by increasing the level of its saturation interval α (2.7) up to a certain magnitude only:

$$\alpha = \begin{cases} 8.03 \cdot 10^{-2} \tilde{f}_p^{1.24} & \text{for } \tilde{f}_p \leq 0.23, \\ 13.2 \cdot 10^{-3} & \text{for } \tilde{f}_p > 0.23 \end{cases} \tag{5.75}$$

(Babanin & Soloviev, 1998a). Here,

$$\tilde{f} = \frac{f U_{10}}{g} = \frac{1}{2\pi} \frac{U_{10}}{c_p} \tag{5.76}$$

is a dimensionless frequency. While the level α is growing and when the spectral densities overcome the threshold value of (5.36), the waves across the spectrum start to break and the rate of breaking should be increasing in response to the growing level and the induced-breaking effects, see Section 5.3.2, Figure 5.29 and associated discussions.

Once the equilibrium limit of α identified in (5.75) is reached, however, the level of the spectrum tail stops growing. As usual in the physics of air–sea interactions, this conclusion is far from comprehensive. The scatter with respect to the constant-α level in (5.75), according to the data shown in Babanin & Soloviev (1998a), does exist, and this scatter is hardly random. Babanin & Makin (2008), for example, listed more than 15 factors other than the wave age which can contribute to the dependences and scatter of the sea drag (3.8) and therefore of the equilibrium level α. There is experimental evidence that the temperature of the water, which affects the spectral density of short waves and aerodynamic roughness z_0 in (3.19), can be another factor to influence level α under what would seem otherwise the same wind-forcing and dynamic conditions (Brown, 1986; Pierson *et al.*, 1986; Bortkovskii, 1997).

As seen in Figure 5.27, the wave-breaking rates across the spectrum in such conditions remain approximately constant also, but only until the wind speed achieves some new limit. In the Lake George scenario shown, when the wind speed exceeds this limit of

$$U_{10_{\text{limit}}} \approx 14 \text{ m/s}, \tag{5.77}$$

it appears that further growth of the wind forcing results in abrupt changes of the breaking probability across the spectrum's equilibrium interval. Here, it is interesting to note that transition of the wave-growth regime at $\tilde{f}_p = 0.23$ in the measurements of Babanin & Soloviev (1998a) parameterised by (5.75), is predicted theoretically by Stiassnie (2010) based on considering the fetch-limited wave growth due to Miles' (1957) and Phillips' (1957) wind-input mechanisms, and predicted exactly at the same wave age.

It should be remembered, however, that connections of the spectral density and the breaking probabilities above are indicative, but are far from being unambiguous. As was discussed in detail in Section 5.3.2, the breaking rates at scales small relative to the peak are affected by the induced breaking and their dependence on the local spectral density is smeared. Rather, they should depend on the integral of the spectrum over longer scales (e.g. (5.41), by analogy with the dissipation function (5.40)).

These scales include the spectral peak. If so, Figure 5.42 provides a reasonable explanation of the observed wind effect in terms of the induced breaking. The two spectra that exhibit higher breaking rates of short waves are the ones which correspond to much higher dominant waves. For these two spectra the peak enhancement above the others is very noticeable. This is particularly apparent in the linear scale (right), that is, the peak spectral density of the 15 m/s wind (crosses) is 1.5 times higher than the rest of the group, and for the 20 m/s (circles) it is twice as high. If so, modulation of the short waves by the larger dominant waves should be stronger and cause more frequent breaking in the corresponding spectral range.

Therefore, the observed response of the breaking rates across the spectrum to the higher wind speeds, under the virtually unchanged local spectral density, can in fact be a result induced by larger dominant waves rather than by the wind directly. The cause of the high dominant waves, however, still rests with the wind, which in this case managed to pump up the spectral peak while the equilibrium level stayed unchanged. One way or another, wind forcing is of course the major player in the dynamics of the fields of wind-generated waves with a continuous wave spectrum, and its multiple roles in wave-breaking behaviour are still in need of further understanding and quantifying (see also discussion of wind effects on the breaking severity in Chapter 6).

To conclude this chapter, and to some extent related results of Chapter 4, we would like to say that the topic of breaking probability has enjoyed the close attention of the ocean-wave community over the past two decades or so, and at the time of writing is still going through a stage of active progress. Not only have some important physical features such as breaking threshold, cumulative effect and limiting breaking steepness been formulated and understood, they have been quantified too. It is instructive to notice that, while breaking onset due to linear focusing was scrutinised much earlier than the details of onset because of modulational instability, focusing-breaking probability is yet to be parameterised in terms of the background wave-field conditions. Many other behaviours and characteristics of breaking probability are still to be identified, appreciated and described. These are first of all related to the breaking occurrence in spectral, and particularly in directional environments.

6

Wave-breaking severity

On many occasions earlier in this book, it has been mentioned and emphasised that knowledge of breaking severity is as important as is understanding the physics driving the breaking occurrence. While the latter, however, has received a lot of attention from the wave-research community lately, our information on breaking strength, its variability, environmental dependences and physics remains limited and fragmental.

If the breaking strength is defined as energy loss in a single breaking event (Section 2.7), then the breaking severity coefficient s can be identified in a number of ways, through measurement of the individual breaking wave (2.24), of the group where the breaking occurred (2.32), of spectra of the respective groups before and after the breaking (2.38) and of short waves modulated by the longer wave only (2.42). The magnitude of such a coefficient varies greatly, from $s = 10\%$ (Rapp & Melville, 1990, or even less as seen in Figure 6.3 below) up to 99% (2.31) based on the Black Sea estimates (see also Bonmarin, 1989; Babanin *et al.*, 2010a, 2011a).

Such a range of change of course cannot be disregarded or substituted with some mean value in applications that involve the breaking severity. A typical application is the wave-energy dissipation function S_{ds} employed in wave forecast models (2.21), (2.61) and (5.40). As defined in (2.21), it can in principle be directly determined as a product of the breaking probability and breaking severity, but since more or less definite parameterisations of the latter are not available a set of inventive indirect methods have been elaborated to estimate the dissipation function (see e.g. and The WISE Group, 2007, for a review).

In other applications, such as, for example, engineering aspects of the wave-breaking impact, the breaking-strength magnitude needs to be known explicitly rather than as an element of the overall energy dissipation. More than that, statistical values such as those involved in obtaining the averaged dissipation term may not be helpful in this regard, as extreme or individual events may be sought. Therefore, quantifying and parameterising the breaking severity is an important outstanding task of wave-breaking studies.

Chapter 6 is shorter than other chapters in this book for obvious reasons. There are not many experimental dependences and no theoretical approaches to discuss. In the two sections, indicative laboratory (Section 6.1) and field (Section 6.2) results will be outlined in addition to those already mentioned in the definition Section 2.7.

6.1 Loss of energy by an initially monochromatic steep wave

Features of the severity of breaking which resulted from linear frequency focusing have been investigated in a number of laboratory studies, and the reader is referred to the comprehensive paper by Rapp & Melville (1990) and to other studies on this topic discussed in Section 2.7, for details. With respect to the breaking strength Rapp & Melville (1990) concluded that

"The loss of excess momentum flux and energy flux was measured and found to range from 10% for single spilling events to as much as 25% for plunging breakers".

The severity of wave breaking that occurs due to modulational instability varies in a much greater range, and appears to be a gradually changing property rather than a characteristic with approximately set values. Since, according to discussions in Section 5.3.3 such breaking is likely to be more frequent in field conditions, some respective results obtained in the course of laboratory experiments described in Babanin *et al.* (2010a) and Galchenko *et al.* (2010) will be discussed here in more detail.

In Figure 6.1, an example of a spectral distribution of the breaking severity due to a single event is demonstrated. This record was part of the experiment in the ASIST wave tank described in Section 5.1. Spectra of the time series of Figure 6.2 are plotted in Figure 6.1a and their ratio in Figure 6.1b. In the top panel, the solid line signifies the pre-breaking spectrum and the dashed line the after-breaking spectrum.

The spectral distribution of this breaking-severity event was discussed in Section 2.7. In short, we will mention that while it is the main wave that is breaking, the energy is lost from all the harmonics too. The peak is reduced by a factor of 5, and the other harmonics almost completely disappear. Across the entire frequency range, the average ratio of the two spectra is 1.8 which translates into the overall spectral severity $s_{\text{spectral}} = 45\%$ (2.38).

In Figure 6.2, wave series before and after the breaking are compared in physical space. In this figure, the solid line shows the waves of IMF $= 1.8\,\text{Hz}$, IMS $= 0.30$, $U/c = 0$ at the second probe (10.53 m from the wavemaker), as in Figure 5.2, and dashed line – on the third probe (11.59 m from the paddle). Breaking of the three incipient breakers seen at probe 2 happened between the two probes. The wave that is seen following the incipient breaker on the second probe also broke between the two probes. This consistent double-breaking, with a small time delay, is again in agreement with field observations (e.g. Donelan *et al.*, 1972). These breaking processes happened within a period of 1.2 s, the time required by the 1.8 Hz waves to travel the distance between the probes at their phase speed. Therefore, the records made by the third probe are time-shifted by 1.2 s in an attempt to superimpose what should have been the same waves, if the breaking had not taken place. Note that the group velocity is different to the phase speed and therefore the individual waves are travelling through the wave envelope seen in the figure. Therefore, in the absence of breaking, the match of the two shifted time series does not necessarily have

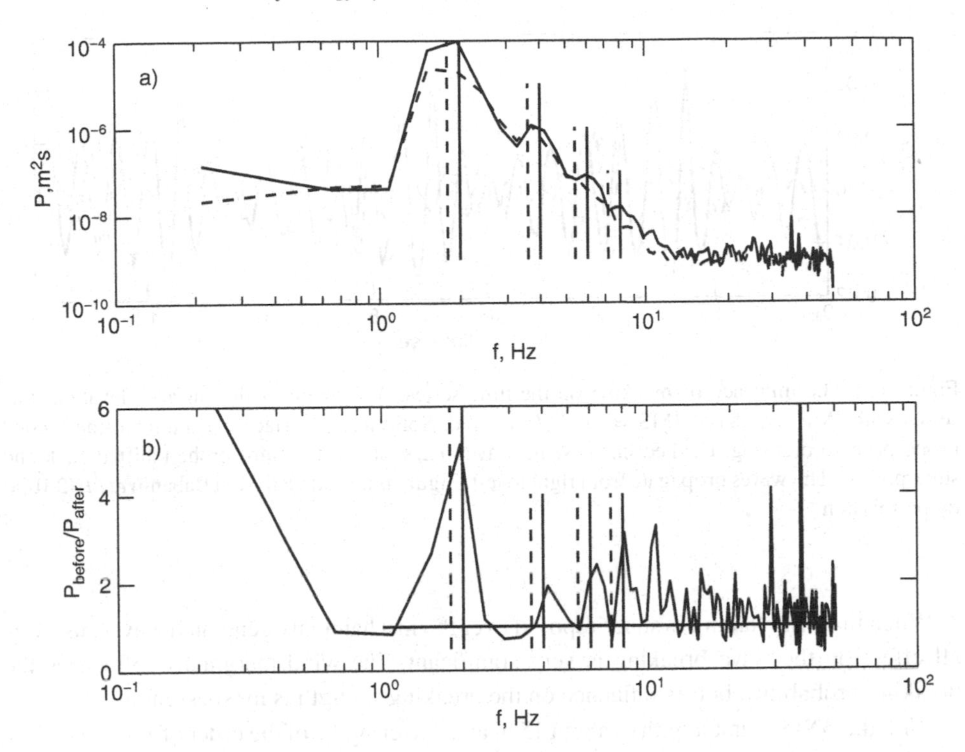

Figure 6.1 a) Spectra P of the time series of IMF $= 1.8$ Hz, IMS $= 0.30$, $U/c = 0$. Solid line corresponds to the pre-breaking spectrum measured at the second probe (see solid-line time series in Figure 6.2). Dashed line is post-breaking spectrum measured at the third probe (dashed time series in Figure 6.2). Multiples of IMF are shown with dashed vertical lines. Multiples of the incipient-breaker 2 Hz frequency are shown with solid vertical lines. b) Ratio of the pre-breaking and post-breaking spectra. Solid horizontal line signifies ratio of 1, vertical lines have the same meaning as in the top panel. Figure is reproduced from Babanin *et al.* (2010a) by permission

to be exact, depending on the phase of individual waves with respect to the envelope (see also Section 2.7 for corresponding discussion of this figure).

With that in mind, we still see that the two waves that broke practically disappeared, as well as the entire modulation. The only waves that can still be tracked are the second and the third ones after the incipient breaker at probe 2. An overall severity of 45%, integrated over the spectrum above, is quite significant.

The number of waves in the segment has also changed. Between the three incipient breakers at probe 2, one can count 17 waves, and at probe 3 this number is closer to 16, i.e. frequency downshifting has occurred (see also Melville, 1982; Su *et al.*, 1982; Bonmarin, 1989; Reid, 1992; Tulin & Waseda, 1999; Babanin *et al.*, 2010a, 2011a). This is how Bonmarin (1989) describes the downshifting as a result of plunging breaking:

"the potential energy of the original breaking wave finally becomes wholly dissipated in the midst of the successive splash cycles" to a point "as if a wave crest had never existed at this location".

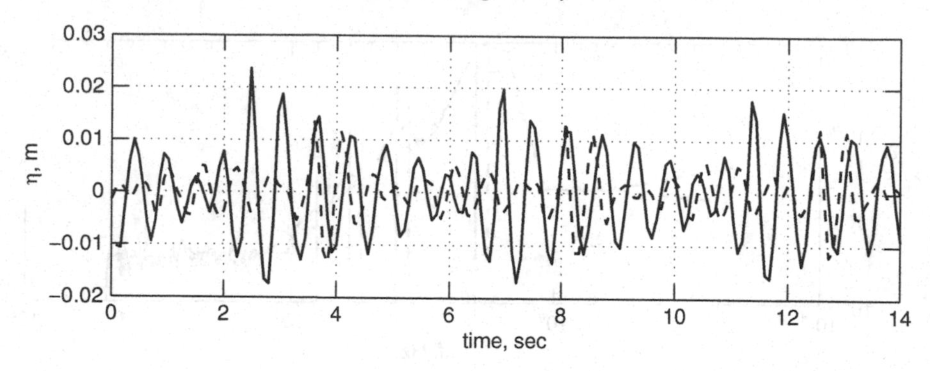

Figure 6.2 The influence of breaking on the time series. A segment of the surface elevation time series with IMF = 1.8 Hz, IMS = 0.30, U/c = 0. Solid line – surface elevations at the second probe prior to breaking. Dashed line – same waves 1.2 s later at the third probe (shifted back and superposed). The waves propagate from right to left. Figure is reproduced from Babanin *et al.* (2010a) by permission

When the wind forcing is superimposed over the mechanically generated waves, its overall effect on the wave breaking is very significant. The wind marginally influences the breaking probability, but its influence on the breaking strength is most essential.

With the ASIST tank length of about 15 m and wavelengths of the order of $\lambda \approx 0.5$–0.7 m, the fetch is short and the wind capacity to change the steepness of mechanical waves is limited, even at strong forcing. Therefore, one would expect only a weak dependence of the distance to breaking on the wind speed, which is what is seen in Figure 5.17. The wind, however, does alter this distance. In the experiment, gradual reduction of the initial monochromatic steepness caused the waves to break further and further from the wavemaker, all the way to the beach at the opposite end of the tank. If then the wind was switched on and its speed gradually increased, the breaking-onset point was brought somewhat back.

The breaking point, however, would not come too far from the beach, certainly not even to the middle of the tank; this is not due to the lack of wind power. The ASIST facility is capable of producing hurricane-force winds. The reason was the breaking severity. As the wind forcing was increasing and the distance to the breaking decreasing, so was the breaking strength diminishing. In the end, already between probes 2 and 3 which were 10.53 m and 11.59 m from the wavemaker, the breaking became a mere toppling of the very crest and effectively disappeared when the wind speed was increased further.

This effect is demonstrated in Figure 6.3 which is analogous to Figure 6.2 except the wind forcing is now superimposed and IMF = 1.5 is different. The initial monochromatic frequency had to be reduced in order to make the waves break between the same probes 2 and 3. As just mentioned, in the presence of the wind forcing it would take fewer wavelengths to the breaking point and therefore longer waves had to be used.

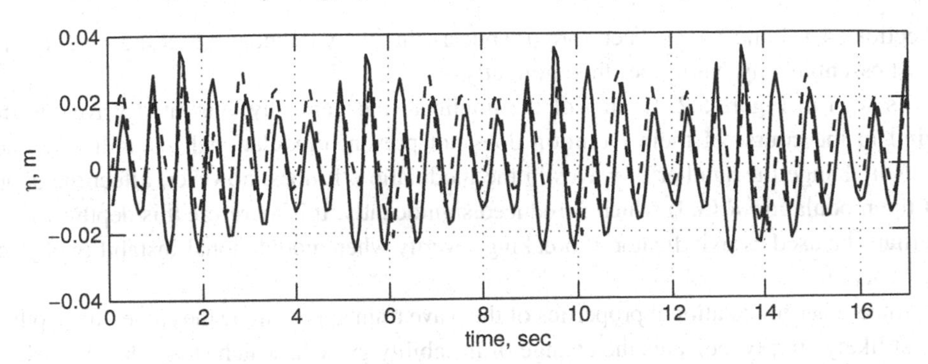

Figure 6.3 As in Figure 6.2, with wind forcing. A segment of the time series with IMF = 1.5 Hz, IMS = 0.30, $U/c = 3.9$. Solid line – surface elevations at the second probe prior to breaking. Dashed line – same waves 1.04 s later at the third probe. The waves propagate from right to left (shifted back and superposed). Figure is reproduced from Babanin *et al.* (2010a) by permission

Again, the wave time series are compared immediately before and after breaking. The solid line shows the waves of IMF = 1.5 Hz, IMS = 0.30, $U/c = 3.9$ at the second probe and the dashed line at the third probe. Breaking of the four incipient breakers seen at probe 2 occurred between the two probes. The breaking was very gentle when observed visually, and there was no double-breaking. With IMF = 1.5 Hz, the time necessary to travel the distance between the two probes is estimated as 1.04 s and therefore the record made on the third probe is shifted back correspondingly.

As was analysed in Section 2.7, because of the difference of the phase speed and group velocity each of the solid-lined waves would move approximately one position ahead within the envelope, and the highest wave should become the second highest (in front of it) in the dash-lined group. Because of this, the height of the wave which was the highest at the previous probe would be significantly reduced even without breaking, but the overall picture of the wave group would remain the same.

Breaking, even though gentle, however, happened and comparison of the breaking impacts in Figures 6.2 and 6.3 is quite instructive. In Figure 6.2, the incipient breaker and the wave following it practically disappeared, as well as the entire modulation. In Figure 6.3, they are all present and each wave in the modulation can be tracked at the third probe. The energy loss was minimal and could not have been quantified within the confidence limits of the respective spectra.

In contrast to Figure 6.2, after the breaking in Figure 6.3 the number of waves did not change and no downshifting is visible. As seen in the figure, the breaking resulted in some truncation of the crest of the highest wave in the group, and in smoothing of the modulation.

The latter, we believe, is a most essential observation. The fact that the wind influence reduces the modulational-instability growth and smears the modulation has already been noticed (i.e. Figure 5.2, Section 5.1.3). It was also seen that the wind impact on the wave breaking as such, at the wave-breaking time scale, is always small or even negligible

(Sections 4.1.3 and 5.1.3). Therefore, it appears that the wind does affect the severity in a most essential way, but it does this indirectly.

As seen in Figure 5.2, forcing of the nonlinear mechanically generated waves by the wind in the course of their evolution does not prevent modulational instability, neither does it change the number of waves in the modulation. It does, however, affect the depth of the modulation R (5.3) which is reduced significantly. It is therefore this depth that can perhaps be used as an indicator of breaking severity when modulational-instability physics is involved.

Since other modulational properties of the wave train appear to be the same, the depth R most likely simply indicates the change of instability growth, which slows down. Such an effect was predicted theoretically by Trulsen & Dysthe (1992) based on numerical experiments on a nonlinear Schrödinger equation with an added linear-growth term to simulate the wind forcing.

The waves break, however, not because the wave group reaches some limiting value of R, but because individual waves within this group reach the limiting steepness (2.47). Growth of the individual-wave steepness is determined both by the growth of the modulation of the group and by direct energy input from the wind to these individual waves. Thus, even though the instability development slowed down, the waves still reach the limiting steepness and break. The severity of the breaking, however, is very different under the circumstances of lower instability-growth rates. Thus, the wind action alters the modulation depth and instability growth on a longer time scale, and the hydrodynamics related to the modulation itself controls the breaking strength at the short scale of the duration of the breaking process.

If so, the breaking severity, like other wave-breaking features, depends mostly on hydrodynamics rather than on air–sea interactions. In order to investigate this dependence in pure hydrodynamic conditions, Galchenko *et al.* (2010) conducted a laboratory experiment where wind action was removed and different depths of the modulation at the breaking onset were achieved by imposing various sidebands, both in terms of their steepness and frequency separation of the primary wave and the sideband.

The experiment was carried out in a wave tank of the Department of Hydraulic and Ocean Engineering of the National Cheng Kung University, Taiwan. A programmable piston wavemaker was used to produce the wave trains with primary waves of steepness ϵ_1, sideband of steepness ϵ_2, and bandwidth $\Delta\nu$ such that the number of waves in initial modulation was $N_m = 6.9$, close to that typical of ocean waves (see (2.12)). Surface elevations were measured and recorded by means of a one-dimensional array of six capacitance-type wave gauges deployed along the main axis of the tank. Thus, when a breaking happened within the array, it was possible to estimate the wave energy within the group where the breaking occurred, before and after the breaking. The severity s_g was defined for the wave groups according to (2.32).

In Figure 6.4, s_g (denoted as S in the figure) is plotted versus R where the modulation depth is estimated immediately before breaking. The scatter is large, but the positive trend is robust. As anticipated, the severity appears to be a function of modulation depth:

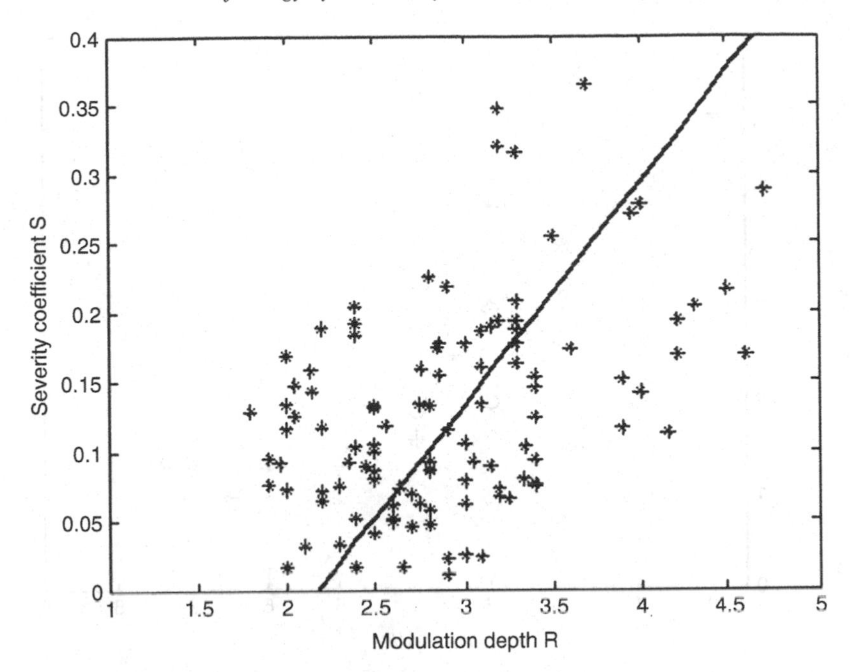

Figure 6.4 Severity coefficient s_g (2.32), denoted as S in the figure, as a function of modulation depth R (5.3). Solid line is parameterisation (6.1). Figure is reproduced from Galchenko *et al.* (2010) © American Meteorological Society. Reprinted with permission

energy loss grows with the modulation depth. The figure includes 109 points, and the linear correlation coefficient between R and S is $K = 0.44^{+0.13}_{-0.17}$ (for a 95% confidence interval). Energy loss varies very significantly, from 2% to 35%. Note that this is energy lost from a group consisting of several waves, that is, the severity in terms of the energy of individual breaking waves is much greater (see Section 2.7). A linear function in Figure 6.4 is

$$S = (0.16 \pm 0.03)R - (0.35 \pm 0.06). \tag{6.1}$$

where 95% confidence limits are shown.

The observed linear trend gives credit to the suggestion made above that it is the hydrodynamics that primarily drives the breaking severity in the case of instability-caused breaking, and the depth of modulation is its indicator. The large scatter, however, points out that such a straightforward picture may be oversimplified. Different depths of the modulation signify different degrees of development for the modulation at the stage of the wave breaking, but it appears that this degree is not the only property to affect the severity even in the absence of the wind.

The role of the wind is also not simple and not limited to variation of this degree only. Stronger winds do reduce modulation depth. This was observed experimentally in self-modulated wave trains by Babanin *et al.* (2010a), and it was confirmed numerically by Galchenko *et al.* (2010). In Figure 6.5, modulation depth R immediately before breaking is

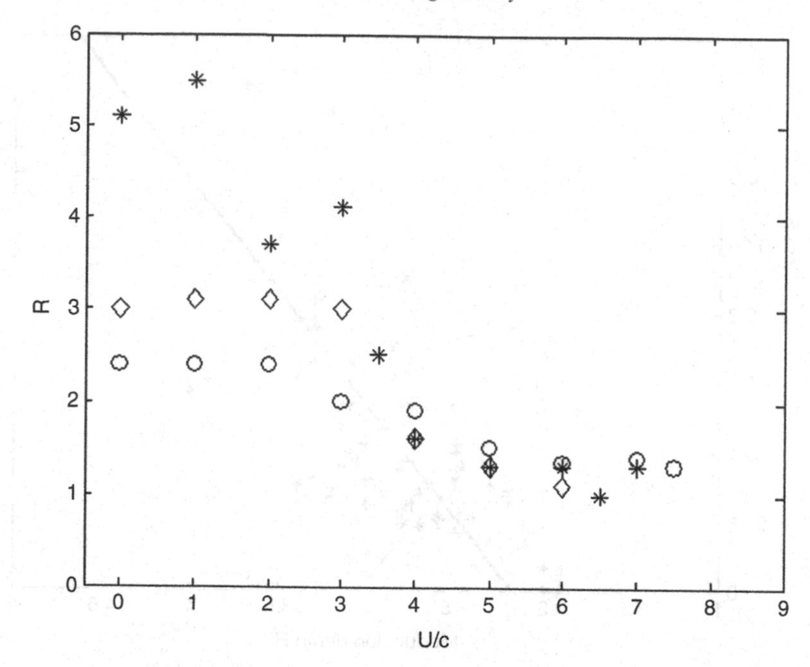

Figure 6.5 Modulation depth R as a function of wind forcing U/c for wave trains with three different steepnesses of $\epsilon_1 = 0.17, \epsilon_2 = 0.005$ (asterisks), $\epsilon_1 = 0.19, \epsilon_2 = 0.005$ (diamonds) and $\epsilon_1 = 0.21, \epsilon_2 = 0.006$ (circles). Here, ϵ_1 is steepness of primary waves, ϵ_2 is sideband steepness, U is characteristic wind speed in the model. Figure is reproduced from Galchenko *et al.* (2010) © American Meteorological Society. Reprinted with permission

plotted versus wind forcing U/c where U is the characteristic wind speed in the CS model which was used for the numerical simulations (see Section 4.1.1). Three different combinations of primary and sideband steepnesses are shown: $\epsilon_1 = 0.17, \epsilon_2 = 0.005$ (asterisks), $\epsilon_1 = 0.19, \epsilon_2 = 0.005$ (diamonds) and $\epsilon_1 = 0.21, \epsilon_2 = 0.006$ (circles).

At the very-strong-wind end, in Figure 6.5 there is a hint of the modulational depth increasing again. This is consistent with laboratory observations of Waseda & Tulin (1999) who found that the wind forcing decreases sideband growth if weak and stimulates their growth when strong. A slower sideband growth would mean, over the same wave fetch, a less-developed modulation and its depth, and vice versa.

In general, Waseda & Tulin (1999) found that the imposed wind can have two effects on modulational instability: alteration of the instability growth and change of the natural bandwidth in (2.12). The latter also depends on the wind speed relative to the wave phase speed, that is, the wave age in the general sense. This interesting observation may mean that relative spectral-peak width cannot be constant, as it is in the JONSWAP spectrum (2.7). That is, if the peak width for the waves with a continuous spectrum is determined by this instability, wave-age-dependent parameterisations of the spectral peak (e.g. Donelan *et al.*, 1985) are more realistic, but that would have to be verified qualitatively and quantitatively.

Coming back to the effects of wind on the modulational instability, Waseda & Tulin (1999) concluded that

"both effects combined will determine whether the modulational instability is enhanced or suppressed".

From Figure 6.4 and (6.1), one can see that on average the severity approaches zero when $R \approx 2.2$, certainly there is no breaking for $R < 1.8$. This makes it possible to conclude that $R \approx 2.2$ represents a lower limit to modulation depth, below which the probability of breaking in the absence of wind significantly decreases. In the presence of wind, however, the modulation depth immediately before breaking can be as low as $R \sim 1$ – a value that signifies no modulation by definition (5.3). This result highlights the fact that the actual role of the wind in wave breaking in general and in regulating breaking severity is not that unambiguous and in the latter case does not reduce to a mere controlling of the rate of modulation development.

The modulational depth, in more general terms, if applied to field conditions, is a property difficult to measure and hardly possible to employ in spectral applications as there is no obvious way how information on such a feature can be obtained from a wave spectrum. In spectral models, perhaps, some combination of characteristic steepness and bandwidth, which determine the extent of modulation, and some combination of steepness and wind forcing, which determine the rate of modulation growth, could be used as parameters for the breaking-severity dependences. The severity of wave breaking in conditions of a continuous wave spectrum is the subject of the next section.

6.2 Dependence of the breaking severity on wave field spectral properties

As with breaking probability discussed above, the spectral and directional distribution of the breaking severity bears the principal uncertainty of how this property, which is not a continuous sequence of instantaneous values, can be converted into the Fourier space. If the number of individual breaking events is counted and energy loss in each of them is measured, then the average value of severity at frequency f in direction θ will depend on the choice of frequency bin $f \pm \Delta f$ and directional bin $\theta \pm \Delta\theta$, particularly in the case of a function rapidly changing along frequency/direction (see Sections 2.5 and 5.3.3 for more discussion).

With this uncertainty in mind, we can discuss the spectral distribution of the breaking severity in the same way as we did with the breaking probability in Section 5.3.2. In order to obtain a distribution similar to that for the probability in Figure 5.27, we need the count and the strength-measure of individual breaking events in the spectral and directional bins. The bubble-detection technique described in Section 3.5 can provide both.

This technique employs the sound emitted by individual bubbles when they are created in the course of breaking. As seen in Figure 3.9, detected bubbles can be identified with a period (frequency f) of the overlying wave. Information on the breaking strength is carried by the bubble size R_0, if detected simultaneously. This is illustrated in Figure 3.12

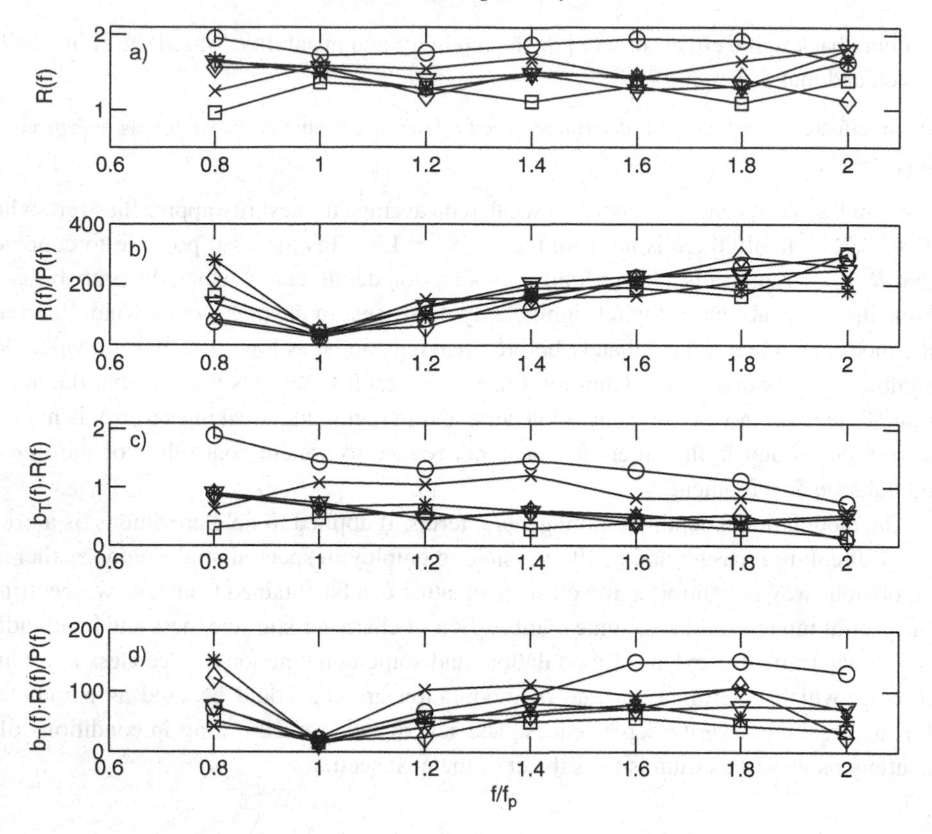

Figure 6.6 Breaking severity analyses versus wave frequency f normalised by the peak frequency f_p. a) Breaking strength assumed as bubble radius $R_0(f)$ in mm (denoted as R in the figure); b) Breaking strength $R_0(f)$ divided by the spectral density $P(f)$; c) Product of breaking strength $R_0(f)$ and breaking probability $b_T(f)$; d) Product of breaking severity $R_0(f)$ and breaking probability $b_T(f)$ normalised by the spectral density $P(f)$. Squares: 12.8 m/s; *: 12.9 m/s; ∇: 13.2 m/s; diamonds: 13.7 m/s; ×: 15.0 m/s; circles: 19.8 m/s. The records are from Table 5.2

for breaking waves in a uniform train. Similar calibration in a spectral environment is yet to be conducted.

In Figure 6.6, the four subplots demonstrate a dependence of the severity-related parameters on wave frequency for the six wave records of Table 5.2 depicted previously in Figure 5.27. The mean-bubble-size distribution with wave frequency $R_0(f)$ is shown in the top panel (R_0 is the radius of the surface bubbles in mm). Since the bubble size does not depend on how many waves broke and it is quite uniform across the frequency range for each of the records, this means that the energy loss across the frequency, at least up to the relative frequency of $2f_p$ shown in the figure, is approximately constant. Since the waves away from the peak, and certainly at the double peak frequency are significantly smaller, this observation implies that the relative strength of the breaking, i.e. severity coefficient s (2.24), goes up at higher frequencies/smaller scales.

This is clearly illustrated in the second panel. A surrogate dimensionless severity here is plotted as a distribution of the ratio of $R_0(f)$, which stands for the energy loss, and spectral density $P(f)$, which is a measure of wave energy – across the frequency. The growth is very consistent and large, about six times on average between f_p and $2f_p$. Once again, we should stress that this estimate and the trend can only be treated qualitatively, since bubble size is not a linear proxy of the energy loss and the calibration in Section 3.5 was only done for a monochromatic wave.

Some interesting observations with respect to this surrogate severity can, however, be made. Firstly, unlike in the other subplots of this figure, all the records are bundled together. This fact signals that an increase of the severity across the spectrum is mainly a function of the relative frequency, with respect to the spectral peak, rather than of any environmental factors and forcings as in the other panels. Secondly, in the Lake George finite-depth environment, where the downshifting is apparently restricted or even blocked by the lower-frequency breaking, the severity of this breaking is greater than at the peak. In fact, breaking severity at the peak is the smallest.

It is interesting to notice here that the breaking-caused directional distribution of the dissipation is also smallest in the main direction of wave propagation (Figure 7.9, Section 7.3.6). That is, the breaking tends to make the spectral peak narrower, both in the frequency and in the directional domains.

Below the peak, the weakest severity corresponds to the highest breaking rates (see Figure 5.27). This is consistent with the observation made in Section 6.1 that, in the case of breaking due to modulational instability, a stronger wind forcing makes the breaking more frequent but less powerful. We should remember, however, that the lower-frequency breaking at Lake George is affected by the bottom proximity, i.e. influenced by a combination of causes.

Since the larger bubbles correspond to more severe breakers, the product of the mean bubble size $R_0(f)$ and the breaking rate $b_T(f)$ can be treated as a surrogate dissipation rate (2.20) at frequency f. Again, the surrogate dissipation must only be interpreted qualitatively, and not even in terms of proportionality.

Here, it has to be mentioned that another method, also based on bubble size, was suggested by Garrett *et al.* (2000) to connect the bubble-size spectra and the dissipation. It is based on an expression for the maximal bubble radius, for which bubble breakup is prevented by the surface tension, and observations of the bubble spectra. The energy dissipation is estimated from a dimensional energy-cascade argument. The energy brought in by the breaking is partially spent on breakup of the bubbles and cascading the bubbles by their size. The cascade is stopped at the bubble size whose further breakup is prevented by the surface tension which opposes the turbulent inertia forces. At this size, the bubbles will be accumulated and, if this radius is measured, it gives an indication of the energy initially produced by the breaker.

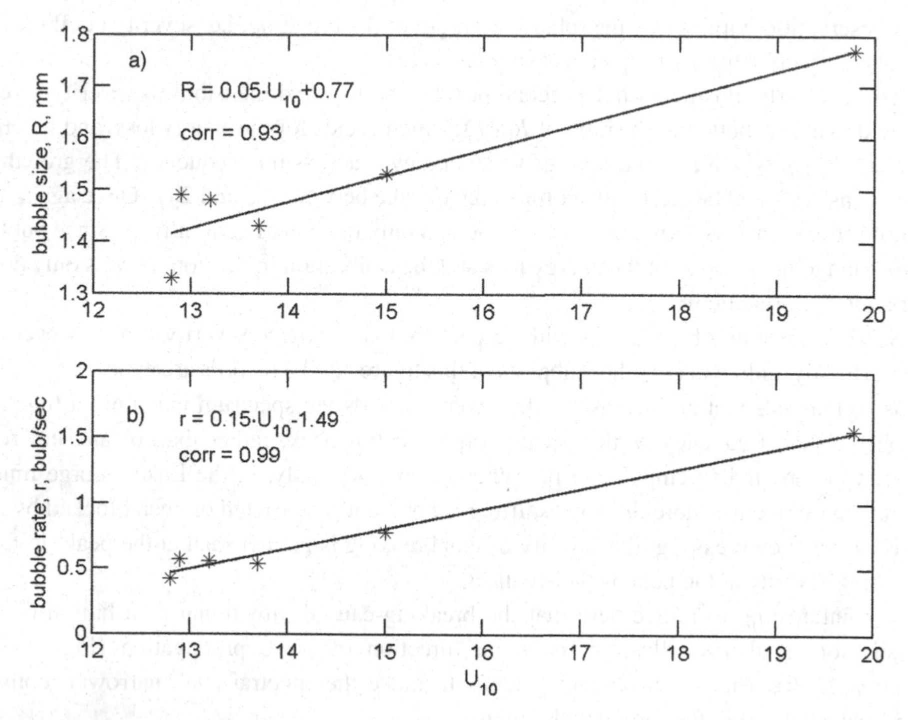

Figure 6.7 Trends with wind speed U_{10}. a) Mean bubble radius R_0 (denoted as R in the figure); b) Count rate. Combined data sets of Figure 6.6 are used. Figure is reproduced from Manasseh *et al.* (2006) © American Meteorological Society. Reprinted with permission

Back to our distribution of $b_T(f)R(f)$, it is shown in Figure 6.6c, and $b_T(f)R(f)$ normalised by the spectral density $P(f)$ in Figure 6.6d. These distributions exhibit features similar to those of $b_T(f)$ in Figure 5.27, including the cumulative effect due to induced breaking at smaller scales.

Further observations can be made in Figure 6.6a with respect to the wind effect on spectral-breaking strength. The mean bubble size $R_0(f)$ shows that the largest bubbles were produced by breakers at the highest wind speeds and the smallest bubbles were produced under the lightest winds. This is also illustrated in Figure 6.7, based on the bubble-detection technique both for breaking severity and bubble-generation rates. The increase in average bubble size with wind speed is now clear (Figure 6.7a). An increase in bubble-production rate with wind speed can be seen in Figure 6.7b. There is an apparent ordering of the wind speeds with respect to both the bubble rate and the mean radius, with quite high correlation.

In summary, higher wind speeds in this Lake George observation generate breaking events more frequently, and the bubbles at higher wind speeds are larger. We should remember, again, that R_0 is a proxy of dimensional energy loss rather than of the dimensionless severity coefficient s of Section 2.7. In Section 6.1, severity was reduced if mechanically generated wave trains of a certain height were forced by the wind. In Figure 6.7, energy loss increases when the wind forcing grows, but here stronger winds also correspond to higher waves.

In the field, this feature is responsible for a change of the ambient noise level (e.g. Manasseh *et al.*, 2006). As a result, this level depends on many factors., i.e. wind speed, wave height, breaking rate of dominant waves etc., and it is hardly possible to set a universal acoustic threshold to distinguish individual breaking events in the field by acoustic means (e.g. Babanin *et al.*, 2007b).

These details expand and reinforce the conclusion already made in Section 6.1 that, although correlations between the wind and the breaking strength can be well-traced in a particular circumstance, generally speaking wind influences are indirect. These influences must be coming through the wave dynamics which is imparted by the wind on temporal scales greater than the wave-breaking event duration. Such dynamics can differ between the two-dimensional modulated wave trains and fully spectral directional oceanic environments, even if both are subject to the same wind forcing. A similar conclusion has also been made with respect to the breaking probability (Section 5.3.1 and Figure 5.19). It should also be noted that both breaking probability and severity can themselves affect the wind input and therefore the wind (Section 8.3).

To conclude this breaking-severity chapter, we will summarise what is unknown rather than what is definite knowledge, because the former certainly outweighs the latter. What does the severity depend on? Will it be constant, even if on average, in different wave fields with the same background steepness, for instance? Does the breaking strength change across the spectrum, i.e. is the relative energy loss different for the dominant waves and the shorter waves away from the peak? For the latter waves, does their severity depend on whether they break due to inherent reasons or their breaking is induced by larger waves? Is the severity dependent on the wind (see Stolte, 1992; Babanin *et al.*, 2010a, in this regard)? How do wave directionality and shortcrestedness impact the breaking strength, if they do? Also, for example, how much of the energy loss is spent on work against buoyancy forces while entraining the bubbles into the water, and what fraction is passed on to water turbulence? These and other questions should be answered and parameterised for practical application, and this topic represents very interesting physics too.

7
Energy dissipation across the wave spectrum

In previous chapters, we have considered in some detail dissipation of wave energy in the course of individual wave-breaking events. This makes clear physical sense as breaking is an intermittent rather than continuous process, and the breaking rates in realistic circumstances are of the order of a few percent. That is, only a few waves out of a hundred break at any given spot or any given time, and that is sufficient to keep the energy balance in a wave system that experiences a consistent and uninterrupted energy input from the wind.

As discussed in Chapter 6 and throughout the book, it is often the case that more than one wave is affected in a single breaking occurrence. The dynamics of these waves is locked and coupled in many ways, and as far as the wave-energy dissipation is concerned, these energy losses are impossible to separate. Therefore, it usually makes sense to look at the properties of the wave group where the breaking took place rather than investigate the dynamics of some individual wave.

One way or another, there are no theoretical or even experimental approaches that would allow us to describe the breaking-dissipation process as such, in terms of some decay rate, as, for example, gradual energy decline due to the action of viscosity in fluid flows. First of all, unlike most if not all other dissipation causes, breaking dissipation has a start and an end. The former, as discussed throughout this book, is most likely due to the water surface becoming too steep and collapsing, and the reason for the end of breaking at this stage is perfectly unclear.

Then, gradual decline, i.e. constant dissipation rate, is perhaps not an option as a breaking-dissipation characteristic either. Plunging breaking has been portrayed above as a series of jets and splash-ups, and similar, although perhaps less intermittent behaviour should be applicable to spilling breakers with their multiple-jet structure.

Apart from the collapse of the water system, whose dynamics can apparently proceed in a number of different ways since it leads to very different outcomes in terms of energy loss as discussed above, breaking is also accompanied by aeration of the water, ejection of spray, generation of turbulence and other features. These would require complex two-phase-fluid modelling approaches, capable of reproducing rotational motions and multiple exchanges across the interface at various scales. The modelling should be conducted in three dimensions for two principal reasons: the breaking starts at the centre of the wave crest and then spreads laterally, and the turbulence produced by the breaking is essentially three-dimensional. Such fluid-mechanics approaches, in principle, are available, but dedicated efforts are needed to combine and further develop them for the wave-breaking application.

Therefore, experimental estimates of the breaking dissipation, both for monochromatic and spectral waves, have so far mostly concentrated on bulk energy loss or energy/momentum flux change, by measuring waves before and after breaking commenced and finished. These are purely empirical approaches, and they also need some empirical time scale in order to convert the results into dissipation rates. Such methods are necessary and very helpful in practical requirements, particularly if parameterised in statistical terms, but they reveal little about the physics of breaking in progress, and correspondingly about the control mechanisms of energy dissipation.

Theories on wave-breaking dissipation, correspondingly, have also primarily focused on interpreting properties of the wave fields by means of their state before and after the breaking takes place, usually in statistical terms. These will be summarised in Section 7.1 for the spectral models and in Section 7.2 for phase-resolvent models. Section 7.1.2 describes an interesting approach which is a combination of both. Experimental measurements of the wave-breaking dissipation in wave trains and fields with a continuous spectrum are not many, and they will be depicted in Section 7.3. Section 7.4 is dedicated to the implementation of spectral-dissipation functions in wave-forecast models. In this regard, that is, as far as modelling the wave energy dissipation is concerned, dissipation does not cease in the wave field if the breaking is not present. Therefore, we have found it relevant to outline non-breaking dissipation mechanisms in the last section of this chapter.

7.1 Theories of breaking dissipation

This section will provide a brief review of analytical theories of dissipation in wave fields with a continuous wave spectrum. Older theoretical models, based on the probability, quasi-saturated and negative-input approaches, have received substantial attention in the literature, and here we have combined them into a single subsection (7.1.1) and will accommodate the most recent updates, but will largely follow revisions by Donelan & Yuan (1994), Babanin *et al.* (2007e) and Babanin (2009). A new kinetic–dynamic approach will be described in a separate subsection (7.1.2).

In a general case, the spectral dissipation $S_{ds}(f, k, \theta)$ in (2.61) is a function of frequency f, wavenumber k and direction θ, but the theories so far only deal with omni-directional frequency-dependent dissipation which is assumed to be a function of spectrum $F(f)$:

$$S_{ds}(f) \sim F(f)^n. \tag{7.1}$$

Even in this relatively simple setup, the diversity of outcomes is such that proposed exponents n vary in the range $n = 1$–5, that is from the dissipation being a linear function of the wave spectrum to being highly nonlinear.

Donelan & Yuan (1994) broadly classified theoretical models of spectral dissipation into three types: probability models, quasi-saturated models and whitecap models. In Babanin *et al.* (2007e), we also added a turbulent model class to this classification, based on work by Polnikov (1993).

Polnikov (1993) proposed a dissipation model where residual features of after-breaking turbulence could supposedly be interpreted. He argued that, no matter what the cause of the breaking, the result is turbulence in the water. In this approach, the rate of wave-spectrum dissipation would be governed by the effective turbulent viscosity. Therefore, to describe the wave energy dissipation in a wave spectrum form, it was necessary to find a link between the wave spectrum and the water-turbulence spectrum.

This link was postulated, but was never further elaborated. Polnikov (1993) assumed a simplified representation of wave dynamics equations, and applied that to monochromatic waves. The assumptions in general may or may not be suitable, but even then, treating wave motion in terms of eddy viscosity is questionable. Most importantly, however, spectral waves of different scales interact, and the turbulent vortices of particular scales are not only generated as a result of dissipation of counterpart waves of the same scales, but also as a result of collapse of larger vortices (Kolmogorov cascade). Besides, the ocean is always turbulent, and not only is the spectrum of this background turbulence unrelated to the wave spectrum, but also a large proportion of such turbulence is produced by sources other than the waves. Therefore, the basic postulate appears unfeasible in principle, and this kind of model will be omitted here.

7.1.1 Probability, quasi-saturated and whitecap models

Historically, theories of breaking dissipation started with the work of Longuet-Higgins (1969a). This was an analytical probability model that considered the waves prior to breaking. It was further developed by Yuan *et al.* (1986, 2009) and Hua & Yuan (1992). The approach was described in detail in Section 3.8, and here we will only provide the most relevant summary.

The approach employs the probability distribution of surface elevations, assuming waves to be linear, in order to predict the likelihood of wave heights, or surface accelerations, or surface velocities exceeding those characteristic of the Stokes limiting wave. Such waves are unstable and will break, and as a result their height, as well as surface accelerations and velocities will reduce. The difference can be interpreted in terms of energy loss, and the resulting dissipation function appears to be a linear function of the wave spectrum.

The model is sensible in a physical sense, and in our view in this regard is the most robust out of the three approaches described in this subsection. There can be qualitative and quantitative concerns brought forward, but they are not of principal concern and can potentially be addressed in the future.

Indeed, as has been discussed in Sections 2.9, 5.1, and throughout the book, the Stokes limiting criteria mostly hold, even if with small variations and even though the wave at the breach of breaking onset does not actually appear to be the classical Stokes wave. The main uncertainty in the probability-based dissipation theories is the difference between the wave state before and after breaking.

The existing analytical approaches presume some constant lower level of wave height, at which the breaking should stop. These levels are different in different models, some are more and some are less realistic as described in Section 3.8. Essentially, however, in reality there is no constant lower wave-height level for the breaking to cease. As described in Section 2.7 and Chapter 6, the wave-height loss can range from 100% to 0%, i.e. the wave can completely disappear as a result of breaking, or can just marginally lose a top of its crest, with all possible variations in between. Therefore, even though conceptually attractive, the probability models, as they have been derived, are not quantitatively plausible at this stage.

The second type of analytical dissipation model is what Donelan & Yuan (1994) called quasi-saturated models (Phillips, 1985; Donelan & Pierson, 1987). These models rely on the equilibrium range of the wave spectrum, where some sort of saturation exists for the wave spectral density. Phillips (1985) argued that in this region the wind input, the wave–wave interactions and the dissipation should balance each other. Therefore, at each wave scale (wavenumber), any excess energy contributed by combined wind input and nonlinear interaction fluxes does not bring about spectral growth, but results in wave breaking and can be interpreted as the spectral dissipation local in wavenumber space. Donelan & Yuan (1994) found additional support to Phillips' assumption in the experiments of Pierson *et al.* (1992) where the breaking energy-loss was found to associate with the peak of propagating wave groups rather than being localised in the physical space. Thus, adjustments to the spectrum are focused in the wavenumber and frequency domain.

Phillips (1985) found that such dissipation is cubic in terms of the spectral density, i.e. $n = 3$ in (7.1). Donelan & Pierson (1987) added consideration of wave directionality to the energy balance of the equilibrium range, arguing that a simple balance between wind input and dissipation is not observed at large angles to the wind. They also separated dispersive (gravity and capillary) waves and non-dispersive (gravity–capillary) waves, as the nature of breaking differs for them because of different speeds of propagation relative to wave groups. Donelan & Pierson (1987) obtained a local-in-wavenumber-space dissipation function, similar to that of Phillips (1985), but their exponent n depends on the wave spectrum $F(k)$ and on the wavenumber k as such. According to them, n can vary significantly: $n = 1$–5. In most ranges of interest, however, $n \sim 5$ – both for gravitational and for capillary waves.

This model type has multiple shortcomings. The most important one is that, even if all the physics implied is true and correct, the results are valid at the spectrum tail and not valid at the spectral peak where a balance between the source functions cannot be expected. If so, the dissipation function based on the quasi-saturated approach is only applicable to a part of the spectrum, and is certainly not applicable to the most energetic waves which, when they break, produce most of the dissipation in absolute values.

In terms of the physics, however, there are essential issues too. Donelan & Yuan (1994) based their argument in favour of the local-in-wavenumber dissipation on observations of breaking in propagating wave groups. This wave-group argument, however, is strictly valid at the spectral peak only. The peak of the wind-generated spectrum is finite-width but narrow, which allows us to explain both the existence and properties of linear groups of dominant waves (e.g. Longuet-Higgins, 1984) and the development and instability of nonlinear wave groups in a narrow-banded environment (i.e. Zakharov, 1966, 1967; Benjamin & Feir, 1967). The tail of the spectrum is not narrow-banded and does not have a characteristic bandwidth in principle, and therefore the nature and very existence of the short-scale groups is unclear and questionable.

Thus, the quasi-saturated argument may be appropriate if it is at the spectrum tail, whereas the local-in-wavenumber-space dissipation is pertinent at the spectral peak, that is both reasons are not applicable at the same time. Furthermore, the quasi-saturated concept of equilibrium interval has been subject to various doubts, but even if it exists, the Phillips saturation level is not constant and depends on environmental conditions (e.g. Babanin & Soloviev, 1998a). Also, none of the source terms that shape the spectral balance are known explicitly and accurately enough to provide a reliable determination of the dissipation as a residual sink term. And finally, as discussed in Sections 5.3.2, 6.2 and throughout the book, there is a growing understanding that dominant waves and the breaking of dominant waves affect waves and their dissipation at smaller scales (Longuet-Higgins & Stewart, 1960; Phillips, 1963; Banner *et al.*, 1989; Meza *et al.*, 2000; Donelan, 2001; Young & Babanin, 2006a; Donelan *et al.*, 2010). If so, the dissipation in the saturation interval should consist of both functions local and non-local in wavenumber space. Therefore, the quasi-saturated models, with all the limitations outlined above, would still be only part of the story.

The whitecapping model was introduced by Hasselmann (1974) and, with a variety of empirical modifications (see e.g. Ardhuin *et al.*, 2007, for an overview), is the dissipation theory most frequently utilised in wave-forecast models, or at least most frequently referred to. This model relies on the distribution of well-developed whitecaps situated on the forward faces of breaking waves. According to Hasselmann (1974), once there is an established random distribution of such whitecaps, it does not matter what caused the waves to break: the whitecaps on the forward slopes exert downward pressure on upward-moving water and therefore conduct negative work on the wave. In a way, this mechanism is similar to the physics of wind-to-wave input which is determined by the downward pressure on the rear wave faces, just with the opposite sign. This dissipation model is a linear function of the wave spectrum, i.e. $n = 1$ in (7.1).

Two main assumptions of the whitecap model are that the dissipation, even if it is strongly nonlinear locally, is weak in the mean, and that the whitecaps and the underlying waves are in geometric similarity. Both assumptions are not strictly accurate. For example, Babanin *et al.* (2001) investigated wave fields with over 10% dominant breaking rates, and Young & Babanin (2006a) examined a 60% dominant-breaking case. It is hardly that the weak-in-the-mean approach is still applicable in such circumstances, which are apparently a regular feature of wind seas.

The geometric similarity is also a questionable approximation for real unsteady breakers. The whitecapping commences at some point on the incipient breaking crest and then spreads laterally and longitudinally (e.g. Phillips *et al.*, 2001) and may or may not satisfy the similarity assumption even in the mean. Therefore, both assumptions need experimental verification which has not been done. Most essentially, however, even the very concept of correlation of the whitecap-exerted pressure with the forward wave slope has never been verified experimentally. We should also point out that, before the distribution of established whitecaps is formed and they commence the negative work on the wave, some energy is already lost from the wave, which cannot be accounted for by such a model.

7.1.2 Kinetic–dynamic model

Zakharov *et al.* (2007) proposed a new analytical approach to the dissipation-function problem which, here, we will conventionally call the kinetic–dynamic model. Unlike the whitecap model in Section 7.1.1, which assumes the dissipation to be weak in the mean, this approach principally relies on co-existence of the so-called 'weak turbulence' and strong and rapid isolated dissipation events.

The weak turbulence comprises weakly-nonlinear background wave fields whose resonant four-wave interactions can be described by the kinetic equation for waves (Hasselmann, 1962) and result in energy flux across the spectrum from large to small scales where they eventually dissipate to viscosity (Zakharov & Zaslavskii, 1982). The strong turbulence is associated with wave breaking:

> "Even if the weak turbulent resonant interaction effects dominate in the greater part of space, strongly nonlinear effects could appear as rare localized coherent events... they could be catastrophic, in which case they are wave collapses... Even rare sporadic collapse events can essentially affect the physical picture of wave turbulence"
>
> (Zakharov *et al.*, 2007).

The coexistence of wave collapses and weak turbulence was verified and confirmed by means of numerical solutions of the primitive dynamic equations for the wave fields (Dyachenko *et al.*, 1992; Dias *et al.*, 2004). In Zakharov *et al.* (2007), the authors further simulated the wave evolution numerically be means of both the Hasselmann kinetic equation and basic Euler equations for the three-dimensional potential flow with a free surface. The first evolution is for a statistical ensemble and, as said above, is due to weakly nonlinear interactions only, and the second evolution is for dynamic variables whose statistics is then compared with the outcomes of the first method.

Qualitatively, agreement between the statistical and dynamic models was good and such key features of the wave evolution as spectral-peak downshifting, directional-spread broadening and ω^{-4} tail were observed in both simulations. Quantitatively, however, the dissipation due to breaking had to be introduced in the statistical model for the outcome to conform with the dynamic numerical experiment.

Thus, the quantitative estimate of whitecapping dissipation was obtained based on first principles. The analytical model, however, does not provide an explicit expression for this dissipation as a function of wave spectrum, i.e. an expression in the standard form (7.1) employed in relevant applications of this function, or some other formulation. Therefore, the authors tested a number of empirical parameterisations in the search for a dissipation term capable of accommodating the observed difference.

They demonstrated that standard dissipation terms used in wave-forecast models, i.e. WAM (Gunther *et al.*, 1992; Komen *et al.*, 1994), essentially overestimate the rates of spectral energy loss in wave fields with moderate background steepness. In order to achieve a reasonable quantitative agreement between the kinetic and dynamic approaches, a power function as strong as $F(f)^{12}$ had to be used. The authors concluded that this result supports the experimental observations of the threshold behaviour of wave breaking by Banner *et al.* (2000) and Babanin *et al.* (2001).

We fully agree with this interpretation. The new method appears to be a powerful analytical means for fundamental investigations of the spectral wave dissipation, capable of investigating and verifying both qualitative and even quantitative properties of the dissipation term. For the last part, however, it remains dependent on the choice of empirical function to be substituted as this term, and this choice needs to be made carefully.

Indeed, $S_{ds}(f) \sim F(f)^{12}$ seems unrealistic. Such a function may help to eliminate the dissipation below the threshold, where the whitecapping contribution is in fact expected to be zero or negligible, but it will provide extremely strong dissipation rates above the threshold where no other empirical or theoretical approach has suggested anything like $n > 5$. In our view, a dissipation function which directly incorporates the threshold behaviour, such as that in (5.40), can be tested in this regard. It would be interesting to see whether the 'moderate steepness' of Zakharov *et al.* (2007) corresponds to the experimental threshold of Babanin *et al.* (2007d, 2010c) described in Section 5.3.2.

7.2 Simulating the wave dissipation in phase-resolvent models

From the point of view of wave-energy dissipation, the phase-resolvent models can be subdivided into two large groups. The first group, the models that simulate the water and air sides of the wavy surface separately, such as the study of water–wave evolution by Zakharov *et al.* (2007) described in Section 7.1.2 above, or an air–sea model which includes the full set of wind–wave interactions (Chalikov & Rainchik, 2011) – have to incorporate the dissipation implicitly. In order to be able to describe the breaking dissipation explicitly, the second group have to be two-phase models, such that they allow for the dynamics of air bubbles injected into the water side and of water droplets ejected into the air side. A significant number of such models are available now (Abadie *et al.*, 1998; Zhao & Tanimoto, 1998; Chen *et al.*, 1999; Watanabe & Saeki, 1999; Mutsuda & Yasuda, 2000; Christensen & Deigaard, 2001; Grilli *et al.*, 2001; Guignard *et al.*, 2001; Tulin & Landrini, 2001; Hieu *et al.*, 2004; Song & Sirviente, 2004; Zhao *et al.*, 2004; Iafrati & Campana, 2005; Dalrymple & Rogers, 2006; Lubin *et al.*, 2006; Liovic & Lakehal, 2007; Iafrati, 2009; Dao *et al.*, 2010; Janssen & Krafczyk, 2010; Lakehal & Liovic, 2011, among others).

Implicit simulation of the dissipation in one-phase-medium models is handled by means of what Zakharov *et al.* (2007) call pseudo-viscosity. Techniques can be different, but the idea is to take the extra energy from the system by artificial means which have the same overall and long-term effect as would the dissipation due to breaking and other effects.

Such a technique is described in detail in Chalikov & Sheinin (2005) and Chalikov & Rainchik (2011). The authors employ two types of dissipation, what they call 'tail dissipation' and 'breaking dissipation'. The tail dissipation in a way mimics viscosity which absorbs the energy fluxes from large to small scales (see e.g. Zakharov & Zaslavskii, 1982, and Section 7.1.2). If not attended to, this energy can accumulate at large wavenumbers, close to the truncation number M, and corrupt the numerical solution.

In order to prevent this, in Chalikov & Rainchik (2011) 'tail dissipation' was added to the right-hand sides of (4.7) and (4.8)/(4.14) in Fourier space:

$$\frac{\partial \eta_k}{\partial \tau} = -\mu_k \eta_k, \tag{7.2}$$

$$\frac{\partial \phi_k}{\partial \tau} = -\mu_k \phi_k. \tag{7.3}$$

Here, μ_k is a damping coefficient such that

$$\mu_k = \begin{cases} rM\left(\dfrac{|k| - k_d}{M - k_d}\right)^2 & \text{if } |k| > k_d, \\ 0 & \text{if } |k| \leq k_d. \end{cases} \tag{7.4}$$

Tuning parameters $k_d = M/2$ and $r = 0.25$ were chosen in Chalikov & Rainchik (2011), and little sensitivity of the outcomes to reasonable variations of these parameters was found. Such a model of tail dissipation is quite straightforward and physical; an increase of the truncation number M shifts the dissipation towards higher wavenumbers, and in any case modes at $|k| < k_d$ are not affected.

Modelling explicitly the dissipation due to breaking, however, represents a considerable challenge. As discussed in Section 4.1, there are both mathematical and physical reasons why potential models cannot go far beyond the breaking-onset point in time or space. But the breaking is a regular occurrence at the scale of tens of wave periods/lengths or less in a wave train/field with background steepness $ak \geq 0.1$, and this is the typical steepness of natural wave trains. Therefore, some realistic breaking-dissipation means have to be developed in order to model phase-resolvent evolution of wave trains at the scale of hundreds/thousands of periods, which is the typical scale for such evolution in deep water in the field.

Chalikov & Sheinin (2005) and Chalikov & Rainchik (2011) used the concept of preventing instability, borrowed from atmospheric models of free convection, and developed an algorithm of breaking parameterisation based on smoothing the interface when indications are that it is becoming too steep and the instability can develop. The differential parameterisation employs diffusion-type terms which are added to the right-hand sides of (4.9) and (4.10) as

$$\eta_\tau = J^{-1}\frac{\partial}{\partial \xi}B\frac{\partial \eta}{\partial \xi}, \tag{7.5}$$

$$\phi_\tau = J^{-1}\frac{\partial}{\partial \xi}B\frac{\partial \phi}{\partial \xi}. \tag{7.6}$$

Their diffusion coefficient B depends on the second derivative of the surface:

$$B = \begin{cases} C_b = \left(\Delta\xi \dfrac{\partial^2 \eta}{\partial \xi^2}\right)^2 & \text{if } \dfrac{\partial^2 z}{\partial \xi^2} > s, \\ 0 & \text{if } \dfrac{\partial^2 z}{\partial \xi^2} \leq s \end{cases} \tag{7.7}$$

where the tunable parameters were chosen as coefficient $C_b \approx 0.1$ and the second derivative as $s \approx 300$.

The algorithm (7.5)–(7.7) obviously does not affect the solution in the absence of breaking. When active, it does not change the volume of the water, but reduces the momentum and energy of the waves which are assumed to be passed on to the motions beyond the wave model, i.e. mean current and turbulence. Such a transfer can be attended to separately outside the potential approach (e.g. Chalikov & Belevich, 1993).

Thus, on the macroscopic level, the differential diffusion-type parameterisation plays the role of the wave breaking in the wave system, and the authors showed this behaviour to be quite realistic. Within the potential model, the parameterisation of the rotational phenomenon prevents development of instability if the second derivative s is becoming too large and thus prevents development of the breaking as such. Therefore, this is not in itself a physical model of the breaking, it is simulating the physical consequences of the breaking necessary for modelling longer-term wave evolution. Like any parameterisation, it has to be applied with caution, and the authors outline the limits of such applications – i.e. the scheme may fail to prevent the breaking occurrence within the model if the initial steepness or the wind energy input are too high.

In this regard, explicit modelling of the breaking in progress, from the start to end of breaking, is an interesting task for fluid mechanics. Such models, capable of handling two-phase flows, have existed for over a decade now (e.g. Abadie *et al.*, 1998; Zhao & Tanimoto, 1998; Chen *et al.*, 1999; Watanabe & Saeki, 1999; Mutsuda & Yasuda, 2000, among the earlier ones), and these days they can even include the effects of viscosity, surface tension, air and water compressibility (i.e. acoustic effects), and three-dimensionality.

These models reproduce the behaviour and multiple observed features of the breaking in progress very realistically, including the formation of turbulence, bubble injection, production and break down, spray emission – key features of the dissipation, i.e. dynamical processes on which the wave spends its lost energy and momentum. In the past, the serious limitation for this kind of model was the extremely high computational demands, and this is still the case, but with the modern rate of development of computing facilities essential two-phase modelling efforts become increasingly feasible. Besides, the breaking in progress as such is only short, a few periods at most including all the turbulent, spray and bubble consequences. Long is the evolution from background conditions to the breaking onset, and that part is simulated very well by fully nonlinear potential models whose computational cost is much less (e.g. Dommermuth *et al.*, 1988; Chalikov & Sheinin, 1998, 2005). Thus, the whole breaking-dissipation process can be covered if the potential-model outputs are connected with the two-phase model input.

Lagrangian models of this type (Tulin & Landrini, 2001; Dalrymple & Rogers, 2006; Dao *et al.*, 2010) have been mentioned and described in some detail in Section 4.2. Lagrangian models are very efficient in handling the nonlinearities of flow processes involved, but have to deal with the conservation issues for fundamental physical properties. Those issues appear to be solvable and therefore it should be a matter of time and dedicated effort to test and use this model for wave-energy dissipation studies.

A number of elaborate non-Lagrangian methodologies were also developed in computational fluid dynamics, and then applied to modelling breaking in progress (Abadie *et al.*, 1998; Zhao & Tanimoto, 1998; Chen *et al.*, 1999; Watanabe & Saeki, 1999; Mutsuda & Yasuda, 2000; Christensen & Deigaard, 2001; Grilli *et al.*, 2001; Guignard *et al.*, 2001; Hieu *et al.*, 2004; Song & Sirviente, 2004; Zhao *et al.*, 2004; Iafrati & Campana, 2005; Lubin *et al.*, 2006; Liovic & Lakehal, 2007; Iafrati, 2009; Janssen & Krafczyk, 2010; Lakehal & Liovic, 2011, among others). Some allow us to combine an Eulerian model with Lagrangian free-surface tracking (e.g. Grilli *et al.*, 2001; Guignard *et al.*, 2001; Janssen & Krafczyk, 2010). Others use direct numerical simulations combined with sophisticated techniques for capturing the interface (e.g. Abadie *et al.*, 1998; Zhao & Tanimoto, 1998; Chen *et al.*, 1999; Song & Sirviente, 2004; Iafrati & Campana, 2005; Lubin *et al.*, 2006; Liovic & Lakehal, 2007; Iafrati, 2009; Lakehal & Liovic, 2011). Some assume the grid size to be refined enough to take into account scales as small as necessary (e.g. Abadie *et al.*, 1998; Chen *et al.*, 1999; Grilli *et al.*, 2001; Guignard *et al.*, 2001; Song & Sirviente, 2004; Iafrati & Campana, 2005; Iafrati, 2009; Janssen & Krafczyk, 2010). Others combine direct numerical simulations with special treatment of the small-scale processes by means of the large eddy simulation (LES) method (e.g. Zhao & Tanimoto, 1998; Christensen & Deigaard, 2001; Lubin *et al.*, 2006; Liovic & Lakehal, 2007; Lakehal & Liovic, 2011) or the Reynolds average Navier Stokes approaches (e.g. Zhao *et al.*, 2004). All of these methods have an apparent potential capacity to address the wave-dissipation problem explicitly, and in this regard intercomparison of the outcomes some time down the track would be most instructive.

Here, we will briefly review the study by Iafrati (2009) who, among other physical problems, specifically investigated dissipation rates in the course of breaking progress, i.e. the topic of the present section. This was done for a single two-dimensional wave of steepness $\epsilon = ak$

$$\eta(x) = \frac{a}{\lambda}\left(\cos(kx) + \frac{1}{2}\epsilon \cdot \cos(2kx) + \frac{3}{8}\epsilon^2 \cos(3kx)\right) \tag{7.8}$$

where the initial steepness ranged very broadly, from $\epsilon = 0.3$ through $\epsilon = 0.65$. A Navier–Stokes solver was employed for two-phase flow of incompressible air and water, with a level-set method to capture the surface.

In the paper it was implicitly assumed that the breaking intensity (severity) will depend on the initial steepness. As discussed in Section 4.1, based on the Chalikov & Sheinin (2005) fully nonlinear model, strictly two-dimensional waves should break within one period if steepness $\epsilon \geq 0.3$. Iafrati (2009) finds this steepness to be

$$\epsilon \geq 0.33 \tag{7.9}$$

which is in good accord with the above conclusion and thus can serve as an independent corroboration of his model. The highest steepness of $\epsilon = 0.65$ used by Iafrati (2009) is unrealistic (see Babanin *et al.*, 2007a, 2010a; Toffoli *et al.*, 2010a) unless such a wave is produced by some artificial means, which is perhaps why the outcome of numerical simulations for waves deviate steeply from the trend established for steepnesses in the range $\epsilon = 0.33$–0.60. In the discussion further down, we omit results for the $\epsilon = 0.65$ steepness.

In between, Iafrati (2009) found the breaking to be of a spilling type if steepness is in the range $0.33 \leq \epsilon < 0.37$, and of a plunging type for $\epsilon \geq 0.37$. According to this model, the type of breaking and the spilling/plunging behaviour difference (see Section 2.8) is determined by the role of surface tension:

"For small amplitude breaking waves, the velocity of the jet tip is not strong enough and thus surface tension forces prevent the formation of the jet which is replaced by a bulge developing about the wave crest."

This bulge starts to slide down the wave's front crest and so the spilling breaking develops. For larger amplitudes, the surface tension goes round the tip of the jet, but cannot stop it, and the plunging breaking is produced (see also Iafrati, 2011).

It has to be noted that the steepnesses mentioned above are background values for the initial wave slopes. Iafrati (2009) does not mention what the value of the steepness was at the breaking onset, but he does point out that in the cases that resulted in wave breaking, progressive steepening of the wave profile was observed before the breaking started. As we know (see Sections 4.1.2, 5.1.2), for two-dimensional waves the breaking onset steepness should be $Hk/2 \approx 0.44$.

The dissipation was investigated in terms of the behaviour of the kinetic energy E_K and potential energy E_P:

$$E_K = \frac{1}{2} \int \rho \left(u^2 + w^2\right) dx dz, \tag{7.10}$$

$$E_P = \frac{1}{2} \int \rho z dx dz, \tag{7.11}$$

(here, we have diverged from the original symbols employed by Iafrati (2009) for consistency throughout this book). Decay of both E_K and E_P, as well as of the total energy

$$E_T = E_K + E_P \tag{7.12}$$

were scrutinised.

For non-breaking waves of $\epsilon = 0.2$ and $\epsilon = 0.3$, the total-energy decline is well described by theoretical expectations due to the action of viscosity (e.g. Landau & Lifshitz, 1987). Once the breaking threshold (7.9) for background steepness is exceeded, the energy dynamics becomes much more interesting.

Unfortunately, only one spilling-breaking case ($\epsilon = 0.35$) is analysed in Iafrati (2009), and therefore it is impossible to draw conclusions about the trends and dependences for this class of dissipation. Compared to viscous decline, the rate of energy drop is much larger, and some 25% of additional loss is observed. It is interesting that in dimensional terms, the time history of this spilling-breaking dissipation eventually leads to the same energy level as in the threshold case of no breaking (steepness $\epsilon = 0.3$). If it is confirmed that there is a consistent outcome, whatever the mean steepness, if it results in a spilling breaking, then the energy will be reduced to the (7.9)-threshold level – then parameterising the spilling dissipation would be a relatively straightforward task, for example, by means of probability models described in Section 7.1.1. This guess would need further investigation, particularly important as spilling breaking is the most frequent breaking occurrence at sea, at least for energetic waves (see Section 2.8).

The plunging breaking in Iafrati (2009) is essentially stronger, but dissipation, instead of approaching some constant dimensional level ultimately, with some scatter asymptotes to some constant dimensionless level of approximately 45–55%. Such a result, if confirmed, first of all provides a very apparent distinction between the spilling- and plunging-breaking dissipation. Secondly, the outcome is in excellent agreement with the experimental estimates of Rapp & Melville (1990).

The latter fact, however, may outline limits on the conclusions drawn as a result of breaking simulations by Iafrati (2009). As discussed in Sections 2.7 and 6.1, the certain-percentage loss of initial wave energy to breaking is a feature of breaking caused by the linear superposition of waves. Modulational-instability breaking may result in any amount of loss, from virtually 0% to 100% when the breaker disappears. Since the latter breaking cause appears more likely in field conditions (Babanin *et al.*, 2011a), the topic needs further research, specifically into dissipation due to modulational-instability breaking.

This further research would need to be extended to simulations of breaking within wave groups, rather than of single waves, as the depth of the nonlinear-group modulation appears to be at least one of the properties well correlated with modulational-breaking severity (Section 6.1, Galchenko *et al.*, 2010). Alternatively, the individual waves can be simulated as before, but the initial conditions for the surface elevations and velocity distributions should be taken as those characteristic of the breaking onset brought about by modulational instability. This can be achieved, for example, by coupling the output of the Chalikov & Sheinin (2005) model with the input of the Iafrati (2009) model.

With these potential limitations in mind, we shall return to other very interesting conclusions of Iafrati (2009) regarding the rate of wave-breaking dissipation. In temporal domain t, from start to finish of the plunging-breaking active phase, the total energy appears to follow the dependence of

$$E_T(t)/E_T(0) \sim t^{-1}. \tag{7.13}$$

Again, if confirmed as a general feature of wave-breaking dissipation, such a function can provide a principal transition in understanding and presenting the dissipation function. Like other source/sink functions in (2.61), the dissipation term can then be approached in physical terms of the gradual evolution of a wave spectrum due to such dissipation, rather than in empirical-parameterisation terms of what was with the spectrum/wave-energy before and after a breaking event.

Iafrati (2009) then follows on to elucidate what processes contribute to the observed plunging-breaking dissipation, and to quantify their relative importance. Immediately after the jet impacts the free surface, two additional active subsurface phenomena appear. The first is a strong rotational flow in the water caused by the large air cavity entrapped, and associated viscous dissipation. The second is work against the buoyancy forces (see also Lammarre & Melville, 1991, 1992; Blenkinsopp & Chaplin, 2007). As far as the wave energy dissipation is concerned, these two processes can be regarded as independent and studied separately.

Therefore, the efficiency and magnitude of different terms in the total-energy balance equation are investigated. The viscous dissipation rate, in the most severe cases, grows up to an order of magnitude following the breaking onset. This high dissipation level can last for about one wave period. After that, the viscous dissipation itself adopts a t^{-2} decay rate.

The overall contribution of the viscous dissipation, however, was only about 15% of the initial energy, which is just one third of the 45–55% energy loss (about one half is spent on work against buoyancy as mentioned above). At the opening stage of the wave breaking, i.e. during the first half-wave-period, the viscous dissipation is even less and only accounts for some 2% of the energy loss. The bulk of the energy in these early times is spent on work done in accelerating the air phase of the fluid mixture, which has been still before breaking. For the rest of the dissipation sources, Iafrati (2009) concludes that the role of the normal and tangential stresses at both sides of the interface is essential too, but the model in its presented version was not able to provide reliable estimates in this regard.

In physical space, the dissipation is mainly localised in a narrow layer below the surface around small air bubbles produced by the collapse of the large air cavity captured by the plunger.

"Sharp velocity gradients are induced by the interaction of the rotating structures with the surrounding fluid at rest".

The simulations of Iafrati (2009) were two-dimensional, and analyses of three-dimensional breaking by means of two-phase models are also available. Most of such modelling efforts are dedicated to shallow waters, which are apparently more interesting for coastal engineering applications which drive these efforts, but are a somewhat different, perhaps quite different phenomenon compared to the deep-water breaking considered here.

Lubin *et al.* (2006) dedicated a section of their paper to investigation of the time evolution of the dissipation of a breaking wave with initial height $H = 0.1\lambda$ in a water depth of $d = 0.1\lambda$, where λ is the length of this wave and the energy is presented in (7.10)–(7.12) terms. The wave period was 0.339 s, and for the 0.1 s preceding the breaking onset the total energy was slightly dissipating to viscosity, while the kinetic energy was being converted into potential energy due to steepening and growing of the wave crest. The latter pre-breaking-onset energy behaviour is familiar from simulations by means of potential models (Chalikov, 2009).

Once the jet starts to plunge and accelerate, potential energy transforms back into kinetic energy until it hits the surface in front of the wave. This happens within 0.07 s, i.e. approximately 1/5th of the wave period, which result is in excellent agreement with the conclusion made in Section 2.4 that the developing phase of the breaking constitutes 20% of the active-breaking duration (provided this duration is of the order of a wave period). The jet rebounds, and the kinetic energy keeps increasing until it reaches a maximum in another 0.03 s which corresponds to generation of the first splash-up. By this time, more than 60% of the original potential energy is either turned into kinetic energy or dissipated.

At $t \approx 0.2$ s after the breaking onset, the splash-ups exhaust themselves, and at this stage about 40% of the total pre-breaking wave energy is dissipated. Then, the energy continues to dissipate at a slower rate, and in five wave periods about 35% of the total energy, 40% of the kinetic energy and 30% of the potential energy remain in the wave.

The difference between the dissipation in two-dimensional and three-dimensional plunging breakers, as concluded both by Lubin *et al.* (2006) and Iafrati (2009), becomes essential at these later stages of the breaking. Iafrati (2009) argues that overturning of the jet and the first jet impact are basically two-dimensional processes. With subsequent splash-ups, vortex generation and production of turbulence, three-dimensional effects become essential. Among them, Iafrati (2009) points out that lateral instabilities affect both the air-pocket fragmentation and dynamics of the large vorticity structures.

These effects, not accounted for in two-dimensional models, should lead to larger levels of energy dissipation in a three-dimensional case. Analogies between the deep-water breaking such as in Iafrati (2009) and the shallow-water breaking of Lubin *et al.* (2006) should be drawn with caution, as the dynamics can change significantly between the former case and the latter, but still the examination of the differences between 2D and 3D cases done by Lubin *et al.* (2006) is quite instructive.

Up to some half of the wave period after breaking onset, the two-dimensional and three-dimensional total-energy curves are hardly distinguishable. The difference then appears and grows up to 10% over some two wave periods. After that, the role of the three-dimensional turbulence and other 3D effects apparently subsides, and while a different amount of energy is now left in a wave, the curves remain parallel to each other.

Thus, the phase-resolvent models represent an excellent opportunity, grossly under-used and under-developed at the present stage, for investigating dissipation behaviour in the course of wave breaking. While one-phase models can simulate outcomes of this behaviour quite realistically as described above, the two-phase models are in principle capable of explicitly reproducing the dissipation process. So far, most of them have concentrated on simulations of plunging breakers in shallow waters, which cases are apparently more challenging and more appealing from the point of view of verification and demonstration of the method. And even then less attention was paid to the dissipation as such, among such interesting phenomena of the breaking as capturing and fragmenting the air pocket, formation of the coherent structures, splash-ups and other free-surface behaviours, transfer of the momentum to the currents and of the energy to the turbulence and vorticity. Given all these achievements, investigation of the breaking in progress, and particularly the spilling breaking which is most abundant in field conditions, by such two-phase models is quite feasible and for now should only be a matter of time and concentrated effort.

7.3 Measurements of the wave dissipation of spectral waves

Measurement of the wave dissipation of monochromatic waves, in a way, was discussed with breaking-severity in Chapter 6. Severity, combined with breaking probability described in Chapter 5, provides the dissipation estimate (see also Section 2.7).

Subdivision of the monochromatic cases and spectral cases, however, is quite superficial. As mentioned throughout the book, strictly sinusoidal linear water-surface waves do not exist. At the very least, they immediately turn themselves into Stokes waves, and therefore in the Fourier space will produce bound harmonics at higher frequencies/wavenumbers (see e.g. Chalikov, 2011), which fact has implications for wave breaking as we shall see below. If the waves are not infinitesimally small, they start interacting with background turbulence and generate resonant sidebands, leading eventually to all sorts of nonlinear interactions in the wave system and to spreading into a broader spectrum (see e.g. Yuen & Lake, 1982, for a review).

Most essentially from the point of view of the present book, however, is that the dissipation impact of the breaking is never local in frequency–wavenumber space. Even in the simplest artificial experiment, when a quasi-linear wave is produced by a wavemaker and made to break mechanically (for instance, by means of a subsurface obstacle), the outcome will be spectral – at the very least, the plunger will generate some propagating ripples of frequencies higher with respect to the original breaker.

Spectral impacts of natural breaking are well documented in the literature (e.g. Melville, 1982; Pierson *et al.*, 1992; Tulin & Waseda, 1999; Waseda & Tulin, 1999; Meza *et al.*, 2000; Manasseh *et al.*, 2006; Young & Babanin, 2006a, among others). In this section, we will describe such measurements conducted in the laboratory (Section 7.3.1) and in the field (Section 7.3.3). Different causes may lead to the breaking, such as linear or nonlinear dispersion, modulational instability and strong wind forcing, and it appears that the dynamics of the resulting dissipation and the spectral outcome of the breaking will also be different. The wave breaks because it becomes too steep, and not because of linear focusing or nonlinear instability, but in a way it 'remembers' what made it that steep.

These spectral consequences of the breaking will be discussed in Section 7.3.2. Cumulative effect, an important feature of dissipation in wave fields with full spectrum, is described in Section 7.3.4. And finally, wind-forcing and directional-distribution issues pertinent to the spectral dissipation will be attended to in Sections 7.3.5 and 7.3.6, respectively.

7.3.1 Laboratory measurements

A lot of laboratory effort, if not most of it with respect to wave-energy dissipation, has concentrated on the total energy loss in a wave train where a breaking occurs. These can be trains of monochromatic waves (e.g. Manasseh *et al.*, 2006), or quasi-monochromatic waves (e.g. Babanin *et al.*, 2007a), or spectral waves (e.g. Rapp & Melville, 1990). In this regard, we also see a recent experimental study complemented with an eddy-viscosity model which investigated the dissipation rates, rather than the total loss, and associated temporal and spatial scales in unsteady plunging breaking (Tian *et al.*, 2010).

The output of the breaking, however, as we have just mentioned above, is always spectral. That is, the energy is never lost from a particular Fourier scale, i.e. frequency or wavenumber, but is rather distributed across a range of scales, or at least has an impact at a number of scales.

Experiments dedicated to the investigation of this spectral impact are much fewer, and here we shall highlight two of them. Both were most thoroughly conducted in controlled laboratory conditions, but produced quite different, in fact directly opposite in some regards, conclusions.

In Tulin & Waseda (1999), evolution of nonlinear deep-water wave groups was tested in a 50 m-long tank. Waves were produced mechanically by a sensitive programmable wave generator. Experiments were conducted for waves of 1–4 m in length, with the initial steepness of primary waves varying in the range $\epsilon = 0.10$–0.28. For such wavelengths and steepnesses, the tank is long, but not long enough for the Benjamin–Feir instabilities to grow and lead to the breaking. Therefore, small sidebands were initially imposed over a steeper carrier wave, in order to speed up the development of instability and to study the outcomes of modulational-instability breaking. The sidebands broadly covered the expected instability range which, in terms of the inverse modulational index M_I (2.12), was $\hat{\delta} = \delta\omega/\epsilon\omega = 0.2$–$1.4$. An array of eight high-resolution wire wave probes were deployed along the tank in order to obtain evolution of the spectrum of such wave trains, including the breaking effects.

Without breaking, the energy was dynamically exchanged between the primary wave and sidebands in the course of evolution, but the exchange was reversible and near-recurrence occurred at some stage (see Figure 7.1 reproduced from Figure 18 of Tulin & Waseda (1999)). When breaking happened, it was always plunging. The ultimate dynamics in this scenario was quite different (also in this figure). There was no recurrence, and rather the wave energy was shifted down the spectral scales.

Tulin & Waseda (1999) provide an interesting and detailed account of the generation of both new free components and bound harmonics, as well as the behaviour of the

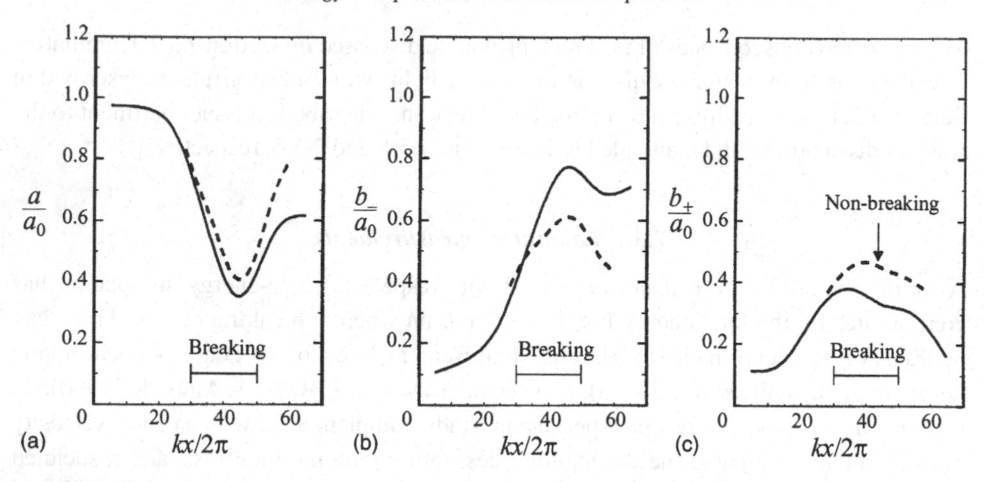

Figure 7.1 Evolution of the wave modes for cases with and without breaking events: (a) the carrier, (b) the lower and (c) the upper sideband waves are presented. The solid line is the evolution of the case with breaking, and the broken line is the evolution of the non-breaking case. Breaking case is $\epsilon = 0.133$, $\hat{\delta} = 0.785$; non-breaking case is $\epsilon = 0.1$, $\hat{\delta} = 0.894$. Figure is reproduced from Tulin & Waseda (1999) by permission

high-frequency spectrum tail. This includes discussion of differences between their observations, which concluded that the newly produced tail spectrum was discrete, and those of Melville (1982) who, also for modulational-instability breaking, obtained a rise in the continuous-spectrum tail.

Here, we will mostly be interested in their conclusions on what happened to the initially existing primary-wave train and two sidebands, as these accounted for the bulk of energy exchange and dissipation (note that this behaviour will be altered in the presence of the wind, i.e. Waseda & Tulin, 1999). Tulin & Waseda (1999) write:

"The end state of the evolution following strong breaking is an effective downshifting of the spectral energy, where the lower and the carrier wave amplitudes nearly coincide"

(see also Reid, 1992). This is illustrated in Figure 7.1: after the breaking, the carrier-wave amplitude a became approximately 60% of the original amplitude a_0, the lower-sideband b_- grew up and stayed at some 70% of a_0, while the upper-sideband b_+ initially rose higher if compared with the recurrence case, but then continued to decrease towards its pre-breaking magnitude.

The most important outcome here for further discussions is that, as a result of the breaking caused by the modulational instability, the primary wave lost the energy, and this loss was substantial. The impact of the breaking was spectral, that is – not only the wave energy was lost from the wave system, but it was also redistributed along the spectrum. The fact that the energy is passed on to lower spectral scales is very instructive too. If such breaking happens in real wave fields, it may be responsible, at least partially, for the spectrum peak downshifting. Since the wind input S_{in} in the radiative transfer equation (2.61) becomes ineffective in providing energy to long waves, whose speed is comparable with the wind speed, the downshifting in RTE is customarily attributed to the work on nonlinear interactions depicted by S_{nl}. As discussed in Chapter 1, this work is slow and takes place at the scale of hundreds or thousands of wave periods. By comparison, the breaking due to modulational instability, if the background wave steepness (5.11) of these long waves can be maintained through the energy sources, is much faster – i.e. it has a temporal scale of tens of wave periods (Babanin *et al.*, 2007a). If so, then the dissipation term S_{ds} may accept responsibility for the downshifting in such wave fields.

In the case of strong modulation, Tulin & Waseda (1999) notice that the breaking is observed not necessarily for the most unstable sideband mode, but in the broad range of initial perturbations. This observation may be related to the dependence of the breaking severity on the modulation depth discussed in Section 6.1 and Galchenko *et al.* (2010). Such an observation also has implications for the instability breaking in wave fields with full spectrum where a continuous set of carrier waves and their perturbations are available.

Tulin & Waseda (1999) then followed on to suggest a theory for their observations of downshifting and sideband behaviour, both in the breaking and non-breaking cases (see also Section 8.1). In particular, in the case of breaking in a wave train with a known bandwidth $\delta\omega/\omega$, the energy difference between the upper and lower sidebands is a function of energy dissipation and momentum loss. For the plunging breaking, if contributions of the secondary sidebands are neglected (which are small), this difference can be parameterised in terms of the dissipation D_b:

$$\frac{\partial}{\partial t}(E_{-1} - E_{+1}) = \gamma \frac{D_b}{(\delta\omega/\omega)} \tag{7.14}$$

where $\gamma = 0.4$ is an empirical coefficient.

The second laboratory experiment to be highlighted here is that by Meza *et al.* (2000). The authors also deal with spectrum measurements before and after the breaking, in carefully controlled conditions, but the breaking was produced by means of frequency dispersion, i.e. linear superposition. Once the superposition leads to a steep wave, nonlinear effects also start playing a role (Brown & Jensen, 2001), but still the wave-breaking causes are very different by comparison with a purely nonlinear dynamics leading to the breaking in the case of modulational-instability development, discussed above.

Meza *et al.* (2000) conducted experiments in a 36 m-long tank, with deep-water waves breaking at a prescribed location either in spilling or plunging fashion depending on the initial steepness.

"Very steep transient wave trains were formed by sequentially generating a series of waves from high to low frequencies that superposed at a downstream location".

Thus, the setup was essentially the same as in Rapp & Melville (1990), with some differences in the form of initial spectra produced. High-resolution resistance wire probes were used to measure the spectra in four locations such that the breaking happened between two of them.

The authors developed a special technique to separate free waves from the bound harmonics, and compared spectra of free-wave components before and after the breaking. One of their main conclusions was that

"the energy loss was almost exclusively from wave components at frequencies higher than the spectral peak frequency. Although the energy density of the wave components of frequencies near the peak frequency is the largest, they do not significantly lose or gain energy after the breaking".

Interestingly, Meza *et al.* (2000) also observed the spectral downshifting as a result of the breaking. They subdivided their spectrum into a low-frequency band of $f/f_p = 0.65$–1 and a high-frequency band of $f/f_p = 1$–3.5. The two bands were clearly separated in terms of the power-spectrum outcome, with the former gaining the energy and the latter losing it. The quantitative estimates varied a lot depending on the type of the breaking (spilling or plunging) and the mean initial steepness, with the average energy gain by the longer waves being 12% (median 8.5%), that is some 10%.

Thus, the two laboratory experiments, where the breaking was created by different physical mechanisms, produced quite different results not so much quantitatively as qualitatively – in terms of their spectral impact. These important differences and their implications will be discussed in the next subsection.

7.3.2 Difference in the spectral distribution of dissipation due to different types of breaking mechanisms

The current subsection is effectively a continuation of Section 7.3.1 above, but because of its significance for further discussions, we have separated it out. Indeed, the breaking-dissipation differences observed in the experiments of Tulin & Waseda (1999) and Meza *et al.* (2000) can be further used for identifying physical mechanisms of the breaking based on indirect observational features. These include field conditions where direct measurements of the breaking are particularly difficult due to its intermittent, random and destructive nature, and therefore the indirect methods are a more effective and feasible way of research. This will be demonstrated in Section 7.3.3 below dedicated to field measurements of the wave-energy dissipation.

In summary, when measuring the spectral consequences of the breaking caused by modulational instability, Tulin & Waseda (1999) found that the energy is lost from the carrier wave, the most energetic in the wave train. For the breaking produced as a result of linear focusing, Meza *et al.* (2000) revealed quite an opposite picture: hardly any energy goes from the most energetic waves, and the observed dissipation is maintained by the energy loss incurred by high-frequency free waves.

Some technical issues have to be mentioned which would need further investigations. In the case of Tulin & Waseda (1999), the initial condition was a line spectrum, whereas Meza *et al.* (2000) worked with a continuous spectrum. Neither, however, was similar to a typical ocean spectrum, e.g. of JONSWAP form (eq. 2.7, Hasselmann *et al.*, 1973) or Donelan *et al.* (1985) form.

The real ocean spectra are continuous and what is the primary component and what are the sidebands in such full-spectrum conditions have to be clarified, as far as the modulational instability is concerned. Studies of this kind are available (e.g. Onorato *et al.*, 2001; Janssen, 2003) and they demonstrate that the modulational-instability analogy should be applicable at the peak of typical narrow-banded wave spectra. Given the fact that the frequency band of this instability, relative to the primary wave, is quite limited (see e.g. Yuen & Lake, 1982; Tulin & Waseda, 1999), the high-frequency components within the continuous spectrum cannot be expected to participate in the nonlinear-interaction dynamics coupled with the dominant waves anyway.

Another conclusion of Tulin & Waseda (1999) and Meza *et al.* (2000), which potentially distinguishes instability breaking from focusing breaking, is the behaviour of low scales in the spectrum (wave components longer than those breaking), in response to the breaking. In both cases, spectral downshifting is observed, i.e. these components in fact gain energy as a result. That is, not all the energy lost is gone from the wave system, at least some part of it is transferred across the spectrum. This, again, highlights the potential of the dissipation term S_{ds} to participate, along with the nonlinear interactions described by S_{in} in (2.61), in the evolution of wind-generated wave fields towards larger waves.

Quantities of the energy gain, however, are very different. In the modulational-breaking scenario, the lower sideband grew on average approximately six times, whereas in linear-superposition breaking, the mean low-frequency gain was some 10%. If applied to the natural wind-generated spectra, however, it should be kept in mind that the energy drops away below the spectral peak very fast (see, for example, JONSWAP parameterisation (2.7)). Therefore, in field conditions such gains may turn out to be quite subtle and difficult to detect, particularly if the swell is also present, which is unrelated to breaking and is a usual occurrence in the open ocean.

Thus, the spectral output of breaking, if detected, can help to distinguish modulational breaking from linear-focusing breaking. What are other possible mechanisms leading to the limiting steepness (2.47), (2.51) and (2.52) and therefore to the breaking?

Babanin *et al.* (2011a) broadly placed such processes into two groups: instability mechanisms and superposition mechanisms. The modulational instability of Tulin & Waseda (1999) and the linear focusing of Meza *et al.* (2000) is only one example for each of the groups. Within these two phenomenological groups, further subdivisions are obvious, which may and perhaps will also correspond to different dynamics in terms of the dissipation outcome. For example, the classical Benjamin–Feir/McLean instability (Zakharov, 1966, 1967; Benjamin & Feir, 1967; McLean, 1982) does result in very tall waves rising within wave groups, provided that the average wave steepness is large enough, but once an individual-wave steepness is higher than $\epsilon = 0.42$, the wave-crest instability (also called second-type instability) becomes important (Longuet-Higgins & Dommermuth, 1997; Longuet-Higgins & Tanaka, 1997).

While the first and second type instabilities, as far as the breaking is concerned, may always be locked together because the breaking starts developing if $\epsilon > 0.42$, separation of the superposition cases is more distinct as each of them can lead to a breaking occurrence independently. These can be subdivided into frequency focusing (e.g. Longuet-Higgins, 1974; Rapp & Melville, 1990; Griffin *et al.*, 1996; Meza *et al.*, 2000, among many others), amplitude focusing (Donelan, 1978; Pierson *et al.*, 1992) and directional focusing (e.g. Fochesato *et al.*, 2007). Spectral outputs of such breaking mechanisms are still to be thoroughly examined and documented, but in the amplitude-focusing observations of Pierson *et al.* (1992), including the cases of breaking waves, it was shown that the loss of energy by dominant waves is accompanied by energy gain both below and above the spectral peak. That is, for amplitude focusing the spectral outcome of dissipation is yet to be different.

Another potential group of mechanisms is external forcing. For example, a very strong wind or a surface current gradient can also lead directly to steepening of the background waves all the way up to the (2.47), (2.51) and (2.52) limit. Usually, the role of the wind in the breaking process is secondary and is limited to gradual growth of the steepness, to stimulating or negotiating development of instabilities or probability of superpositions (Sections 4.1.3, 5.1.3, 5.3.4, Babanin *et al.*, 2010a). If the wind forcing, however, is very strong and can provide a wave growth from background to the Stokes steepness within a few wave periods, or even within one period, then, on the contrary, the other mechanisms become irrelevant. A strong current gradient certainly possesses such capacity too. Whether the breaking limit will still be the same in such circumstances and what will be the spectral outcome of such breakings are still to be verified.

7.3.3 Field measurements

Field measurements of the spectral distribution of wave-breaking dissipation are very few. They are a complicated undertaking for a great number of reasons, ranging from the technical difficulties of breaking measurements as such, through the difficulties of interpretation of the physics of breaking, to the mathematical difficulties of presenting properties of random and irregular events as a spectral dissipation term, whereas the spectrum by definition implies a continuous and uniform distribution of some property. Setting the technical and mathematical difficulties aside, we will draw the attention of the reader to the variety of physical mechanisms for wave-energy dissipation active simultaneously in a single breaking event (see, for example Section 7.2) and to the diversity of the spectral outcomes of wave-breaking dissipation (Sections 7.3.1, 7.3.2).

The latter particularly presents a considerable challenge for field measurements. Indeed, if the dissipation output of breaking of a monochromatic or quasi-monochromatic wave train is spectral, then how can the outcomes of breaking be understood and sorted out in an environment where a continuous spectrum of breaking waves is the initial condition?

In this section, we will present methods that allow us to identify and isolate breaking occurrences by relatively narrow spectral bands where the initially breaking waves belonged. Then, their broader spectral outputs can be interpreted in the context of discussions of Section 7.3.2. That is, attempts can be made to not only quantify the breaking outcome, but also to understand, based on the spectral distribution of the dissipation, what are the physical courses of the breaking in the field.

In this regard, two separate passive acoustic methods were employed (see also Section 3.5). The spectrogram method was applied by Young & Babanin (2006a) to determine the dissipation due to breaking of the dominant waves only. The individual-bubble detection technique of Manasseh *et al.* (2006) allows study of the breaking, and links the dissipation effects to different narrow frequency bands, starting from below the spectral peak and up to double the peak frequency $2f_p$.

The spectrogram method developed by Babanin *et al.* (2001) was used by Young & Babanin (2006a) to investigate differences between pre-breaking and post-breaking wave spectra. Such differences, when attributed to the breaking, should clarify the spectral contribution of the breaking, including the directional distribution of this contribution.

For the analysis, a wave record with an approximately 60% dominant-breaking rate was chosen (record 5 in Table 5.2). This was as close as possible to a 50% rate which would mean that half of the time waves within a single stationary record were breaking and half of the time waves were recovering from the breaking loss. The 50% division of the record into breaking/non-breaking parts enabled Young & Babanin (2006a) to estimate spectra of breaking and non-breaking waves with similar confidence intervals. The waves were stationary (scatter of 1 min standard-deviation surface elevation, relative to the 20 min mean, was less than 10%, with no drift of the mean) under steady $U_{10} = 19.8\,\mathrm{m/s}$ wind, with peak frequency $f_p = 0.36\,\mathrm{Hz}$ and significant wave height $H_s = 0.45\,\mathrm{m}$. The parameterisations of Young & Verhagen (1996) and Young & Babanin (2006b) were used to verify that the waves were fully developed in the bottom-limited environment with a measured depth of $d = 1.1\,\mathrm{m}$. The wave record was 20 min-long and was segmented into five breaking parts and four non-breaking parts.

In this strongly forced situation, approximately half of the waves were actively breaking. It was assumed that those waves not breaking had recently done so, having lost their energy in the breaking process. This assumption seems reasonable in this highly forced but steady-in-the-mean environment and the analysis that follows is predicated on this assumption.

Models of wind–waves, both physical and numerical, implicitly accept a double-scale approach to the wave field (see e.g. Melville, 1994; Lavrenov, 2003). This implicit assumption is important for correct interpretation and treatment of the wave breaking, and we shall discuss it here referring to the radiative transfer equation (Section 2.10) that is routinely employed in the wave spectral modelling.

At long scales of thousands of wavelengths and periods, the waves are assumed to be evolving. In a general case, at this scale the left-hand side derivative in (2.61) is positive as the waves grow under wind forcing. If the wave field is stationary and characterised by constant-depth (or deep) conditions, the evolution along the wave fetch is described by the advective term on the left-hand side of (2.61) which is small (less than 5% according to Donelan (2001)) compared to the wind input S_{in} and the dissipation S_{ds} on the right-hand side. The energy transfer across the spectrum due to the nonlinear term S_{nl} becomes essential at the scales of thousands and tens of thousands of wave periods (Hasselmann, 1962; Zakharov, 1968). In our case of a stationary fully developed constant-depth wave environment, the full derivative is zero and the right-hand-side terms of (2.61) are balanced.

At medium scales of hundreds of wavelengths, the wave fields are usually assumed to be stationary and homogeneous. Time/space series of surface elevations in such waves are used to obtain statistically reliable estimates of wave spectra in experiments, and spectra of this averaging scale are used in wave forecast and research spectral models. Such models have been reasonably successful and this, to some extent, justifies the assumption.

Indeed, the small advective terms and S_{nl} are not capable of bringing about significant changes to the wave spectrum at such time scales (unless changes to the spectrum are sudden and abrupt; see e.g. Young & van Vledder, 1993). For the dissipation term controlled by wave breaking, however, the scale of hundreds of waves is not small. For example, laboratory experiments on unsteady deep-water breaking by Rapp & Melville (1990) and Babanin *et al.* (2010a) show that the breaking is a rapid process of the same order of magnitude in time as the wave period, tens of periods at most, and may cause a major energy loss from the wave group where the breaking occurs. If, hypothetically, within the measurement time-span of hundreds of waves, each wave breaks even once, changes to the spectrum will be very significant. Such dramatic losses of energy, however, are not observed at these time scales since the breaking rates are typically not very large (Holthuijsen & Herbers, 1986; Babanin *et al.*, 2001, among others) and, at this time scale, the wind input is apparently capable of restoring the mean spectrum after breaking. This also means that, at this time scale, energy input by the wind is a slower process than the energy loss from breaking, as the energy is input to every wave in the field, whereas it is only lost from a small fraction of the breaking waves.

There are points of view that the spectral models based on the medium-scale averaging of the wave fields may have reached their limit in the accuracy with which they can simulate realistic wave generation and growth conditions (e.g. Liu *et al.*, 1995). This can be, in part, due to the fact that they average out variations of the wave field at scales of several waves (wave group scale). It is this shorter-scale group structure which plays a major role in intermittent wave breaking (Donelan *et al.*, 1972; Holthuijsen & Herbers, 1986; Babanin, 1995; Banner *et al.*, 2000; Babanin *et al.*, 2007b, 2010a) and modulation of the wind stress (Skafel & Donelan, 1997). There is modulation of the surface roughness at even shorter scales of dominant waves (Hara & Belcher, 2002; Kudryavtsev & Makin, 2002) and over breaking waves (Babanin *et al.*, 2007b), and disregarding this effect in models of wave growth can lead to underestimation of the growth rate parameter by a factor of 2–3 when compared to measured values (see Donelan *et al.*, 2006, for a discussion). The spectral equation (2.61) is not designed for applications at such time scales of individual waves and wave groups.

What happens at the scales of dozens of waves or a hundred waves? In the case of ordinary weak-in-the-mean breaking conditions, there should not be much variation in properties at this scale. In the case investigated here, the strongly forced and frequently breaking dominant waves come in alternating breaking and non-breaking trains from 4 to 120 dominant waves long. The breaking waves are, on average, significantly higher and steeper than those not breaking (Holthuijsen & Herbers, 1986). It is expected therefore that there will be a noticeable difference between the spectra calculated over breaking wave-train segments and the spectra over non-breaking segments.

Since breaking is the only major process to contribute to the rapid dissipation at this time scale (the bottom friction is relatively small and also relatively constant across the breaking/non-breaking segments), the difference can be attributed to dissipation due to the breaking of the dominant waves in the spectrum (as the spectrogram method is able to detect the breaking of dominant waves only – see Section 3.5). This difference will constitute a non-zero term on the right-hand side of (2.61). The main assumption, which was discussed and supported in Young & Babanin (2006a), is that the difference can be attributed to the partial derivative only and the advective term is small. An assumption regarding the advective term was needed as it was not possible to directly estimate its value. The waves were measured using a spatial array of wave probes with the largest separation of 30 cm between the probes (Young *et al.*, 2005). At such distances, the spectral difference along the wave fetch (between the probes) was negligible and could not be detected with a reasonable degree of confidence. Thus, to determine the spectral energy loss due to dominant breaking, it was assumed that it should be sufficient to measure differences between spectra of breaking and non-breaking waves based on measurements of time series at a point.

This difference will be a lower-bound estimate of the dissipation due to breaking. The approach treats the segments of breaking waves as a sequence of incipient breakers. In fact, waves breaking at the measurement point already exhibit some whitecapping and therefore they have already lost some energy prior to arriving at the measurement point. The broken waves in the non-breaking sequence are already gaining energy from the wind, but this energy is still not sufficient, on average, to bring them up to the breaking point. Therefore, the energy difference between the breaking-onset and the just-broken waves should in fact be somewhat larger than the one actually measured by this segmenting method. Also, as pointed out by Ardhuin *et al.* (2010), it should be kept in mind that, since the spectra are different, nonlinear interactions must be different too, and this fact, apart from the dissipation alone, can contribute to the observed spectrum variation as well.

Additionally, it is instructive to observe that the breaking waves are at the same time receiving energy from the wind. This means that the wind-input rate is still slower compared to the dissipation, but wave-growth rates and breaking-dissipation rates are now comparable and differ only by a factor of 2–3. This interesting observation indirectly supports the conclusion made in Donelan *et al.* (2006) that the wave-growth rates depend on the wave steepness.

To summarise the segmenting approach described here, we would mention again that, in the wave record with 60% dominant breaking rate, the trains of dominant breakers are treated as sequences of incipient breakers and the trains of non-breaking waves as sequences of waves which have just broken. This should lead us to a lower-bound estimate of the dominant-breaking impact across the spectrum. Thus, there will be multiple segments of wave record used, from half-a-minute to a few-minutes long, to obtain spectra based on these individual segments. The spectra obtained for the breaking segments and those obtained for the non-breaking segments will then be averaged separately in order to produce reliable estimates of the *incipient-breaking* spectrum and of the *post-breaking* spectrum.

As mentioned above, the spectrogram method developed by Babanin *et al.* (2001) will be used to segment the wave records. Spectrograms of the acoustic noise recorded by hydrophone clearly demonstrate patches of enhanced and lowered noise level, which were shown to be associated with the breaking activity of dominant waves at the wave-measurement spot above the hydrophone (Section 3.5). For example, in Figure 7.2 the first 35 seconds would be an *incipient-breaking* segment and the last 25 seconds – a *post-breaking* segment.

Thus, a mean *incipient-breaking* spectrum $F_i(f)$ and a mean *post-breaking* spectrum $F_p(f)$ were obtained within the record with nearly 50% breaking rate. The two spectra are shown in Figure 7.3. There is a clear difference between the spectra (note that the scale is logarithmic), with $F_p(f)$ having consistently lower spectral density as one would expect.

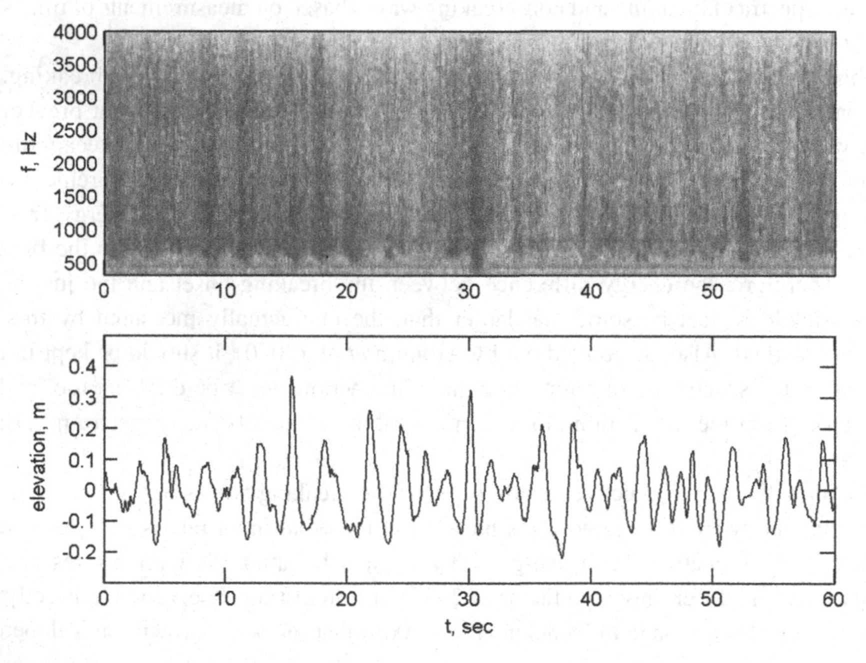

Figure 7.2 (Top panel) Spectrogram of acoustic noise of one minute of the record. Dark crests are associated with breaking waves. (Bottom panel) Synchronous surface elevation record. Figure is reproduced from Young & Babanin (2006a) © American Meteorological Society. Reprinted with permission

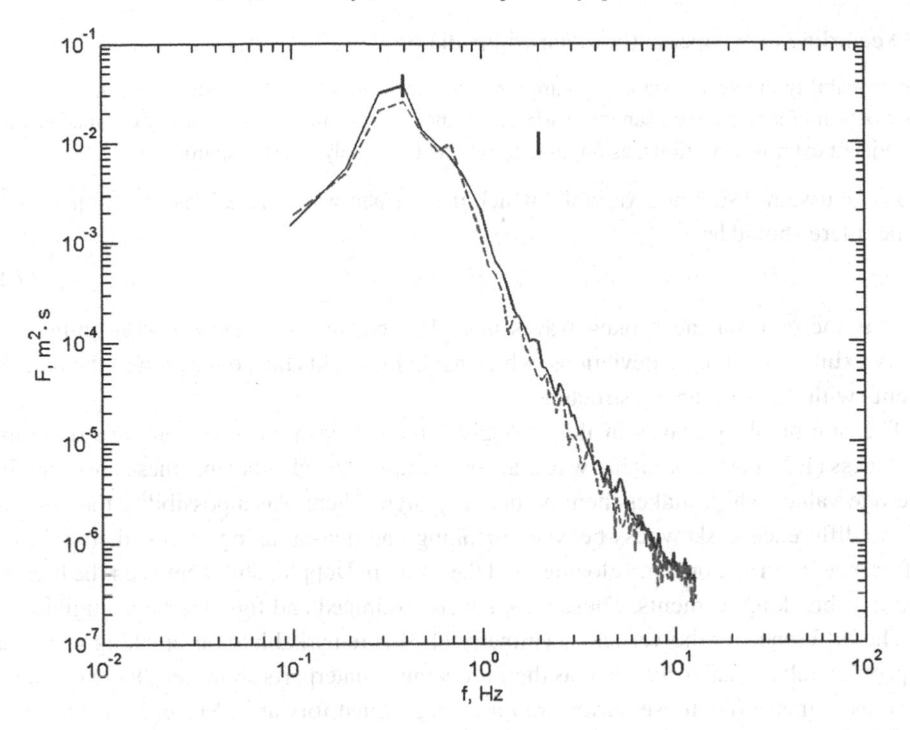

Figure 7.3 Mean power spectrum of *incipient-breaking* (solid line) and *post-breaking* (dashed line) waves. 95% confidence limits are shown. Figure is reproduced from Young & Babanin (2006a) © American Meteorological Society. Reprinted with permission

Confidence limits are very small, and overall the spectra of broken waves lie below the 95% confidence limits of the spectra of non-breaking waves. The higher-order moments also exhibit interesting substantial differences:

$$\begin{aligned} S_{k_i} &= 0.45, \\ S_{k_p} &= 0.31, \\ A_{s_i} &= -0.186, \\ A_{s_p} &= -0.017, \\ K_i &= 3.34, \\ K_p &= 2.96. \end{aligned} \tag{7.15}$$

Here, K_i and K_p are, respectively, *incipient-breaking* and *post-breaking* kurtosis

$$K = \frac{m_4}{m_0^2}. \tag{7.16}$$

According to Wikipedia (http://en.wikipedia.org/wiki/Kurtosis),

> "In probability theory and statistics, kurtosis... is a measure of the "peakedness" of the probability distribution of a real-valued random variable. Higher kurtosis means more of the variance is due to infrequent extreme deviations, as opposed to frequent modestly sized deviations".

For a Gaussian-distributed variable, which the surface-wave elevations are often assumed to be, there should be

$$K = 3. \tag{7.17}$$

This is the case for the broken waves in (7.15), but the *incipient-breaking* trains obviously exhibit the extreme deviations, which again highlights the connection of the breaking events with the wave-group structure.

The non-breaking waves in this strongly forced steep-wave situation have quite high skewness (1.2), but the breaking waves are even more skewed. Note that these are ensemble-average values which makes them particularly high. There was a possibility that, because of the difference in skewness between breaking and non-breaking waves, there will be a difference in surface orbital velocities and therefore in Doppler shifts between the breaking and non-breaking segments. These effects were examined and found to be negligible.

The difference for the vertical asymmetry (1.3) is remarkable: non-breaking waves are, approximately, symmetric, whereas their breaking counterparts show very large mean negative asymmetry (i.e. these waves are on average tilted forward). Since, as has been discussed in Section 5.1.1, at the very point of the breaking onset the individual wave-to-break is nearly symmetric, the negative asymmetry indicates the obvious fact that the waves measured in the *incipient-breaking* trains are already breaking (they were detected through their whitecapping in the first place). Since, however, the average asymmetry here is an ensemble-average, its distinctly negative value for the trains with many breaking waves has an important implication.

Indeed, if the waves do not break, their asymmetry still goes through the oscillations, that is the waves periodically tilt forwards, then recover the symmetric shape and tilt backwards (see Sections 4, 5.1, Agnon *et al.* (2005)). On average, however, the asymmetry of such trains should be zero, as it certainly is in the *post-breaking* train (7.15). When the waves break, they tilt forwards and then do not appear to recover their symmetric form, at least not until the breaking is ceased and a new cycle of the wave re-development starts. Therefore, wave trains which contain the breaking waves will have negative-on-average asymmetry, depending on how many breakers are embedded into the record. Thus, proper calibration of the breaking probability in terms of ensemble-average wave-train asymmetry should allow us to estimate wave-breaking rates in such trains without actually having to observe and detect the breaking events, at least for the dominant waves. This has been a long-standing problem, i.e. judging on the wave-breaking rates in widely available records of surface-wave elevations. Based on linear interpolation of (7.15), we can suggest that

$$\begin{cases} A_{s0\%\text{breaking}} &= 0 \\ A_{s100\%\text{breaking}} &= -0.2, \end{cases} \tag{7.18}$$

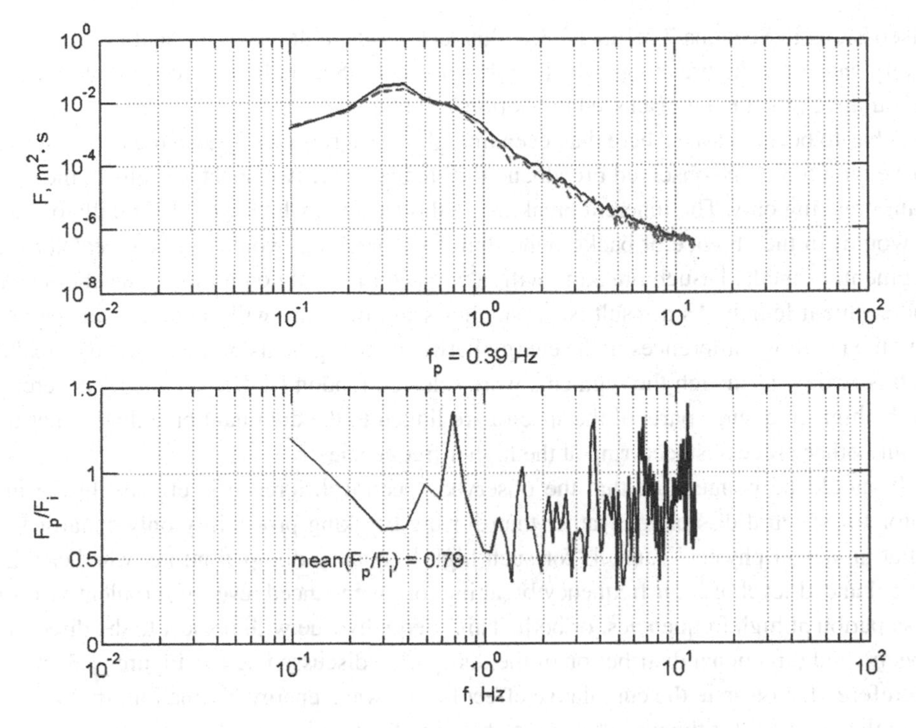

Figure 7.4 (Top panel) Mean power spectrum of *incipient-breaking* (solid line) and *post-breaking* (dashed line) waves. (Bottom panel) Ratio of the spectra shown in top panel. Figure is reproduced from Young & Babanin (2006a) © American Meteorological Society. Reprinted with permission

that is

$$b_T = -5A_s. \tag{7.19}$$

The spectral difference is quantified in the bottom subplot of Figure 7.4, where the ratio of the two spectra is plotted as a function of frequency f. The top subplot duplicates Figure 7.3 to make comparisons easier. Clearly, the loss of spectral density, following the breaking of dominant waves, is spread across almost the entire frequency range. Although there is considerable noise, the longer wave scales appear more affected by the dominant breaking than the shorter scales of $f > 5f_p$. It is possible that these shorter scales recover from the breaking impact at a time scale of a few tens of dominant periods (shorter than the averaging period of the spectra shown here). It is also possible that the broadband impact of the dominant breaking becomes less effective at high frequencies. The strongest attenuation is observed for waves in the $3f_p$–$5f_p$ range.

The mean energy loss across the full measured spectrum is approximately 20%. This is an important observation which will be mentioned in different aspects and discussed many times below. Firstly, this does mean that the dominant wave scales, the spectral peak, do lose substantial energy in the course of the dominant breaking. In the context of the discussions in Sections 7.3.1 and 7.3.2, this fact points to the modulational instability as a likely cause for the breakings which took place here. Secondly, this result underlines the cumulative effect to be discussed in Section 7.3.4 below.

It has to be emphasised here that, even though the short waves do also break as discussed in Section 5.3.2, the observed broadband difference of the two spectra is due to the dominant breaking only. The inherent breaking of short waves, which would naturally occur (if it would) in the absence of background dominant breaking, would not be detected by the segmenting method. Such breaking will take place randomly on a time scale shorter than the segment length. As a result, such smaller-scale breaking will happen in all segments and the resulting differences in the energy between the segments will not identify breaking at this scale – as though the segments were selected randomly. Therefore, any differences in the high-frequency parts of the spectra are linked to the dominant breaking, rather than to any other processes occurring at the higher frequencies.

It should be pointed out that the observed spectral difference is effectively the indicator for spectral dissipation, rather than for the breaking probability only, regardless of what an actual physical cause is for such high-frequency dissipation: i.e. whether this is an enhanced level of high-frequency breaking, or an enhanced level of turbulent viscosity dissipation at high frequencies, or both. This is equally true with respect to the directional spectra and directional distribution of the dissipation discussed below. Figures 7.3 and 7.4, therefore, demonstrate the cumulative effect for the wave-energy dissipation, similar to the cumulative effect for the wave breaking shown in Section 5.3.2.

The conjecture that the difference/dissipation between the observed spectra in Figures 7.3 and 7.4 is due to dominant breaking required quantitative verification. Measurements of the total dissipation in the water column beneath the surface waves were used for this purpose. Estimates obtained by means of such measurements are not necessarily more accurate than the estimates obtained in Young & Babanin (2006a) by the segmenting method, but the first approach is quite well established and provided a good reference value for the new results.

The total dissipation per unit of surface, which is the usual measure of the wave-energy properties including, for example, radiative transfer equation (2.61), can be estimated directly by measuring the volumetric dissipation rate $\epsilon_{dis}(z)$ in the water and then integrating over the water-column depth z. If measurement of ϵ_{dis} is only available at a single distance from the surface, this can be used as a reference point, and for the integration a parameterised depth-dependence would need to be adopted.

Field measurements of the $\epsilon_{dis}(z)$ profiles were discussed in detail in Section 5.3.4, and the respective parameterisation was suggested there (5.74). For the strong-wind Lake George case employed here, the inverse-quadratic dissipation profile of (5.74) was used. During the wave record analysed in our segmenting exercise, ADV measurements synchronised with the wave and sound recordings were carried out at a distance of 20 cm from the surface at the location of the wave array. The ADV velocity spectra of *incipient-breaking* and *post-breaking* periods are shown in Figure 7.5. The orbital velocities of waves around the peak, unlike the peak spectral densities of the power spectra in Figures 7.3 and 7.4, do

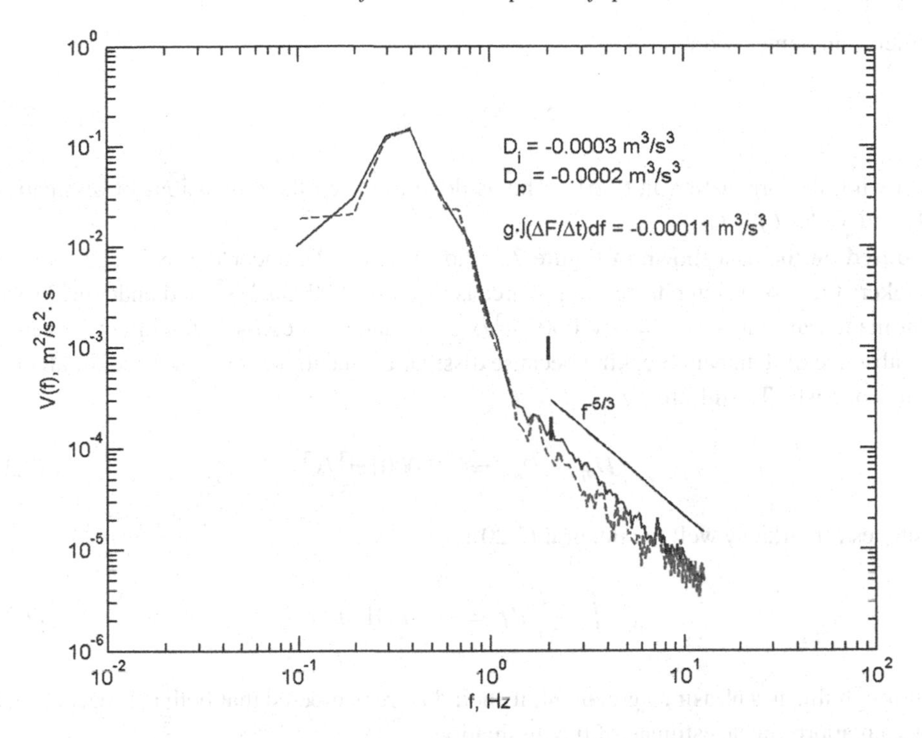

Figure 7.5 Mean velocity spectrum of ***incipient-breaking*** (solid line) and ***post-breaking*** (dashed line) waves. The Kolmogorov interval slope of $f^{-5/3}$ is shown. The text demonstrates values of respective total dissipation rates D_a (5.67, denoted as D in this figure) and the integral of spectral dissipation seen in Figure 7.3. 95% confidence limits are shown with the vertical straight segments. Figure is reproduced from Young & Babanin (2006a) © American Meteorological Society. Reprinted with permission

not appear to differ significantly. The levels of the Kolmogorov intervals at higher frequencies of the two spectra, however, are essentially different. At these frequencies and at the 20 cm depth, the turbulent velocities apparently dominate the orbital velocities. The difference between Kolmogorov intervals can be attributed to the loss of energy from across the wave spectrum seen in Figures 7.3 and 7.4. The total dissipation estimated from the velocity spectra in Figure 7.4 and the wave spectra in Figure 7.3 should match:

$$D_{a_i} - D_{a_p} = g \cdot \overline{\int_f \left(\frac{F(f)_p - F(f)_i}{\Delta t} \right) df} \tag{7.20}$$

where Δt is the time difference between the mean time points of subsequent breaking and non-breaking segments and the overline represents the ensemble average. Here, the spectral dissipation term is multiplied by g:

$$S_{ds}(f) = g \cdot \frac{\Delta F(f)}{\Delta t} \tag{7.21}$$

rather than being used as

$$S_{ds}(f) = \frac{\Delta F(f)}{\Delta t} \tag{7.22}$$

following its formulation in (2.61). This is done to match the dimensions of dissipations D_a and $\int_f S_{ds}(f)df$.

Based on the data shown in Figure 7.5 and using (5.67), under the incipient dominant breakers the dissipation rate per unit of area is $D_{a_i} = -0.0003m^3/s^3$, and under the broken dominant waves it is $D_{a_p} = -0.0002m^3/s^3$. The latter, of course, should not be zero in the absence of dominant breaking because dissipation due to breaking of waves of all other scales persists. The difference

$$D_{a_i} - D_{a_p} = -0.0001m^3/s^3 \tag{7.23}$$

matches remarkably well the integral (7.20):

$$g \cdot \int_f \frac{\Delta F}{\Delta t} df = -0.00011m^3/s^3. \tag{7.24}$$

Although this is a pleasing agreement, it should be remembered that both approaches yield only an approximate estimate of the dissipation.

As has been mentioned above, the spectral difference in Figure 7.3 is a lower-bound estimate of dissipation due to the breaking, because some energy must have already been lost before the spectra were measured. The difference $D_{a_i} - D_{a_p}$ of the total dissipation rates of kinetic energy is also likely to be a lower-bound estimate of the loss of energy from the wave field due to breaking. As has also already been mentioned, some of the energy is expended on work against buoyancy forces whilst entraining bubbles into the water, rather than on generating the turbulence. The fraction of the wave energy dissipated, which is expended on entraining the air, can be from 14% to 50% of the total (Lammarre & Melville, 1991, 1992; Melville *et al.*, 1992; Blenkinsopp & Chaplin, 2007) and this defines the limits of accuracy of the segmenting method. For the laboratory measurements of Lammarre & Melville (1991, 1992) and Melville *et al.* (1992), some of the breaking was of a plunging type. In contrast, most of the Lake George breaking was of the spilling type, which spends less energy on entrainment. Therefore, we would expect that in our case the fraction of energy expended this way is at the lower bound of 14% (Blenkinsopp & Chaplin, 2007) or similar. Nevertheless, the above comparison shows that the proposed method of estimation of wave-breaking dissipation by considering the difference between the *incipient* and *post-breaking* spectra provides a reasonable magnitude of the total dissipation.

Another interesting observation can also be made on the basis of Figure 7.5. Here, one can see that the orbital velocities of the dominant waves did not change between the breaking and non-breaking segments. This implies that the energy, which is obviously gone from the power spectra, must have been the potential energy, rather than kinetic, at least for the dominant waves. What also matters here is that the dominant orbital velocities are the same between the segments, and therefore the bottom friction is the same for the breaking and non-breaking wave trains which is essential for the estimates made.

The second set of field measurements of the spectral distribution of wave-breaking dissipation, mentioned in the beginning of this section, is based on the bubble-detection method of Manasseh *et al.* (2006). This method was described in detail in Section 3.5 and its application to measuring, across the wave spectrum, the breaking probabilities in Section 5.3.2, and the breaking severity combined with the breaking probability (i.e. dissipation), in Section 6.2.

Since measurements of the spectral dissipation by this method have thus already been described in the book, we refer the reader to the above-mentioned sections. In short, the conclusions are that it is only at the spectral peak that the dissipation intensity can be related to the density of the wave power spectrum. As in the estimates done by means of the spectrogram method above, major loss of energy is observed from the peak. In the context of the discussions of Section 7.3.2, this observation, again, points to modulational instability as the likely course for the breaking here.

Neither below nor above the peak does the wave spectrum seem to correlate with the dissipation at those spectral scales. Below the spectral peak, in the finite-depth Lake George conditions development of the longer waves is arrested by the bottom-induced breaking (see Section 3.5; Young *et al.*, 2005; Young & Babanin, 2006b), and this is the most likely reason for divergence of the wave-spectrum and dissipation-spectrum intensities. Above the peak, this divergence is another demonstration of the cumulative effect which is the subject of the next section.

7.3.4 Cumulative effect

Section 7.3.2 is essentially a separated out summary of the most important outcomes of Section 7.3.1, dedicated to the laboratory measurements of the dissipation and similarly, the current section concentrates on the most important conclusion of the previous field-measurement Section 7.3.3. The significance of the cumulative effect for applications which involve spectral distributions of the wave-energy dissipation is hard to overestimate, and yet to date it is missing in the operational forecast models and is only starting to make its way into research models.

The cumulative effect is due to breaking and/or dissipation of short waves being influenced or even directly induced by longer waves. That is, as was shown in Section 7.3.3 (and Young & Babanin, 2006a), when the dominant waves of frequency $\sim f_p$ break, the energy is lost not only from the waves of this frequency, but from the entire spectrum range of $f > f_p$. So if the waves of frequency $2f_p$ will be breaking due to inherent reasons (rather than being induced by $f < 2f_p$ breakings as just described above), they will not affect the dissipation at the peak f_p, but will cause a further associated dissipation at scales of $f > 2f_p$, and so on. Thus, dissipation of waves at a higher frequency, for example, $3f_p$, will be induced every time waves in the range $f < 3f_p$ are breaking, and the higher the frequency the stronger will be the accumulating overall dissipation due to the lengthening lower-frequency band of breaking waves.

Mathematically, the cumulative dependence of the spectral dissipation is described by an integral of the wave spectrum rather than by an algebraic function of this spectrum. One such formulation for the dissipation cumulative term, the original expression by Young & Babanin (2006a), is given in (5.40), and for the cumulative behaviour of the breaking probability at higher frequencies in (5.41). Graphically, this effect is illustrated in Figure 5.30.

It should be pointed out that the cumulative dissipation does not deny or cancel the inherent dissipation, that is the dissipation due to non-forced breaking at scales higher than the spectral peak. The inherent term is present in the $S_{ds}(f)$ formulation (5.40), and at each frequency it is a function of the local-in-frequency spectrum $F(f)$ as with any other dissipation term. That is, the breaking dissipation should consist of two terms, the inherent-dissipation term S_{inh} and the cumulative-dissipation term S_{cum}:

$$S_{br}(f) = S_{\text{inh}}(f) + S_{\text{cum}}(f). \tag{7.25}$$

$S_{\text{inh}}(f)$ is the only breaking-dissipation mechanism at the peak and below the peak, but for shorter waves of $f > f_p$ its role becomes progressively less important as the strength of $S_{\text{cum}}(f)$ accumulates, to the point of being negligible at some scales of $f > 3f_p$ (see Babanin & Young, 2005; Babanin *et al.*, 2007c, and discussions in Section 5.3.2).

The dissipation terms presently employed in operational wave-forecast models (see Section 7.4 for a dedicated discussion of this topic), effectively only include the inherent term $S_{\text{inh}}(f)$ and disregard the cumulative term $S_{\text{cum}}(f)$. Various weighting coefficients have been proposed over the years to help the associated apparent dissipation bias across the spectrum, but since the two terms are functionally different, no amount of tuning can compensate for the missing physics in all circumstances.

It is relevant to mention here that in reality, the wave attenuation consists of very many physical processes, independent or interdependent, and not all of them even relate to the wave breaking. As Babanin & van der Westhuysen (2008) suggested, the overall dissipation should be best described by a sum of those as

$$S_{br}(f) = S_{\text{inh}}(f) + S_{\text{cum}}(f) + S_{\text{turb}} + S_{\text{visc}} + S_{\text{wind}} + \cdots \tag{7.26}$$

where S_{turb} is dissipation due to turbulent viscosity, S_{visc} is due to molecular viscosity, S_{wind} is due to interaction of waves with adverse winds etc. This way, each dissipation term may have a different formulation as dictated by the relevant physics, and any of them may turn to zero as necessary while the wave evolution and the wave energy dissipation will still proceed.

Now, we should step back and point out that the induced breaking/dissipation is not a newly discovered feature which still awaits further experimental and theoretical support and confirmation. It has been observed and elaborated analytically for a substantial period of time (e.g. Longuet-Higgins & Stewart, 1960; Phillips, 1963; Longuet-Higgins, 1987; Banner *et al.*, 1989; Donelan, 2001; Melville *et al.*, 2002; Manasseh *et al.*, 2006). This effect, however, has been overlooked by the wave-forecast modelling community and it is only very recently that the cumulative integral was explicitly suggested as an additional breaking-dissipation term for spectral models (Donelan, 2001; Babanin & Young, 2005; Young & Babanin, 2006a).

Formally speaking, this single expression for the cumulative term is also an oversimplification. Indeed, a number of different physics can lead to the cumulative dissipation, such as the straining action of long waves on shorter waves (e.g. Longuet-Higgins & Stewart, 1960; Longuet-Higgins, 1987; Donelan, 2001; Donelan *et al.*, 2010), the induced breaking of shorter waves caused by larger breakers (e.g. Manasseh *et al.*, 2006; Young & Babanin, 2006a) and the turbulent-viscosity damping of short waves in the wake of large breaking (e.g. Banner *et al.*, 1989). These mechanisms can be concurrent and tied together, or can be disconnected as, for example, in the case of mature dominant waves which do not break themselves, but still cause the strain and breaking of smaller-scale waves (see e.g. Tsagareli, 2009; Babanin *et al.*, 2011a). Therefore, following the logic of Babanin & van der Westhuysen (2008), the cumulative term itself may eventually need to be subdivided into sub-terms

$$S_{cum}(f) = S_{straining}(f) + S_{induced_{breaking}}(f) + S_{turbulent_{wake}} + \cdots, \tag{7.27}$$

depending on the relative significance of the sub-terms and importance of consequences of the separation of the cumulative effects in the physical space if that happens.

7.3.5 Whitecapping dissipation at extreme wind forcing

Many processes in the small-scale air–sea interactions run differently at extreme-wind conditions. As a reference point for the extreme winds, the hurricane-scale classification can perhaps be used: that is a tropical storm becomes a hurricane if the wind speed reaches

$$U \sim 33\,\mathrm{m/s}\ (119\,\mathrm{km/h}). \tag{7.28}$$

It is interesting to note that this is approximately the wind speed U_{10} at which the drag coefficient C_D (3.8) was found to saturate in the field observations by Powell *et al.* (2003) and Jarosz *et al.* (2007) (see also Section 9.1.3).

A further number of experimental and theoretical studies followed and investigated this conjecture (Andreas, 2004; Donelan *et al.*, 2004, 2006; Barenblatt *et al.*, 2005; Makin, 2005; Bye & Jenkins, 2006; Kudryavtsev, 2006; Kudryavtsev & Makin, 2007, 2009, 2010; Black *et al.*, 2007; Stiassnie *et al.*, 2007; Vakhguelt, 2007; Troitskaya & Rybushkina, 2008; Rastigejev *et al.*, 2011; Soloviev & Lukas, 2010, among others). A detailed review of the peculiarities of air–sea interactions in such conditions is far beyond the scope of the present book, but for a shorter review see Section 9.1. What should be emphasised here is that the observed magnitude change of the sea drag and particularly the apparent change of the C_D-growth trend signifies a change of the surface roughness (see Section 3.1). And this issue is certainly relevant for the wave-breaking and wave-dissipation topics in this book.

The surface roughness is largely determined by the waves in general and the wave breaking in particular (see e.g. Donelan, 1990; Komen *et al.*, 1994; Kudryavtsev *et al.*, 2001; Kudryavtsev & Makin, 2002; Babanin *et al.*, 2007c; Babanin & Makin, 2008), and therefore the characteristics of the surface waves are apparently different in such circumstances, and perhaps much different. The interface is smeared as the water is covered with foam and the air is full of spray, and the anecdotal description of the ocean surface in hurricane conditions says that the water is too difficult to breathe, but the air is not dense enough to swim (see Figure 1.9). Obviously, the very notion of short-wave scales loses its physical meaning attributed to waves with a distinct air–sea interface, but the longer waves would still persist and be breaking. The behaviour of the whitecapping dissipation at such extreme wind forcing, however, can and will be altered.

Apparently, there is no sudden change of the wave-surface and boundary-layer conditions at wind speed (7.28). There is a gradual build-up of the change when the moderate-condition features of the air–sea interface are not necessarily cancelled, but the new physical exchanges certainly appear and grow in significance to the point of the new physics becoming dominant.

In Sections 5.3.4 and 6.2, examples of the wave-breaking response to wind speeds as low as $U_{10} > 14\,\mathrm{m/s}$ (5.77) were demonstrated, when the growth of the wind did not cause a further growth of the waves, but rather brought about an increase of the breaking rates and dissipation. Thus, at least one effect of the strong winds on the breaking can be outlined: when the wind-energy input rates at some spectral scales become too strong for the waves of those scales to 'digest' them (that is to redistribute this energy to different scales and directions through nonlinear interactions), the waves respond by more frequent breaking (Figure 5.27) which dissipates the extra-input energy (Figure 6.6), without a noticeable change to the wave spectrum at those scales (Figure 5.42).

This conclusion of Sections 5.3.4 and 6.2, however, was not unambiguous because the results shown in the above-mentioned figures correspond not just to different wind speeds, but to different magnitudes of the dominant waves also. The higher wind speeds are associated with larger waves, and these larger waves, too, in addition to the wind, could contribute to enhanced breaking rates of shorter waves, without having the spectrum altered at those shorter scales.

On the other hand, the conclusion agreed with the independent observation-based deep-water parameterisation (5.75) of Babanin & Soloviev (1998a), also discussed in Section 5.3.4. This parameterisation was obtained in a broad range of spectra and winds, and implies that for younger waves (stronger wind forcing) the level of the wave-spectrum tail remains relatively constant. It certainly does not respond in any consistent way to an increase in wind forcing, that is the shorter waves do not on average grow higher if the wind is stronger, which can only mean that they break more frequently.

The results of Sections 5.3.4 and 6.2 on wind influence on wave breaking and dissipation were obtained for quite strong winds, up to mean wind speeds U_{10} in excess of $20\,\mathrm{m/s}$, but such magnitudes are still very distant from extreme conditions like those hurricanes identified by (7.28).

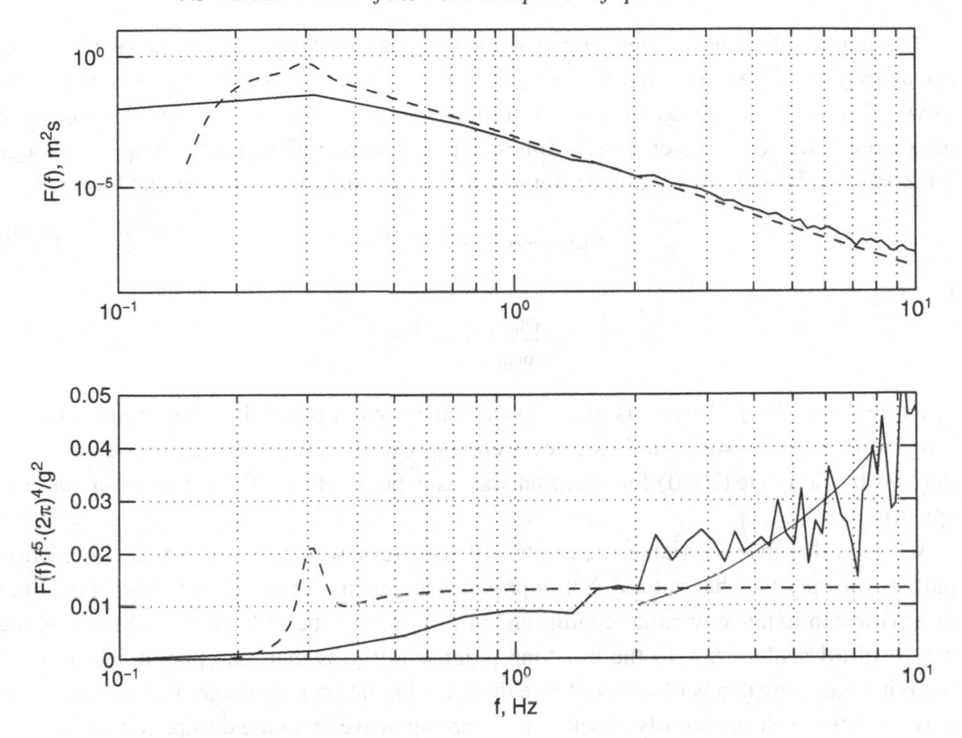

Figure 7.6 Comparison of JONSWAP spectrum (2.7), dashed line, with the helicopter-forced waves, solid line. (top) The spectra; (bottom) spectrum-tail level α. Curved line indicates f^{-4} trend

In Figure 7.6, outcomes of an artificial field test for an extreme wind forcing are shown. The waves were measured by the wave array as described in Babanin & Soloviev (1998a,b), and a helicopter pilot was asked to hover his machine over the array in order to observe what difference to the background conditions that would make.

The background conditions were represented by a light wind of $U_{10} \approx 6\,\text{m/s}$ with the wind–wave spectrum of $f_p \approx 0.3\,\text{Hz}$. The JONSWAP spectrum (2.7) with such a peak frequency and tail-level $\alpha_{\text{Phillips}} = 13.2 \cdot 10^{-3}$, corresponding to the upper limit of the parameterisation (5.75) for comparison, is shown by a dashed line.

The Russian helicopter MI-8 weighs 11 T, and with a blade diameter of 25 m it was estimated that the enforced surface horizontal winds were of the order of $U_{\text{surface}} \sim 30\,\text{m/s}$. Hurricane-like conditions, therefore, were created locally in the vicinity of the wave array. Visually, every wave was breaking.

The measured spectrum of the background waves with such additional localised forcing is plotted in Figure 7.6 with the solid line. The top panel shows the spectra themselves, and the lower panel – the level α of these spectra. Directional spectra at frequencies $f < 1\,\text{Hz}$ were unimodal, i.e. still carried information on the wind which generated them initially, and at higher frequencies they were observed to be isotropic, that is, disconnected from the wind and fully driven by the helicopter.

The impact of the helicopter forcing in the frequency spectrum is clearly visible at frequencies $f > 2\,\text{Hz}$. Note that the tail rise at the high end of the frequency range is not physical, it is noise caused by the wire probes averaging the surface elevations over an area exceeding its diameter several times (e.g. Cavaleri, 1979). In the frequency range $2\,\text{Hz} < f < 7\,\text{Hz}$, the helicopter response is rather flat and the corresponding level α is

$$\alpha_{\text{helicopter}} \approx 24.5 \cdot 10^{-3}, \tag{7.29}$$

that is

$$\frac{\alpha_{\text{helicopter}}}{\alpha_{\text{Phillips}}} \approx 1.9. \tag{7.30}$$

For reference, the f^{-4} trend is also shown in the bottom panel. It is interesting to point out that $\alpha = 0.23$ was found as a level corresponding to the 'limiting' wave spectrum derived by Stiassnie (2010) for maximal wave steepness of $\epsilon = 0.4$ for waves across the spectrum.

We should realise, of course, the limits of this exercise where the wind–wave interaction pattern was anything but natural. Yet, it provides an instructive insight into wave breaking/dissipation in extreme weather conditions. In response to such extreme wind forcing, the waves of all scales grow to the breaking point within less than one period and a 100% breaking-rate situation is observed. Once the breaking rate can no longer be increasing, the wave system must, apparently, react with a more intensive breaking dissipation, in order to balance the wind energy input in conditions when the spectrum level does not (or cannot) grow any more.

Indeed, the wave spectrum responding to the helicopter forcing by the spectral-density increased less than two times. If the dependence of the wind-energy input at those scales is approximately quadratic in terms of the wind speed (Donelan *et al.*, 2006), and if we presume an upper-limit wind speed $U_{10} = 20\,\text{m/s}$ for α_{Phillips} and a lower-limit $U_{10} = 30\,\text{m/s}$ for the measured $\alpha_{\text{helicopter}}$, then

$$\left(\frac{U_{\text{helicopter}}}{U_{\text{Phillips}}}\right)^2 = 2.25, \tag{7.31}$$

quite close to the ratio (7.30). If we choose more realistic midrange values of $U_{10} = 10\,\text{m/s}$ for α_{Phillips} and $U_{10} = 40\,\text{m/s}$ for $\alpha_{\text{helicopter}}$, then the ratio becomes

$$\left(\frac{U_{\text{helicopter}}}{U_{\text{Phillips}}}\right)^2 = 16. \tag{7.32}$$

This is an enormous gap between the increased wind-input rate and the spectral-power response. This gap should be filled in with the greatly increased dissipation which, in such a scenario, should be a function of the growing wind forcing or of the wind speed directly, as was suggested in Sections 5.3.4 and 6.2.

Quantitatively, we should remember that estimates (7.30)–(7.32) are only a guide. As emphasised already, the helicopter experiment does not fully reflect real hurricane-like wind forcing, and also the approximations (7.31)–(7.32) were based on an extrapolation of the dependence of Donelan *et al.* (2006), obtained at moderately strong winds, in extreme conditions. As discussed at the beginning of this section, such extrapolation is not straightforward, since the sea drag most likely tends to saturate (and perhaps even decrease) at extreme weather. Therefore, this topic, very important for adequate sea-state predictions in tropical cyclones, is in serious need of experimental and field observations.

7.3.6 Directional distribution of the whitecapping dissipation

The least known feature of the spectral dissipation function is its directional behaviour, which is the main topic of this subsection. As mentioned throughout the book (see also Section 5.3.3), the waves are directional, as are all the source functions in (2.61). Therefore, some directional shape, usually isotropic, must be assumed for the dissipation term. There is, however, little, if any, experimental validation of this directional shape.

The segmenting method can be used to obtain *incipient-breaking* and *post-breaking* directional spectra, similar to that used to obtain *incipient-breaking* and *post-breaking* omni-directional spectra in Section 7.3.3 and Figure 7.3 above. The maximum likelihood method (MLM) developed originally by Capon (1969) (see also Young, 1994; Babanin & Soloviev, 1987, 1998b; Young *et al.*, 1996) was used to analyse the wave-array data and the wave directional distributions in this respect.

It was noticed that the main wave propagation direction θ_{max} changes from segment to segment (in Figure 7.7 it is shown for the spectral-peak frequency f_p). This scatter around the mean main direction appeared random, not connected to whether the segment consisted of breaking or of non-breaking waves. Therefore, the non-normalised directional spectra $\phi(f_p, \theta)$ were obtained for each of the segments and turned to have the same main direction ($\theta_{max} = 0$ in Figure 7.8). The connection between the non-normalised directional spectra $\phi(f, \theta)$ and normalised directional distributions of $K(f, \theta)$ in (5.34)–(5.33) is

$$\phi(f, \theta) - A(f)K(f, \theta) \qquad (7.33)$$

(see Babanin & Soloviev, 1987, 1998b, for details). Since MLM does not produce dimensional values for the spectral densities, however, the vertical scale in Figure 7.8 is arbitrary.

In Figure 7.8, the solid line designates the mean *incipient-breaking* directional spectrum $\phi_i(f_p, \theta)$ at the spectral peak, and the dotted line – the mean *post-breaking* directional spectrum $\phi_p(f_p, \theta)$. Clearly, the major energy loss occurs at angles oblique to the main propagation direction.

Figure 7.9 shows the ratio of $\phi_i(f_p, \theta)$ and $\phi_p(f_p, \theta)$ spectra. Qualitatively, this ratio reflects the directional behaviour of the dissipation at the spectral peak. Contrary to existing assumptions, the energy loss in the main propagation direction is a minimum, with the loss increasing away from this main direction. Since the dissipation has to start decreasing again

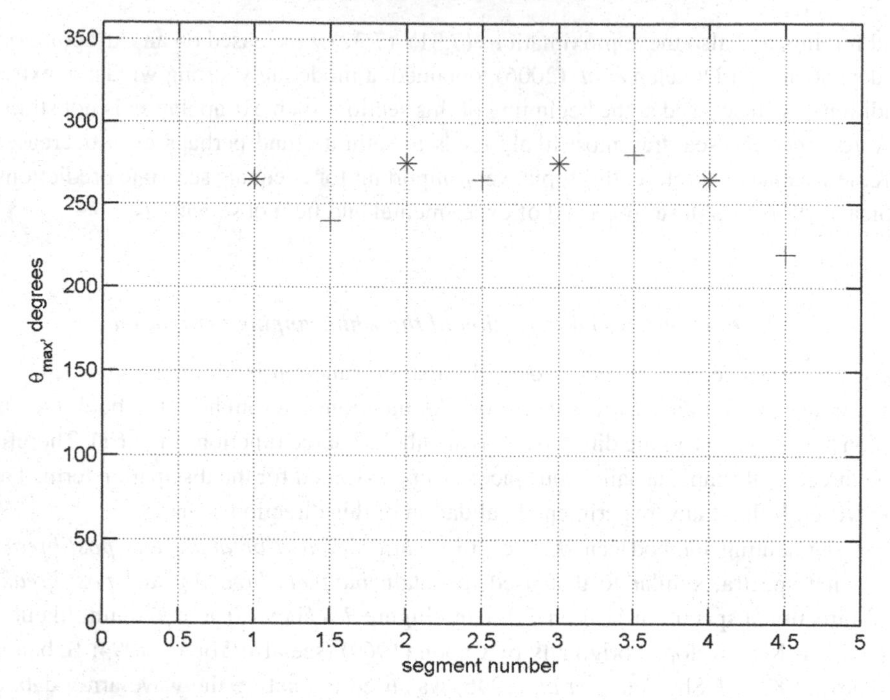

Figure 7.7 Main direction of subsequent breaking-wave (*) and broken-wave (+) segments. Figure is reproduced from Young & Babanin (2006a) © American Meteorological Society. Reprinted with permission

at larger angles, certainly in the half-plane opposite to the wave-propagation half-plane, this result implies that the dissipation function $S_{ds}(f, \theta)$ in (2.61), (2.25) and (5.49) has to be bimodal in the directional space. Unlike the wind-input function $S_{in}(\theta)$ which is unimodal and has a distinct maximum in the main wind direction, $S_{ds}(\theta)$ may have maxima at angles oblique to the wind-input maximum angle.

The impact of the dominant breaking on directional dissipation at $2f_p$ is similar, though less pronounced (Figures 7.10, 7.11). There is a noticeable loss of energy in the primary wind/wave direction, and the oblique peaks are pushed farther away. The former can be explained by the influence of the induced breaking. That is, the dominant waves, whose directional distribution is quite narrow, enforce the short-wave breaking in the wind-predominant direction primarily, and the dip in the centre of Figure 7.11 becomes less pronounced compared to Figure 7.9. The trend of the oblique peaks requires further confirmation, understanding and explanation, as does the very directional bimodality of the f_p waves.

Thus, it appears that the directional dissipation rates at oblique angles are higher than the dissipation in the main wave-propagation direction. With respect to the breaking as such, this conclusion does not necessarily immediately translate into directional properties

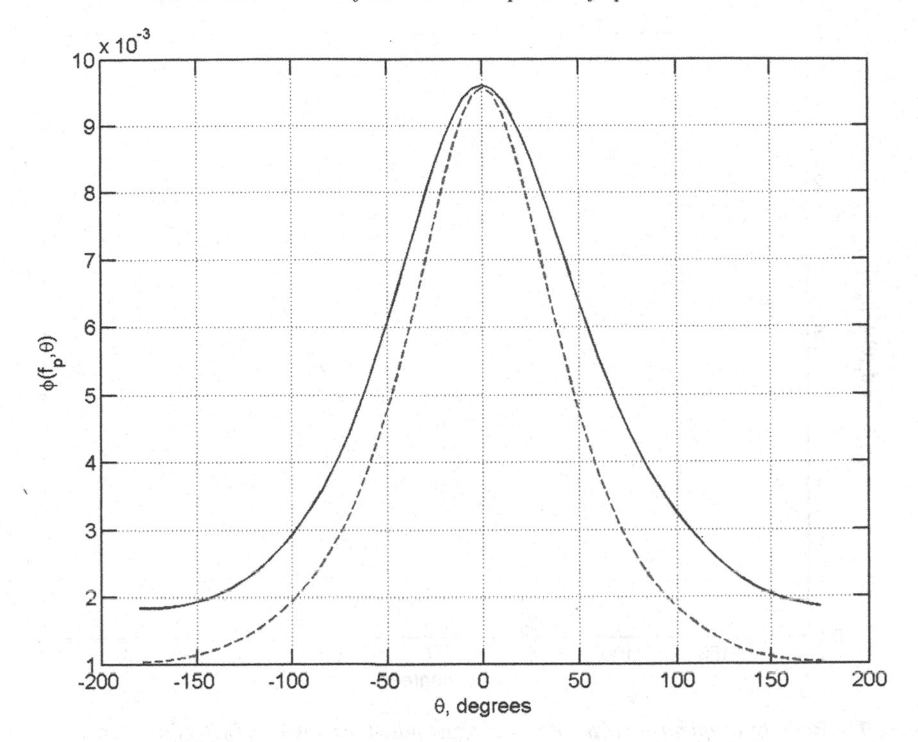

Figure 7.8 Non-normalised *incipient-breaking* $\phi_i(f_p, \theta)$ (solid line) and *post-breaking* $\phi_p(f_p, \theta)$ (dashed line) directional spectrum (7.33) at peak frequency f_p. Units of the MLM directional distributions are arbitrary. Figure is reproduced from Young & Babanin (2006a) © American Meteorological Society. Reprinted with permission

of the breaking occurrence. As has been discussed on a number of previous occasions, the dissipation consists of contributions of both the breaking probability and the breaking severity.

It is interesting to notice here that the frequency distribution of the breaking severity is also smallest at the spectral peak (Figure 6.6 and discussion in Section 6.2). That is, the breaking appears to narrow down the spectral peak, both in the frequency and in the directional domains.

In order to obtain the angular distributions of the breaking occurrence directly, it would be necessary to combine one of the methods able to detect individual wave-breaking events, e.g. one of the acoustic techniques described in Section 3.5, with a method suitable for detecting a direction of propagation of these individual waves, for example, the wavelet directional method (WDM) of Donelan *et al.* (1996) or some of the remote-sensing techniques of Section 3.6. This is, however, a very demanding experimental exercise as it would require very long observations in order to obtain reliable statistics. Even if the directional plane is subdivided into reasonably broad bins, i.e. $\pm 10°$ or even $\pm 20°$, there will be required many hours of continuous stationary wave-breaking records, particularly given

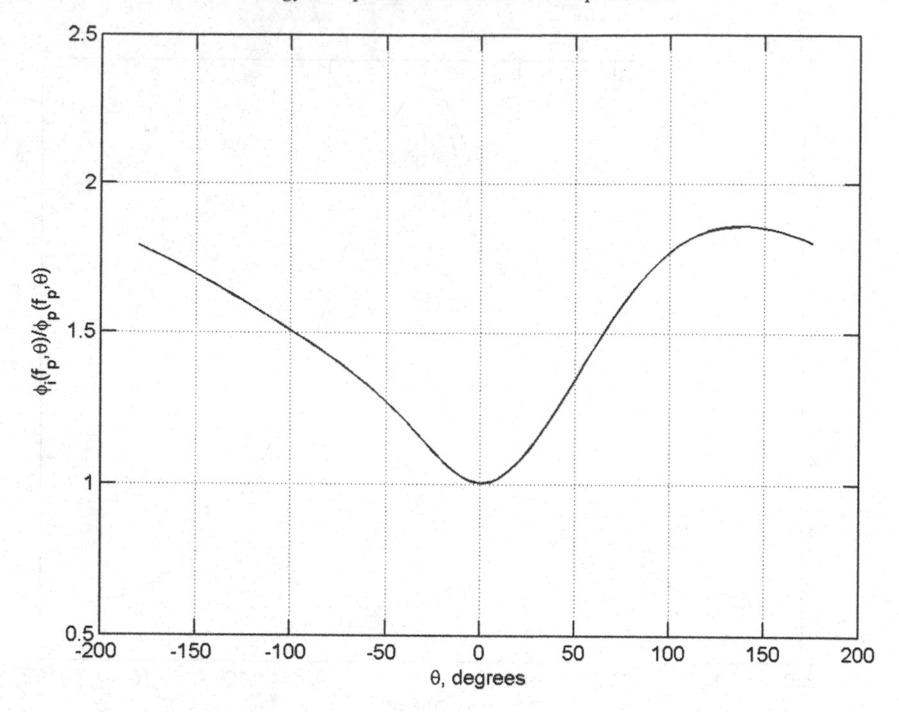

Figure 7.9 Ratio of *incipient-breaking* $\phi_i(f_p, \theta)$ and *post-breaking* $\phi_p(f_p, \theta)$ directional spectra of Figure 7.8 at peak frequency f_p. Figure is reproduced from Young & Babanin (2006a) © American Meteorological Society. Reprinted with permission

the fact that the overall directionally integrated rates are typically only of the order of a few percent (e.g. Babanin *et al.*, 2001).

Such an investigation was conducted by Filipot (2010, personal communication) and revealed an isotropic distribution for the breaking probability. In connection with the bimodal directional function of the dissipation here, this implies a strongly bimodal behaviour of the breaking severity. Such conjecture is still to be confirmed experimentally.

In this regard, a recent study by Kleiss & Melville (2010) has to be mentioned. They investigated the directional distribution of $\Lambda(c, \theta)$, the breaking crest length per unit area (see Section 3.6), which relates both to the breaking probability and to the dissipation. They found it predominantly unimodal and broadening towards smaller scales. For these slow breakers under developing-wave conditions, a bimodal directional distribution of $\Lambda(c, \theta)$ was observed.

7.4 Whitecapping dissipation functions in spectral models for wave forecasting

Dissipation functions which are used to represent the spectral distribution of dissipation in spectral wave-prediction models is a large topic and may well constitute a large review paper in its own right. Many, if not most (to date) such terms are not based on strictly

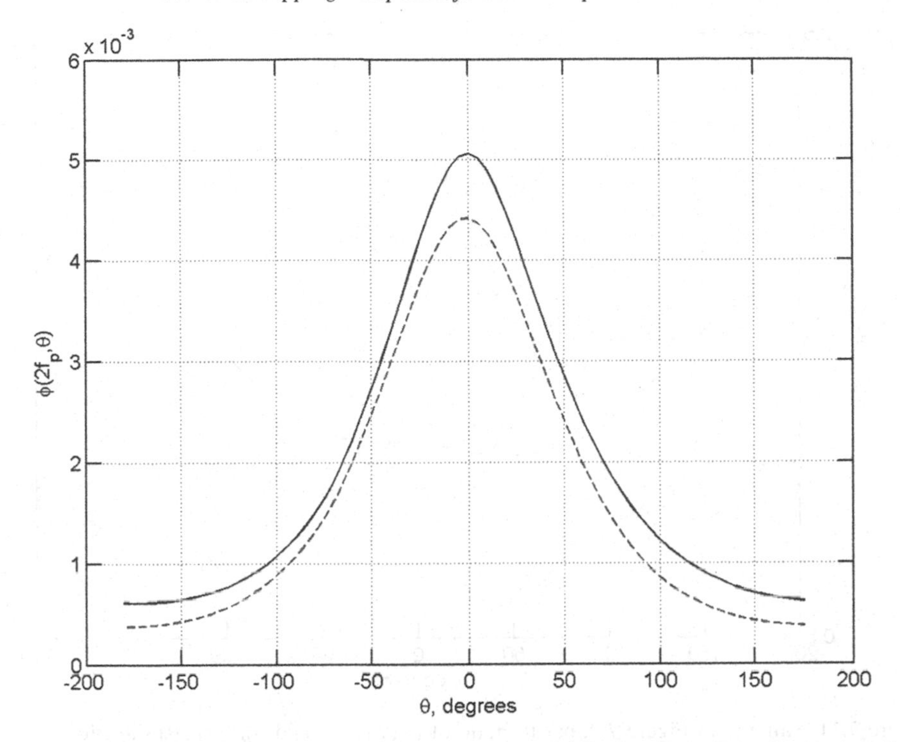

Figure 7.10 Same as in Figure 7.8, but for $\phi_i(2f_p, \theta)$ (solid line) and $\phi_P(2f_p, \theta)$ (dashed line) directional spectrum at double peak frequency $2f_p$. Figure is reproduced from Young & Babanin (2006a) © American Meteorological Society. Reprinted with permission

physical, either experimental or theoretical argument and therefore, technically speaking, are not the subject of this book. We should treat with all due respect the tremendous job the forecast-model developers have done, including their efforts to treat dissipation in a sensible way. In the absence of reliable knowledge and firm understanding of the breaking-dissipation behaviour, such models still perform very well in the majority of circumstances, but this is a topic of forecasting technologies, numerical modelling, data assimilation, applied mathematics and other related disciplines.

The chapter on spectral dissipation would however be incomplete without outlining the dissipation terms presently employed for practical needs, their advantages and deficiencies, and future prospects in this regard. At the time of writing, this particular topic is undergoing rapid development and is subject to concentrated effort by a number of research groups around the world, and therefore we will pay more attention to the prospects rather than to the present state of wave-dissipation modelling. We will largely follow Babanin & van der Westhuysen (2008) and Babanin *et al.* (2010c) in this discussion.

The dissipation term S_{ds} is one of the three most important source functions of the radiative transfer equation (2.61) employed by all spectral wave models to predict the wave spectrum F. Since a major, if not dominant part of S_{ds} is attributed to energy losses

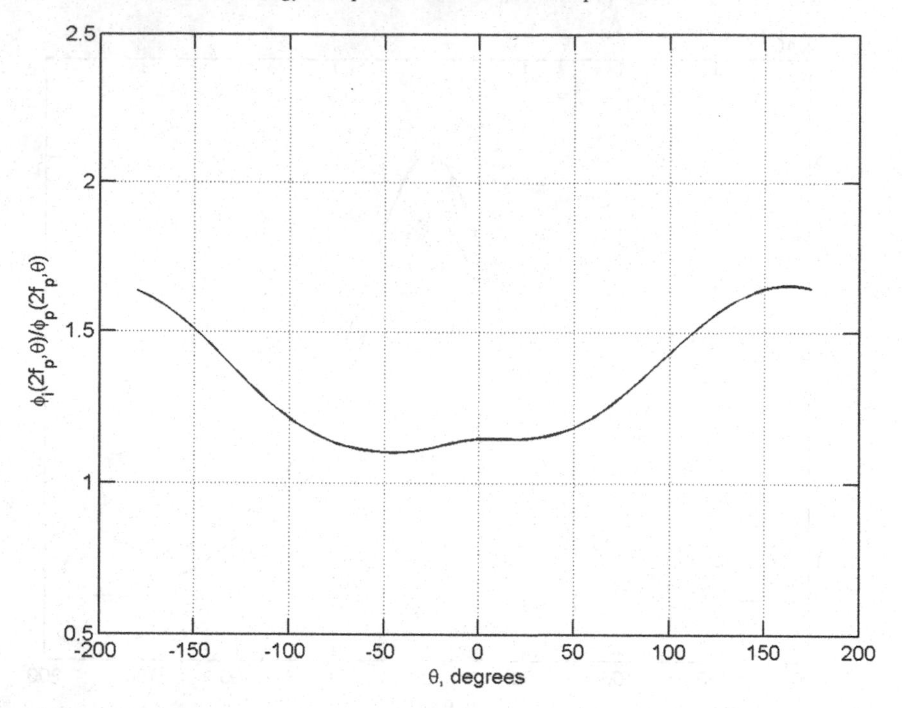

Figure 7.11 Same as in Figure 7.9, but for ratio of $\phi_i(2f_p, \theta)$ and $\phi_p(2f_p, \theta)$ directional spectra of Figure 7.10 at double peak frequency $2f_p$. Figure is reproduced from Young & Babanin (2006a) © American Meteorological Society. Reprinted with permission

due to wave breaking, and the breaking has usually been regarded as a poorly understood and basically unknown phenomenon, formulations of the term have often been loosely based on physics and served as a residual tuning knob. The tradition was laid by Komen *et al.* (1984), who calibrated the dissipation and other source terms together, based on their joint contribution to the integral-wave-growth curves which were the only reliable experimental knowledge at the time, and the approach persisted throughout more than 20 years. Such significant attempts of the past years to improve the S_{ds} parameterisation as Polnikov (1991), Young & Banner (1992), Banner & Young (1994), Tolman & Chalikov (1996), Alves & Banner (2003), Rogers *et al.* (2003) and Bidlot *et al.* (2007), among others, rested firmly within this tradition. More recent studies by Ardhuin *et al.* (2007) and Van der Westhuysen *et al.* (2007), which both highlighted some serious limitations of this approach and resulted in updates of operational wave-forecast models, are still, to an extent, based on the residual tuning. And it is only now that the newly devised dissipation functions and their calibration appeal directly to the recently found and/or understood physics of breaking and dissipation, including its threshold behaviour and cumulative effects (Babanin *et al.*, 2007d, 2010c; Tsagareli, 2009; Ardhuin *et al.*, 2010). Babanin *et al.* (2007d, 2010c), Tsagareli (2009) and Tsagareli *et al.* (2010) went further and suggested a methodology of separate independent calibration of the wind-input S_{in} and dissipation S_{ds} functions.

At present, when modelling equation (2.61), there is almost no flexibility in formulating S_{nl} and some limited flexibility in formulating S_{in}. By contrast, functions to represent S_{ds} still can be chosen rather arbitrarily and are used in models without much objection from the wave-modelling community. There is no consistency and sometimes even little similarity between the terms of Komen *et al.* (1984), Polnikov (1991), Tolman & Chalikov (1996), Babanin *et al.* (2007d, 2010c), Van der Westhuysen *et al.* (2007) and Ardhuin *et al.* (2010), all of which are incorporated in models and used alongside some standard terms for S_{nl} and S_{in}.

The latter two are based on more or less defined physics (e.g. Hasselmann (1962) and Donelan *et al.* (2006), respectively). Obviously, all the formulations for the dissipation S_{ds} refer to some physics too, but theoretical and experimental guidance had been very uncertain in the past.

Existing theories of the wave-breaking dissipation, both their advantages and shortcomings, were analysed in Section 7.1. In short, it is generally assumed that the spectral dissipation $S_{ds}(f, k, \theta)$ in (2.61) depends on the wave spectrum $\Phi(f, k, \theta)$. Since the functional form of this dependence is basically unknown, the dissipation is usually parameterised as an algebraic power function of omni-directional spectrum $F(f)$ (see eq. 7.1). Some directional distribution is then assumed, typically isotropic.

It would be fair to mention that, in spite of a relatively broad choice of theoretical models in Section 7.1, it is the theory by Hasselmann (1974) which is most frequently referred to in S_{ds} formulations. From the very beginning, however (i.e. Komen *et al.*, 1984), this theory was employed only conditionally – that is, speculative properties and parameters were added to meet tuning needs, and in particular to compensate for the missing threshold behaviour and cumulative effect as was discussed, in particular, in Section 7.3.4 above. Over the years, this term has undergone a significant number of similarly speculative alterations and additions, a review of which is available in Appendix A of Ardhuin *et al.* (2007).

Contrary to the theory of dissipation, recent experimental advances in wave-dissipation studies have brought about much more certainty regarding the behaviour of S_{ds}. In our view, the notion that the dissipation function is a great unknown and that any formulation that helps to satisfy the energy balance is considered legitimate, is no longer satisfactory. Over the past decade, many physical features of the dissipation performance have been discovered experimentally and explained as described throughout this book and this chapter.

How are these physics, which are by no means tentative reasoning, but are definite field observations, included in S_{ds} terms? In WAVEWATCH (see Tolman, 2009, for the latest version of this model), two-phase behaviour of the dissipations term is accommodated, i.e. that it is different at the spectral peak and the spectrum tail (although the assumed physics of Tolman & Chalikov (1996) is different to that revealed in the experiments by Babanin & Young (2005), Young & Babanin (2006a) and Manasseh *et al.* (2006)). Rogers *et al.* (2003), in a way, attempted to introduce a threshold-dissipation behaviour by disallowing the breaking of swell. This was done within the SWAN model (Booij *et al.*, 1999). Overall, however, most models still lack these new insights into the physics of wave breaking. Since there appears to be some confusion in this regard, we would like to point out, that, for example, the analytical model of Hasselmann (1974) (see Section 7.1) is neither in contrast nor in support of the above-mentioned threshold behaviour, but is unrelated to it. It predicts the behaviour of the whitecapping dissipation provided that whitecaps already exist and conduct the negative work on the forward face of the wave, i.e. what happens if the waves are above the steepness/spectral threshold and are already breaking.

These observed features need to be accommodated in modern dissipation terms, otherwise the models do not reflect the correct physics and do not adequately describe the reality. It is fast becoming clear that, without incorporating these new features, the models cannot properly forecast complex or non-standard circumstances. Ardhuin *et al.* (2007) showed two such situations: wave growth in the presence of swell and over slanting fetches.

The most apparent non-standard circumstances where failure of the standard-tuned terms is to be expected are extreme and complex wind–wave conditions – which are also of utmost interest from practical points of view (Greenslade, 2001; Babanin *et al.*, 2010d). As discussed in Section 7.3.5 above, the dissipation function is altered under such conditions, and so is the wind input. No amount of good tuning and statistical fitting, as opposed to employing correct physics, will be able to extrapolate source terms, tuned to standard conditions, into such extreme situations (see Cavaleri, 2009, for the most recent update on spectral-modelling limitations in this regard).

Babanin *et al.* (2010d) specifically investigated the modelling capacity of modern third-generation models by means of hindcasting the wave conditions in Typhoon Krosa prior to landfall on Taiwan in October 2007. Two models were used, SWAN (Booij *et al.*, 1999) and WWM (Hsu *et al.*, 2005; Roland, 2009). During this typhoon, the highest wind-generated waves ever measured were recorded, with the trough-to-crest elevation for an individual wave of

$$H_{max} = 32.3\,\mathrm{m} \tag{7.34}$$

and the maximal significant wave height, measured over 10 min duration, of

$$H_{s_{max}} = 23.9\,\mathrm{m} \tag{7.35}$$

(Liu *et al.*, 2007, 2009; Doong *et al.*, 2008).

After careful analysis, Babanin *et al.* (2010d) decided that the measurement did not appear faulty and is physically realistic. Numerous numerical tests were then performed in order to reproduce the observed conditions, and it was concluded that neither SWAN nor WWMII are able to hindcast the extreme observation. Thus, an urgent need for introduction of new and adequate physics and numerics into wave-forecast models is apparent.

One of the first dedicated efforts to incorporate some of the new experimental observations (i.e. the breaking-threshold behaviour, see Section 5.2) into a nonlinear dissipation function was conducted by Alves & Banner (2003). Babanin & van der Westhuysen (2008) later showed that the statement that the form of the Alves & Banner (2003) dissipation term is based on a threshold behaviour of wave breaking appeared to be mistaken. The formulation did not consistently employ experimentally observed threshold dependences, nor did it use experimentally obtained values for the threshold. On the contrary, the magnitude of the dimensionless switch, which imitated the no-breaking threshold, appeared to correspond to the spectra which certainly comprise breaking waves, and these waves break at a typical for moderate conditions rate of some 5% (i.e. Babanin *et al.*, 2001). As a result, the meaning of the switch turned to be to an extent opposite to its interpretation. That is, the switch did not signify a breaking threshold, i.e. a lower spectral limit below which no breaking occurs, but rather a soft upper limit for the spectrum growth, allowed by the formulation itself. Subsequently, the dissipation function, being postulated as strongly nonlinear, worked as quasi-linear. The role of the nonlinear option was mainly limited to quenching the spectrum down if it happened to exceed the switch level; below this level conventional spectral evolution, controlled by linear dissipation and input, took place and thus regular growth rates were well reproduced.

As such, however, the Alves & Banner (2003) intention was an important step towards adequate description of wave-field dynamics. It has been followed by a number of studies employing similar so-called 'saturation-based' dissipation terms S_{ds} (i.e. formulations based on a non-dimensional 'saturation' spectrum (5.31), see Van der Westhuysen *et al.* (2007) and Ardhuin *et al.* (2010)).

Van der Westhuysen *et al.* (2007) initially based their simulations by means of the SWAN model (Booij *et al.*, 1999) on the Alves & Banner (2003) formulation, but incorporated appropriate threshold limitations and a wind-forcing dependence for the dissipation function. As a result, in the physical sense it is essentially a new dissipation term.

First, the threshold parameter was qualitatively and quantitatively returned to its original meaning of the bottom limit for wave breaking. Secondly, Van der Westhuysen *et al.* (2007) reviewed the scaling of the major source terms and made them consistent, i.e. exponents of the dissipation and the wind-input functions now matched (in Alves & Banner (2003), the dissipation was declared strongly nonlinear in terms of the wave spectrum, but then linear counterpart wind-input functions were used to reproduce regular wave-growth dependences). Besides, gradual transition from the strongly forced sea-state cases to the mature seas was implemented which was a novel physically sound feature. The Van der Westhuysen *et al.* (2007) function is indeed nonlinear, and this nonlinearity varies in the course of wave evolution. Also, depth dependence for the dissipation was introduced.

Ardhuin *et al.* (2010) further extended the saturation-based formulations. They added the cumulative term and an S_{turb} dissipation term (see eq. 7.26), responsible for wave-turbulence interactions and based on the Ardhuin & Jenkins (2006) mechanism. Their inherent-breaking term has an isotropic part and a direction-dependent part. The latter allows us to use the dissipation function to control the directional spread of the wave spectrum (see Section 7.3.6 for a discussion on this issue). The dissipation is essentially linear in terms of the wave spectrum $F(f, \theta)$.

A separate dissipation term was reserved for swell, which does not break and exhibits physical properties, different to wind-seas, in other regards as well. Ardhuin *et al.* (2010)

argued that most of the swell dissipation comes from the negative momentum/energy flux from the swell field to the wind, and they formally included this kind of wave-energy dissipation as an additional negative wind-input term (see also Harris, 1966; Volkov, 1970; Makin & Chalikov, 1980; Donelan, 1999; Drennan *et al.*, 1999; Grachev & Fairall, 2001; Grachev *et al.*, 2003; Lavrenov, 2004; Babanin & Makin, 2008; Hanley *et al.*, 2010; Soloviev & Kudryavtsev, 2010, among others).

Here, we should note that swell dissipation, at least at initial stages of swell propagation while it is still steep, can also be explained through generation of the wave turbulence, unrelated to breaking (see Babanin, 2006; Babanin & Haus, 2009; Dai *et al.*, 2010, see also Section 7.5). The relative strength of the swell friction against the air at the air–sea interface and the swell's role in producing the wave-induced turbulence are unclear at this stage. In Ardhuin *et al.* (2010), the swell dissipation was calibrated on the basis of satellite observations of swell attenuation when propagating large distances across the ocean (Ardhuin *et al.*, 2009a).

Thus, Ardhuin *et al.* (2010) took dissipation-function modelling to very new heights. While the formulations for the different sub-terms of the dissipation function are still largely empirical and the tuning was still needed, the authors made an extraordinary effort to accommodate most of the known up-to-date experimental features of the spectral dissipation, and, where possible, used theoretical justification and experimentally observed dependences for calibrations.

A variety of tests and validations were conducted. These covered both the standard set of steady-wind fetch-limited growth according to Komen *et al.* (1984) and hindcasts of real-life storms, on regional as well as global scales. The regional scales included such diverse situations and water bodies as young waves in Lake Michigan and Hurricane Ivan in the Gulf of Mexico. The hindcasts were conducted by means of WAVEWATCH-III^{TM} (Tolman, 2009) and demonstrated a potential for improvements in performance.

Such improvements do not necessarily come hand-in-hand with updating the physics of models, which had been well-tuned to predicting standard situations over years. Incorporation of new dissipation features is more complex than a mere replacing of one dissipation term with another. For example, if a cumulative integral is simply added to the breaking-dissipation term, then local-in-wavenumber-space balance can no longer be satisfied at smaller scales where the cumulation is significant, and re-formulations and adjustments of the wind input function and potentially of the entire model will also be required. Therefore, an essentially modified version of the Janssen (1991) input was employed by Ardhuin *et al.* (2010) which, in turn, led to a markedly different spectral balance of the source terms.

Another new implementation of the observation-based physics into source functions of spectral wave models is that by Babanin *et al.* (2007d), Tsagareli (2009), Tsagareli *et al.* (2010) and Babanin *et al.* (2010c). In this case, both wind-input S_{in} and dissipation S_{ds} terms in (2.61) are new and observations based. Mathematical expressions for these functions are not based on any of the above-mentioned parameterisations presently employed in wave models, and they were initially designed to accommodate new features observed in the course of the Lake George experiment (see Section 3.5) which was specifically designated to field investigations of the source functions under moderate-to-strong wind forcing.

The new wind-input parameterisation of Donelan *et al.* (2006), for example, follows the theoretical form proposed by Jeffreys (1924, 1925), based on the wind-sheltering argument, and incorporates two newly observed features of wind–wave coupling, i.e. dependence of the growth increment on wave steepness (which makes the input function nonlinear in terms of the wave spectrum) and full air-flow separation in strong wind-forcing situations (which leads to a relative reduction of the wind input). Additionally, it accommodates enhancement of the wind input over breaking waves (Babanin *et al.*, 2007b, see also Section 8.3), which becomes essential when the breaking rates are high. The parameterisation was able to reconcile experimental outcomes for fractional wind–wave growth rates $\gamma(f)$, which previously appeared incompatible, that is low values of the sheltering coefficient in well-developed oceanic conditions (Hsiao & Shemdin, 1983) and two-and-a-half-times as high magnitudes of the sheltering for strongly forced and steep young waves in the laboratory (Donelan, 1999).

As was mentioned above, Tsagareli *et al.* (2010) for S_{in} and Babanin *et al.* (2010c) for S_{ds} employed additional constraints when testing the source functions, which constraints allowed calibration of these functions separately, before they are put together in a wave model to reproduce overall wave growth and overall wave-spectrum evolution. For S_{in}, this is the total integrated wind input which must agree with independently observed or known magnitudes of the wind stress. Within this approach, a new technique – a dynamic self-adjusting routine – was developed for correction of the wind-input source function S_{in} at each step if the constraint was not observed. This correction involves a frequency-dependent adjustment to the growth rate $\gamma(f)$, based on extrapolations from field data. The model results also showed that light winds require higher-rate adjustments of the wind input compared to strong winds.

Even though this book is dedicated to the dissipation, we shall briefly review the wind-input constraint and its implementation in Tsagareli *et al.* (2010). It is important, first of all, as the first step to the dissipation-calibration constraint. Secondly, it is a key item of the new methodology of separate source-function calibrations, intended to replace the bundled-calibration method according to Komen *et al.* (1984).

So, the momentum flux τ across the water surface (3.7) was considered the key boundary parameter for calibrating the wind-input source function. The generation of surface waves by the action of wind is due to work done by the wind stress exerted on this surface. Wind stress is a result of the air–sea interaction, i.e. 'friction' of the air flow against the water surface, and reflects the strength of this interaction. Physically, it is the drag force per unit area exerted on the interface by the adjacent layer of the airflow. Therefore, wind stress determines the exchange of momentum between the atmosphere and the water.

Significant stresses arise within the near-surface atmospheric boundary layer because of the strong shear of the wind between the slowly moving air near the water surface and the more rapidly moving air in the layer above (see e.g. Komen *et al.*, 1994). Close to the surface, the total wind stress τ can be represented by three components: turbulent stress τ_{turb}, wave-induced stress τ_{wave} and viscous or tangential stress τ_ν (e.g. Kudryavtsev *et al.*, 2001):

$$\tau = \tau_{\text{turb}} + \tau_{\text{wave}} + \tau_\nu. \tag{7.36}$$

The atmospheric turbulent momentum flux decreases to zero at the surface itself where the turbulence vanishes. Therefore, directly at the surface the total wind stress is a combination of stress τ_{wave} induced by the ocean waves and the viscous stress τ_ν which generates surface currents.

Among the different contributions to the total stress (7.36), the wave-induced stress is directly related to the momentum exchange between the wind and the waves, and it is this part which is used by Tsagareli *et al.* (2010) as the constraint for the wind-input source term. It is estimated at the air–sea interface as:

$$\tau_{\text{wave}} = \tau - \tau_\nu. \tag{7.37}$$

On the other hand, the wave-induced stress is determined by the wind momentum input as:

$$\tau'_{\text{wave}} = \int_{f_{\text{min}}}^{f_{\text{max}}} M(f)df, \tag{7.38}$$

where $M(f)$ is the momentum-input function, and the integration limits signify minimal and maximal frequency f which still contribute to this input. The momentum-input function $M(f)$ can be obtained from the wind energy-input source term $S_{in}(f)$ in (2.61):

$$M(f) = \rho_w g \frac{S_{in}(f)}{c(f)} df. \tag{7.39}$$

Therefore,

$$\tau'_{\text{wave}} = \rho_w g \int_{f_{\text{min}}}^{f_{\text{max}}} \frac{S_{in}(f)}{c(f)} df. \tag{7.40}$$

This constraint is apparent from the physical point of view. In the practical sense, satisfying this criterion determines the credibility of a parameterised form for the wind-input source term S_{in}, and sets the main physical framework for investigation of the behaviour of this parameterisation and its validation, before it is employed in a model together with other sources.

Wave-induced stress is dependent on the upper limit of the integral in (7.40). Contribution of short-wave scales to the total stress is significant, and therefore the higher the upper limit of the integral, the more precise is the estimate of the wave-induced stress. Thus, the upper limit of $f_{\text{max}} = 10\,\text{Hz}$ was selected by Tsagareli *et al.* (2010) which signifies the shortest waves in the capillary range still involved in air–sea coupling.

Computation of the wave-induced stress using (7.37) required knowledge of the viscous stress. Here, the viscous-stress contribution to the total stress was estimated according to

Banner & Peirson (1998). Substituting their $\tau_v = \rho_a C_V U_{10}^2$, where C_V is the viscous drag coefficient, along with (3.7) into (7.37) yields:

$$\tau_{\text{wave}} = \rho_a U_{10}^2 (C_D - C_V). \tag{7.41}$$

Banner & Peirson (1998) demonstrated a qualitative trend of the viscous stress as a function of the wind speed, but did not present a quantitative dependence. Tsagareli *et al.* (2010) digitised the data of Banner & Peirson (1998) and parameterised the viscous drag as a function of wind speed U_{10}:

$$C_V = -5 \cdot 10^{-5} U_{10} + 1.1 \cdot 10^{-3}. \tag{7.42}$$

Combined with a known dependence for sea drag C_D, this now provides an integral constraint which the wind-input spectral distribution $S_{in}(f)$ must satisfy.

A variety of parameterisations exist to choose from for the sea drag. Most common are dependences for the drag coefficient C_D as a function of wind speed U_{10}; some also introduce the wind-forcing parameter U_{10}/c_p (see e.g. Guan & Xie, 2004; Babanin & Makin, 2008, for a review). Babanin & Makin (2008) also showed that there are more than a dozen other properties that the drag can depend on. As a result, the scatter of measurements with respect to any of the C_D parameterisations is large, of the order of tens of percent, but as Tsagareli *et al.* (2010) demonstrated errors due to absence of a normalisation of the wind input by the sea drag can be of the order of hundreds of percent or even more.

Specifically, Tsagareli *et al.* (2010) used C_D parameterisation by Guan & Xie (2004), which is an attempt to unify 25 relevant experimental dependences obtained by different researchers in the period from 1958 to 2003. It is a C_D-versus-U_{10} expression which also accommodates wave-age:

$$C_D = (0.78 + 0.475 f(\delta) U_{10}) \cdot 10^{-3} \tag{7.43}$$

where

$$f(\delta) = 0.85^B A^{1/2} \delta^{-B}. \tag{7.44}$$

The wave age dependence is included through the deep-water wave steepness

$$\delta = H_s k_p = H_s \frac{\omega^2}{g}. \tag{7.45}$$

The empirical parameters $A = 1.7$ and $B = -1.7$ are chosen such that C_D in (7.43) is in agreement with the latest results on this topic by Drennan *et al.* (2003).

The computational range of operational spectral wave models is usually limited by a relatively low-frequency upper cutoff in the vicinity of the spectral peak, and therefore in order to verify the constraint (7.37) in operational models a parametric tail should be added for the integration. Tsagareli *et al.* (2010) argue that such a tail has to be $S(f) \sim f^{-5}$, as in JONSWAP parameterisation (2.7). A similar parametric tail was also argued for before by the authors of the WAM model (Komen *et al.*, 1994).

On the other hand, the $S(f) \sim f^{-4}$ tail is also a reality, both measured and predicted theoretically (see Sections 5.3.2 and 8.2 for this discussion). There are well-justified parameterisations of the wave spectrum based on such behaviour of the small spectral scales (Donelan *et al.*, 1985). As concluded in Tsagareli *et al.* (2010) and Babanin *et al.* (2010c), both subintervals may in fact exist in real spectra, and in the case of their co-existence the f^{-5} tail is situated at higher frequencies which then lets the integral in (7.37) converge. Co-existence of the two subintervals is also supported by measurements (e.g. Forristall, 1981; Evans & Kibblewhite, 1990; Kahma & Calkoen, 1992; Babanin & Soloviev, 1998b; Resio *et al.*, 2004).

In order to accommodate both types of tail behaviour for calibration purposes of the source functions, Tsagareli (2009) and Babanin *et al.* (2010c) suggested a combination of the JONSWAP spectral parameterisation form (2.7) with that of Donelan *et al.* (1985), so that both subintervals of f^{-4} and f^{-5} are present in the equilibrium interval:

$$F(f) = \begin{cases} \alpha g^2 (2\pi)^{-4} f_p^{-1} f^{-4} \exp\left[-\frac{5}{4}\frac{f}{f_p^{-4}}\right] \cdot \gamma^{\exp\left[-\frac{(f-f_p)^2}{2\sigma^2 f_p^2}\right]} & f \leq f_t, \\ \alpha g^2 (2\pi)^{-4} f_t f_p^{-1} f^{-5} \exp\left[-\frac{5}{4}\frac{f}{f_p^{-4}}\right] \cdot \gamma^{\exp\left[-\frac{(f-f_p)^2}{2\sigma^2 f_p^2}\right]} & f > f_t \end{cases} \tag{7.46}$$

where f_t is the transition frequency. Similar spectral shapes, which can accommodate either type of spectral tail, had been suggested before (e.g. Young & Verhagen, 1996). Here, they were unified and, following Tsagareli (2009), this combined spectral-shape parameterisation is called the Combi spectrum.

The transition is typically located at $f \sim 3f_p$ (5.42) and has great importance for calibration of the dissipation function. As discussed in Section 5.3.2, it signifies transition from the inherent-breaking spectral region to the part of the spectrum which is dominated by the cumulative dissipation.

As far as the separate calibration of the dissipation term S_{ds} is concerned, the main constraint states that the dissipation-function integral must not exceed the total wind input:

$$\int_0^{f_\infty} S_{ds}(f)df = R_D \int_0^{f_\infty} S_{in}(f)df \tag{7.47}$$

where ratio $R_D \leq 1$. The ratio of the two integrals as a function of the wave development stage U_{10}/c_p is known experimentally (e.g. Donelan, 1998), and therefore the dissipation term can also be studied and tuned individually.

According to Donelan (1998), this ratio R_D stays within the range of 95–100% for most stages of wave development, reaching 100% at the Pierson-Moscowitz full-development limit (Pierson & Moskowitz, 1964). It is only at the very early stages that the total wind input can be significantly larger than the total dissipation.

Donelan (1998) did not provide an explicit quantitative dependence for the ratio R_D. For practical purposes, in Babanin *et al.* (2010c) his Figure 6 was segmented, digitised and parameterised here as following:

$$R_D = \begin{cases} -0.12\dfrac{U_{10}}{c_p} + 1.52 & 4.5 < \dfrac{U_{10}}{c_p} \le 5.8, \\ 0.0031\dfrac{U_{10}}{c_p} + 0.96 & 1.5 < \dfrac{U_{10}}{c_p} \le 4.5, \\ -0.052\dfrac{U_{10}}{c_p} + 1.043 & 0.83 < \dfrac{U_{10}}{c_p} \le 1.5, \\ 1 & \dfrac{U_{10}}{c_p} \le 0.83. \end{cases} \tag{7.48}$$

The upper wind-forcing limit of Donelan (1998) was $U_{10}/c_p = 4.5$, and parameterisation (7.48) also includes the range of very young dominant waves $U_{10}/c_p = 4.5$–5.8. For this range of wave ages, the dissipation ratio R_D was determined on the basis of consistency between the model results for the variance of the energy-density spectra and the experimental data of Babanin & Soloviev (1998a). It was found that for very young waves the dissipation ratio is relatively small compared to the ratio for mature waves. For very young waves of $U_{10}/c_p = 5.8$, it was of the order of $R_D = 0.82$ (see Section 4.3 of Tsagareli (2009) for more details).

The segmented parameterisation (7.48) produces discontinuities of the derivatives and such sharp transitions are undesirable in numerical modelling. Therefore, for application within a spectral model, the relationship was further smoothed and used in the following form

$$R_D = \begin{cases} 0.97 - 0.07 \cdot \left(1 + \tanh\left(3\left(\dfrac{U_{10}}{c_p} - 5.2\right)\right)\right) & 2 < \dfrac{U_{10}}{c_p} \le 5.8, \\ 0.97 + 0.015 \cdot \left(1 - \tanh\left(5\left(\dfrac{U_{10}}{c_p} - 1.1\right)\right)\right) & 0.9 < \dfrac{U_{10}}{c_p} \le 2, \\ 1 & 0.83 \le \dfrac{U_{10}}{c_p} \le 0.9. \end{cases} \tag{7.49}$$

The dissipation term tested in Babanin *et al.* (2010c) is that proposed by Babanin & Young (2005) and Young & Babanin (2006a) and written out in (5.40); it has already been outlined in Section 5.3.2. In short, it consists of the inherent-dissipation term $T_1(f)$ and the cumulative dissipation $T_2(f)$:

$$S_{ds}(f) = T_1(f) + T_2(f) \tag{7.50}$$

where

$$T_1(f) = -a_1 \rho_w g f (F(f) - F_{\text{threshold}}(f)) \tag{7.51}$$

and

$$T_2(f) = -a_2 \rho_w g \int_{f_p}^{f} (F(q) - F_{\text{threshold}}(q))dq. \qquad (7.52)$$

As discussed in Section 5.3.2, normalisation by the directional spread $A(f)$ (5.33) can be omitted which is done here in (7.51)–(7.52).

The expression uses dimensional wave spectrum $F(f)$, rather than its dimensionless counterpart $\sigma_{\text{Phillips}}(f)$ (5.31) like the saturation-based functions discussed in this section above. The rationale behind the saturation use is that it allows nonlinear formulations for the spectral dissipation, but the investigations of Tsagareli (2009) led to a conclusion that linear whitecapping dissipation with the cumulative term is most suitable. If the (5.40) or (7.50)–(7.52) term is to be made nonlinear, it can be done of course, but $F(f)$ has to be normalised by a relevant spectral-density distribution first in order to satisfy the dimensional argument (Rogers *et al.*, 2011).

As discussed in Section 5.3.2, since the saturation $\sigma_{\text{Phillips}}(f)$ (5.31) is the fifth moment of the spectrum, experimental investigations of the breaking probabilities versus such a parameter, for field-observed irregular spectra, led to formidable scatter (see Figure 5.23). This is why it was preferred to keep the dissipation a function of the spectrum $F(f)$ itself; dependences of the breaking probabilities across the spectrum in terms of the spectral density $F(f)$ are quite robust and therefore can be investigated experimentally (see Babanin & Young, 2005; Babanin *et al.*, 2007c).

The dimensionless threshold $\sigma_{\text{threshold}}(f)$ (5.36), on the contrary, is a steady value. In the dimensional formulation of the dissipation function, this value can then be converted into dimensional threshold at each frequency algebraically (5.37).

The weighting coefficients a_1 and a_2, of the inherent (7.51) and cumulative (7.52) dissipation terms, were introduced by Babanin & Young (2005) and Young & Babanin (2006a) on the basis of a single extreme-breaking record analysed. They obtained $a_1 = 0.0065$ and assumed the same value for a_2, but obviously in a general case this issue needed to be revisited in Babanin *et al.* (2010c).

Computations of the spectral dissipation function (7.50) with such weighting coefficients, performed for a Combi spectrum (7.46) of moderately-forced waves with $U_{10}/c_p = 2.7$ and wind speed $U_{10} = 10$ m/s, are presented in Figure 7.12. The inherent breaking term $T_1(f)$ (7.51) and the forced-dissipation term $T_2(f)$ (7.51) are also shown. The shape of the wave-dissipation source function is the result of superposition of these two terms in (7.50).

As implied in (5.40), the long-scale waves down to the size of dominant waves do not experience induced dissipation. The contributions of the forced dissipation to the total start at the spectral peak and then increase towards higher frequencies until it saturates. This gradual transition is due to the integral of the forced dissipation term T2(f) (7.52).

Obviously, the high-frequency waves, which have reached the saturated-dissipation limit, mostly experience forced dissipation due to the influence of longer waves, and their inherent-breaking dissipation can be neglected. The coefficients a_1 and a_2 are crucial in

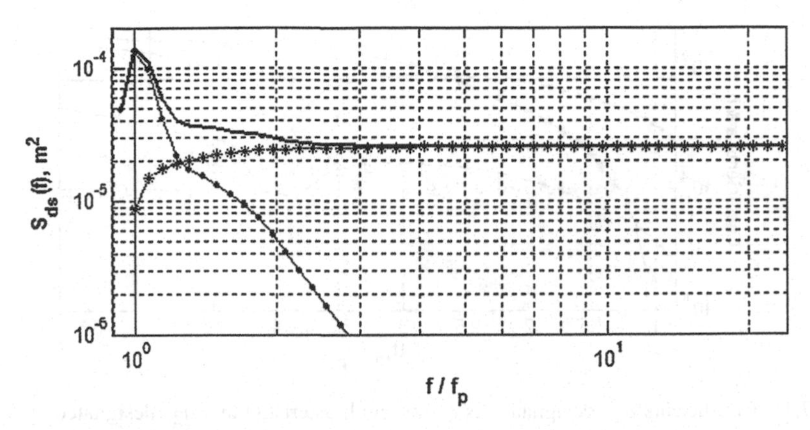

Figure 7.12 Spectral dissipation function $S_{ds}(f)$ (7.50) (bold line) with the two terms T_1(f) (7.51) (line with dots) and T_2(f) (7.51) (line with asterisks), with coefficients $a_1 = a_2 = 0.0065$. Computations were performed for a Combi spectrum with $f_p = 0.13$ Hz, $U_{10}/c_p = 2.7$, $U_{10} = 10$ m/s. Figure is reproduced from Babanin *et al.* (2010c) © American Meteorological Society. Reprinted with permission

determining the saturation level, as well as the relative contributions of the two different types of dissipation. Therefore, to achieve the correct level of wave dissipation in the model, it is necessary to calibrate these coefficients carefully.

Because of the two-phase behaviour of the spectral dissipation, i.e. combination of terms both depending on the wave spectrum $F(f)$ directly and depending on the integral of $F(f)$, the induced-dissipation term $T_2(f)$ (7.52) is zero below the peak, and thus the entire dissipation in this region was attributed to the inherent breaking in order to estimate the coefficient a_1(note that this is an upper estimate of the dissipation because the nonlinear transfer could have contributed towards spectrum growth below the peak too). As the waves approach full development, the dissipation $T_1(f)$ becomes negligible, and therefore $a_1 \to 0$ in such conditions. Once the coefficient a_1 is then calibrated, the quantitative dependence for a_2 can be obtained based on the constraint (7.47).

The behaviour of the coefficients a_1 and a_2 obtained this way is shown in Figure 7.13 (designated as a and b, respectively) as a function of the wave-development stage U_{10}/c_p. The magnitude of these coefficients is compared with the experimental value $a_{\text{exp}} = 0.0065$ estimated by Young & Babanin (2006a).

As shown in Figure 7.13, values of both coefficients decrease as waves develop, and the rate of decrease accelerates as the waves approach full development. This trend, in particular, indicates a reduction of the relative contribution of the dominant waves into wave-energy dissipation, as the wave system develops.

For young waves ($U_{10}/c_p > 2$), the coefficients a_1 and a_2 differ by an order of magnitude, a_1 being larger. Regardless of this, as can be seen in Figure 7.14, the cumulative term, the magnitude of which is effectively determined by the coefficient a_2, dominates the dissipation at smaller scales of the wave spectrum (that is the tail of the dissipation function

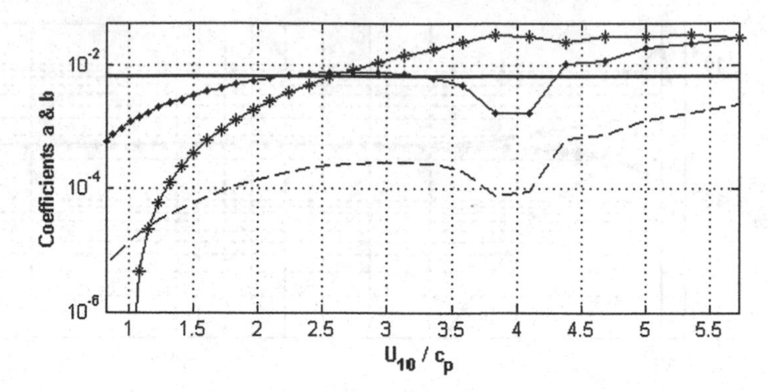

Figure 7.13 Coefficients a_1 (designated as a, line with asterisks) and a_2 (designated as b, dashed line) as functions of the wind forcing parameter U_{10}/c_p computed for Combi spectra at wind speed $U_{10} = 10$ m/s. Line with dots shows coefficient a_{2_0} after the frequency correction was applied. The experimental coefficient $a_{\text{exp}} = 0.0065$ (Young & Babanin, 2006a) is shown with a bold line. Figure is reproduced from Babanin *et al.* (2010c) © American Meteorological Society. Reprinted with permission

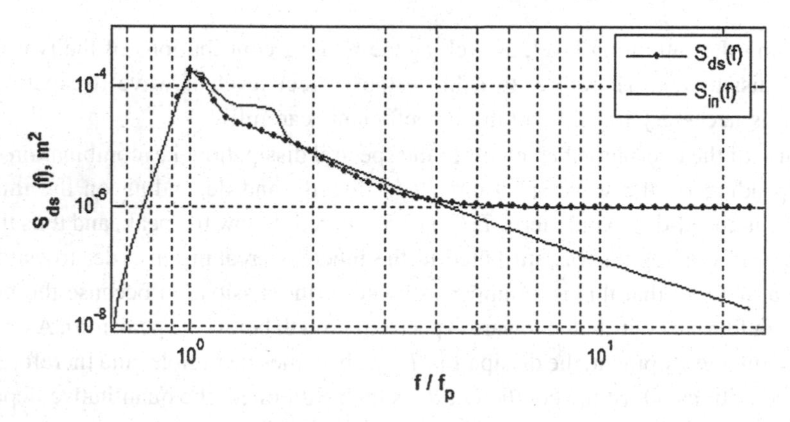

Figure 7.14 The spectral dissipation function $S_{ds}(f)$ (line with dots) with coefficients a_1 and a_2 of Figure 7.13. Computations were performed for a Combi spectrum with wind forcing $U_{10}/c_p = 2.7$ and wind speed $U_{10} = 10$ m/s. The corresponding wind input $S_{in}(f)$ is shown as the plain line. Figure is reproduced from Babanin *et al.* (2010c) © American Meteorological Society. Reprinted with permission

saturates due to the cumulative term according to Figure 7.12). This fact highlights the significance and importance of the cumulative term $T_2(f)$ which, no matter how small it is in absolute value, cannot be disregarded in the spectral sense.

For older waves, the difference between the coefficients reduces; and for a mature stage of development, $U_{10}/c_p \sim 1.2$, the relationships cross: $a_1 = a_2$. For even older waves, approaching the full-development, the magnitude of coefficient a_1 drops rapidly, signifying absence of wave breaking at the spectral peak (Banner *et al.*, 2000). The coefficient a_2

continues to decrease gradually, but unlike a_1 does not become zero. Even at these stages of development, the induced dissipation persists, even though it slows down as the dominant waves become less steep. This is due to, for example, stretching and compressing of short waves by underlying longer waves, thus causing these shorter waves to break (e.g. Longuet-Higgins & Stewart, 1960; Phillips, 1963; Longuet-Higgins, 1987; Donelan, 2001; Donelan *et al.*, 2010).

Figure 7.13 also compares the coefficient a_1 obtained by means of the constraint (7.47) and experimental data (bold line). The only experimental estimate of $a_1 = 0.0065$ is available from Young & Babanin (2006a) based on their analysis of a single wave record when wind forcing was quite extreme: $U_{10}/c_p \sim 6.5$. In Figure 7.13, the value of $a_{\text{exp}} = 0.0065$ is achieved at $U_{10}/c_p = 2.6$, and at higher values of wind forcing the magnitude is somewhat greater. Qualitatively, this is consistent with the experiment. Young & Babanin (2006a) stressed that their estimate is a lower-bound approximation of the actual value since they measured the dissipation by comparing the difference in energy of wave trains which were already breaking to wave trains which had completed breaking and were once again gaining energy from the wind. By definition, this approach will underestimate the energy loss. In this regard, the quantitative agreement between the calibrated values of a_1 and the measurement is encouraging.

Figure 7.14 shows the dissipation source function $S_{ds}(f)$ (5.40) based on the coefficients a_1 and a_2, computed for the Combi spectrum (7.46) with $U_{10}/c_p = 2.7$ and wind speed $U_{10} = 10\,\text{m/s}$. In the figure, the corresponding wind energy-input source function $S_{in}(f)$ is also plotted, computed for the same wind-forcing conditions as described above in this section (see also Tsagareli *et al.*, 2010, for more detail). The integrals of the two source functions are consistent according to the physical constraint (7.47), but the shape of the dissipation function at high frequencies raises further questions about the calibration of $S_{ds}(f)$.

The most striking feature of the comparison of $S_{ds}(f)$ and $S_{in}(f)$ is the difference at high frequencies, where the cumulative term $T_2(f)$ dominates, by up to two orders of magnitude. Mathematically, this feature is apparent if coefficients a_1 and a_2 in (5.40) are frequency independent, i.e. only vary as a function of wave age, as shown in Figure 7.13. Physically, however, such a difference is difficult to justify because, in order to maintain the high dissipation rates, a very strong energy flux from the lower-frequency part of the spectrum by means of, for example, the nonlinear interaction term S_{nl} in (2.61) would be necessary.

This raises a question as to whether the coefficient a_2 is scale independent. It is likely that a frequency-dependent correction is required to ensure that the magnitude of the wave dissipation $S_{ds}(f)$ remains comparable with the wind-input function $S_{in}(f)$.

To define the frequency-dependent form for coefficient a_2, it was decided to apply a dimensionless correction function $Z(f)$ such that as a result $S_{ds}(f) \sim S_{in}(f)$. Note that this correction is only approximate as the nonlinear transfer S_{in} should also play some role (e.g. Young & van Vledder, 1993). It should be pointed out that such a correction does not affect the principal physical constraint (7.47) as it is applied in the frequency region where values of the dissipation function are two or more orders of magnitude less than at the peak and therefore their contribution into the integral is negligible.

A function of the following form was chosen:

$$Z(f) = \left(\frac{f}{f_p}\right)^{\mu} \tag{7.53}$$

where μ is an exponent fitted to the high-frequency tail of the wind-input source term $S_{in}(f)$ at each stage. The resulting coefficient for the cumulative dissipation term $T_2(f)$ can now be represented as:

$$a_2 = a_{2_0} Z(f). \tag{7.54}$$

The magnitudes of the coefficient a_{2_0} computed at different stages of wave development are shown in Figure 7.13 by the line with dots. These new values of a_{2_0} are now significantly larger than the previous values of a_2 (dashed line). They are also closer to the experimental value a_{exp}, which is what was implied for coefficient a_2 in the experimental paper by Young & Babanin (2006a).

This new coefficient was then tested for different types of wave spectra and for different wind speeds $U_{10} = 7\,\text{m/s}, 10\,\text{m/s}, 15\,\text{m/s}, 20\,\text{m/s}$ and $30\,\text{m/s}$. For further technical details, we refer the reader to Babanin *et al.* (2010c), and here we will mention that in order to keep the cumulative term positive, an upper limit was imposed on the coefficient a_1 such that the condition of $a_2 > 0$ is always satisfied. In technical terms, the correction function $Z(f)$ was also applied to coefficient a_1, but in the high frequency range only:

$$a_1 = \begin{cases} a_0 & f \leq f_p, \\ a_0 Z(f) & f > f_p. \end{cases} \tag{7.55}$$

As a result of this correction, since coefficients a_1 and a_2 are not independent as discussed above, values of coefficient a_1 decreased somewhat and coefficient a_2 correspondingly increased and now stayed positive in all circumstances.

Figure 7.15 shows the wave-dissipation source function $S_{ds}(f)$ with the frequency-corrected coefficients a_1 and a_2 according to (7.54)–(7.55), at different stages of wave development U_{10}/c_p. At intermediate wind forcing of $U_{10}/c_p = 2.7$, the dissipation exceeds values of the wind input in the spectral tail, which implies an additional influx of energy due to nonlinear transfer in this spectral region. Other than this region, the dissipation is always smaller than the input across the spectrum.

In Figure 7.16, the spectral dissipation function is shown computed for $U_{10}/c_p = 2.7$ at different wind speeds of $U_{10} = 7, 10, 15, 20$ and $30\,\text{m/s}$. The figure clearly demonstrates growth of the dissipation level as the wind speed increases. Furthermore, the contribution of the forced (induced) cumulative wave breaking to the total wave dissipation slightly increases with increasing wind speed as indicated by reduction of the slope of the high-frequency tail of the wave-dissipation function.

In Babanin *et al.* (2010c), an attempt was also made to model the least known feature of the spectral dissipation function, its directional behaviour. Previously, isotropic

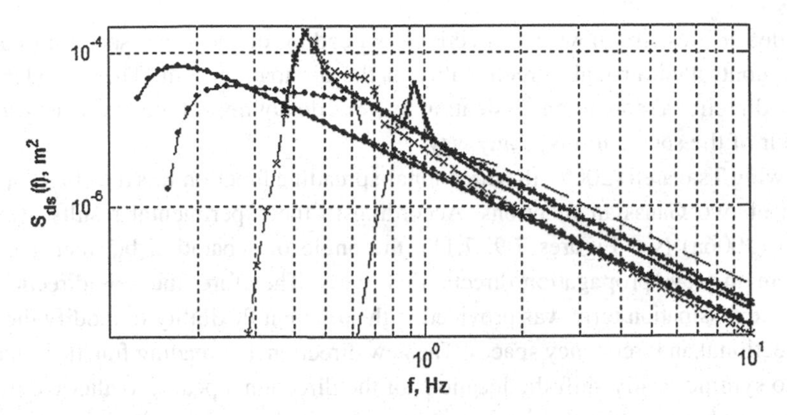

Figure 7.15 Spectral dissipation source function $S_{ds}(f)$ (5.40) computed with coefficients a_1 (7.55) and a_2 (7.54). Computations were performed for Combi spectra (7.46). Different stages of wave development: $U_{10}/c_p = 5.7$ (bold line), 2.7 (bold line with crosses), 0.83 (bold line with dots), for wind speed $U_{10} = 10\,\mathrm{m/s}$. Respective wind-input source functions $S_{in}(f)$ are also shown with plane lines marked with symbols corresponding to the dissipation functions. Figure is reproduced from Babanin *et al.* (2010c) © American Meteorological Society. Reprinted with permission

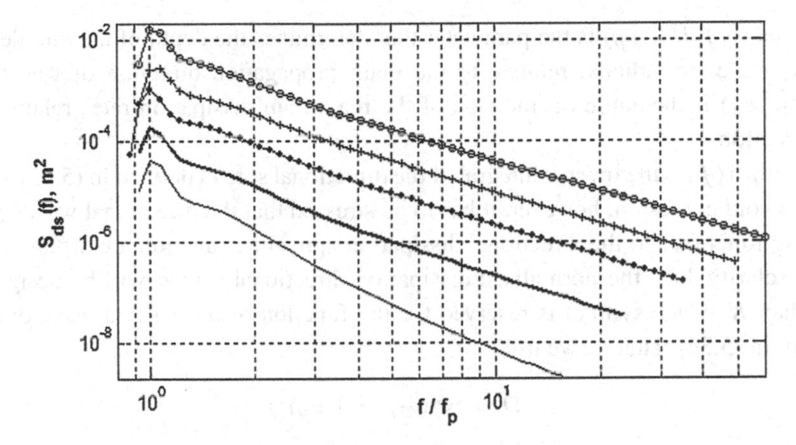

Figure 7.16 Spectral dissipation source function $S_{ds}(f)$ (5.40) computed with coefficients a_1 (7.55) and a_2 (7.54). Computations were performed for Combi spectra (7.46). Waves at $U_{10}/c_p = 2.7$ for wind speeds $U_{10} = 7\,\mathrm{m/s}$ (plain line), 10 m/s (bold line), 15 m/s (line with dots), 20 m/s (line with crosses) and 30 m/s (line with circles). Figure is reproduced from Babanin *et al.* (2010c) © American Meteorological Society. Reprinted with permission

or unimodal directional shapes were assumed for the dissipation source term (see also a recent study by Ardhuin *et al.*, 2010). However, the Lake George experiments (Young & Babanin, 2006a, see Section 7.3.6) revealed that the dissipation function may have symmetric maxima at angles oblique to the main wave-propagation direction. In terms of spectral modelling, this fact can be interpreted as a bimodal shape of the directional spreading for the three-dimensional dissipation spectrum. Note that this is a feasibility study only, which is intended to describe how the directional spreading of the wave spectrum can potentially be controlled through directionality of the dissipation term. These moderate alterations of directional-spreading dissipation features do not impact the frequency-distributed behaviour of the source terms in any way.

Following Tsagareli (2009), the directional-spreading function was developed as a superposition of two Gaussian functions. According to the experimental results of Young & Babanin (2006a) (see Figures 7.9, 7.11), the angle of separation between the maxima and the main wave-propagation direction can vary. Therefore, the new directional function for the dissipation term was provided with sufficient flexibility to modify the shape in both directional and frequency spaces. The new directional-spreading function includes the ability to symmetrically shift the locations of the directional peaks, to alter the magnitude of the trough between the maxima, to change the cross-sectional shapes as a function of frequency, and to vary with different wind-forcing conditions.

The spreading function V takes the form

$$V\left(\theta, f, \frac{U_{10}}{c_p}\right) = \begin{cases} A(f)\exp(-p(\theta+\theta_p)^2) & \theta < 0, \\ A(f)\exp(-p(\theta-\theta_p)^2) & \theta \geq 0 \end{cases} \tag{7.56}$$

where $p = p(f, U_{10}/c_p)$ is the parameter that determines the depth of the middle trough, θ is the angle (in radians) relative to the main propagation direction of waves, $\theta_p = \theta_p(f, U_{10}/c_p)$ is the angle (in radians) of the maximum dissipation rates relative to this main direction.

The term $A(f)$ is the inverse integral of the directional spread defined in (5.33) above. In order to avoid confusion, however, it has to be stressed that the directional wave-spectrum spreading function and the directional dissipation-spreading function are different properties. For clarity, here the normalised dissipation directional spread will be designated D rather than K which symbol is reserved for the function of normalised wave directional spectrum in (5.34). That is, we used

$$D = \exp(-p(\theta+\theta_p)^2) \tag{7.57}$$

in (7.56). The normalisation condition $\int_{-\pi}^{\pi} V(f,\theta)d\theta = 1$ is then satisfied; and if the angle $\theta_p = 0$, the directional spreading has a unimodal shape.

Now, as the angle θ_p increases, the lobes of the directional spreading function move further apart, enhancing the depth of the trough between them. Increasing the parameter p reduces the width of the lobes, increasing the depth of the trough at the same time. As a result, variations of parameters p and θ_p lead to different directional spreading of the dissipation which allows the model to control the directional wave spectrum, a property which has proved difficult in previous modelling tests (see e.g. Banner & Young, 1994). Figure 7.17 demonstrates the resulting directional spreading function $V(\theta, f, U_{10}/c_p)$ (7.56) with different values for the parameters p and θ_p. For convenience of comparison, all the directional spreading functions were normalised by the maximum value at the angle θ_p, i.e. $D = V(\theta_p, f, U_{10}/c_p) = 1$.

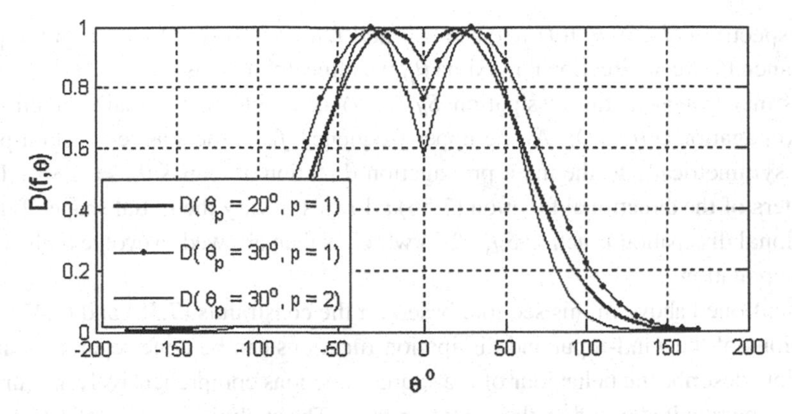

Figure 7.17 Normalised directional spreading function $D(\theta)$ (7.57), for different values of θ_p and p. Figure is reproduced from Babanin *et al.* (2010c) © American Meteorological Society. Reprinted with permission

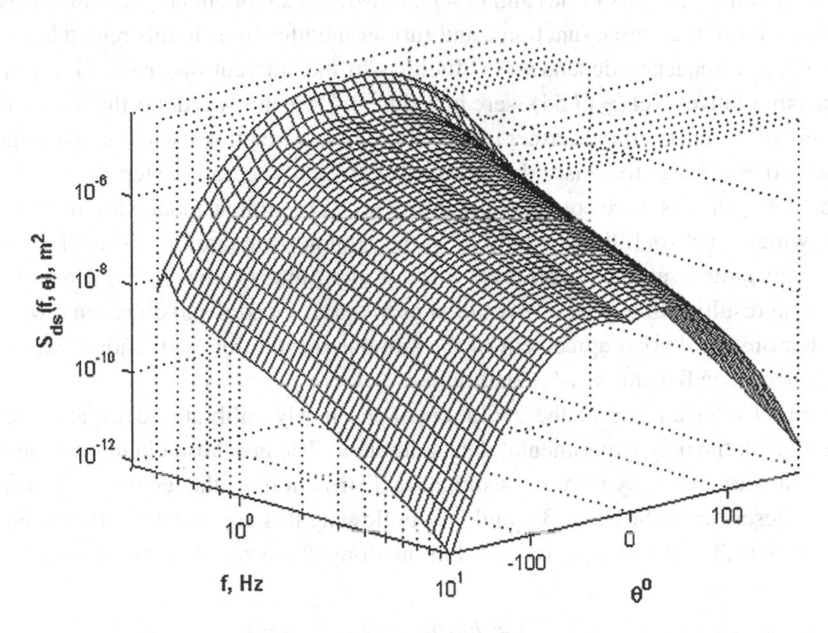

Figure 7.18 Three-dimensional dissipation function $S_{ds}(f, \theta)$ with the bimodal directional spreading $V(f, \theta)$ (7.56). Figure is reproduced from Babanin *et al.* (2010c) © American Meteorological Society. Reprinted with permission

For further technical details of implementing the bimodal directional dissipation distribution we refer the reader to Tsagareli (2009) and Babanin *et al.* (2010c). Here, Figure 7.18 shows an example of an initial model setup of the three-dimensional spectral dissipation function $S_{ds}(f, \theta)$ with such a spreading function $V(\theta, f, U_{10}/c_p)$ (7.56). In the figure, at the spectral peak it was assumed $\theta_p = 20°$, $p = 1$, and the initial setup was done for a

Combi spectrum (7.46) with $U_{10}/c_p = 2.7$ and $U_{10} = 10\,\text{m/s}$. Away from the peak f_p, the distance between directional maxima θ_p was lineally increased.

A distinct trough in the dissipation $S_{ds}(f, \theta)$ is visible in the main direction of the wave propagation at $\theta = 0°$. At the upper frequency $f_{\max}$, the source-function peaks are located symmetrically to the main propagation direction at angles $\theta_p = \pm 30^o$. Thus, the parameters of the directional function (7.56) allow us to vary the initial shape of the three-dimensional dissipation function $S_{ds}(f, \theta)$ which is then allowed to evolve in the course of wave propagation.

As mentioned above in this section, based on the constraints (7.38) and (7.47), primary calibrations of the wind-input and dissipation functions can be done separately, and if the constraints describe the behaviour of the source functions comprehensively, no further tuning would be required based on the evolution tests. The evolution curves, of course, have to be satisfied in any case. This way they become verification, rather than tuning/calibration means.

The integral constraints (7.38) and (7.47), however, cannot attend to details of the spectral behaviour of the source functions, and further modifications in this regard have proved necessary, i.e. frequency-dependent coefficients of the inherent-dissipation level a_1 (7.55) and the cumulative level a_2 (7.54) were introduced. Once this was done, the wave-evolution tests were performed without further tuning of the source functions, with the self-adjustment routine active in order to satisfy the constraints at each integration step.

The evolution tests were conducted with research two-dimensional wave model WAVE-TIME, which employs full computations of the nonlinear integral S_{nl} (Van Vledder, 2002, 2006). The latter uses the Webb–Resio–Tracy algorithm (Webb, 1978; Tracy & Resio, 1982). The resulting time-limited evolution of integral, spectral and directional wave properties demonstrated good agreement of the simulated evolution with known experimental dependences (see Babanin *et al.*, 2010c).

For direct comparisons of the evolution of the newly calibrated dissipation term S_{ds} (7.50)–(7.52), the only experimental data suitable to date are those obtained, based on the dimensional argument, by means of estimating distributions of the length $\Lambda(c)$ of breaking crests as described in Sections 3.4 and 3.6. Following this argument (3.30) and empirical dependence (3.31), the dissipation distribution along phase speeds c can be expressed as

$$S_{ds}(c) = b_{br}\rho_w \frac{c^5}{g} \Lambda(c) \left(\frac{10}{U_{10}}\right)^3 \tag{7.58}$$

(note that the quantitative empirical dependence is not dimensionally consistent). $S_{ds}(c)$ can then be converted into customary $S_{ds}(f)$:

$$S_{ds}(f) = \frac{g}{2\pi} \frac{1}{f^2} S_{ds}(c). \tag{7.59}$$

As discussed in Sections 3.4 and 3.6, in the case of a steady breaker, b_{br} in (7.58) is simply a proportionality coefficient. However, in the field the breaking is principally unsteady. Measurements within realistic unsteady-breaking conditions in the laboratory and the field found b_{br} varying by up to four orders of magnitude, which therefore makes dissipation formulations based on such a coefficient quite unreliable.

Besides, the connection between dissipation in a breaking wave with its phase speed is clearly not applicable in the case of induced breaking (i.e. the breaking of short waves caused and carried by the large waves as in Sections 5.3.2 and 7.3.4). Therefore, it is only the inherent term of dissipation (7.51) which should be considered in (7.59) for comparison purposes. Thus, a dissipation formulation of the type shown in (7.58) is applicable at the spectral peak region or below, where the cumulative effect is negligible.

Therefore, to reduce the uncertainty, values of coefficient b_{br} were compared at the spectral peak. The spectral dissipation function $S_{ds}(f)$ (5.40) was computed for the modelled Combi spectrum (7.46) of mature waves with $U_{10}/c_p = 1.3$ and at the wind speeds of $U_{10} = 7.2, 9.8$ and $13.6\,\mathrm{m/s}$ (i.e. same as in Melville & Matusov (2002)), and compared with the dissipation (7.58)–(7.59). Reasonable agreement was achieved at $b_{br} \approx 0.01$ (see Figure 7.19). This quantity is close to the value $b_{br} = 8.5 \cdot 10^{-3}$ used by Melville & Matusov (2002).

In Figure 7.19, the parameter b_{br} of (7.58) is plotted as a function of inverse wave age U_{10}/c_p, obtained from the dissipation function (5.40) at the spectral peak of Combi spectrum (7.46) for different wind speeds: $U_{10} = 7.2\,\mathrm{m/s}$ (plain line), $9.8\,\mathrm{m/s}$ (line with dots), $13.6\,\mathrm{m/s}$ (line with asterisks), $15\,\mathrm{m/s}$ (line with circles), $20\,\mathrm{m/s}$ (line with crosses) and $30\,\mathrm{m/s}$ (line with diamonds). The straight line signifies the value obtained in Melville & Matusov (2002).

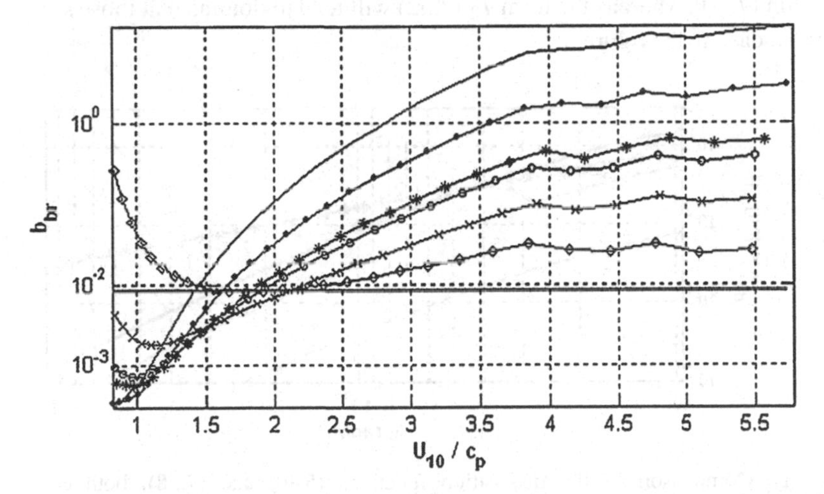

Figure 7.19 Dependence of the parameter b_{br} in (7.58) on the sea state U_{10}/c_p, based on the dissipation function (5.40), computed at the spectral peak of Combi spectra (7.46) for different wind speeds: $U_{10} = 7.2\,\mathrm{m/s}$ (plain line), $9.8\,\mathrm{m/s}$ (line with dots), $13.6\,\mathrm{m/s}$ (line with asterisks), $15\,\mathrm{m/s}$ (line with circles), $20\,\mathrm{m/s}$ (line with crosses) and $30\,\mathrm{m/s}$ (line with diamonds). The straight solid line is the value obtained in Melville & Matusov (2002). Figure is reproduced from Babanin *et al.* (2010c) © American Meteorological Society. Reprinted with permission

Even at the peak, where the induced breaking is absent and the dimensional hypothesis of Duncan (1981) can be expected to hold, the proportionality coefficient varies by many orders of magnitude depending on the wave age and, for the same wave age, on the wind speed. It is only in the range of the wind forcing conditions $1.5 < U_{10}/c_p < 2$ that the coefficient b_{br} is close to the value of $8.5 \cdot 10^{-3}$ of Melville & Matusov (2002) for most wind speeds.

For wind speeds $U_{10} < 20\,\text{m/s}$, as waves develop, the coefficient b_{br} exhibits a reducing trend, reaching down a value of the order of 10^{-3} at full development (10^{-2} at $U_{10} \sim 20\,\text{m/s}$). For very strong winds, $U_{10} = 30\,\text{m/s}$ in the figure, coefficient b_{br} exhibits significant growth towards full development, after reaching a minimum close to the Melville–Matusov value.

Comparison of the dissipation function (5.40) with the dissipation function based on phase speed c only (7.58), is shown in Figure 7.20. For convenience of the comparison, both functions were converted into wavenumber space, and the dissipation (5.40) was weighted by $(10/U_{10})^3$. Computations were performed for Combi spectra (7.46) with $U_{10}/c_p = 1.3$ at wind speeds of $U_{10} = 7.2\,\text{m/s}$, $9.8\,\text{m/s}$ and $13.6\,\text{m/s}$. For the dissipation (7.58), a number of values of coefficient $b_{br} = 2 \cdot 10^{-5}$, $1 \cdot 10^{-3}$ and $8.5 \cdot 10^{-3}$ were employed.

As has already been pointed out by Babanin & Young (2005) and Babanin *et al.* (2007c), (7.58) is expected to underestimate the dissipation at smaller scales away from the spectral peak because, even if there was a universal constant value for the proportionality coefficient b_{br}, it would only be applicable to the inherent dissipation term T_1 (7.51) of the total dissipation (7.50), whereas the term T_2 (7.52) will tend to dominate at those scales. This is clearly the case in the figure.

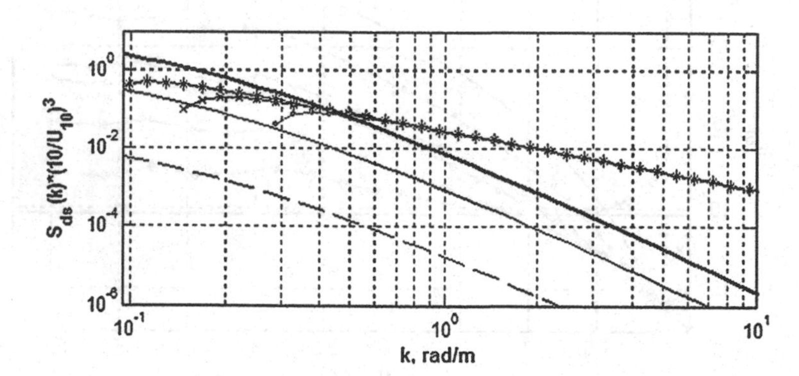

Figure 7.20 Comparison of the dissipation functions (5.40) and (7.58), both converted into wavenumber space. Dissipation (5.40) is weighted by $(10/U_{10})^3$, computations are performed for Combi spectra (7.46) at $U_{10}/c_p = 1.3$ for wind speeds $U_{10} = 7.2\,\text{m/s}$ (line with dots), $9.8\,\text{m/s}$ (line with crosses) and $13.6\,\text{m/s}$ (line with asterisks). For dissipation (7.58), different values of coefficient $b_{br} = 2 \cdot 10^{-5}$ (dashed line), $1 \cdot 10^{-3}$ (plain line) and $8.5 \cdot 10^{-3}$ (bold line) are used. Figure is reproduced from Babanin *et al.* (2010c) © American Meteorological Society. Reprinted with permission

At this stage of wave development ($U_{10}/c_p = 1.3$), agreement of the two dissipations at the spectral peak is reasonable for the Melville–Matusov value of $b_{br} = 8.5 \cdot 10^{-3}$. The value of $b_{br} = 2 \cdot 10^{-5}$ suggested by Gemmrich (2005) gives much lower dissipation values across the entire spectral band. As discussed by Tsagareli (2009), however, there is a dimensional issue in comparing the b_{br} coefficients of Melville & Matusov (2002) and Gemmrich (2005), due to the normalisation by $(10/U_{10})^3$. If this issue is accounted for, the agreement between the experimental outcomes of the two studies at the peak of the dissipation function (7.58) at $U_{10}/c_p = 1.3$ is quite good (not shown here).

Thus, to summarise this section on the whitecap-dissipation function for spectral models, we can conclude that new developments are being seen in almost every regard lately. The substantial experimental progress which led to direct estimates of the breaking dissipation and finding new features of its behaviour in recent years, has produced a number of different formulations and parameterisations. These are converging qualitatively and quantitatively, if compared within ranges of their expected performance and applicability. And the most encouraging fact is that newly advanced and proposed spectral-dissipation source terms are now being adopted in spectral models, all the way to operational wave-forecast modelling.

It is also worth mentioning in conclusion that there is a growing discussion not only on what the model-dissipation has to damp, but also on what it should not and what it may in fact produce. For example, Lavrenov (2004) showed that, if the dissipation function in wave models is not forced to suppress the low-frequency spectral energy, this may result in return energy fluxes from the waves into the atmospheric boundary layer, up to a quarter of the total wind-to-wave flux in magnitude. Such a considerable additional source of energy for the atmosphere may prove a significant factor in weather and climate forecasts (see also Hanley *et al.*, 2010).

An interesting feature which the dissipation can also produce is downshifting of the spectral energy. Such a downshift does appear a consistent outcome of the breaking due to modulational instability (see Section 8.1) and attempts have been made to attribute this behaviour to the dissipation function (e.g. Schneggenburger *et al.*, 2000, and papers presented at WISE meetings by Donelan and Meza).

7.5 Non-breaking spectral dissipation

This book is dedicated to wave breaking and the dissipation of ocean waves, and as far as the latter is concerned most attention is paid to whitecapping dissipation, i.e. that due to wave breaking. There is a good rationale for this of course, as once the waves break, most of the dissipation in such wave trains/fields is due to the breaking.

There are many other sources of dissipation, however, and some of them have already been outlined in Section 7.3.4 (see eq. 7.26). Compared to the whitecapping dissipation, their efficiency may be small or even negligible, but once the wind forcing drops down or ceases, or in case of swell, the breaking subsides or stops, these become a major sink of energy in the wave system.

Ardhuin *et al.* (2009a) investigated propagation of swell across the oceans by means of space-borne synthetic aperture radar data. They measured energy e-folding scales along wave-propagation distances 3000 to 30000 km and provided unique material for investigating and estimating dissipation of wave energy in field conditions due to causes other than breaking.

Ardhuin *et al.* (2009a) conducted their analysis based on what they call the linear energy e-folding scale α such that

$$dF(f,\theta,\phi) = \alpha RF(f,\theta,\phi). \tag{7.60}$$

Here, R is the radius of Earth, F is the wave frequency-directional spectrum, which also depends on ϕ, the separation angle from the source storm of the swell on the Earth's sphere. They compare the measurements with academic linear decay α_i which corresponds to an idealistic scenario of swell propagating over the global oceans, without even viscosity included. Such a scenario would be observed in the case of no energy lost by the wave field, i.e. the decay is due to the spatial expansion of the energy front only.

The first obvious energy sink in any fluid motion is that due to viscosity. The viscosity of water, however, is too small to have any significance in dissipating the energy of dominant wind-generated waves directly (e.g. Lamb, 1932, although indirectly its role is very important in promoting wave-induced turbulence as will be discussed later in this section). For the swell, Ardhuin *et al.* (2009a) found that indeed dissipation of the small-amplitude swells can be explained by the viscous dissipation with measured α corresponding to the scales of 20000 km or more.

However, for a swell steepnesses of

$$s = \frac{H}{\lambda} > 0.005 \tag{7.61}$$

α is not constant and depends on steepness. As a result, decay of swell energy is always larger, sometimes by several orders of magnitude, than that expected from the theory for molecular viscosity, under ambient temperature and pressure. "Steep swells lose a significant fraction of their energy, up to 65% over a distance as short as 2800 km". In terms used in this book, steepness (7.61) can be converted into

$$\epsilon = ak >\approx 0.016 \tag{7.62}$$

and is a threshold indicator in the swell-attenuation behaviour.

Other mechanisms of ocean-wave decay unrelated to breaking, i.e. such that can persist both in the presence and absence of breaking, are due to interactions of waves with turbulence, both in the water and in the air (see e.g. and The WISE Group, 2007). These can either be background turbulence or wave-induced turbulence, that is turbulence that owes its existence to the mean wave motion itself.

In the laboratory, it is possible to investigate wave motion and propagation in a turbulence-free water environment (e.g. Babanin, 2006; Babanin & Haus, 2009; Dai *et al.*, 2010), but the ocean is always turbulent (see e.g. Thorpe, 2005). A number of mechanisms of interaction of waves with such a background turbulence have been proposed (e.g. Kinsman, 1965; Kitaigorodski *et al.*, 1983; Cheung & Street, 1988; Jiang *et al.*, 1990; Benilov *et al.*, 1993; Drennan *et al.*, 1997; Teixeira & Belcher, 2002; Ardhuin & Jenkins, 2006, among others). These result in passing the energy to the turbulence and mean circulation which is therefore, as far as the wave fields are concerned, dissipation of wave energy. The mechanisms include, for example, scattering the waves by water turbulence or stretching the small-scale turbulence by the waves.

A separate large group of wave-turbulence interactions is generation of Langmuir cells. These can be interpreted as a large-scale turbulence produced by the surface waves (e.g. Langmuir, 1938; Craik & Leibovich, 1976; Smith, 1992, 1998; McWilliams *et al.*, 1997; Melville *et al.*, 1998; Phillips, 1998, 2001, 2002, 2003, 2005; Thorpe, 2004; Sullivan & McWilliams, 2010).

Interactions of waves with water turbulence can certainly be energetic, to the point of being an essential dissipation sink for, for example, short waves in the intensive surface-turbulence wake of larger breakers (e.g. Banner *et al.*, 1989, see also discussion in Section 7.3.4). For longer waves, with periods in excess of 10 s, however, such mechanisms appear not to be intensive enough to explain the observed wave-energy decay of swell. Ardhuin & Jenkins (2006) compared swell-attenuation rates measured in ocean observations, as well as those empirically inferred in operational wave models, with viscous and wind-caused (Kudryavtsev & Makin, 2004) damping and concluded that wave interaction with oceanic turbulence is too weak to justify the observed decay.

Rather, Ardhuin *et al.* (2009a) presumed that the decay is due to the swell interacting with the turbulent boundary layer on the atmospheric side. They found that estimates of such decay, based on analogy with the turbulent boundary layer, are not inconsistent with the measurement.

Wind–wave interaction can be due to work done by normal (pressure) or tangential (shear) stresses imparted by the waves on the air flow in the wave boundary layer, the bottom sublayer of the atmospheric boundary layer (see e.g. recent publications by Kudryavtsev *et al.*, 2001 and Chalikov & Rainchik, 2011, among many others). It is generally accepted that, as far as the energy input from the wind to the waves is concerned, the normal stresses drive the interaction (see e.g. Donelan *et al.*, 2006 and The WISE Group, 2007, for recent reviews of this topic).

For the swell, which gives the energy/momentum back to the air, Ardhuin *et al.* (2009a) found no apparent connection between swell energy decay and the wind speed or swell age c/U_{10}. Therefore, they argued that the normal-stress interaction can be neglected, and they considered only the shear-stress modulations imposed by the wave orbital motion. Here, we should mention that the swell-induced wind is necessarily small in absolute magnitude, even according to the most extreme estimates of its significance (see e.g. Donelan, 1999; Lavrenov, 2004), and therefore such a correlation would be difficult to measure in the field where background winds, unrelated to the swell, would almost always be expected to prevail. This wind, however, is by far not insignificant (Hanley *et al.*, 2010). Whatever the role of the normal stresses is in this regard, it does not discard of course the swell-induced shear stresses.

Measurements of air flow over swell are quite rare (see e.g. a recent study and review by Soloviev & Kudryavtsev, 2010), and Ardhuin *et al.* (2009a) relied on analogy with the wall boundary layer. They argued that the air–sea boundary layer should have the same properties if the waves are treated as surface undulations, but magnitudes of the orbital velocity u_{orb} and displacement a_{orb} should be doubled in the respective estimates of Reynolds numbers, compared to the fixed boundary (Collard *et al.*, 2009). For the boundary-layer flow to be turbulent then, the Reynolds numbers have to be:

$$\text{Re} = \frac{4u_{\text{orb}}a_{\text{orb}}}{\nu_a} > 10^5 \tag{7.63}$$

where ν_a is the air viscosity (e.g. Jensen *et al.*, 1989).

If the flow is laminar, strong shear above the interface makes the air viscosity important, and the dissipation coefficient is expected to be (e.g. Dore, 1978; Collard *et al.*, 2009):

$$\alpha_\nu = 2\frac{\rho_a}{\rho_w}\frac{1}{gc_g}\left(\frac{2\pi}{T}\right)^{5/2}\sqrt{2\nu_a}. \tag{7.64}$$

At ambient conditions ($\nu_a = 1.4 \cdot 10^{-5} m^2 s^{-1}$), α_ν is a function of T only, and for observed swell periods of $T = 13$–19 s it is expected to vary in the range from $2.2 \cdot 10^{-8}\ \text{m}^{-1}$ down to $5.8 \cdot 10^{-9}\ \text{m}^{-1}$, respectively.

For the turbulent boundary layer, the energy decay can be estimated based on knowledge of the temporal dissipation coefficient

$$\alpha_{\text{time}} = c_g\alpha = \frac{\rho_a}{\rho_w}\frac{4\pi^2}{gT^2}f_e u_{\text{orb}}. \tag{7.65}$$

Here, f_e is an analogy of the smooth-wall friction factor f_w and is expected to be of the order of $f_e = 0.002$–0.008 (e.g. Jensen *et al.*, 1989).

Thus, Ardhuin *et al.* (2009a) interpreted their observations of swell decay in terms of such a mechanism of swell friction against the air. The ratio of observed values of α over α_ν (7.64) ranged from 1 to 28, indicating that both laminar and turbulent behaviours were present. For the latter, curve fitting needed the friction constant in the range $-0.001 \le f_e \le 0.019$ which is therefore, with the median value of 0.007, close to the expectations for a smooth surface. Transition occurs at

$$\text{Re} \approx 5 \cdot 10^4 \tag{7.66}$$

which is another swell-attenuation threshold to keep in mind, in addition to the steepness (7.61)–(7.62).

There is, however, another phenomenon which also has laminar-to-turbulent transition and also leads to dissipation of energy of waves, including swell. That is, it may also be effective in the observed attenuation of swell, provided the relevant Reynolds numbers are above their critical value. This is turbulence induced by the mean wave orbital motion in the water, rather than produced by breaking or by friction against the air interface. Below, we will follow Babanin (2006) and Babanin & Haus (2009) to introduce the phenomenon of this turbulence.

Turbulence can be generated by a sheared fluid motion if a certain inertia-to-viscosity ratio limit is overcome (e.g. Reynolds, 1883). While this should be true in a general case (e.g. Kinsman, 1965), the motion due to surface water waves is generally regarded to be irrotational (e.g. Young, 1999) and thus does not produce shear stresses and turbulence directly. The conjecture of irrotationality in such theories is a consequence of the initial assumption that the waves are free, that is have no viscosity and surface tension, and therefore cannot cause shear stresses (e.g. Komen *et al.*, 1994). Albeit small and negligible from the point of view of many applications, water viscosity, however, is not zero. In the presence of a strong exponential vertical gradient of the wave orbital velocity, and this velocity being one or two orders of magnitude larger than the other velocities in the water column usually attributed with the shear stresses, wave-caused shear stresses are unavoidable. Thus, the existence of wave-induced turbulence has been suggested (Babanin, 2006). In the ocean, which is always turbulent, the instability of pre-existing, even if negligibly small vortices can also take energy from potential waves (Benilov *et al.*, 1993).

Babanin (2006) proposed the concept of a wave-amplitude-based Reynolds number that indicates a transition from laminarity to turbulence for the mean wave orbital motion. Estimates of the critical wave Reynolds number (see (7.70) below for the definition) provided an approximate value of

$$Re_{wave_{critical}} \approx 3000. \tag{7.67}$$

Babanin (2006) tested this number on mechanically-generated laboratory waves and it was confirmed. Once this number is used for ocean conditions when mixing due to heating and cooling is less important than that due to the waves, quantitative and qualitative characteristics of the ocean mixed layer depth (MLD) were shown to be predicted with a good degree of agreement with observations. Testing the hypothesis against other known results in turbulence generation and wave attenuation, including swell propagation across the Pacific, was also conducted. Later, in further laboratory experiments it was demonstrated that, intermittently, the wave-induced turbulence can appear at even lower Reynolds numbers, i.e. $Re_{wave} \sim 1000$ (Babanin & Haus, 2009; Dai *et al.*, 2010).

In the discussion below, the linear wave theory will be both employed for estimates and criticised for not including viscosity. Therefore, some clarifications should be mentioned at this stage to avoid what may seem a contradiction.

The wave motion considered is that due to the wind-generated waves at the ocean surface. Nonlinear corrections to the mean orbital velocities and amplitudes are unimportant from the point of view of turbulence generation, and the linear wave theory is used here to scale the mean wave motion as a function of water depth. Although such an approach is routinely used and is regarded to be reliable on average, it is necessary to comment that very significant deviations from the linear theory predictions, both overestimations and underestimations of anticipated depth-dependent wave characteristics, have been reported (see, for example, Cavaleri *et al.*, 1978, for a review).

At this stage, a few further words of caution with regard to the linear theory have to be mentioned. As said above, ocean waves of lengths $\lambda > 1$ m, if described by the perturbation theory based on the Euler equation, are considered free, with no viscosity and surface tension (e.g. Komen *et al.*, 1994). Such assumptions further lead to conjecture that the waves happen to be irrotational. Although the former assumptions are a mere approximation and justified in most cases (with noticeable faults in other cases, as mentioned above, though), the latter feature of irrotationality imposes a serious limitation on the wave motion because it basically bans wave-induced turbulence (although some theoretical mechanisms of generating the turbulence by irrotational waves are still possible, provided some background turbulence exists, as discussed above in this section).

There is, however, accumulating evidence, both direct and indirect, that such turbulence does exist. For example, as far as 40 years ago, Yefimov & Khristoforov (1971) concluded that their measurements provide

> "a basis for assuming that small-scale turbulence is generated by the motion of waves of fundamental dimensions".

They did not mention explicitly whether those waves were breaking or not, and Babanin (2006) conducted estimates based on the breaking-threshold criteria of Babanin *et al.* (2001). He concluded that, for the two records analysed by Yefimov & Khristoforov (1971) in their Fig. 5, breaking rates of dominant waves were 0.4% and 0.01%. Both rates are marginal, the second one being negligible, and we have to conclude that the observed substantial levels of wave-associated turbulence could not have been brought about by wave breaking, but were induced by the mean wave motion.

Cavaleri & Zecchetto (1987) in their dedicated and thorough measurements of wave-induced Reynolds stresses gave explicit accounts for wave breaking. One set of their data corresponds to active wind-forcing conditions (many breakers present are mentioned), whereas the other set describes steep swell (no breaking). Non-zero vertical momentum fluxes in the absence of breaking are evident. The magnitude of the fluxes appears to depend quadratically on the height of individual waves which is consistent with the wave-amplitude-based Reynolds number hypothesis (7.70) below. Cavaleri & Zecchetto (1987) concluded that

> "in the water boundary layer there can occur an additional mechanism of generation of turbulence... full, correct description of the phenomenon is still lacking".

Later, Babanin *et al.* (2005) conducted simultaneous measurements of the wind-energy input rate and the wave-energy dissipation through a water column at Lake George (see Sections 3.5 and 5.3.4). They showed that, at light winds and in the absence of wave breaking, turbulence persisted through the entire water column, not only in the shear boundary layers near the surface and bottom. In the finite-depth lake, with no currents and internal waves, the only source of the turbulence was surface waves, and the intensity of turbulence indeed correlated with the magnitude of these waves.

A need for wave-induced turbulence has also been felt by the ocean modellers in their search for mechanisms to fill the gaps in explanations of upper-ocean mixing. Jacobs (1978), Pleskachevsky *et al.* (2001, 2005), Qiao *et al.* (2004, 2008, 2010) and Gayer *et al.* (2006) all brought in wave-induced turbulent viscosity and applied it in their circulation models. The ocean sites ranged from the finite-depth North Sea (Pleskachevsky *et al.*, 2001, 2005; Gayer *et al.*, 2006) through the open ocean (Jacobs, 1978) to global applications (Qiao *et al.*, 2004, 2008, 2010). In all cases, it was either impossible to describe the observed mixing without wave turbulence, or introducing such turbulent diffusion essentially improved the correlation between the data and numerical simulations.

The above-mentioned Ardhuin & Jenkins (2006) mechanism does not exactly belong to the pure class of wave-induced turbulence. The approach remained within the frame of zero-vorticity for wave solutions. The source of their turbulence is Stokes drift, the zero-frequency solution, which then interacts with the background turbulence. The same can be said about the mechanism of Benilov *et al.* (1993), except it is the mean potential wave motion that causes the unstable infinitesimal vortices to grow. If the viscosity is allowed in theory, however, a viscous fluid can further promote shear instability in such a system (Balmorth, 1999).

Here, we will be using the linear theory to find the wave-based Reynolds number, to describe its distribution along the water depth and to approach a possible upper-ocean mixing mechanism due to such waves (in Section 9.2.2 below). We would like to emphasise that this is not a compromise with the criticism of the limitations of the linear theory above. To estimate the Reynolds numbers we only need a scaling of mean wave orbital motion at different depths which the linear theory approximates well enough.

The hypothesis of wave Reynolds number has three important consequences. First, the wave motion should be able to generate turbulence even in the absence of wave breaking. As discussed above, this is not a completely unexpected conclusion as such turbulence has been observed for a while. What has not been appreciated, however, is the potential significance of such a turbulence source as the waves in the ocean being ever present, unlike currents or Langmuir cells, and wave-caused speeds of water motion are at least an order of magnitude greater than those of shear currents and Langmuir circulations which are usually held responsible for the turbulence supply as mentioned above.

A second consequence is decoupling of the wave-induced non-breaking turbulence from analogies with the wall-layer law tradition which are often employed to approach wave turbulence (e.g. Soloviev *et al.*, 1988; Agrawal *et al.*, 1992). According to the wave-turbulence hypothesis, the principal difference of such a turbulence is the existence of a characteristic length scale (radius of the wave orbit) as opposed to the wall turbulence which does not have a characteristic length other than distance to the surface.

Third, such a wave-induced turbulence would enhance the upper-ocean mixing on behalf of the normal component of the wind stress. The wind stress plays a dual role in the upper-ocean dynamics. A tangential component of the stress generates the surface shear currents which further induce turbulence and promote mixing. Under moderate and strong winds, however, a normal component of the wind stress dominates, which is supported by the momentum flux from wind to waves (e.g. Kudryavtsev & Makin, 2002). The latter means that, before this momentum is received by the upper ocean in the form of mean currents, and thus enters the further cycle of air–sea interaction, it goes through a stage of surface wave motion. This motion can directly affect or influence the upper-ocean mixing and other processes, and thus skipping the wave phase of momentum transformation undermines the accuracy and perhaps the validity of approaches based on the assumption of direct mixing of the upper ocean due to the wind. This should also be a sink of wave energy, small by comparison with the breaking dissipation, but perhaps essential for the swell decay.

The wave-Reynolds-number hypothesis thus attempts to link together three oceanic features which are routinely treated as separate properties: the wind–waves, the near-surface turbulence and the upper-ocean mixed layer. Mechanisms of MLD deepening are believed to be affected by a number of ocean properties and processes: i.e. wind stress, heating and cooling, advection, wave breaking, Langmuir circulation and internal waves, with the surface wind forcing being the major factor in many circumstances (e.g. Martin, 1985). In accordance with this hypothesis, the role of the wind stress, acting at the ocean surface, may need to be reconsidered in terms of the mixing throughout the water column due to wind-generated wave orbital motion. This mixing can be accounted for through the wave energy spent on generation of turbulence, i.e. wave-turbulence dissipation (Pleskachevsky *et al.*, 2011).

The hypothesis of the wave Reynolds number, as mentioned above, refers to the mean surface wave elevation η, propagating in time t and space x with amplitude a_0:

$$\eta(x,t) = a_0 \cos(\omega t + kx). \tag{7.68}$$

This propagating elevation has two characteristic length scales: wave length λ and wave amplitude a. In deep water, the amplitude a decays exponentially away from the surface:

$$a(z) = a_0 \exp(-kz) \tag{7.69}$$

where z is the scalar vertical distance from the mean water level.

Wavelength λ does not depend on depth z and defines the horizontal scale over which the wave oscillations change phase. It also characterises the depth of penetration of the wave oscillations (the distance from the surface where the oscillations can still be sensed is approximately $\lambda/2$). This scale, however, does not comprise the physical motion of the water particles. The motion of physical particles involved in the wave oscillations is depicted by the other scale, a, which is also the radius of wave orbits.

The wave Reynolds number (hereinafter WRN) is

$$\mathrm{Re}_{\mathrm{wave}} = \frac{a u_{\mathrm{orb}}}{\nu} = \frac{a^2 \omega}{\nu} \tag{7.70}$$

where orbital velocity is $u_{\mathrm{orb}} = a\omega$ and ν is the kinematic viscosity of the ocean water. $\mathrm{Re}_{\mathrm{wave}}$ indicates the transition from laminar orbital motion to turbulent. It is interesting to notice that WRN can have the velocity scale eliminated if the dispersion relationship (2.17) is used, and can be expressed in terms of the two length scales:

$$\mathrm{Re}_{\mathrm{wave}} = \frac{a^2\sqrt{gk}}{\nu} = \sqrt{2g\pi}\frac{a^2}{\nu\sqrt{\lambda}}. \tag{7.71}$$

Thus, according to (7.69), for a given wavelength λ (wave frequency ω) Reynolds number $\mathrm{Re}_{\mathrm{wave}}$ decays rapidly as a function of depth:

$$\mathrm{Re}_{\mathrm{wave}_\lambda}(z) \sim a(z)^2 \sim \exp(-2kz). \tag{7.72}$$

At the surface ($z=0$, $\exp(-2kz)=1$), longer waves of the same amplitude a_0 will produce smaller Reynolds numbers.

Let $\mathrm{Re}_{\mathrm{wave}_{\mathrm{critical}}}$ be the critical value of WRN (7.70)–(7.71). If near the surface $\mathrm{Re}_{\mathrm{wave}_\lambda} > \mathrm{Re}_{\mathrm{wave}_{\mathrm{critical}}}$, then the corresponding wave orbital motion will be turbulent. At some depth z_{critical}, dependence (7.72) will lead to $\mathrm{Re}_{\mathrm{wave}_\lambda} = \mathrm{Re}_{\mathrm{wave}_{\mathrm{critical}}}$, and from that depth down the orbiting will become laminar, or, in the turbulent ocean, will stop generating turbulence. Depth z_{critical}, therefore, will define a mixed layer depth – the depth of the upper-ocean layer mixed due to the turbulence generated by orbital movement produced by the surface waves. Obviously, in reality the convection, advection, heating and other processes can alter this value. Also, the background ocean waters are nearly always turbulent, and therefore z_{critical} is not the depth below which turbulence is absent, but is rather a depth below which we do not expect the presence of wave-induced turbulence.

From (7.70)–(7.71), the Reynolds number at a given $\lambda(\omega)$, as a function of z, is

$$\mathrm{Re}_{\mathrm{wave}} = \frac{\omega}{\nu}a_0^2\exp(-2kz) = \frac{\omega}{\nu}a_0^2\exp\left(-2\frac{\omega^2}{g}z\right). \tag{7.73}$$

Therefore, if the critical Reynolds number $\mathrm{Re}_{\mathrm{wave}_{\mathrm{critical}}}$ were known, the critical depth, which is also wave-induced MLD, would be readily available:

$$z_{\mathrm{critical}} = -\frac{1}{2k}\ln\left(\frac{\mathrm{Re}_{\mathrm{wave}_{\mathrm{critical}}}\nu}{a_0^2\omega}\right) = \frac{g}{2\omega^2}\ln\left(\frac{a_0^2\omega}{\mathrm{Re}_{\mathrm{wave}_{\mathrm{critical}}}\nu}\right). \tag{7.74}$$

As seen, for a wave frequency ω, if the wave height grows, MLD will increase. If a few waves of the same height but different scales are present at a time, the mixed layer will mostly be determined by the lowest frequency ω (longest length λ) as its z_{critical} will be the largest.

The real wind-generated waves are spectral, and, apart from the cases of pure swell, multiple wave scales are superposed at the ocean surface. The wave spectrum, however, has a sharp peak, with the spectral density (wave height) dropping very rapidly away from the peak frequency ω_p both towards smaller scales (higher frequencies) and larger scales (lower frequencies) – see, for example, JONSWAP parameterisation (2.7). Therefore, ω_p and associated wave height can be chosen as characteristic wave scales which determine $MLD\text{-}z_{\mathrm{critical}}$ in the case of a wind–wave spectrum. As the representative amplitude of spectral waves, half of the significant wave height $a_s = H_s/2$ (2.39) will be used.

Babanin (2006) performed qualitative and quantitative verifications of the hypothesis, and consistency checks by testing conclusion (7.74) on the basis of known values of MLD (z_{critical}) for different water bodies and different wave conditions. Additionally, a laboratory feasibility check was conducted.

Based on the Black Sea depth of mixed layer ($z_{\mathrm{critical}} \approx 25$ in April (Kara *et al.*, 2005)), we will determine the value of critical WRN $\mathrm{Re}_{\mathrm{wave}_{\mathrm{critical}}}$ according to (7.73) by having use of the typical extreme values of peak frequency and amplitude of the wind-generated waves in this sea in April. This dimensionless number should be universal for the wave motion, according to the hypothesis (7.70), and therefore be equally applicable to such outermost extremes as the deep ocean and much smaller laboratory mechanically-generated waves. Thus, once this critical Reynolds number is known, we should be able to predict the transition of non-forced wave motion from laminarity to turbulence in the laboratory and to predict MLD in the ocean for different wave circumstances on the surface in cases when wind stresses (waves) dominate over other mixing mechanisms.

April in the Northern hemisphere was chosen for our estimates because the combined effect of surface cooling and heating on MLD is expected to be minimal in early spring (e.g. Martin, 1985), and therefore April data will provide the cleanest material for investigation of mixing due to wind (waves). An extensive wind–wave data set collected in a deep water region of the Black Sea throughout April by Babanin & Soloviev (1998a) was used to determine the relevant wave climate. The minimal value of the peak frequency $f_p = 0.175\,\mathrm{Hz}$ is encountered three times in Table I of Babanin & Soloviev (1998a) and was thus chosen as representative of the typical extreme wave conditions in the Black Sea in April. Corresponding values of variance m_0 (2.10) ranged from $0.379\,\mathrm{m}^2$ to $0.500\,\mathrm{m}^2$.

Water temperatures recorded at the measurement site in April were around 10° C, and therefore, for the Black Sea whose salinity is half of that in the ocean, the kinematic viscosity was chosen as $\nu = 1.35 \cdot 10^{-6}\,\mathrm{m}^2/\mathrm{s}$. The wave amplitude used was $a_s = H_s/2 = 2\sqrt{m_0}$. Finally, (7.73) leads to the critical Reynolds number in the range $\mathrm{Re}_{\mathrm{wave}_{\mathrm{critical}}} =$ 2602–3433. Given the approximate nature of the estimates, we chose $\mathrm{Re}_{\mathrm{wave}_{\mathrm{critical}}} = 3000$ as was already indicated in (7.67), close to the centre of the range. This is the critical Reynolds number for wave orbital motion.

Such a Reynolds number is in very good accordance with the critical numbers for other fluid flows. Typical Reynolds numbers for a great variety of engineering applications outside the boundary layer are $\mathrm{Re}_{\mathrm{critical}} =$ 2000–4000 (e.g. Cengel & Cimbala, 2006).

A laboratory test was conducted in the wave tank of Monash University to check the feasibility of the calculated critical number. Regardless of its relation with the upper ocean mixing, critical WRN, $\mathrm{Re}_{\mathrm{wave}_{\mathrm{critical}}} = 3000$ should be able to predict the onset of turbulence for a simple case of monochromatic waves in the tank. Mechanically-generated gently sloped ($\epsilon = ak = 0.035$–0.075) waves were run which therefore involve no additional forcing superposed over wave orbital motion.

A single set of measurements was conducted for waves of 0.667 Hz frequency, with the wave amplitude gradually being changed from 2 cm up to 4 cm, then down to 2 cm and up to 4 cm again. For such a frequency and water depth of approximately $d = 1$ m, the waves are in a finite depth environment ($kd = 1.9$), and the wave orbits are somewhat elliptical. For simplicity, the vertical amplitude of the wave orbit was chosen to estimate the Reynolds number. Traces were injected into the water in the centre of the tank at 10 cm depth below the surface ($\exp(-kz) = 0.83$), far away from the bottom, surface and wall boundary layers.

At the lower margin of $a_0 = 0.02$ m, WRN (7.70) is $Re_{wave} = 1150$ and the orbital motion was expected to be fully laminar. At the other end of $a_0 = 0.04$ m, $Re_{wave} = 4600$ and fully turbulent motion was expected. Transition was anticipated at $a_0 \approx 0.03$–0.035 m where $Re_{wave} = 2600$–3500. Note that, according to (7.72), at different distances from the surface z transition is expected at different surface wave amplitudes a_0.

The experiment is illustrated in Figure 7.21. At $a_0 = 0.02$ m, the motion remained clearly laminar, and patterns of injected ink, while moving along the orbits, stayed unchanged for minutes. At $a_0 = 0.03$ m, some vortexes became visible which eroded the

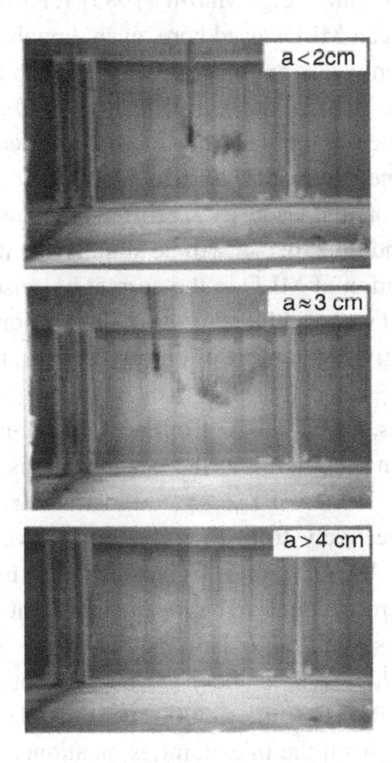

Figure 7.21 Wave-tank experiment with wave-induced turbulence. $f = 0.667$ Hz, $d = 1$ m, ink is injected for waves of amplitude $a_0 = 2$ cm (top), $a_0 = 3$ cm (middle), $a_0 = 4$ cm (bottom)

upper parts of the ink patterns. At $a_0 = 0.04$ m, the motion was obviously turbulent, with the ink being completely dissolved within seconds after injection. When the amplitude was reduced down to $a_0 = 0.02$ m again, the laminar behaviour of the traces was immediately restored as the source of turbulence was apparently removed. Note that the time interval between the measurements is minutes, that is wave turbulence completely dissipates at this time scale, and certainly cannot persist for hours and tens of hours as it is implicitly assumed in ocean-mixing schemes which rely on wave-breaking turbulence. The wave-breaking turbulence is directly injected down to the depths of the order of wave height (e.g. Agrawal *et al.*, 1992; Babanin *et al.*, 2005; Gemmrich, 2010), and to mix the ocean at the scale of MLD ~ 50 m, as, for example, if the wave-breaking turbulence scheme of Craig & Banner (1994) is employed, would have to be diffused down and survive for hours and tens of hours (e.g. Martin, 1985). Back to our experiment, on the return back to $a_0 = 0.04$ m, the onset of turbulence was observed at approximately the same wave amplitude as previously.

Now, let us apply $Re_{wave_{critical}} = 3000$ to find z_{cr} (7.74) in the ocean. This will only be an approximate value as, for the ocean wave conditions it is just mean (rather than mean extreme) magnitudes of wave height and peak period that are available to us (wave atlas of Young & Holland, 1996). Mean extremes would need a designated definition for this kind of estimate anyway, even if the raw wave data of interest were available, because those would have to be not only high waves, but also waves persisting long enough (normally tens of hours within the month – e.g. Martin (1985) for MLD to settle. Besides, readings of mean waves and mean MLDs used here, even though taken for the same month of April, were done in different years and may be out of synch to some degree due to interannual variability. Also, the thermal, advective and other effects, although assumed to be relatively small, may not be negligible. And importantly, sea surface temperature (SST) is used here to estimate kinematic viscosity which is further applied in (7.73) and (7.74). The temperature and therefore the viscosity are different below MLD (temperature is lower and viscosity is greater), and thus Re_{wave} will be slightly smaller down there. This fact can lead us to some overestimation of MLD in the current exercise. Such precision details are beyond our scope, and what we would like to see in this section is an approximate quantitative and reasonable qualitative agreement of our predictions, based on (7.73)–(7.74), with ocean observations.

For the ocean estimations, we will begin from well-documented directly measured April values of MLD and SST in the Pacific at the ocean stations *November* (140°W, 30°N) and *Papa* (145°W, 50°N). Values of $z_{critical} \approx 104$ m and $z_{critical} \approx 75$ m were read for the stations N and P respectively from Figs. 5 and 6 of Martin (1985). Corresponding surface temperatures were 19.6°C and 5.3°C, which lead us to $\nu \approx 1.07 \cdot 10^{-6}\,\mathrm{m^2/s}$ and $\nu = 1.58 \cdot 10^{-6}\mathrm{m^2/s}$, respectively. The mean significant wave height and mean peak period, provided by the atlas of Young & Holland (1996) for April, were $H_s = 2.33$ m and $f_p = 0.084$ Hz at N, and $H_s = 3.44$ m and $f_p = 0.097$ Hz at P.

Estimates, obtained for $Re_{wave_{critical}} = 3000$ using (7.74), are $z_{critical} \approx 95$ m for N and $z_{critical} \approx 78$ m for P. Given the uncertainties mentioned above, they are in excellent quantitative (9% and 4% deviations, respectively) and very good qualitative agreement with the measurements. One and a half times higher waves at *Papa* did not produce a deeper mixed layer because the excessive height was compensated by a more rapid depth attenuation of the wave motion, due to the higher peak frequencies.

To investigate these agreements further, we conducted calculations for a transect across the Pacific and Atlantic Oceans in April at the 30°N latitude which provides a significant variety of wave conditions (Figure 7.22b and c). Wave climatology was taken from Young & Holland (1996), and MLD climatology from the US Naval Research Laboratory (NRL) site http://www7320.nrlssc.navy.mil/nmld/nmld.html. The latter was

"constructed from the 1° monthly-mean temperature and salinity climatologies of the World Ocean Atlas 1994 (Levitus & Boyer, 1994; Levitus *et al.*, 1994) using a method for determining layer depth that can accommodate the wide variety of temperature and density profiles that occur within the global ocean."

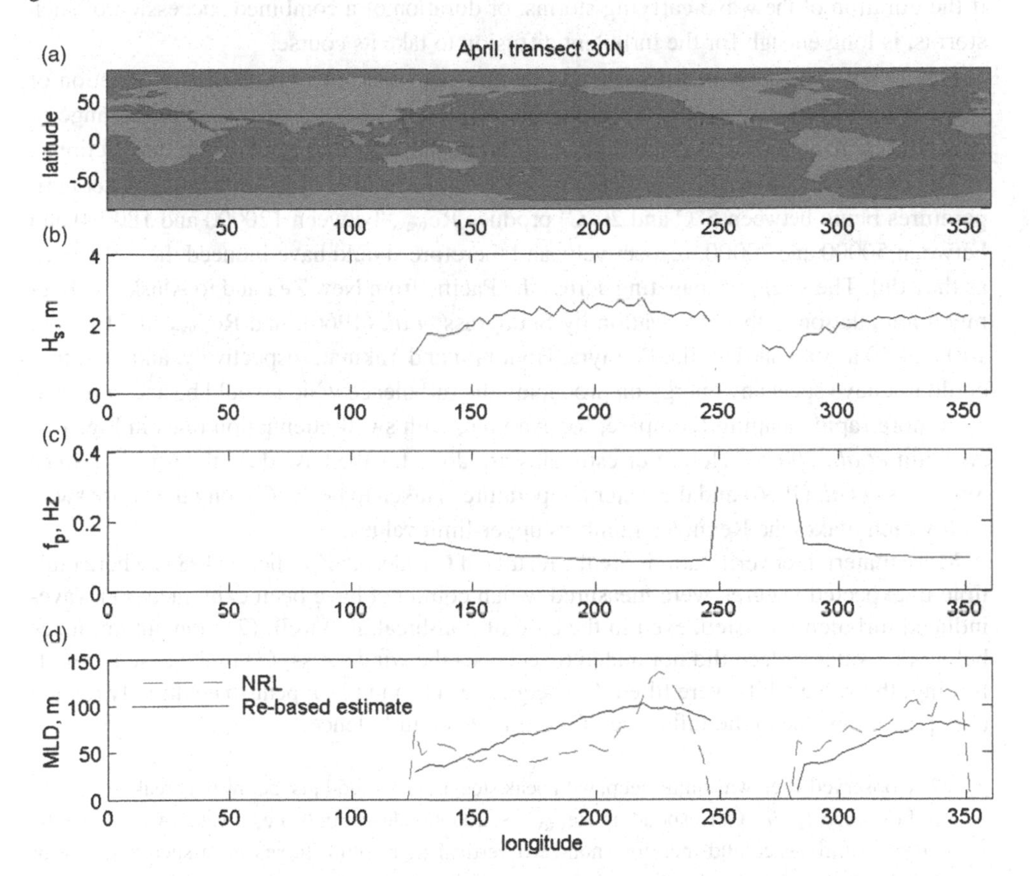

Figure 7.22 (a) World map. The transect latitude 30°N is shown with a solid line. (b) Mean April significant wave height H_s along the 30°N latitude. (c) Mean April peak frequency f_p along the 30°N latitude. (d) Reynolds-number-based estimate of MLD (solid line) and NRL estimate of MLD (dashed line) along the 30°N latitude. Figure is reproduced from Babanin (2006) by permission of American Geophysical Union

The satellite-measured SST for April were taken from US NOAA site http://www.osdpd.noaa.gov/PSB/EPS/SST/al_climo_mon.html. Because variations were small, we assumed an approximately constant value of 25°C which thus gives $\nu \approx 0.96 \cdot 10^{-6}\,\mathrm{m}^2/\mathrm{s}$ at 30°N.

The results are shown in Figure 7.22. Along the transect, significant wave height varies four times, from 0.65 m to 2.7 m, and so does the peak frequency which varies from 0.08 Hz to 0.3 Hz (Figures 7.22b and c, respectively). Comparison of the NRL results with the present Reynolds-number-based estimate of MLD is shown in Figure 7.22d. The Re-based estimate reproduces the trend very well, and given the fact that both methods are results of modelling rather than direct *in situ* measurements, the quantitative agreement is also very good. It should be pointed out that possible rates of development of the mixed layer cannot be addressed on the basis of this hypothesis: the predicted MLD will be achieved if the duration of the wave-carrying storms, or duration of a combined succession of such storms, is long enough for the turbulent diffusion to take its course.

Obviously, prediction of MLD is not the only potential outcome of the introduction of wave-induced turbulence. It is therefore interesting to check our results across a range of oceanographic applications, including swell propagation discussed above in this section.

The two records described by Yefimov & Khristoforov (1971) (assuming the water temperatures being between 5°C and 20°C) produce $\mathrm{Re}_{\mathrm{wave}}$ between 120000 and 180000 and between 50000 and 70000, respectively, and therefore should have induced the turbulence as they did. The swell, propagating across the Pacific from New Zealand to Alaska without much attenuation in the observation by Snodgrass *et al.* (1966), had $\mathrm{Re}_{\mathrm{wave}}$ of 310, 260, 140 and 50 at stations Tutuila, Palmyra, Honolulu and Yakutat, respectively, and therefore could not have spent any energy on producing the turbulence which would be accompanied by a more rapid damping (compare, for example, with swell-attenuation rates in Fig. 1 of Ardhuin *et al.*, 2009a). Here, our estimates are done for the case described in Fig. 30 of Snodgrass *et al.* (1966) and the water temperature is taken to be 20°C along the entire swell path which makes the Reynolds numbers upper-limit values).

More material for verifications are the results of Cavaleri & Zecchetto (1987) where multiple unexpected features were measured which could not have been explained: (1) wave-induced turbulence existed, even in the case of non-breaking swell; (2) momentum fluxes below the water surface did not match (exceeded) the wind stress; (3) in the case of wind-forcing, the wave orbits were tilted. Consequences (1) and (2) appear to be directly related to expectations due to the influence of wave-induced turbulence.

(1) The observed swell was quite steep, with peak steepness of 0.054 just below the breaking threshold of 0.055 (5.19). Corresponding $\mathrm{Re}_{\mathrm{wave}} \sim 200000$ clearly indicates a value well-above the onset of turbulence and therefore non-zero vertical momentum fluxes from such a source of turbulence are to be expected.
(2) Turbulent stresses in the water and in the air do not have to match in a general case, as the wave turbulence can be generated even in the absence of the wind if the swells are steep enough. Obviously, the two-order magnitude difference observed in the paper under conditions of active generation is mostly due to the phase shift between the horizontal–vertical components

of the orbital velocities, but the present wave-induced turbulence would also contribute towards excessive vertical momentum fluxes.

(3) Tilting of the orbits does not appear to relate to the wave-induced turbulence as it was only present in the case of the wind-generated waves. Since those waves were actively breaking (dominant-breaking rate of some 13% according to our estimate based on (5.25)), we would suggest that the tilting was perhaps due to asymmetry of the wave shape which is a prominent feature of the breaking waves (see (7.19) and Caulliez, 2002; Young & Babanin, 2006a).

We must emphasise again that the suggested critical WRN only signifies onset of turbulence generation and does not bear information on rates of this generation. Such rates, which would be of most interest for the turbulence, dissipation and mixing modelling, cannot be inferred from the Reynolds number alone and will be approached by experimental means below.

The measurements were conducted in the air-sea interaction salt water tank (ASIST) of the University of Miami (see Section 5.1 for tank description). The tank has a fully programmable wave maker able to produce waves with a predefined spectral form, including monochromatic waves. These waves are dissipated at the opposite end of the facility by a minimum-reflection beach.

In the experiment, in order to avoid any ambiguity due to wind-caused shear stresses and spectral wave superpositions, a simple setup was realised. Unforced (no wind) mechanically generated monochromatic deep-water two-dimensional wave trains of 1.5 Hz were generated. Particle image velocimetry (PIV) measurements were conducted at 5.1 m from the wavemaker in the centre of the tank directly under the troughs of the deepest waves, i.e. always at the same distance from the still surface level which was 38.8 cm. The distance of 5.1 m had been shown to be sufficiently far away from the paddle for the wave shape to be well established, but shorter than required for intensive modulational instability to develop (Babanin *et al.*, 2007a, 2010a). Measurements were made after the first few waves passed the measurement region, but before there could be any reflections off the dissipative beach which was 9.2 m from the measurement region. The tank has a smooth acrylic bottom and sidewalls that lead to a very thin wave boundary layer (<0.005 m). Observations above the bed and near the sidewalls under the wave conditions of interest revealed no upward/outward eddy propagation above the bottom/side boundary layers. Possible contaminations by wave-induced secondary circulations (Groeneweg & Battjes, 2003) were also investigated. The PIV system was positioned to observe the long-tank and cross-tank velocities in a horizontal plane at a level just below the lowest wave troughs (38.5 cm). Measurements of the cross-tank mean flow, for the 1.5 Hz waves studied, showed that it was very small (<2.5 mm/s) and spatially incoherent.

The PIV system employed here operated at a sampling frequency of 15 Hz. It was a double-pulsed system and the time between the image pairs ranged from 3 to 4 ms, depending on the wave amplitude. The tank was seeded with 50 μm polyamide seeding particle beads with a specific gravity of 1.06, and a 1-megapixel camera was positioned to look from the side of the tank. The laser was projected from the bottom to provide an illuminated plane in the long-tank and vertical dimensions at the cross-tank centre. The image size was 17 cm (1600 pixels) alongtank by 12 cm (1186 pixels) in the vertical for the first series of measurements. In the second set of measurements, the camera was set farther back and the image size/pixel resolution was correspondingly larger. Velocities were computed using an adaptive correlation technique in which the initial interrogation area was 256×256 pixels and in three successive refinement steps the final horizontal resolution was 1.2 mm. Based on the image pixel size and elapsed time between images, the uncertainty of each velocity observation was ± 0.03 m/s (Hyun *et al.*, 2003).

Fifteen wave periods were recorded for each of the wave trains whose wave amplitudes a_0 ranged from 6 mm through 32.5 mm (at this amplitude waves started to break at and before the measurement point, and therefore this measurement was not taken into consideration here). Wave amplitudes were recorded at the measurement location using a digital laser elevation gauge.

The PIV system provided spatial distributions of the u (horizontal, parallel to the tank's axis) and w (vertical) components of the two-dimensional velocity, and thus avoided the usual problem of converting the frequency velocity spectra into wavenumber spectra $P(k)$ (see Section 5.3.4 and eq. 5.66), in the oscillatory wave flow where mean velocity is close to zero and therefore there is no steady advective velocity to scale turbulence frequency ω into turbulence wavenumber k (Lumley & Terray, 1983; Agrawal *et al.*, 1992). If isotropic turbulence is generated, the velocity spectra, at the scales where its intensity dominates over the orbital oscillations, are expected to exhibit a $k^{-5/3}$ Kolmogorov interval at small scales (high wavenumbers) $k \gg k_s$ where k_s characterises the source of energy in the dynamic system (e.g. Monin & Yaglom, 1971). In our case of the monochromatic waves, $k_s \approx k_{1.5hz} = 9.82$ rad/m, and therefore the wavenumber scales of $k = 20$–2600 rad/m, resolved by the PIV frame, provided a sufficiently broad band.

WRN (7.70) for 1.5 Hz waves of surface amplitudes $a_0 = 6$–25.5 mm and water of 20°C ranged from 330 to 5900 and therefore the transition to turbulence could be expected. This transition depends strongly on the vertical distance z according to (7.72). Directly at the surface, however, PIV measurements were not possible as, in order to obtain the wavenumber spectrum, the entire horizontal layer of measured velocities has to be in the water. For consistency of comparisons, the layer at 30 mm from the still surface was chosen as a compromise which was close enough to the surface, but below the wave troughs for all recorded non-breaking waves.

It was expected that the turbulence, subjected to periodic forcing, will exhibit complex and modulated behaviour (e.g. Bos *et al.*, 2007). Close to the critical Reynolds number, the turbulence observed was also highly intermittent in both the space and time domains. At the surface wave amplitude of $a_0 = 22.5$ mm at each of the ten recorded phases of the wave period, the Kolmogorov interval in the velocity spectra appeared from 0 to 3 times over the duration of 15 periods. Most frequently, it was observed at the rear face of the wave profile, close to the instant of zero down-crossing, and therefore this phase of the wave was chosen for further analysis.

Examples of the wavenumber spectra recorded during the down-crossing, with the turbulence, are demonstrated in Figure 7.23. Here, a few $P_u(k)$ spectra exhibiting the

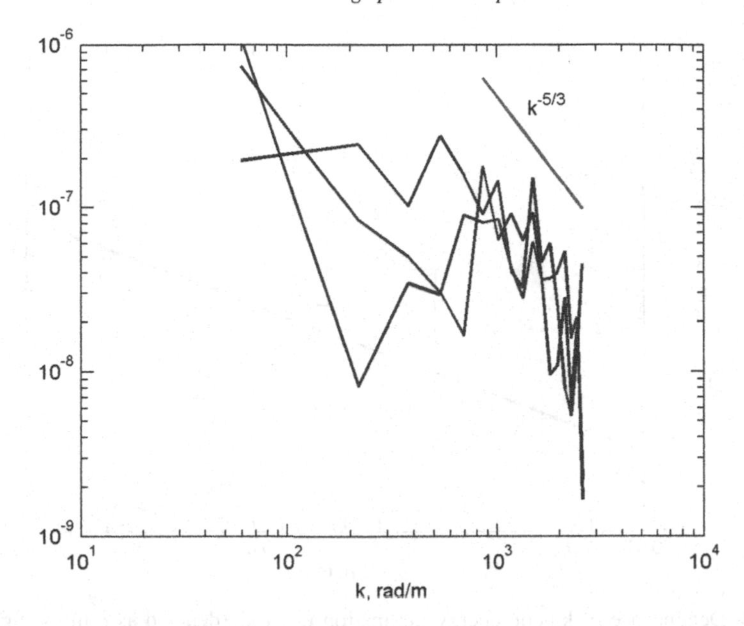

Figure 7.23 Velocity spectra exhibiting the Kolmogorov interval are shown as solid lines. Straight line indicates the $k^{-5/3}$ slope. Figure is reproduced from Babanin & Haus (2009) © American Meteorological Society. Reprinted with permission

Kolomogorov interval are shown. The spectra were obtained as a Fourier transform of the velocity u space series across the 30 mm layer, with averaging over four wavenumbers. Because the size of the image is less than the wavelength, the monochromatic-orbital-motion peak cannot be seen in the spectrum. The spectra show a clear Kolmogorov subinterval at wavenumbers $k > 800$ rad/m as should be expected for small turbulence scales far away (in the Fourier space) from the energy source. The Kolmogorov-interval slope is indicated with the straight line.

As was explained in Section 5.3.4, the level of the Kolmogorov interval contains information about the volumetric kinetic energy dissipation rate ϵ_{dis} (e.g. Veron & Melville, 1999). Since the wavenumber, rather than frequency spectra were measured with the PIV system, the connection is simpler than (5.66) and does not have to rely on the rms advection hypothesis of Lumley & Terray (1983):

$$P(k) = \frac{18}{55} \left(\frac{8\epsilon_{dis}}{9\alpha} \right)^{2/3} k^{-\frac{5}{3}}. \tag{7.75}$$

Using (7.75) to define ϵ_{dis}, a dependence on the wave amplitude a_0 was observed (Figure 7.24). Given the large confidence intervals of the individual estimates and the highly-intermittent behaviour of the observed turbulence, the experiment was repeated in order to verify the consistency of the result. Data points denoted with circles were obtained, for the same experimental setup, a year later than the asterisk data points.

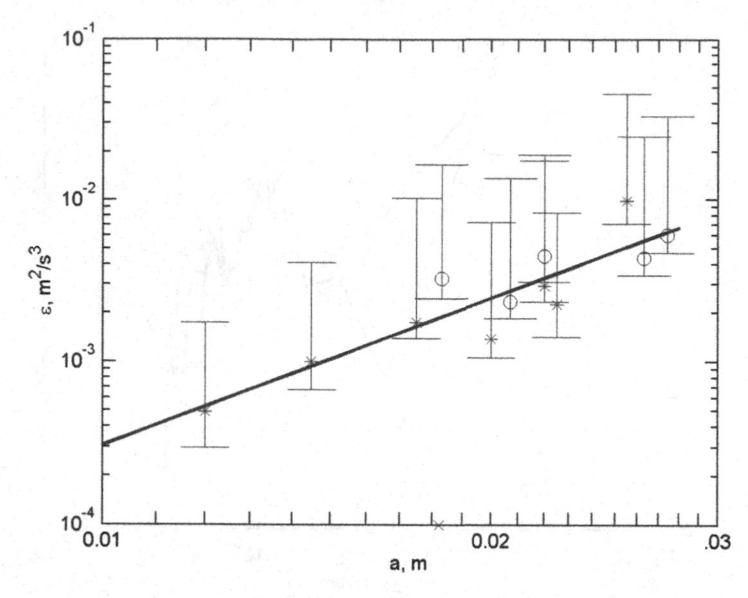

Figure 7.24 Dependence of kinetic energy dissipation rate ϵ_{dis} (denoted as ϵ in the figure) versus wave amplitude a_0 (denoted as a). Dependence (7.76) is shown with a solid line. Data points designated with asterisks and circles were obtained in two separate experiments in the same facility. The $\times$-mark on the bottom axis indicates value of a_0 which corresponds to $Re_{wave_{critical}} = 3000$, and the vertical bars show the standard deviations. Figure is reproduced from Babanin & Haus (2009)

Direct fit (solid line in Figure 7.24) provides the dependence

$$\epsilon_{dis} = 300 a_0^p. \tag{7.76}$$

Values of ϵ_{dis} and a_0 in this formulae of Babanin & Haus (2009) are dimensional. Therefore, the coefficient 300 is dimensional too, and in the form presented in (7.76) it is only suitable for the waves of 1.5 Hz considered in the experiment (see (7.85) for a non-dimensional version of this parameterisation). Statistical uncertainty of this coefficient (the intercept of dependence (7.76)) is large, but for the exponent p in (7.76) the 95% confidence limit places it between 2 and 4: $p = 3.0 \pm 1.0$. This is close to the expectation (Babanin, 2006) that, since the force due to the turbulent stresses is proportional to a_0^2, the energy-dissipation rate should be $\epsilon_{dis} \sim a_0^3$.

Before applying the results of (7.76) and Figure 7.24 to estimation of possible swell dissipation in the ocean, which would associate with the wave-induced turbulence, it should be emphasised again that the ϵ_{dis} estimates obtained by Babanin & Haus (2009) are not sustained dissipation rates. They are instantaneous values incurred intermittently at the rear-face phase of the wave below the level of the wave trough. Even for the steepest waves in the experiment, such turbulence occurred (that is occupied the entire 30 mm level of the camera image) not more than once out of the recorded ten phases per wave period. This is due to the experiment being designed to avoid wave breaking and thus it dealt with small-steepness waves and relatively low WRN, close to the critical number (7.67). As a result, if averaged over the wave period, the estimates of ϵ_{dis} in (7.76) have to be divided at least by a factor of 10 and perhaps more.

It is interesting to notice that such intermittent turbulence, which apparently belongs to the transitional fluid state between fully laminar and fully turbulent conditions, does not reveal a threshold value of wave amplitude, below which it does not occur, i.e. the critical WRN (7.67), and the data fit shows a correlation of 89% for the dependence which goes through the origin (see also Section 9.2.2 and Figure 9.10). The value of $a_0 = 0.0182$ m which corresponds to $Re_{wave_{critical}} = 3000$ for the 1.5 Hz waves is marked in Figure 7.24 with the ×-mark on the bottom axis. The lowest wave amplitude, at which the turbulence was still observed, i.e. $a_0 = 0.012$ m, corresponds to $Re_{wave} = 1300$. If approximately scaled to the free surface $z = 0$ by assuming $\exp(-kz)$ attenuation of the wave amplitude, the lowest observed turbulence-producing Reynolds number would thus be $Re_{wave} = 2300$. This is close to the lower-margin estimate of the critical Reynolds number of Babanin (2006), and this margin for the transitional wave turbulence is even lower in experiments and simulations of Dai *et al.* (2010), down to $Re_{wave} \sim 1000$. Such scaling of the lowest transitional WRN, however, is quite speculative, if talking about the measurements just below troughs of steep waves. Therefore, for the reference estimates, we chose to show the fit going through the origin rather than trying to produce a dependence based on an imposed threshold value for the amplitude.

Levels of the turbulent rates of $\epsilon_{dis} \sim 10^{-3}$ are quite high and comparable with dissipation rates measured in the presence of wave breaking (Agrawal *et al.*, 1992; Young & Babanin, 2006a). Thus, locally, the wave-induced turbulence can be quite intensive, but its average rates have to be scaled down over the wave period due to the intermittency of the observed turbulence as described above. It is also worth mentioning that in a recent study of turbulence and energy dissipation associated with wave breaking (Gemmrich, 2010) it was found that in the absence of wave breaking the turbulent dissipation persists with energy dissipation levels of $\epsilon_{dis} \sim 5 \cdot 10^{-3}$. Such levels are in close agreement with the wave-breaking onset levels measured in this study (i.e. Figure 7.24). Experimental approximation (7.76) is also consistent with $\epsilon_{dis} \sim a_0^3$ dependence implied by Qiao *et al.* (2004, 2008, 2010) who introduced additional wave-induced turbulent mixing in ocean circulation models. The very significant advances in performance of such models can be interpreted as an indirect support of the experimental results and dependence (7.76) presented here.

Although treating the results obtained in a particular case of short and steep laboratory waves has to be done with caution, this may still indicate a significance of the non-breaking wave-induced turbulence even in a more general case. When comparing breaking and non-breaking rates directly, our measurements revealed quite close values in the case of steep waves. The instantaneous dissipation rates of the $a_0 = 32.5$ mm, 1.5 Hz waves at the rear face, where the non-breaking turbulence was found in this study, and at the front face, where the breaking-in-progress turbulence develops, show magnitudes of $\epsilon_{dis} \approx 22.5 \cdot 10^{-3}$ and $\epsilon_{dis} \approx 14.6 \cdot 10^{-3}$, respectively.

Now, let us try to apply the estimates of transition to turbulent wave motion (7.67) and of the wave-turbulence dissipation rates (7.76) to the swell-attenuation measurements of Ardhuin *et al.* (2010). As said above, these observations provide a unique data set for testing theories and verifying mechanisms of surface-wave dissipative phenomena which are weak and slow compared to wave breaking.

First of all, we will check whether the steepness (7.61)–(7.62) is consistent with WRN (7.70) indicative of the turbulent wave motion. It should be noticed that $\mathrm{Re}_{\mathrm{wave}}$ cannot be unambiguously converted into the wave steepness, as for waves of the same steepness but different wavelengths or amplitudes WRN can be different.

The kinematic viscosity ν of the sea water, with three significant digits of precision, can be calculated as follows:

$$\nu = ((0.659 \cdot 10^{-3} \cdot (T_s - 1) - 0.05076)(T_s - 1) + 1.7688) \cdot 10^{-6} \tag{7.77}$$

where T_s is the surface water temperature in degrees of Celsius (this empirical formula is taken from http://ittc.sname.org/2002_recomm_proc/7.5-02-01-03.pdf). Ardhuin *et al.*'s (2009a) data (their online auxiliary material) cover swell periods of $T = 13$–$19\,\mathrm{s}$. For waves with $T = 16\,\mathrm{s}$, in the middle of this range, $k = 0.0157\,\mathrm{rad/m}$ and, if we choose constant $T_s = 10°$ over the swell path, then $\nu = 1.37 \cdot 10^{-6}\,\mathrm{m^2/s}$ and, based on the critical WRN (7.67), the critical amplitude is $a_0 = 0.102\,\mathrm{m}$. This leads us to

$$a\dot{k} = 0.0016, \tag{7.78}$$

way below the steepness (7.62). Therefore, for such a wavelength and for the steepness (7.62), WRN is quite high: $\mathrm{Re}_{\mathrm{wave}} \approx 29000$ and thus the wave motion is certainly turbulent and should lead to an excessive swell dissipation compared to the linear case (7.60).

Before estimating this dissipation, we should look in more detail into the issue of the critical steepness (7.61)–(7.62). In Figure 7.25 (top), a swell-decay rate α of (7.60) is plotted versus swell steepness $s = H/\lambda$ based on the online data of Ardhuin *et al.* (2009a).

When analysing these data points, the best correlation was observed not for the linear fit (corr = 0.86), but for a quadratic dependence with the intercept at

$$s = 0.0028 \tag{7.79}$$

rather than at $s = 0.005$ as in (7.61):

$$\alpha = 2.9 \cdot 10^{-3}(s - 0.0028)^2. \tag{7.80}$$

This dependence is shown with the solid line in Figure 7.25 (top). To obtain dependence (7.80), the obvious outlier at $s \approx 0.004$ was removed, but in any scenario, the correlation is higher for the quadratic expression.

This correlation, with intercept (7.79) is corr ≈ 0.932, but it is only reduced marginally to corr ≈ 0.930 if the intercept is set to 0, that is

$$\alpha = 1.6 \cdot 10^{-3} s^2 \tag{7.81}$$

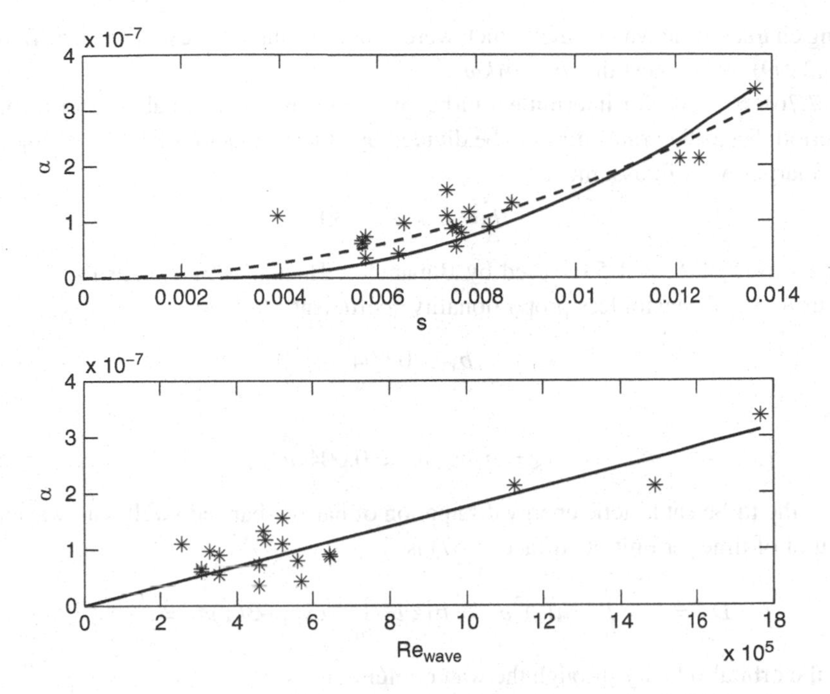

Figure 7.25 Swell decay rate α (7.60) versus (top) steepness $s = H/\lambda$. Solid line is expression (7.80), dashed line is expression (7.81); (bottom) Wave Reynolds number $\mathrm{Re}_{\mathrm{wave}}$ (7.70). Solid line is expression (7.82)

as shown with the dashed line in Figure 7.25 (top). In fact, the dashed line is visually preferable.

Absence of a clear threshold in terms of steepness is not that unexpected, and the dependence of ϵ_{dis} on steepness a_0 (7.76) also goes through the origin. If the transition is due to the wave-orbital motion becoming turbulent, then it should have a threshold in terms of WRN, which does not convert into a steepness unambiguously. In this regard, steepness (7.78) is not in contradiction with the observational data of Figure 7.25 (top).

Figure 7.25 (bottom) demonstrates the dependence of α versus $\mathrm{Re}_{\mathrm{wave}}$. The best correlation corresponds to the linear fit. The threshold $\mathrm{Re}_{\mathrm{wave}_{\mathrm{critical}}} = 3000$ is not apparent in the plot, due to overall values of WRN being high, but it is there and it is imposed in the dependence

$$\alpha = 1.77 \cdot 10^{-13}(\mathrm{Re}_{\mathrm{wave}} - 3000) \tag{7.82}$$

shown with the solid line. All the data points correspond to $\mathrm{Re}_{\mathrm{wave}} > \mathrm{Re}_{\mathrm{wave}_{\mathrm{critical}}}$, and therefore the mechanism of turbulence production by the waves should be active.

In order to estimate the rate of turbulence production by long ocean swells of Ardhuin *et al.* (2009a), the laboratory short-wave dependence (7.76), i.e. $\epsilon_{\mathrm{dis}} = b \cdot a_0^3$ should be converted into dimensionless form first. Here, b has dimensions of $\mathrm{m}^{-1}\mathrm{s}^{-3}$, and therefore, if using characteristic wave scales which were kept constant in the experiment of Babanin & Haus (2009), we suggest that $b = b_1 k\omega^3$.

In (7.76), $b = 300$ for intermittent turbulence, and, as discussed above, for mean-over-the-period dissipation rates it has to be divided by at least a factor of 10. Therefore, at this upper margin, we will assume

$$b = b_1 k\omega^3 = 30. \qquad (7.83)$$

For the waves of $f = 1.5$ Hz used by Babanin & Haus (2009) in their experiment that leads us to the dimensionless proportionality coefficient

$$b_1 = 0.004, \qquad (7.84)$$

that is

$$\epsilon_{\text{dis}} = b_1 k\omega^3 a_0^3 = 0.004 k u_{\text{orb}}^3. \qquad (7.85)$$

Now, the turbulent kinetic energy dissipation of narrow-banded swell with wavenumber k per unit of time per unit of surface (5.67) is

$$D_a = b_1 k \int_0^\infty u(z)^3 dz = b_1 k u_0 \int_0^\infty \exp(-3kz) dz = \frac{b_1}{3} u_0^3. \qquad (7.86)$$

Here, the orbital velocity through the water column is

$$u_{\text{orb}}(z) = u(z) = u_0 \exp(-kz) \qquad (7.87)$$

and u_0 is the orbital velocity of the surface water particles.

D_a is dissipation of wave energy E (2.22), normalised by the water density, per unit of time:

$$D_a = \frac{\partial(gE)}{\partial t} = \frac{1}{8}\frac{\partial(gH^2)}{\partial t} = c_g \frac{\partial(gE)}{\partial x} = c_g D_x. \qquad (7.88)$$

Dissipation per unit of propagation distance x, therefore, is

$$D_x = \frac{1}{c_g} D_a = \frac{b_1}{3} 2 \frac{k}{\omega} u_0^3 = \frac{2}{3} b_1 k\omega^2 a_0^3 = \frac{2}{3} b_1 g k^2 a_0^3. \qquad (7.89)$$

(7.88) and (7.89) lead us to differential equation

$$\frac{g}{2}\frac{\partial(a_0(x)^2)}{\partial x} = \frac{2}{3} b_1 g k^2 a_0(x)^3, \qquad (7.90)$$

that is

$$\frac{\partial(a_0(x)^2)}{\partial x} = \frac{4}{3} b_1 k^2 a_0(x)^3 = B a_0(x)^3. \qquad (7.91)$$

The solution yields

$$a_0(x)^2 = \frac{4}{B^2} x^{-2} = \frac{9}{4 \cdot b_1^2 k^4} x^{-2} = \frac{9}{64} 10^6 k^{-4} x^{-2}. \qquad (7.92)$$

The online data of Ardhuin *et al.* (2009a) provide swell height H_1 at distance 4000 km from the slow-moving limited-area storm and the attenuation rate α measured over the subsequent 11000 km. Expression (7.92) does not have distance to the storm which is effectively implied to be a point-source and have an infinite power at $x = 0$. Therefore, (7.92) was first used to estimate distance x_0 from this fictitious point-source such that the resulting wave height is H_1:

$$x_0 = 2\frac{3}{8}10^3\frac{1}{k^2 H_1}. \tag{7.93}$$

Such inferred distance is then used to obtain what would be the wave height 11000 km down the path, based on (7.92):

$$H_B = \frac{3}{4}10^3\frac{1}{k^2(x_0 + 11000000)}. \tag{7.94}$$

This wave height is then compared to the estimate obtained from H_1 and the decay rate α:

$$H_A = H_1 \exp(-11000000\alpha). \tag{7.95}$$

The results of (7.94)–(7.95) are compared with each other in Figure 7.26 (asterisks). Attenuation rates of (7.94) are, on average 2.5 times larger than those of (7.95) estimated by Ardhuin *et al.* (2009a). In the top panel, H_B points are always below the one-to-one

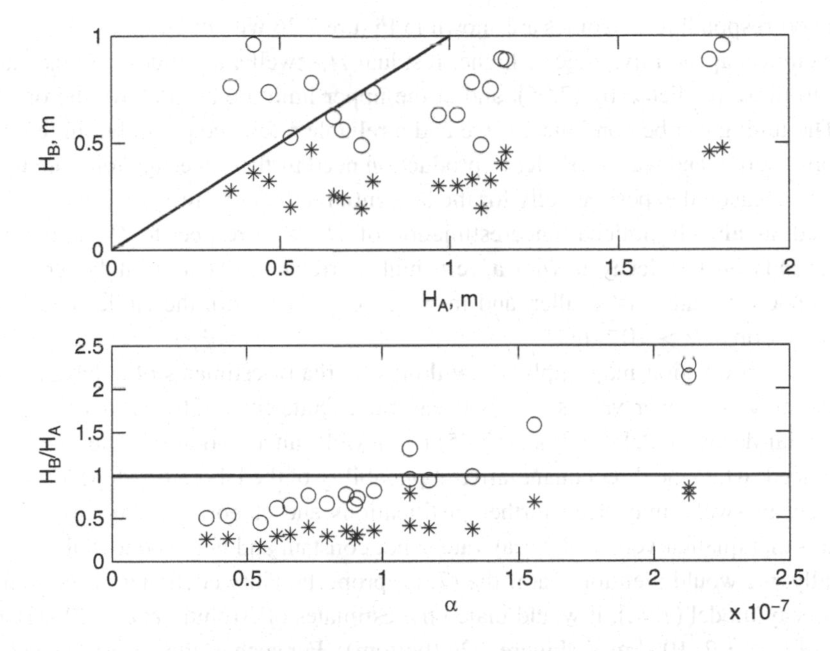

Figure 7.26 (top) Swell height H_B (7.94), estimated by means of decay described by (7.92), versus height H_A (7.95) based on the experimental decay rate α (7.60) of Ardhuin *et al.* (2009a); (bottom) Ratio H_B/H_A versus α. In both subplots, asterisks correspond to the empirical coefficient (7.84) and circles to (7.96), and the solid line indicates the one-to-one ratio

line, and the mean ratio is $H_B/H_A = 0.41$. Here, we have excluded and do not show the single data point which produced a ratio of $H_B/H_A = 3.1$ and is an obvious outlier.

This ratio, however, is not a constant value and depends clearly on α of Ardhuin *et al.* (2009a): H_B/H_A versus α is plotted in Figure 7.26 (bottom). The ratio grows as a function of α, with a very high linear correlation of 96%.

Before discussing the implications of these results, we should notice that the dissipation rate of (7.85) and (7.94) is perhaps too strong and overestimates the observed swell-height decay. This conclusion, however, highlights an important conjecture that the dissipation due to wave-induced turbulence is a significant player in observed attenuation of swell. Not only do the dissipation rates identified due to such mechanisms explain the swell decay, in the present form they over-predict it.

The present form includes empirical coefficient b_1 whose value was adopted as $b_1 = 0.004$ at the upper margin of its possible values, by assuming that the intermittent turbulence in the non-breaking experiment of Babanin & Haus (2009) appears at one out of the ten wave-phases measured. In fact, as mentioned above, this is only true for the steepest waves, and for the smaller waves of the same length this intermittency rate is lower for most data points of Figure 7.24. If dependence (7.76) is divided by 20, rather than by 10, in order to estimate mean-over-wave-period turbulence production, then

$$b_1 = 0.002, \tag{7.96}$$

and the corresponding outcomes are shown in Figure 7.26 with circles.

The circles, apparently, indicate higher residual H_B swell amplitudes, and they are now closer to those predicted by (7.95), and at the upper limit are even above the one-to-one line. The tuning can be continued, if we had a reliable reference point, but in reality both swell-air friction and wave-turbulence production need further investigation. The former is still to be measured experimentally for the tangential turbulent stresses.

Indeed, as already noticed, underestimation of H_B with respect to H_A is not random and depends on the decay α with a very high correlation. That is, at larger values of α, the underestimation is smaller, and in the case of b_1 (7.96), the circles even indicate overestimation at $\alpha > 10^{-7}\,\mathrm{m}^{-1}$.

Such an observation may imply a few things. Perhaps, estimates of α have a progressive bias towards lower values of this decay rate. Quite likely also is that the constant-exponential-decay model (7.60) and (7.95) is not well suited to describe the swell attenuation. Indeed, while the direct quantitative applicability of the laboratory dependence (7.76) to the ocean swells may need further justifications and elaborations, in any scenario it indicates that qualitatively such decay rate is not constant and not exponential.

Finally, we would mention that if the (7.96) proportionality coefficient is chosen in the swell-decay model (7.92), it would match the estimates of Ardhuin *et al.* (2009a) at their values of $\alpha \approx 1.2 \cdot 10^{-7}\,\mathrm{m}^{-1}$ (Figure 7.26 (bottom)). For such α, the observed swell decay can be fully explained by the waves spending their enery on generating turbulence in the upper ocean. If $\alpha \approx 2 \cdot 10^{-7}\,\mathrm{m}^{-1}$, H_B is approximately twice as high as H_A, that is the wave-induced turbulence alone is not a sufficient dissipation source. In such a case, some compatible energy sink, like, for example, the swell friction against the air discussed above in this section, should be present.

To conclude this section, we should stress again that, apart from the wave-breaking dissipation, a number of other dissipation mechanisms certainly exist in the wave system. In the presence of breaking, they may be small and negligible, but they are certainly most essential when describing propagation of swell whose forecast is one of the most difficult practical hydro-meteorological applications to date.

We should also remember that dissipation of waves is a source of energy for other mechanisms in the atmosphere–ocean system. Small it may be as a dissipation, but its role can become significant and certainly not negligible as this source. Two such secondary-to-breaking dissipation terms were identified as significant in this section: the wave-induced turbulence and the wave-to-air fluxes.

In this context, for example, if the powerful dissipation breaking only injects turbulence at the scale of wave height, the gentle dissipation mean-wave-orbit-induced shear stresses generate turbulence directly through the water column at the scale of wave length. This way, it is an essential mechanism of the upper-ocean mixing (e.g. Babanin *et al.*, 2009b; Pleskachevsky *et al.*, 2011).

The wave-against-air friction is essential in modifying the lower part of the atmospheric boundary layer (e.g. Kudryavtsev, 2004, see also Section 7.3.6 for discussion and further references). Collard *et al.* (2009) demonstrated that it has a significant effect for wind speeds in excess of 7 m/s (see also Hanley *et al.*, 2010).

In this section, we mainly concentrated on the dissipation phenomena which could be identified when observing swell propagation over large distances. There are, however, other potential dissipation mechanisms whose role may be essential both in wind-generated wave fields, i.e. small but not negligible by comparison with wave breaking, and in swells. One of them is the loss of energy by large waves due to work spent on the modulation of short waves (see e.g. and The WISE Group, 2007). In the case of swell, such process can be active when swell propagates through wave fields generated by adverse wind.

This phenomenon is suggested as a combination of the maser mechanism (Phillips, 1963; Longuet-Higgins, 1969b) and a theory of exchange of the potential energy between long and short waves (Hasselmann, 1971). Ardhuin & Jenkins (2005) demonstrated that such dissipation is greater than the action of viscosity. Moreover, Hasselmann (1971) neglected associated modulations of the wind stress on the atmospheric side, which may be significant and enhance the effect (Garrett & Smith, 1976; Ardhuin & Jenkins, 2005, 2006).

Thus, the analytical theory of this particular dissipation phenomenon is well developed, but experimentally its significance is still to be investigated and quantified. The considerable challenge in such experiments and observations of this and similar dissipation effects is due to the fact that they are small in the presence of breaking and in the course of swell propagation they are patchy. The interaction of swell with short waves, for instance, will only occur if the swell encounters counter-propagating wind seas, which may or may not have taken place in satellite observations similar to those of Ardhuin *et al.* (2009a).

On a broader scale, in this section, and throughout the book, we concentrate on deep-water waves which are typically treated in isolation from other oceanic processes. In and The WISE Group (2007), however, interactions of surface waves with the vertical structure of the upper ocean are outlined as another source of the dissipation. These can be interaction of surface waves with internal waves (e.g. Gargett & Hughes, 1972), or mixing through the thermocline conducted by the surface waves (e.g. Qiao *et al.*, 2006), among many others.

To summarise the entire chapter, we should say that it is one of the most important chapters of the book. The book is dedicated to wave breaking, and while the breaking is an interesting physical phenomenon in its own right, relevant across a broad range of scientific and practical applications, the wave energy dissipation is one of its most significant outcomes and applications. Such dissipation occurs non-uniformly across the wave spectrum, and its spectral distribution is still poorly understood, but is included in most of the models which are used for forecast of the waves and plays a role in describing interaction of the waves with the upper ocean and the atmospheric boundary layer.

In this chapter, analytical theories of the spectral dissipation are described in Section 7.1. They are broadly classified into probability, quasi-saturated, whitecap and kinematic–dynamic model types in the respective subsections.

Phase-resolvent numerical models are outlined in Section 7.2. They deal with a challenging task of simulating the wave-breaking phenomenon explicitly, based on first principles, and therefore have to take into account the strongly turbulent nature of fluid flow, with air bubbles embedded on the water side and spray emitted and suspended in the air. With the advance of fast computing, this class of models has been rapidly expanding over the past two decades, and a variety of clever mathematical and numerical means of treating this two-phase dynamic medium are available these days. Potentially, this way of investigating the wave breaking and associated wave-energy dissipation, which is not based on limiting assumptions and conditions like the analytical theories, has a great future.

While analytical theories drive the understanding of physics and progress of numerical simulations of phenomena, the ultimate judgment of validity, applicability, significance and relative importance of the theories always belongs to the experiment and observations. Section 7.3 is dedicated to measurements of the spectral dissipation. These are considered both in the laboratory, where components of the phenomenon can be isolated and studied separately, and for the real waves in the field. New features of the total-dissipation and spectral-dissipation behaviour found, or rather re-discovered recently, are attended to in detail. These are, first of all, the wave-breaking threshold and the cumulative effect. The effects that the wind imparts on the dissipation process and its strength, as well as the directional behaviour of the dissipation, are still in a blurred area at the present stage, but they are important and highlighted in respective subsections.

This book is on the physics rather than applications of the wave-breaking and wave-dissipation knowledge, but without mentioning these applications it would be incomplete. Therefore, a substantial Section 7.4 outlines past experience, present attempts, and updates on the potential future progress with respect to the dissipation terms employed in wave-forecast models. The newly-found characteristics and properties of the wave-breaking dissipation are still slow in making their way into operational forecast, but they are a consistent feature of research wave models and hindcasting exercises these days, and new perspectives of the operational modelling are also very encouraging.

Finally, the present Section 7.5 is dedicated to dissipation processes and their estimates, which are unrelated to the breaking. These are many, but they are often disregarded because in the presence of wave breaking their relative efficiency is small. It is not small, however, for the swell, and without essential energy inputs and nonlinear interactions, the swell-energy evolution is mainly driven by such dissipation processes. Two of them, which are secondary in the presence of breaking, but primary in its absence are analysed in detail: the production of turbulence by the mean wave-orbital motion, and the friction of surface waves against the turbulent air. As also discussed in the section, the significance of such dissipation processes extends far beyond the associated wave-energy loss and provides a link between the surface waves, the ocean below and the atmosphere above. This loss may be small, but the role of the lost energy in the upper-ocean mixing and in the dynamics of the near-surface atmospheric boundary layer can be large.

8

Non-dissipative effects of breaking on the wave field

One of the main roles of wave breaking in the ocean-wave fields is providing the dissipation sink of wave energy (see Chapters 1 and 7). The waves do not grow unlimited in height, and breaking is the main process which restricts this growth even in conditions of continuous wind-energy input over unlimited wave fetches.

Being the loss from the wave system, the breaking dissipation provides a link for the wind-generated waves with other phenomena in the lower atmosphere, upper ocean and on the ocean interface itself. It is a source of energy and momentum for infra-gravity waves, near-surface currents, upper-ocean mixing and the atmospheric boundary layer (e.g. and The WISE Group, 2007).

Some features and physical processes in the air–sea system, related to the breaking, however, are not necessarily of dissipative nature or can be described in terms of the wave-energy dissipation as such, and these are the phenomena to which the current chapter will be dedicated. Some of them have already been mentioned or discussed throughout the book. It was needed and essential to highlight them in order to understand the described features of breaking and whitecapping dissipation.

Others, for example, the spectral-peak downshift which is the topic of the next section, 8.1, have been mentioned as those which contribute to spectral dissipation (see Section 7.3.1). There is no contradiction here. In the context of the spectral distribution of the dissipation, energy lost from any part of the wave spectrum is interpreted as dissipation of waves of the respective scales. On the total-energy scale, however, this energy is not necessarily lost to the wave system in general, and can be transferred to the shorter waves (high-frequency tail of the spectrum) or to the longer waves (for instance, contribute to the downshift of the spectral peak). On an even broader picture of ocean waves, it can be converted into energy of infra-gravity waves at the scales of wave groups (e.g. Smith & Mocke, 2002). These are still surface waves, even though outside the range customarily attributed to wind-generated waves, and therefore the question of whether this energy was dissipated from the wave field or not is a matter of definition.

Thus, in this chapter, Section 8.1 is dedicated to the downshift of the spectral wave power and the next section, 8.2 to the roles wave breaking plays in maintaining the level of the spectrum tail. As discussed in that section, these roles are few. Some of them are indeed not due to the loss of wave energy as such. Others are definitely dissipative in nature, but here we would like to stress the fact that breaking shapes the spectrum tail and its level, and to concentrate on this character of the breaking rather than on the associated dissipation. This section is relatively short as a large proportion of its content serves as a summary of relevant outcomes and conclusions which have been discussed throughout the book in a different context.

With respect to the dissipation, the content of Section 8.3 is quite unique. It discusses the issue of enhancement of the wind energy input due to wave breaking. In other words, in a way in this sense the breaking provides a negative dissipation because it instigates higher energy and momentum fluxes from the wind to the waves when it happens.

8.1 Spectral peak downshift due to wave breaking

In the context of the radiative transfer equation (2.61) used in spectral models to forecast the evolution of wave fields, downshift of the spectral peak is typically attributed to the weak resonant four-wave interactions, i.e. term S_{nl} in RTE. This is another long-lasting tradition in wave-modelling applications, analytically based on the theory initially suggested by Hasselmann (1962) and further developed by Zakharov (1968), Krasitskii (1994) and Zakharov (2010) for homogeneous wave fields and recently by Gramstad *et al.* (2010) for non-homogeneous conditions. For this mechanism to be active, the wave fields have to be directional, with continuous distribution of wave energy along frequency and wavenumber vectors, and exact resonance of the four waves defined by the dispersion relation (2.17) has to occur. While the continuous spectrum is a natural appearance of the wind-generated waves, the exact resonance is a more limiting condition. First, the real waves are always nonlinear and non-steady, and this fact causes deviations from the linear dispersion (2.14), which can be as large as of the order of 10% (see Section 5.1.2). Secondly, downward acceleration of the short waves is strongly modulated by the dominant-wave motion at a scale as large as the magnitude of the gravitational acceleration itself (see e.g. Sections 3.2 and 3.7).

In the meantime, it is generally known that downshifting of the spectral energy of nonlinear wave trains, propagating in a dispersive environment, is also an outcome of modulational instability. This dispersive environment does not have to be surface water waves. In nonlinear optics, for example, it is the so-called Raman effect which transfers energy from high to low frequencies and causes a continuous red shifting of the mean frequency of pulses propagating in optical fibres (e.g. Gordon, 1986). It is described by the nonlinear Schrödinger equation (NLS, Zakharov, 1968), the same as used for water waves (see Section 4), which not only predicts the effect, but is also able to describe the magnitude of the frequency shifting quantitatively.

For water waves, the downshifting which associates with the modulational instability has also been observed and described, both experimentally (e.g. Lake *et al.*, 1977) and theoretically (e.g. Segur *et al.*, 2005). Breaking, if it happens in the course of these growing instabilities, significantly modifies and accelerates such downshifting (e.g. Melville, 1982; Su *et al.*, 1982; Reid, 1992; Tulin & Waseda, 1999; Hwung *et al.*, 2009; Babanin *et al.*, 2010a; Hwung *et al.*, 2010).

This question was first investigated by Trulsen & Dysthe (1990) by means of a modified nonlinear Schrödinger equation (Dysthe equation). The effect of breaking was simulated by adding a dissipation term to this equation which was solved numerically. As mentioned in Section 4, NLS and its modifications work in the physical space for the complex wave envelope, and this dissipation was expressed as a function of wave steepness ϵ:

$$D_{\text{NLS}} = \frac{1}{\tau_r}\epsilon\left(\left(\frac{\epsilon}{\epsilon_{\text{thr}}}\right)^r - 1\right)H(\epsilon - \epsilon_{\text{thr}}). \tag{8.1}$$

This term has two very important features. These are the breaking limiting steepness ϵ_{thr} and the Heaviside unit-step function H. Here, r and τ_r define relaxation of the wave steepness back to its limiting value once the critical steepness is exceeded, that is the breaking.

Nowadays, we do know that the limiting steepness exists (see Section 5.1.2, Babanin *et al.* (2007a); Toffoli *et al.* (2010a)). Once the wave exceeds this steepness, it breaks, but while breaking, its steepness is typically reduced to a level much lower than the critical steepness. This issue, however, is of secondary importance here and can be easily modified in the approach (see also Section 7.1.1). The significance of the Heaviside function will be outlined with respect to the model of Hwung *et al.* (2009) and Hwung *et al.* (2010) discussed below.

Typical behaviour of the modulational instability in a uniform wave train is growth of two resonant sidebands, both below and above the primary mode, that leads to strong modulation of the wave train, which sidebands then subside and the initial near-uniform train recurs. In the case of wave breaking happening in the course of such a cycle, however, as was shown by Trulsen & Dysthe (1990) theoretically and Tulin & Waseda (1999) experimentally, the grown lower-sideband becomes disconnected from the cycle and thus a permanent downshift of the wave energy takes place.

The topic has been consistently developed further by Trulsen and his colleagues. Trulsen & Dysthe (1992) demonstrated modification of this process subject not only to the dissipation, but also to strong energy input (i.e. in a way simulating the wind forcing, see also Sections 5.1 and 6.1). Trulsen & Dysthe (1997) extended the research into three-dimensional wave fields. When using the Dysthe equation without dissipation, no permanent downshifting was observed for two-dimensional waves, but if oblique perturbations were allowed, the permanent downshift of the spectral peak did occur, with the energy shifted to the oblique modes. Trulsen *et al.* (1999) confirmed such significant downshift to the transverse modes experimentally, in a directional wave tank.

The most detailed and dedicated study of the downshifting due to breaking of waves was conducted by Tulin & Waseda (1999) and Waseda & Tulin (1999). This was done in a flume for two-dimensional waves, with and without the wind, and a theory for explaining the results was also proposed (see discussion in Section 7.3.1).

The modulational instability in the experiments was, mostly, seeded, that is the fastest-growing sidebands were imposed to accelerate development of the instability. While substantial shift of the energy to the lower sideband was observed at the peak of the modulation, it was near-recurrent without breaking.

The breaking would change the dynamics in a most principal way. Transfer of the energy to the lower sideband would increase, and, most essentially, it would become irreversible.

> "The end state of the evolution following strong breaking is an effective downshifting of the spectral energy, where the lower and the carrier wave amplitudes nearly coincide; the further evolution of this almost two-wave system was not studied..."

Here, for more details, we will follow Hwung *et al.* (2009) and Hwung *et al.* (2010), who combined the Trulsen & Dysthe (1990) and Tulin & Waseda (1999) theories with the most recent observations of the wave breaking and dissipation in a two-dimensional super tank of the National Cheng Kung University of Taiwan.

The super tank is 300 m-long, with a cross-section of 5 m × 5.2 m. In such a long tank, there was no need to seed the instability sidebands and therefore the naturally developing evolution of nonlinear groups was investigated, with the lower and upper sidebands growing from the background noise (e.g. Reid, 1992; Babanin *et al.*, 2007a, 2010a). The waves were recorded with 66 high-resolution capacitance-wire probes along the fetch.

The long tank also gave the possibility of studying evolution of the wave system past the breaking occurrence, and additional features of this subsequent evolution of the wave trains were revealed (Hwung *et al.*, 2004; Hwung & Chiang, 2005). Once the energy was lost to breaking and nonlinearity could no longer lead to a new breaking, the train restabilised asymmetrically with respect to the two-wave system which was the final stage of the Tulin & Waseda (1999) observations. The lower sideband grew up further and finally periodic modulations were observed, without further energy loss and with the lower sideband as the new carrier wave.

Measurements were then compared with the theory. Hwung *et al.* (2009) and Hwung *et al.* (2010) derived the initial evolution equations for energy density $E = ga^2/2$ and group velocity c_g, following the variational approach of Tulin (1996) and Tulin & Li (1999):

$$\frac{\partial E}{\partial t} + \frac{\partial c_g E}{\partial x} = I - D_b; \tag{8.2}$$

$$\frac{\partial c_g}{\partial t} + c_g \frac{\partial c_g}{\partial x} = \frac{1}{4} k \frac{\partial E}{\partial x} + \frac{c_g^2}{8k^2} \frac{\partial}{\partial x} \left(\frac{a_{xx}}{a} \right) + \frac{c_g}{E} \gamma D_b. \tag{8.3}$$

Here, I is the wind-input energy rate, and the nonlinear group velocity is used (see (2.14)–(2.19)):

$$c_g = \frac{c}{2} \left(1 - \frac{(ak)^2}{4} - \frac{a_{xx}}{8ak^2} \right). \tag{8.4}$$

Dissipation rate D_b and momentum-loss rate M_b are such that

$$\gamma D_B = c M_b - D_b > 0 \tag{8.5}$$

where γ is an empirical coefficient of the order of 1, which in numerical simulations described here was varied in the range $\gamma = 0.3$–0.8 (in Tulin & Waseda (1999), it was chosen $\gamma \approx 0.4$, see also (7.14)).

The positive last term in the evolution equation for group velocity (8.3) simulates the wave breaking. It signifies a continuous increase of the group velocity, that is the frequency downshift at the rate which depends on the breaking energetics.

Hwung *et al.* (2009) and Hwung *et al.* (2010) combined (8.2)–(8.3) into a single equation for the complex amplitude B:

$$\frac{\partial B}{\partial t} + c_g \frac{\partial B}{\partial x} + i\omega k^2 |B|^2 B = B \left(\frac{I - D_b}{g|B|^2} - i4\gamma \int \frac{k D_b}{g|B|^2} dx \right) \tag{8.6}$$

where

$$B = a e^{i(\theta + \alpha)} \tag{8.7}$$

and the phases are

$$\theta = k_0 x - \omega_0 t \tag{8.8}$$

for the carrier waves, and α is the fluctuating phase such that (see (8.4)):

$$\frac{1}{2k} \frac{\partial \alpha}{\partial x} = \frac{(ak)^2}{4} + \frac{a_{xx}}{8ak^2}. \tag{8.9}$$

In the computations, input I was assumed $I \sim g|B|^2$, and therefore another empirical proportionality coefficient was introduced. The dissipation D_b, following Tulin (1996) parameterisation based on the fetch laws, was adopted as

$$D_b = 0.1 E\omega \cdot (ak)^2. \tag{8.10}$$

To compare the theory with the experiments, a solution was sought for a wave system consisting of the respective carrier wave and the two resonant sidebands. Simulations were performed for the range of initial carrier steepness of $\epsilon_0 = a_0 k_0 = 0.15-0.25$.

The comparison was very good for the evolution leading to the breaking, including the breaking event itself. Initial exponential growth rates of the resonant waves, asymmetrical growth of the sidebands, distance to the location where the amplitudes of the carrier waves and the sub-harmonics were level, redistribution of energy between the harmonics both at the growing and breaking-in-progress stage – all of these features were correctly reproduced quantitatively.

Past the breaking occurrence, however, the train stabilisation was not described by the model. Rather, it continued to quench the wave amplitude as the wave-breaking term in the model was still active.

To help this apparent shortcoming of the model, the Heaviside function of Trulsen & Dysthe (1990) from (8.1) was incorporated into (8.6):

$$\frac{\partial B}{\partial t} + c_g \frac{\partial B}{\partial x} + i\omega k^2 |B|^2 B = B \left(\frac{I - D_b}{g|B|^2} - i4\gamma \int \frac{k D_b}{g|B|^2} dx \right) \cdot H \left(\frac{|B_X|}{B_{\text{thr}}} - 1 \right). \tag{8.11}$$

Here, B_{thr} is the limiting value of steepness and $B_X = \sum(a_i k_i)$ is the combined steepness of wave components i at distance X along the tank. The threshold $B_{\text{thr}} = 0.32$ was chosen and perhaps it needs to be updated to $B_{\text{thr}} = 0.44$ in the future (see Section 5.1.2).

Now, the dissipation/input are disallowed past distance X and ordinary nonlinear wave evolution proceeds. The model, updated this way, correctly reproduces the entire evolution of the wave train in the super tank. The waves break in the prescribed location, permanent downshift occurs, followed by restabilisation and steepness oscillations with the new carrier frequency. Hwung *et al.* (2009) and Hwung *et al.* (2010) conclude that the post-breaking wave system 'forgets' its pre-breaking initial conditions and further evolution is described by the nonlinear behaviour of the past-breaking three wave components.

Thus, we can conclude that downshift of the wave energy is a robust feature of the breaking due to modulational instability. If such instability is present in the real wave fields, then the dissipation term in RTE (2.61) should be attributed with the downshifting. Recent experiments have demonstrated that the modulational instability is indeed active in typical directional fields with typical background wave steepness (Babanin *et al.*, 2011a, see Section 5.3.3). Since the breaking happens at the scale of tens of wave periods, whereas the weak nonlinear interactions at the scale of thousands of wave periods (see Babanin *et al.*, 2007a, Section 5.1.2 and discussion in Section 7.3.3), then S_{ds} term in RTE perhaps can be held responsible for the downshifting, in addition to that provided by S_{nl}. Such models have been attempted (e.g. Schneggenburger *et al.* 2000; WISE-meeting presentations by Donelan and Meza), and this issue needs further research and clarification.

8.2 Role of wave breaking in maintaining the level of the spectrum tail

Mention of the role of wave breaking in maintaining the level of the spectrum tail is scattered throughout the book. We, however, thought that it is helpful to briefly summarise and update the effects which this role imparts on the spectrum shape. As has been discussed in Sections 5.3.4, 6.2 and 7.3.5, the equilibrium level of this spectrum, even though with some variations, remains remarkably stable given the very large magnitudes of differences in wind forcing. These differences occur both in absolute values, that is in terms of the wind speed which in this book varies from near-zero to some 30^+ m/s, and in relative terms across the wind–wave scales, that is if $U_{10}/c_p = 1$ for waves with $f_p = 0.1$ Hz, then at the 10 Hz tail end it will be $U_{10}/c_p \sim 100$ (neglecting the surface-tension contribution for clarity). Here, we will largely follow Babanin (2010) in discussing the role of breaking at the tail of the wave spectrum.

The first analytical justification for the equilibrium range of the spectrum, based on a dimensional argument, was proposed by Phillips (1958). The idea of the equilibrium spectrum tail was already in the air, and a number of experimentalists had suggested their dependences for such a spectral interval. The first parameterisation was by Neumann (1954) for a 'fully-developed spectrum',

$$F(\omega) = C\omega^{-6} \exp\left(\frac{-2g^2}{\omega^2 U^2}\right) \tag{8.12}$$

where $C = 1.4 \cdot 10^3\,\mathrm{cm/s^3}$ is a dimensional constant. The parameterisation of Burling (1955) was measured in the range of wave fetches $X = 400$–$1350\,\mathrm{m}$ and wind speeds $U_{10} = 5$–$8\,\mathrm{m/s}$ and stated the fact that, even though the spectra themselves were clearly developing, with the spectrum peak and the total energy evolving, the tails

"obtained in different conditions very nearly coincide and become apparently independent of the fetch and of the strength of the wind":

$$F(\omega) = 7.0 \cdot 10^3 \omega^{-5}. \tag{8.13}$$

Bretschneider (1958) produced a parameterisation which is already remarkably familiar (see (2.7)):

$$F(\omega) = \alpha g^2 \omega^{-5} \exp\left(-0.657\left(\frac{2\pi}{T\omega}\right)^4\right) \tag{8.14}$$

where T is

"a 'mean wave period' which depends upon wind speed and the state of development of the sea".

Bretschneider (1958) found that constant $\alpha = 7.4 \cdot 10^{-3}$, obtained from observations of Burling (1955), was consistent with his spectra.

Phillips (1958) rightly argued that for the spectrum parameterisation to be physically sound, its dimension has to be correct. He produced a rather long list of physical dimensions to be considered, i.e. air and water densities ρ_a and ρ_w, friction velocity u_*, surface roughness length z_0, acceleration g, water surface tension σ and viscosity ν, together with the relevant spectral scales k and ω.

In different parts of the spectrum, different combinations of the dimensional properties may be valid and/or significant, but when it comes to the gravitational range of surface waves, it was u_* which represented the wind stress and gravitational acceleration g that remained. If talking about the spectrum tail, which clearly exhibited some level of saturation in the spectra of Burling (1955), Phillips (1958) argued that this saturation signifies some limiting shape of the wave crests which is apparently due to the waves breaking and not being able to grow any further.

With respect to the breaking, Phillips (1958) argued that

"the geometry of the limiting shape near the sharp crests is determined by the condition that the downward acceleration should not exceed g, so that the asymptotic forms... would not be expected to involve u_*. An increase in u_* ... should not influence the geometry of such a sharp crest itself".

Phillips (1958) even made a footnote remark that

"we have excluded a different possible type of surface instability in which the sharp crests may be 'blown off' by very high winds".

We must notice that this is exactly the view of the breaking onset that we have these days. As discussed in detail in Chapters 4 and 5, the breaking onset is defined by the limiting steepness (2.47), (2.51) and (2.52), and it is only in extreme conditions that the wind can reduce this steepness, and even then only marginally.

If so, the original argument of Phillips (1958) is right and the only relevant dimensional parameter for this part of the spectrum is g. This led to the frequency spectrum of the saturated interval

$$F(\omega) = \alpha g^2 \omega^{-5} \tag{8.15}$$

and wavenumber spectrum

$$\Phi(\mathbf{k}) = \alpha_k(\theta) k^{-4}. \tag{8.16}$$

Following (8.13), Phillips (1958) defined his constant as $\alpha = 7.4 \cdot 10^{-3}$. Expression (8.16) defines spectral densities in different directions of wavenumber vector **k**, and the dimensionless coefficient α_k depends on this direction θ.

Parameterisations (8.15)–(8.16) must have been the first physical and even quantitative account for wave-breaking effects on the behaviour of the wave system. By limiting the growth of waves, they define the shape of the spectrum tail and even set its level.

Many more measurements of the equilibrium interval have been conducted since 1958. The level α proved to be not that constant. The early observations of Burling (1955) were conducted at what we would now classify as relatively light winds, and for stronger winds even Phillips (1977) himself reconsidered this value. As shown in Babanin & Soloviev (1998a), α indeed remains approximately constant for the best part of the wave-spectrum development, but for mature waves (which would usually accommodate the light-wind conditions too) it starts dropping towards the value of Phillips (1958). This is expression (5.75) in this book.

As discussed in Section 5.3.4, the tail level α responds to the wind forcing if $U_{10}/c_p <$ 1.45 (i.e. $\tilde{f}_p \leq 0.23$). At such forcing, apparently, the average Phillips' 'geometry of the limiting shape' of the wave crests can change due to, perhaps, changing rates of the breaking occurrence in the high-frequency spectral bands.

Once it reaches, however, $\alpha \approx 13.2 \cdot 10^{-3}$ at $U_{10}/c_p \geq 1.45$, the level of the f^{-5} tail does not grow up further on average even if the wind forcing increases very significantly (Figure 5.42). As shown in Figure 5.27, such a situation does not actually mean that the breaking rates of short waves reach 100%, they stay essentially below this ultimate percentage. Therefore, this condition can only signify a change of the breaking severity. Thus, for the younger waves, the breaking reacts to the changing wind forcing by altering the dissipation rates rather than the rate of breaking occurrence, in its role of maintaining the spectrum-tail level (Figure 6.6).

At hurricane-like winds, that is for the very young waves, with the 100% breaking rate of the short waves occurring, the tail is still roughly at the same level range (see Section 7.3.5 and Figure 7.6). This fact further highlights the significant role of breaking in the formation of the wave spectrum as we know it. Now, the forcing is most extreme, but the spectrum level has only changed by less than two times (7.30). That is, the wave-breaking control still maintains approximately the same tail shape and balances the much grown wind input (7.31)–(7.32) by adjusting its severity up: that is the extra wind input is lost locally to the dissipation, but the spectrum tail more or less holds its level.

While talking about the saturation level, it must be mentioned of course that using the same dimensional argument as Phillips (1958), but assuming the role of the wind still to be essential, Toba (1973) obtained a different parameterisation for the spectrum tail:

$$F(\omega) = \beta u g \omega^{-4}, \tag{8.17}$$

i.e. he argued in favour of a quasi-equilibrium interval, whose level also depends on some characteristic wind speed u. Well before that, Zakharov & Filonenko (1966) achieved the exact analytical solution of the Hasselmann equation for the spectrum defined by interactions in the system of nonlinear waves, which also produced an ω^{-4} tail:

$$F(\omega) = c_K p^{1/3} g \omega^{-4}. \tag{8.18}$$

Here, the spectral level varies as a function of energy flux p across the spectrum, and c_K is the Kolmogorov constant (see Zakharov, 2010).

The observations were divided and produced both f^{-5} and f^{-4} parameterisations, e.g. JONSWAP spectrum ((2.7), Hasselmann *et al.*, 1973) and the Donelan *et al.* (1985) spectrum, respectively. The f^{-4} interval required more attention as the initial formulations (8.17) and (8.18) seemed contradictory: the first one associated the equilibrium with the external wind forcing and the second one with the internal nonlinear flux. The contradiction was basically resolved by Resio *et al.* (2004) who concluded, based on a comprehensive collection of field data sets, that

$$u = (U_{10}^2 c_p)^{1/3} - u_0 \tag{8.19}$$

where the first term defines the external energy flux. This flux can now be related to the internal flux p, and the contradiction is answered. Here, u_0 is a dimensional offset of the experimental dependences, which complicates the theoretical argument, but is apparently the robust feature of the experimental analysis by Resio *et al.* (2004).

In the meantime, based on experimental evidence and theoretical argument, there came understanding that both the f^{-4} and f^{-5} subintervals are present at the tail, the first one being closer to the peak (Forristall, 1981; Evans & Kibblewhite, 1990; Kahma & Calkoen, 1992; Babanin & Soloviev, 1998b; Resio *et al.*, 2004). Moreover, it became clear that the total wind input, integrated over the entire spectrum $F(f)$, cannot converge to the total wind stress known independently, unless the f^{-5} subinterval does exist (Tsagareli *et al.*, 2010; Babanin *et al.*, 2010c). For the oceanic waves, the transition was observed at frequency $\omega_t \approx (2.5 - 3)\omega_p$ (Forristall, 1981; Evans & Kibblewhite, 1990). Kahma & Calkoen (1992), based on a 'grand average' of many measurements of the saturation spectra from different data sets, indicated a dimensional value for such a frequency:

$$\omega_t \approx 5 \frac{g}{U_{10}}. \tag{8.20}$$

which can alternatively be expressed as

$$\tilde{\omega_t} = \frac{U_{10}}{c_p} \approx 5. \tag{8.21}$$

Now, with the view of (8.19) above, the transition frequency can be parameterised in terms of the peak frequency and characteristic wind speed. Eqs. (8.15), (8.17) and (8.19) can now be combined to give a dependence of the transitional frequency on peak frequency and wind speed:

$$\omega_t = \frac{\alpha}{\beta}\frac{g}{u} = \frac{\alpha}{\beta}\frac{g}{(U_{10}^2 g/\omega_p)^{1/3} - u_0}. \tag{8.22}$$

α can be defined by (5.75), and $\beta = 6.09 \cdot 10^{-3}$ according to Resio *et al.* (2004) .

In Figure 8.1, ratio ω_t/ω_p is plotted versus $f_p = \omega_p/(2\pi)$ (top panel) and versus the wind forcing U_{10}/c_p in the bottom panel. The range of changes of the latter corresponds to the peak frequencies of the top subplot, and the wind speeds shown are U_{10} = 10, 15, 20, 30 and 40 m/s (top to bottom).

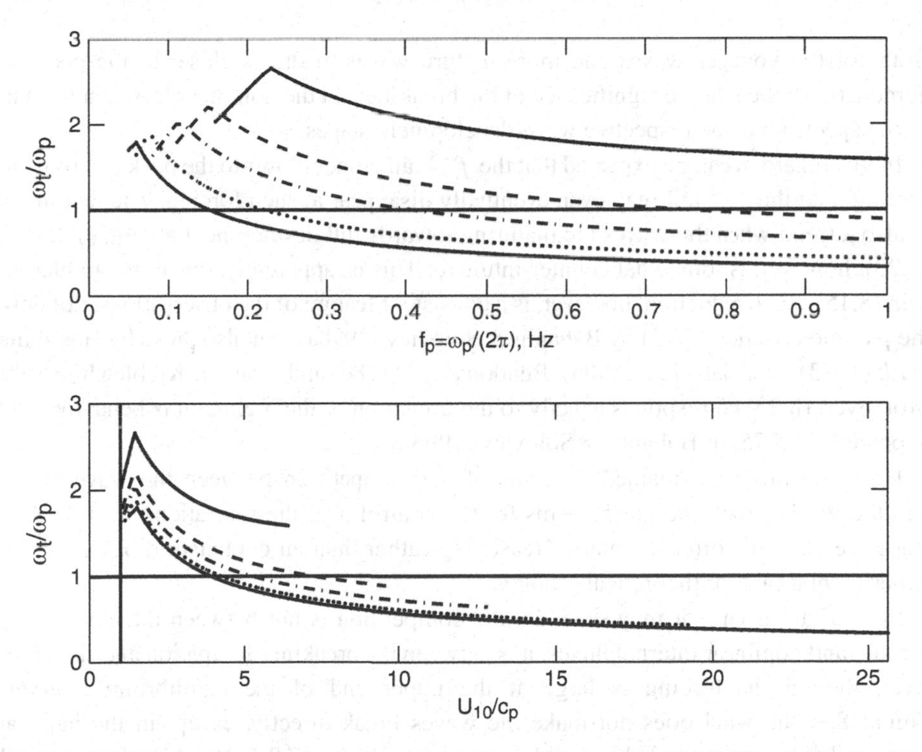

Figure 8.1 In both panels, top to bottom wind speeds are U_{10} = 10, 15, 20, 30 and 40 m/s. (top) Ratio of transitional and peak frequencies ω_t/ω_p versus peak frequency $f_p = \omega_p/(2\pi)$. (bottom) Ratio of transitional and peak frequencies ω_t/ω_p versus wind forcing parameter U_{10}/c_p

At typical oceanic winds of $U_{10} \sim 10\,\mathrm{m/s}$ and typical wave age $U_{10}/c_p \sim 1.2$, (8.22) produces the transition at $\omega_t \approx 2.2\omega_p$ which is close, even though slightly below the values of Forristall (1981) and Evans & Kibblewhite (1990) mentioned above. If dependence (8.20) is used, then for $U_{10} \sim 10\,\mathrm{m/s}$ and $\omega_t = 2.2\omega_p$ it gives $f_p = 0.35\,\mathrm{Hz}$ which is exactly what the corresponding line in the top subplot of Figure 8.1 indicates.

The highest transitional frequency ω_t, for individual cases over the five wind speeds, is in the range of $\omega_t \approx 1.8$–$2.7\omega_p$. The lowest transitional frequency is usually an asymptotic hypothetical value below the peak ($\omega_t < \omega_p$) which fact basically means that in those circumstances the f^{-5} interval will extend all the way to the spectral peak, i.e. there will be no f^{-4} subinterval.

Such condition (i.e. no f^{-4} subinterval) is never reached for wind speeds $U_{10} = 10\,\mathrm{m/s}$ (top subplot), that is at such wind the spectrum will always have the two subintervals. At all the other wind speeds, for wind forcing stronger than $U_{10}/c_p \approx 3.9$–7.2, there will be only the f^{-5} spectrum tail, and therefore no condition where the spectrum shape is controlled by the nonlinear fluxes (8.18) will take place.

In accordance with (5.75), the maximal transitional frequency ω_t will occur at

$$U_{10}/c_p \approx 1.45. \tag{8.23}$$

Both for the younger waves and more mature waves, it draws closer to the peak. This demonstrates the relative significance of the breaking and the nonlinear fluxes in formation of the spectrum at the respective wave-development stages.

In this regard, it can be expected that the f^{-5} tail comes closer to the peak at strong wind forcing, and the f^{-4} tail may even eventually disappear at such forcing. The fact that this is also a trend when the waves are maturing towards full development at U_{10}/c_p less than (8.23), however, is somewhat counter-intuitive. This is, apparently, due to the reduction of α in (8.15). Such reduction, however, is a consistent feature of the observations, not only of the parameterisation (5.75) by Babanin & Soloviev (1998a), but also those by Hasselmann *et al.* (1973), Donelan *et al.* (1985), Bandou *et al.* (1986) and Evans & Kibblewhite (1990). Moreover, (8.23) corresponds exactly to the transition of the regime of α behaviour in the dependence (5.75) of Babanin & Soloviev (1998a).

How can this be explained in terms of the competition between the wave-breaking and the weak-turbulence mechanisms for the control over the saturation interval? At this stage we can only offer a tentative reasoning rather than an explanation based on direct experimental facts or theoretical grounds.

First of all, even if it may seem so, the competition is not between the external wind forcing and nonlinear internal fluxes, it is between the breaking/dissipation and such fluxes. Even though the forcing is large at the upper end of the equilibrium interval in Figure 8.1, the wind does not make the waves break directly, except in the hurricane-like conditions mentioned above and described in Section 7.3.5. Breaking happens either due to hydrodynamic reasons, i.e. wave superposition or modulational instability, or is induced.

For the spectrum tail, the induced breaking, due to various influences of the large waves, dominates (see Section 5.3.2). The transition between the prevalence of induced breaking over the inherent breaking happens at approximately the same relative frequency $\omega_0 \sim 3\omega_p$ (see discussions in Sections 2.7 and 5.3.2) as the transition ω_t in this section (see also (7.46)). This means that most of the breaking which we see to control the f^{-5} interval is induced, and whatever the wind does to stimulate this breaking is done mostly indirectly through altering the long-wave conditions in the field.

On the other hand, this induced breaking for younger waves reacts to the changing wind-forcing conditions by adjusting the breaking strength, whereas the breaking rates as such do not necessarily change (see also Section 5.3.2 and current section, above). This is for $U_{10}/c_p > 1.45$. For the mature waves of $U_{10}/c_p < 1.45$, as was already discussed in this section, the induced breaking is likely to simply alter the rate of occurrence because of the wave age.

Why and how the dominant waves do that to the induced breaking of the short waves is yet to be understood, but one way or another, at both sides of the $U_{10}/c_p \approx 1.45$ wave age, the extent of the f^{-4} interval starts shrinking. This shortening leads to complete disappearing of this dynamic spectral range for younger waves, but for the mature waves such an interval controlled by the nonlinear fluxes always exists. Even at full development, $\omega_t = 1.6$–$2.1\omega_p$ depending on the wind speed (bottom panel of Figure 8.1). Transitional frequency $1.6\omega_p$, however, which occurs at full development at wind speed $U_{10} = 40$ m/s is very close to the spectral peak whose half-width is approximately 1.3–$1.5\omega_p$. Therefore, at very strong winds, the width of the f^{-4} interval, even when it exists, is quite short.

To conclude this section, we would also briefly mention that wave breaking can make other, perhaps even somewhat unexpected contributions to the spectrum tail. Willemsen (2002), for example, showed by means of the deterministic modelling of directional waves that the power of the dissipation function, if altered, influences the asymptotic wave tail. Breaking of large waves can actually generate short waves and ripples (e.g. Pierson *et al.*, 1992; Hwang, 2007, see also Section 2.7). Such a contribution is interesting, but is perhaps small and is yet to be quantified. Control of the level of the f^{-5} interval, which is achieved both through the breaking frequency and strength, and its competition with the nonlinear fluxes for the shape of the spectrum tail, is, however, a very important role of the breaking in wave fields with a continuous spectrum.

8.3 Wind-input enhancement due to wave breaking

The third subsection in this chapter dedicated to non-dissipative effects of wave breaking indeed deals with the phenomenon which is not associated with dissipation in any way (or, rather in any direct way, since any additional input in the balanced wave system ultimately leads to an additional dissipation in some part of the wave spectrum, see Section 9.2.1 below). This section is about the energy input to the wave system due to the breaking. This energy, of course, comes from the usual source, that is from the wind, but the input rates are essentially enhanced due to the presence of breaking. In this section, we will mostly follow the paper by Babanin *et al.* (2007b).

The topic focuses on the long-standing problem of the aerodynamic consequences of wave breaking on wind–wave coupling and was investigated in the more general context of the Lake George field experiment (Young *et al.*, 2005, see also Section 3.5). In this experiment, direct measurements of the influence of wave breaking on the wave-induced pressure in the air flow over water waves, and hence on the energy flux to the waves, were conducted. As described in Section 3.5, the forcing covered the range $U_{10}/c_p = 3$–7. The propagation speeds of the dominant waves were limited by the water depth and the waves were correspondingly steep (Young & Babanin, 2006b). These measurements allowed an assessment of the magnitude of any breaking-induced enhancement operative for these field conditions and provided a basis for parameterising the effect.

Overall, appreciable levels of wave breaking occurred for the strong wind-forcing conditions prevailing during the observational period. Associated with these breaking wave events, a significant phase shift in the local wave-coherent surface pressure was observed. This produced an enhanced wave-coherent energy flux from the wind to the waves, with a mean value of two times the corresponding energy flux to the non-breaking waves. Thus, it was proposed that the breaking-induced enhancement of the wind input to the waves can be parameterised by the sum of the non-breaking input and the contribution due to the breaking probability.

This aerodynamic effect had been investigated before on the basis of laboratory measurements (e.g. Banner & Melville, 1976; Reul, 1998; Giovanangeli *et al.*, 1999; Banner, 1990) and numerical simulations (Maat & Makin, 1992; Kudryavtsev & Makin, 2001; Makin & Kudryavtsev, 2002), but detection and quantifying the wind-input enhancement in field conditions was first reported following the Lake George observations (Young & Babanin, 2001; Babanin & Young, 2006; Babanin *et al.*, 2007b). It was expected that local air-flow separation accompanies wave breaking, and causes a phase shift of the wave-induced pressure, and that this significant modification to the near-surface aerodynamics can result in enhanced wave-coherent momentum and energy fluxes from the wind to the waves.

Instrumentation and measurement techniques for such a fine phenomenon is quite a complicated issue which was described in great detail in a special paper by Donelan *et al.* (2005), and will not be repeated here. In short, in order to measure microscale oscillations of induced pressure above surface waves, a high precision wave-follower system was developed at the University of Miami, Florida. The principal sensing hardware included Elliott pressure probes, hot film anemometers and Pitot tubes. The wave-follower recordings were supplemented by a complete set of relevant measurements in the atmospheric boundary layer, on the surface and in the water body. The precision of the feedback wave-following mechanism did not impose any restrictions on the measurement accuracy in the range of wave heights and frequencies relevant to the problem. Thorough calibrations of the pressure transducers and moving Elliott probes were conducted. It was shown that the response of the air column in the connecting tubes provides a frequency-dependent phase shift, which was then accounted for to recover the low-level induced pressure signal.

A subsequent paper by Donelan *et al.* (2006) described prevailing environmental conditions and presented new results on the physics and parameterisation of the spectral wind-input source function for the wave field. One of the major outcomes reported in this paper was the finding that the customary exponential growth rate parameter (fractional energy increase per radian) depended on the mean steepness of the waves (i.e. the wind-input term S_{in} in *RTE* (2.61) is a nonlinear function of the wave spectrum, with growth rates going up if the steepness was increasing).

Another major finding arose in the context of very strong forcing of steep non-breaking waves, where a condition of 'full' separation was observed. It was argued that the full separation occurs when the local surface curvature at the crest becomes too large for the pressure gradient normal to the wave to be able to balance the centrifugal acceleration of the wind layer in contact with the water surface. During the full separation of the wind flow over a steep wave crest, the streamlines detach near the steep crest and do not reattach until well up the windward face of the preceding wave towards its crest. The consequence is that the shear layer, which is normally adjacent to the surface, detaches and moves upwards to leave a 'dead zone' in the trough region between the crests. Thus, the external flow passes over the wave troughs and the imposed pressure pattern is weaker than in the usual case of non-separated flow. In a way, the wind 'does not know' how deep the wave troughs are and responds to what 'it thinks' are much smaller amplitude waves. The phase shift of the pressure maximum towards the re-attachment point on the windward face of the wave, however, becomes larger. It was not immediately obvious whether the combined effect would cause enhancement or reduction of the dimensionless wind input, but the quantitative estimates exhibited a significant reduction, compared to the extrapolated input at the same wave frequencies for the same wind forcing if the full-separation effects were not taken into account.

It should be specifically stressed that the full separation of Donelan *et al.* (2006) and the breaking-caused separation discussed in this section lead to different consequences. In this regard, the flow separation due to wave breaking considered here does not correspond to the full separation: it does somewhat increase the phase shift of the induced-pressure maximum with respect to the wave trough, but the flow does not pass over the wave troughs altogether as in the case of the full separation. As a result, there is an enhancement rather than reduction of the wave-induced pressure magnitude, plus the increased phase shift, and the flow separation due to breaking was always found to lead to enhancement of the wind input in the field conditions, as described below. Qualitative comparison of the two separation effects is sketched in Figure 8.2.

Things can prove different in the laboratory, that is in conditions when the wind forcing is strong and the waves are steep and also very short. The separation due to breaking can cause skipping of the wave troughs altogether, that is to bring about the full separation effect due to the breaking in such conditions. This is how Kudryavtsev & Makin (2007) explain the laboratory experiment of Donelan *et al.* (2004) with the surface-drag saturation. And this is why laboratory measurements of wind input S_{in}, as a function of wave steepness, indicate negative rather than positive correlation with the high wave

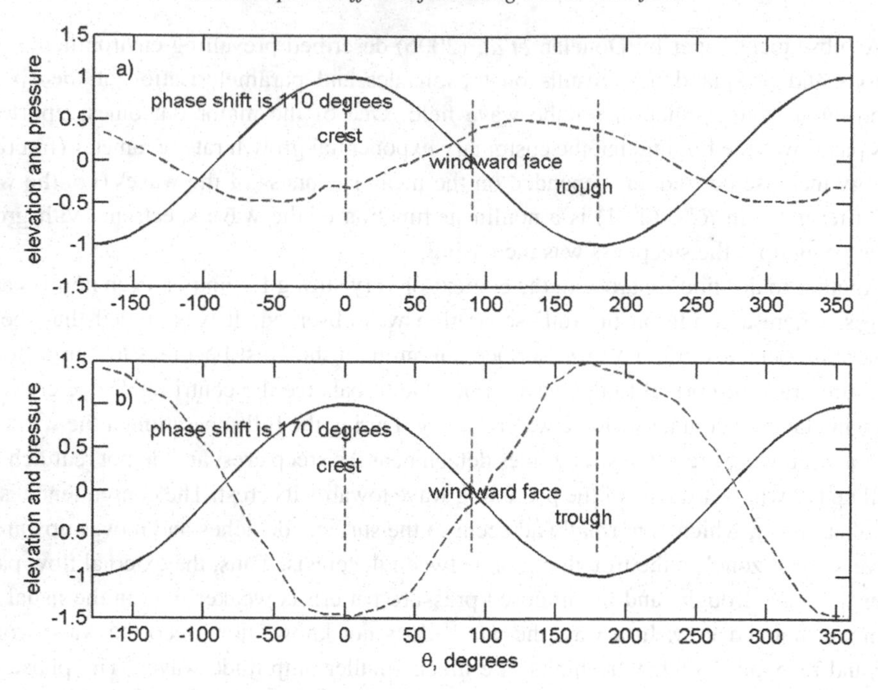

Figure 8.2 Sketch. Illustration of qualitative differences between the full flow separation (a) and separation due to wave breaking (b). Solid line signifies the surface water wave. Positions of crest, trough and windward face of this wave are indicated with vertical lines. Dashed line demonstrates the induced pressure wave. Vertical scales are arbitrary, vertical dimensions of the pressure waves in the two subplots are scaled relative to the water wave. Figure is reproduced from Babanin *et al.* (2007b) © American Meteorological Society. Reprinted with permission

steepnesses (Peirson & Garcia, 2008; presentation of Savelyev *et al.* at WISE-16, Ensenada, Mexico, 2009; see also Section 9.1.3). In discussion of Figure 8.4 below in this section, it is also demonstrated that, relatively, the enhancement effect decreases once the full flow separation starts to occur.

The wind-input source function proposed in Donelan *et al.* (2006) was parameterised by wave steepness and degree of separation, in addition to the traditional wind-forcing properties. This formulation was shown to be in agreement with, and in fact to be able to reconcile, previous field and laboratory data obtained for a variety of conditions in terms of wind forcing and wave steepness. Hence the steady-state strong-forcing conditions during the Lake George experiment provided a possibility to define a generalised wind-input source function S_{in} in (2.61) that is suitable for parameterising wave amplification through wind action for a wide range of conditions.

Since wave breaking can be very frequent (up to 60% in Babanin *et al.* (2001)), the breaking-induced separation has the potential to produce a noticeable enhancement of such atmospheric input to the waves and the function S_{in}. Apart from the wave growth as such,

Table 8.1 *Environmental conditions of records used. Here,* U_{10} *is the wind speed at* 10 m *height,* f_p *is peak frequency,* H_s *is significant wave height.*

Run	U_{10}, m/s	f_p, Hz	H_s, m
4	6.6	1.33	0.05
8	11.9	0.60	0.16
9	12.0	0.52	0.13
10	8.1	0.77	0.08
11	10.6	0.57	0.08
14	8.2	1.12	0.06
15	7.3	0.60	0.07
24	6.4	1.22	0.05

this issue is important for the alterations of the air–sea exchanges in these extreme situations (see Section 9.1.3). Indeed, Kudryavtsev & Makin (2001) and Makin & Kudryavtsev (2002) have parameterised breaking into their phase-resolvent model for ocean wind–waves and they attribute up to 50% of the total wave-induced stress to the breaking. Clearly, such model estimates need to be validated observationally.

The measurements were conducted at the Lake George field experimental site during active wind-generating situations. The prevailing environmental conditions for the set of records analysed here are summarised in Table 8.1.

The breaking-detection methodology relied on detecting enhanced acoustic noise at three bottom-mounted pressure sensors attached to the base of the wave-gauge array frame. The setup is described in detail in Donelan *et al.* (2005). The breaking waves generate an enhanced acoustic pressure at high frequencies which was sensed clearly by the collocated hydrophone (e.g. Babanin *et al.*, 2001; Manasseh *et al.*, 2006, where two different methodologies for breaking detection were investigated, see also Section 3.5). The same pressure was also detected by the pressure probes, and Babanin *et al.* (2007b) relied on a third method that uses these probes.

Thus, the breaking-detection procedure was based on 'hearing' the breaking, but it was also verified by visual means – i.e. by 'seeing' the breaking. This was done using video records of waves at the measurement location. The video record was taken at the rate of 25 images per second, the same as the sampling frequency of most of the other measurements. All these measurements were synchronised. Once a breaking event was registered, a zero-crossing analysis was used to identify that whole wave as a breaker and to measure its relevant properties (i.e. period, height and steepness).

Figure 8.3 illustrates the phenomenon, which is quantified in the subsequent figures in this section. Figure 8.3a displays the surface elevation η, measured by the wave resistance wire. In the segment shown, the waves around the 2nd, 7th, 12th and 14th seconds were identified as breaking by repeated viewing of the video record. Figure 8.3b demonstrates

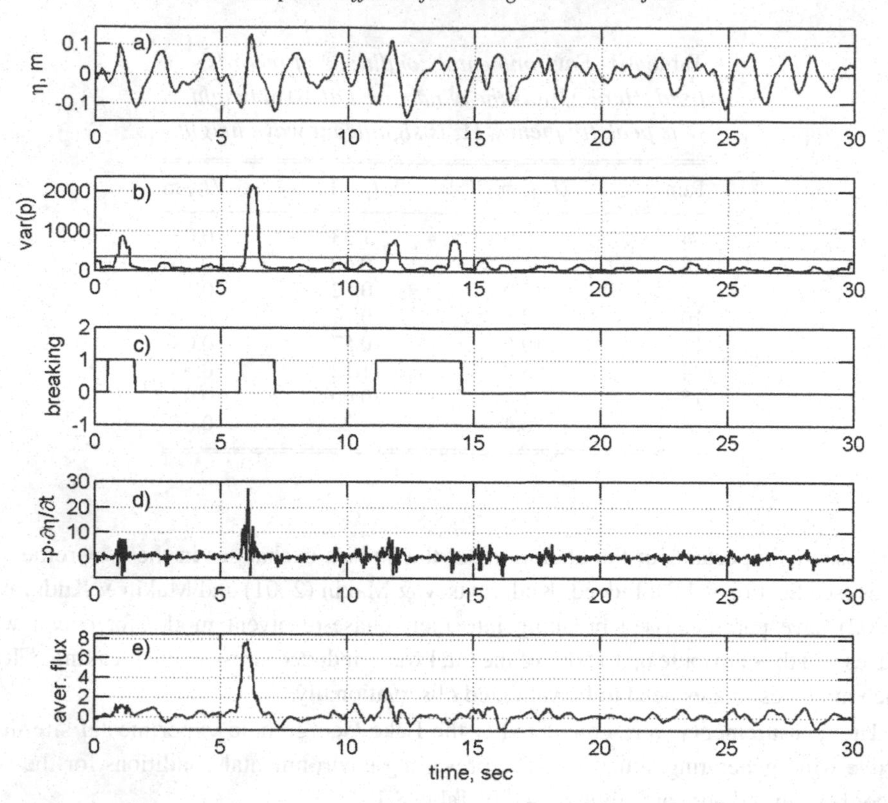

Figure 8.3 Data illustrate the breaking enhancement. a) Typical signal of the surface elevation η, measured by the wave resistance wire. b) Running average of the instantaneous pressure variance var(p) (high-pass-filtered pressure squared), based on an averaging interval of $0.25/f_p$. Breaking threshold, taken as a factor $b_t = 2.5$ times the run-averaged bottom-pressure-squared signal mean(var(p)) is shown as the straight line. c) Unit-step-function breaking indicator – from trough to trough of the wave in the panel a). 0 and 1 correspond, respectively, to 'no breaking' and 'breaking'. d) Instantaneous energy flux $p\frac{\partial \eta}{\partial t}$. e) Running average of the energy flux, based on the averaging interval of $0.25/f_p$. Figure is reproduced from Babanin *et al.* (2007b) © American Meteorological Society. Reprinted with permission

the ability of the bottom pressure probes to detect these same events. In this panel, the running average of the instantaneous pressure variance var(p) is plotted. This is the square of the high-pass-filtered pressure signal. The averaging interval employed to smooth the instantaneous property was chosen as $0.25/f_p$. The bursts in the bottom-transducer high-frequency pressure at appropriate moments are clearly evident, but require setting a relevant threshold to distinguish them above the background pressure/acoustic-noise in order to analyse the breakers routinely. This breaking threshold, taken as a factor b_t times the run-averaged bottom-pressure-squared signal mean(var(p)) is employed in the middle subplot, Figure 8.3c. This panel is a unit-step-function breaking indicator – from trough to trough of the wave in the top panel, where 0 indicates no breaking and 1 indicates breaking.

In Babanin *et al.* (2007b), b_t was taken as 2.5 throughout the analysis. This choice of b_t will be proved below.

In Figure 8.3d, the synchronous instantaneous energy flux $p\frac{\partial \eta}{\partial t}$ is plotted, where p is the instantaneous pressure detected just above the moving surface and $\frac{\partial \eta}{\partial t}$ is the partial time derivative of the elevation η. The wave-height signal, which in this measurement was sampled at $f_s = 50$ Hz, was smoothed using a 5 Hz low-pass filter. Again, the bursts are evident, but require a formal averaging procedure to quantify the integral enhanced-pressure effect.

This is done in the last panel, Figure 8.3e. To highlight the enhancement effect, this subplot shows the running average of the energy flux, based on the same averaging interval of $0.25/f_p$ as above. It is clear that the flux is enhanced over the second and third breakers, somewhat enhanced over the first one, but hardly at all over the last breaker.

Thus, the approach adopted by Babanin *et al.* (2007b) was based on measuring the flux over breaking waves, the breakers being detected on the basis of the acoustic noise emitted. The capability of the breakers to emit noise, however, depends strongly on the phase of the wave breaking (not to be confused with the phase angle of the wave). Classification of these phases was proposed by Liu & Babanin (2004) and discussed in detail in Chapter 2. There are four phases: incipient breaking, developing breaking, subsiding breaking and residual breaking. At the incipient stage, the water surface becomes unstable and the breaking starts, but little if any whitecapping is produced and therefore the acoustic or visual methods will not detect such a breaker. They will detect it at the developing and subsiding stages when the whitecaps are actively formed, the latter stage being characterised by the originally steep waves having lost much of their height and their steepness having dropped below the mean-steepness level. During the last, residual, stage of breaking, whitecaps are left behind, and this stage is not relevant here. Obviously, the first three breakers in Figure 8.3a are developing while the last one is subsiding, its steepness being no different to that of non-breaking waves.

The enhancement effect is expected to be due to flow separation over the steep breakers and hence will exhibit itself at the incipient and developing breaking stages, but not at the subsiding and residual stages. Therefore, it is not unexpected that there is no enhancement evident over the fourth breaker in Figure 8.3e. This kind of breaker will, however, be routinely detected by means of the acoustic-based technique and will tend to lower the overall enhancement value compared to the integrated energy flux over the entire wave-breaking set. Additionally, the acoustic technique will not detect the incipient breakers that may or may not produce a separated flow. If they do, energy fluxes over such waves will be integrated into the contribution of non-breaking waves and thus will lead to underestimation of the breaking-induced enhancement. Therefore, estimates made here have to be regarded as a lower bound.

The basis for the choice of the bottom-pressure threshold b_t for registering breaking events is justified in Figure 8.4. The data sets used to illustrate this are from three different wind-speed cases of Table 8.1: $U_{10} = 6.6$ m/s (record 4, circles), 8.1 m/s (record 10, x-symbols) and 11.9 m/s (record 8, plus-symbols). The bottom-pressure signal was processed as described above.

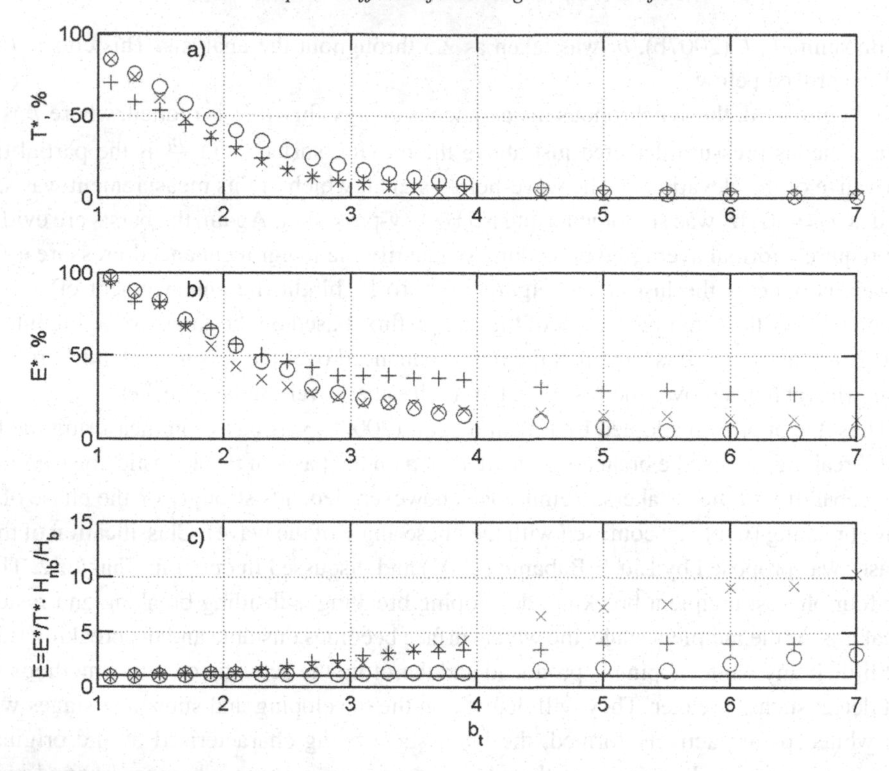

Figure 8.4 Basis for the choice of bottom-pressure threshold for registering breaking events. b_t (bottom scale) is the factor used as a multiplier to the run-averaged bottom-pressure-squared signal mean(var(p)). Data are from record 4 ($U_{10} = 6.6$ m/s, circles), record 10 (8.1 m/s, x-symbol) and record 8 (11.9 m/s, pluses) of Table 8.1. a) T^*, the observed ratio of breaking duration to total duration. b) Corresponding dependence of E^*, the relative contribution of breaking waves to the total energy flux. c) Normalised energy-flux enhancement ratio, defined as the ratio of $E^* = E_{br}/E_{tot}$ to $T^* = T_{br}/T_{tot}$. Figure is reproduced from Babanin *et al.* (2007b) © American Meteorological Society. Reprinted with permission

As already mentioned, the threshold b_t was sought as a multiplier for the run-averaged bottom-pressure-squared signal, mean(var(p)). When calculated, the product $b_t \cdot$ mean(var(p)) identifies a critical value for the running average of the high-pass-filtered bottom-pressure signal. If this value is exceeded at any instant, the synchronously recorded wave is considered breaking, its physical properties are determined by the zero-crossing analysis as described above, and the total energy flux over such a local wave is obtained by integration of the instantaneous flux $p\frac{\partial \eta}{\partial t}$ from the preceding trough to the following trough. We should mention again that the threshold was chosen such that waves above the critical-pressure property were definitely breaking, whereas waves below could possibly have been breaking or non-breaking. In carrying out this analysis, an iterative procedure was used to suppress the contribution of breaking events to the mean bottom pressure.

The subplots of Figure 8.4 demonstrate different wave and wind properties as functions of the magnitude of the chosen b_t value. Figure 8.4a shows such a dependence for $T^* = T_{br}/T_{tot}$, the ratio of observed breaking duration to the total duration. If b_t is chosen very low, all waves will be identified as breaking and the ratio will approach 100%. If b_t is excessively high, no breaking waves will be detected in the record and T^* will asymptote to zero. For high values of b_t well above the threshold, but below the ultimate value, the method will select particularly severe breakers, depending on the strength of their acoustic impact.

The acoustic method demonstrates the expected behaviour for the breaking duration in general, with noticeable differences between the three wind-forcing situations at low-to-intermediate values of b_t. For example, for $b_t = 1$, the method would identify some 80% of waves as breaking at the two lighter winds, but only 70% at the 11.9 m/s wind. The latter signifies a relatively smaller contribution of particularly 'noisy' events to the total noise, which is most likely caused by a higher level of ambient acoustic noise at strong winds.

The second subplot, Figure 8.4b, shows the corresponding dependence of $E^* = E_{br}/E_{tot}$, the relative contribution of breaking waves to the input energy flux to the waves. It immediately exhibits the expected effect of the enhancement of the wind input. If there were no enhancement, the ratio E_{br}/E_{tot} would follow the $T^* = T_{br}/T_{tot}$ dependence, tending to asymptote to zero for severe but rare events. It is clearly not the case for strong breakers with b_t greater than about 2.5, which means that the relative contribution of those events to the total flux is greater than their relative duration.

The bottom panel in Figure 8.4 is the most informative subplot in this figure, and underpins our choice of threshold level $b_t = 2.5$. In this panel, plotted as a function of the threshold property b_t, is the ratio $E = E^*/T^*$. This ratio should be 1 for the case when the flux E_{br} had no enhancement compared to the flux that would occur during the period T_{br} if the waves were not breaking. Since the breaking waves usually have larger amplitudes than those not breaking, to avoid any influence of wave steepness on the instantaneous energy fluxes, E_{tot} has been normalized by the significant wave height H_s of the non-breaking waves H_{nb}, and E_{br} by the significant wave height of the breaking waves H_b.

The ratio E is seen to be close to or below 1, up to a threshold value of b_t around 2.5, and corresponds approximately to the stage where the threshold starts detecting the breakers and not detecting the non-breakers. Furthermore, for $b_t = 2.5$, all waves with a bottom pressure exceeding this threshold certainly break. This was verified by viewing the synchronised video imagery for a representative subset of the records. Therefore, this bottom pressure threshold was adopted throughout the analysis to register the breaking waves.

Finally, with respect to this subplot, it should be noted that waves with acoustic signatures below the bottom-pressure threshold of $b_t = 2.5$ may or may not break – the bottom pressure-sensing system did not detect breakers reliably near the threshold. That is why the enhancement curves can depart from unity. These data were not taken into account in order to deal with genuinely breaking waves only.

It should also be pointed out at this stage that the T^* property of Figure 8.4a, taken at $b_t = 2.5$, should not be interpreted as the breaking rate. First of all, as noted above, waves falling below the threshold may or may not break and thus may or may not contribute to the rate. Secondly, T^* is related to a relative duration of breaking 'ringing' (over the wave period) rather than to a number of breakers, the former being an unknown function of environmental properties such as breaking severity, wind speed, perhaps water temperature and others (see e.g. Bortkovskii, 1997, and Section 3.1 for further discussions). And finally, if T^* is attempted to compare breaking rates for different records, the spectral distribution of breaking events may become an issue. For example, if for $U_{10} = 6.6\,\mathrm{m/s}$ it is mostly peak waves that are breaking and for the other records these are waves above the spectral peak, there will be different duration T^* for the same breaking rate.

Had a threshold level b_t above 2.5 been adopted, it would have significantly reduced the breaking occurrence-rate statistics. As may be seen in Figure 8.4b, the number of breaking waves with higher steepness decreases dramatically as the breaking threshold is increased. Thus, the threshold level of 2.5 is a purely empirical level and may only be applicable to this particular experimental setup with the prevailing mean water depth. In deeper water, a different value of b_t could have been needed to detect the breakers reliably if the acoustic-induced pressure above the background noise is reduced for the deep-water breakers.

It should be mentioned that other methods of deriving the breaking-detection threshold were attempted in the search for a universal high-frequency pressure property which would characterise the breaking. In particular, the rms level (averaged over the local wave period) of the high-frequency pressure fluctuations was considered as an indicator of the average breaking intensity over a local wave. It was expected that, if the rms background level of fluctuations at these frequencies in the certain absence of breaking (during light winds, for example) is subtracted, the remaining property would unambiguously identify the breaking. The background level was obtained from the high-frequency-noise histograms where it should have a high probability. It was found that the extreme values of the high-frequency-noise rms histograms clearly depend on the wind speed, i.e. (in arbitrary units) 2.4 for $U_{10} = 6.6\,\mathrm{m/s}$, 13.1 for $U_{10} = 8.1\,\mathrm{m/s}$, 35.8 for $U_{10} = 11.9\,\mathrm{m/s}$. Mean values of the noise also depend on the wind. Apparently, the background ambient noise at different wind speeds changes due to the presence of small-scale breakers and the detection threshold would have to be determined for each wave record individually.

It is instructive to highlight some consequences of Figure 8.4 that are further related to the topics other than the wave-breaking wind-input enhancement. Figure 8.4c shows qualitatively the dependence of the enhancement effect on the breaking severity, since higher thresholds imply that only more severe breakers are detected. It is apparent that the contribution to the ratio E in Figure 8.4c increases for more severe breaking events, implying a dependence of the enhancement on the breaking severity. Another interesting feature is that the enhancement effect itself increases for stronger winds (record 10 of Table 8.1), but reduces for the fully separated case (record 8 of Table 8.1 was classified as fully separated in Donelan *et al.* (2006)). Hence, there are some additional dependences underlying the mean value of O(100%) enhancement indicated in this section.

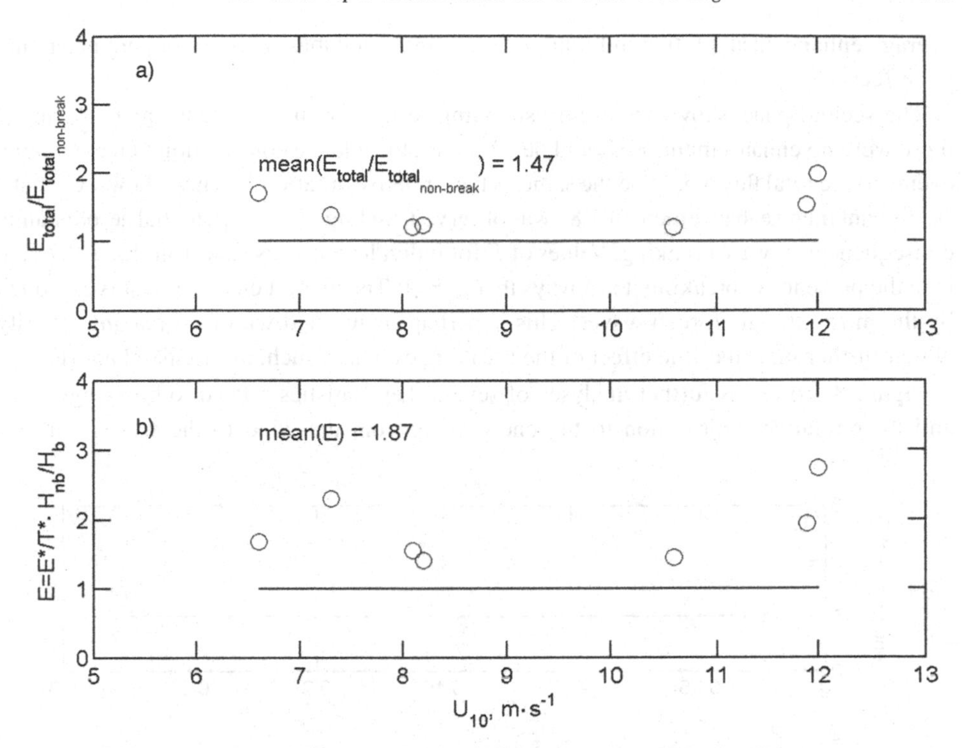

Figure 8.5 Measures of breaking enhancement due to the wave-coherent energy flux from the wind. a) Ratio of the total energy flux from the wind to the waves, to this flux in the absence of breaking, as a function of U_{10}. b) Corresponding results for the breaking enhancement to the wind input E. $E = 1$, shown with solid line, means no enhancement. Figure is reproduced from Babanin *et al.* (2007b) © American Meteorological Society. Reprinted with permission

Figures 8.5 through 8.7 of this section demonstrate different analyses of the flux-enhancement effect. In Figure 8.5, the average enhancement for the eight records of Table 8.1 is shown.

In the first panel, the ratio of the total energy flux from the wind to the waves T_{non-br}, to this flux in the absence of breaking, is plotted as a function of the wind speed U_{10}. To determine this ratio, the non-breaking part of the record was effectively 'stretched' to the whole length of the record:

$$E_{total_{non-br}} = E_{non-br} \frac{T_{tot}}{T_{non-br}} \tag{8.24}$$

where T_{non-br} is the total duration of non-breaking segments and E_{non-br} is the total measured energy flux during the T_{non-br} period. Values of the ratio for individual records vary from 1.2 to 2.0, with no dependence on the wind speed. Obviously, this ratio will depend on the wave-breaking rates which are a complex function of the wave spectrum and wind speed (e.g. Babanin & Young, 2005; Manasseh *et al.*, 2006; Babanin, 2009). Therefore, the average enhancement plotted in Figure 8.5b is a more suitable measure for parameterising the effect.

The second panel shows these corresponding results for the enhancement E. Again, if there were no enhancement, E should be 1 because the relative contribution of the breaking events to the total flux would be the same as their relative duration. We note, however, that a significant mean enhancement of 1.87 was observed, highlighting the potential aerodynamic consequences of wave breaking. Values of E for individual records vary from 1.4 to 2.7, that is in the presence of breaking it is always that $E > 1$. The highest enhancement is exhibited by the most-strongly forced waves. This is perhaps connected with the breaking severity which further magnifies the effect of the breaking events as such, as discussed above.

Figure 8.6 presents further analyses of several key statistics related to breaking waves and their relative contribution to the energy flux from the wind to the waves. Since a

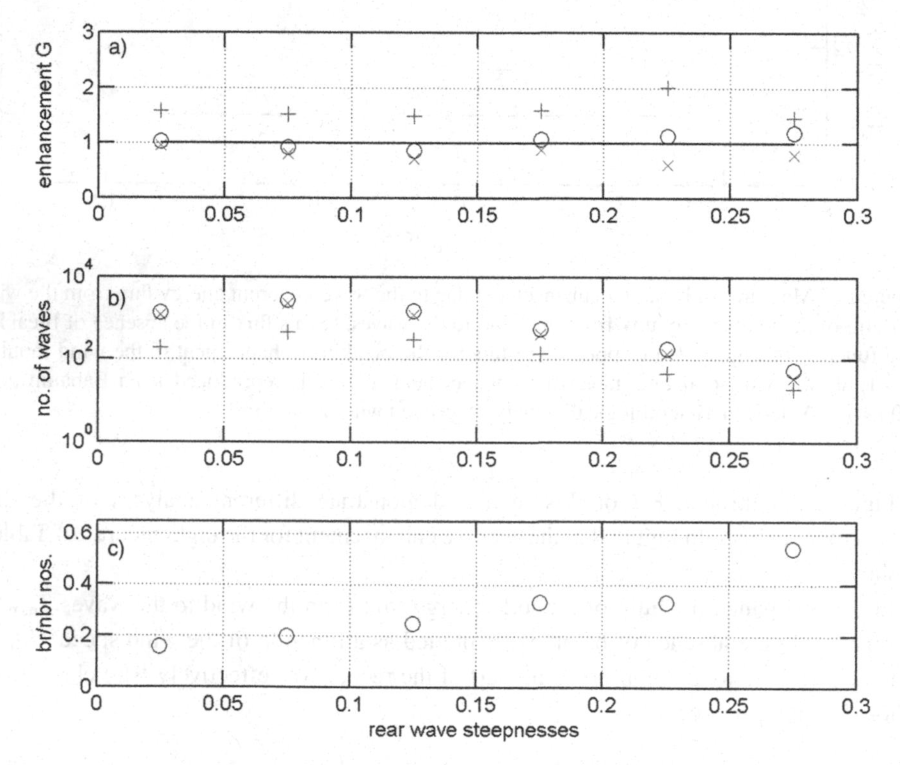

Figure 8.6 Key breaking-wave statistics and their relative contributions to the energy flux from the wind to the waves. In the top two panels, symbols denote breaking (pluses), non-breaking (x-symbols) and non-segregated (circles) waves. a) Mean energy flux from the wind as a function of the rear-face steepness. Energy flux to each individual wave was normalised by its steepness. b) Statistics derived from counting the waves in each steepness group. c) Ratio of counted breaking to non-breaking waves as a function of the steepness of the windward (rear) face of individual waves. Figure is reproduced from Babanin *et al.* (2007b) © American Meteorological Society. Reprinted with permission

dependence of the enhancement on the breaking severity was evident, a possible dependence of the breaking enhancement on wave steepness was also investigated. Individual waves were identified by their zero-crossings and their windward (rear) face steepness was calculated as

$$\epsilon_{rear} = \frac{H}{L} \tag{8.25}$$

where H is the rear crest-to-trough height and L is the rear crest-to-trough length (see also Section 2.9). L was determined from the time series as

$$L = \frac{t_L}{T}\lambda \tag{8.26}$$

where t_L is the rear crest-to-trough duration of the wave of period T. The wavelength λ was approximated from the period T on the basis of linear wave theory.

Contributions of the individual waves to the local mean energy flux $p\frac{\partial\eta}{\partial t}$ were then estimated. The energy flux to each individual wave was normalised by its rear-face steepness. To calculate the enhancement G for individual waves, this energy flux was divided by the mean energy flux for the entire record, normalised by the significant wave steepness defined as $\epsilon_s = \frac{H_s}{\lambda_p/2}$ where λ_p is the length of waves at the spectral peak:

$$G = \frac{E_{ind}}{E_{mean}}\frac{\epsilon_s}{\epsilon_{rear}}. \tag{8.27}$$

The influence of normalising by wave height was also examined. The results were, however, not sensitive to this choice.

Next, the individual waves were segregated into groups according to their steepness. The waves were separated into breaking, non-breaking and non-segregated categories, and then grouped according to their rear-face steepness. The energy-flux enhancement for each of these groups was estimated and averaged. This was done for dominant waves from all the available records. In total, this analysis included 6347 individual waves, 1132 of which were breaking.

The energy-flux enhancement is plotted as a function of the steepness for the full ensemble in Figure 8.6a. The result of this analysis demonstrates that the enhancement does not depend noticeably on the steepness. Thus, once an individual wave of a certain steepness breaks, the mean flux over that wave increases by approximately a factor of 2 compared to the flux over a non-breaking wave of the same steepness.

A further statistic derived from grouping and then counting the waves, according to their steepness and breaking/non-breaking characteristics, is shown in Figure 8.6b. A semi-logarithmic scale was used because the number of waves with large steepness decreases by two orders of magnitude. The distribution according to steepness of Lake George waves has a maximum in the steepness range 0.05–0.1 (2689 waves), and the number of waves rapidly drops towards higher steepness. It is interesting, however, that a significant number (54) of very steep waves, even as steep as 0.25–0.3, were detected.

A variant of this approach, the ratio of counted breaking to non-breaking waves as a function of the steepness of the windward (rear) face of individual dominant waves, is shown in Figure 8.6c. This figure shows the relative probabilities of breakers, conditioned on their rear face steepness. As one would expect, the larger the wave steepness, the more frequently they break. The ratio reaches 54% for waves in the steepness range 0.25–0.3 (percentage of breaking waves, if defined as the ratio of the number of breakers to the total number of waves, is 35%).

Next, a phase-average technique was used to analyse links between surface waves and pressure oscillations induced by them in the case of breaking and no-breaking. This phase-average technique was described and widely utilised in the wind-input study by Donelan *et al.* (2006). It employed the average pressure conditionally sampled on the phase of the surface elevation, in order to obtain a statistically robust relation between phases of the surface and pressure waves.

Phase-averaging techniques have been used for a variety of applications (see e.g. Hristov *et al.*, 1998). Here, this conditional-averaging method employs time series of the phase of a reference signal (surface wave) at a particular frequency to obtain an average profile of various flow variables sampled on the phase of the reference (e.g. pressure). For example, if the mean profile of the wave at a particular frequency (e.g. 1 Hz) is sought out of a non-filtered wave record, the record should be bandpass-filtered in a narrow band around the chosen frequency and then used to obtain time series of the phases of the 1 Hz surface elevations, by means of a Hilbert transform or wavelet analysis (see Section 3.7 for a description of those). In other words, at each instant we now know what was the phase of the 1 Hz surface oscillations.

The phase record can then be used to conditionally sample the original record to choose values of surface elevations, pressure, velocity, etc., in selected phase bins. That is, for each phase bin, for example, from 175° to 185°, whenever a value of the wave-phase belongs to this bin in the phase time series, we notice a respective value of the pressure at the same moment, from the pressure time series, and then add up and average these pressure readings. The mean and standard deviation within the phase bins then yield the conditionally-averaged pressure-flow variable on the phase of the wave component chosen.

This method provides an interesting and instructive insight into the behaviour of wave-induced pressure fluctuations relative to the surface waves here. It is a powerful data-analysis tool, operative at frequencies and signal-to-noise ratios well beyond the limit where co-spectral analysis fails to find any correlation between two related signals.

Figure 8.7 shows the breaking enhancement of the wind input to the waves from a phase-averaged perspective. In all the subplots, the upper lines are phase-average for 1132 breakers, the lower lines – for 5215 non-breaking waves, and the middle lines – for the non-segregated 6347 waves.

A Hilbert transform was used to determine the phase of the individual dominant waves. This required bandpass filtering the wave height signal around the spectral peak f_p in the spectral band $f_p \pm 0.1 f_p$. We note that bandpass filtering changes the wave height and steepness significantly, and this is important here, particularly for the breakers, which are

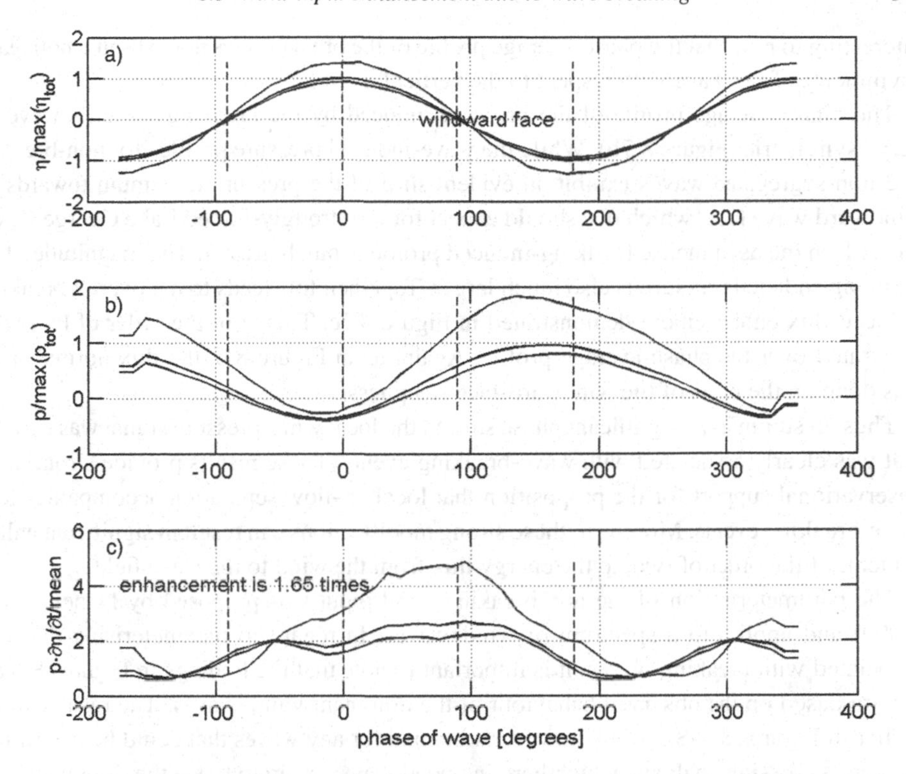

Figure 8.7 Phase-averaged breaking enhancement of the wind input for all the records of Table 8.1 (6347 waves of which 1132 were breaking). Each plot shows the distribution for all waves (middle line), for the breakers (upper line) and for non-breaking waves (lower line). a) Mean phase-resolved wave profile, b) surface-pressure distribution, c) energy-flux distribution – all obtained for the 36 phases resolved. Figure is reproduced from Babanin *et al.* (2007b) © American Meteorological Society. Reprinted with permission

strongly asymmetric and hence are smoothed the most by the bandpass filtering. Therefore, after determining the phases, the original wave records were used rather than the bandpass-filtered signals.

Once the time series of phases of individual dominant waves had been obtained, the same zero-crossing methodology as above was applied to single out the individual waves. For each wave, the instantaneous phases over the wave profile were placed into one of 36 groups, covering the entire 360°, and the instantaneous flux, wave elevation and pressure were registered for each phase group. Analysing all individual waves, the distribution of the average energy flux, the average wave profile and the average air pressure were obtained for the 36 phases.

The top panel of Figure 8.7 shows the mean phase-resolved wave profile for the 36 phases resolved. As expected, the average breaking wave is significantly higher and steeper than the average wave whereas the average non-breaking wave is marginally lower. It is interesting to note that the phase-average profile of the breaker does not exhibit a noticeable asymmetry of the wave with respect to the vertical.

The phase-average profile of the pressure induced by the breaking wave, however, is very asymmetric (Figure 8.7b). While the wave-induced pressure profiles for non-breaking and non-segregated waves exhibit an evident shift of the pressure maximum towards the windward wave face, which one should expect for the strongly-forced Lake George waves, the shift in the asymmetric breaking-induced profile is much greater. The magnitude of the breaking-induced pressure is also much larger. Together, this leads to the overall breaking-induced flux enhancement demonstrated in Figure 8.7c. This is of the order of two when integrated over the phase-average profile. Again, as in Figure 8.6, the flux normalisation was done on the basis of the windward-face steepness.

Thus, in summary, a significant phase shift in the local wind-pressure signal was detected that was clearly associated with wave-breaking events. These results provide strong field-observational support for the proposition that local air-flow separation accompanies local wave-breaking events. Moreover, these strong modifications can result in significant enhancement, of the order of two, to the energy flux from the wind to the wave field.

The parameterisation of the non-breaking wind input was proposed by Donelan *et al.* (2006) and applies to a spectrum of wind–waves. In regard to parameterising the input associated with breaking waves, it is important to note that the findings in Figures 8.6 and 8.7 are based on the observed behaviour of the dominant wind–waves. The results shown earlier in Figures 8.4–8.5 were, however, obtained for any waves that could be determined by a zero-crossing analysis, which here included waves of up to twice the dominant wave frequency (see Manasseh *et al.*, 2006).

While the present study has shown clearly that air-flow separation is operative for such breaking waves, it would be desirable to verify directly that this same effect is also operative for short breaking waves riding on much longer non-breaking waves, a commonly observed occurrence at sea. This will require an open-ocean version of a wave-following, near-surface aerodynamic-pressure measurement system, which is a particularly challenging measurement to make successfully.

If we adopt the standard mean value of two times the mean flux for the energy-flux enhancement, then this breaking-induced enhancement of the wind input to the waves can be parameterised as the product of the non-breaking input with this factor of two. This contribution then needs to be weighted by the breaking probability for these waves. Babanin *et al.* (2007b) proposed such parameterisation as

$$\gamma(f) = \gamma_0(f)(1 + b_T(f)) \tag{8.28}$$

where $\gamma_0(f)$ is the spectral wave-growth rate increment in the absence of wave breaking and $b_T(f)$ is the associated breaking probability (2.3). Here, $\gamma(f)$, $\gamma_0(f)$ and $b_T(f)$ are spectral functions (see Donelan *et al.*, 2005, 2006; Babanin *et al.*, 2001, for detailed definitions, respectively). In this regard, see also Section 5.3.2 for discussions of the problem of wave breaking in the spectrum based on field breaking-wave observations over a range of spectral scales.

Thus, to summarise the entire chapter, we described three effects in the wave system which relate to or, rather, are defined by the wave breaking and extend beyond the usual physical-oceanography applications of the breaking such as whitecapping dissipation. More effects which extend beyond such applications will be attended to or outlined in the next chapter, 9, dedicated to the atmospheric boundary layer and the upper ocean. Here, we only concentrated on those directly pertinent to the field of surface waves themselves.

Section 8.1 describes downshifting of the spectral wave energy in the course of wave breaking brought about by the modulational instability. Customarily, such downshifting in the wave system is attributed to the weak nonlinear interactions, but the instability-caused breaking can certainly contribute to this feature, if not dominate it. Indeed, as discussed in this section, the breaking happens at a much shorter time scale than the resonant four-wave interactions. The theory of the downshifting for two-dimensional waves, which is able to describe qualitatively and quantitatively evolution of the nonlinear groups, including breaking and post-breaking periods, is described in the section. So, eventually, it comes down to the question of whether the modulational instability is active in realistic directional (i.e. three-dimensional) wave fields, and the recent experimental progress on this topic is quite positive.

The next section, 8.2, attends to the oldest known other-than-dissipation effect due to wave breaking, i.e. the role of breaking in maintaining the level of the spectrum tail. Not only the breaking defines the shape of the wave spectrum at small scales, i.e. its well-known f^{-5} behaviour, it also holds a major responsibility for the magnitude of the spectral power in this frequency band. For younger waves, the tail level is approximately constant, and it is achieved through the breaking varying its severity in response to changing wind-forcing magnitude. For mature waves, starting from a wave age defined in the section, the level starts decreasing towards the full-development value. The wave breaking controls this by altering the breaking-occurrence rates. This wave age also potentially signifies change in the wind-input regime (see Stiassnie, 2010), and within the spectrum this dimensionless frequency defines the transition from the f^{-4} to f^{-5} interval, from inherent breaking to induced breaking (see Section 8.2).

In controlling the shape and level of the spectral tail, the breaking has to compete with another mechanism, nonlinear fluxes across the spectrum owing to energy input by the wind. This leads, except in very strongly-forced wind–wave fields, to the co-existence of the f^{-4} and f^{-5} subintervals above the spectral peak. In Section 8.2, the expression for the transitional frequency between these subintervals is obtained.

And the final section of this chapter, the current section, is dedicated to the energy-input rather than energy-dissipation capacity of the breaking. It is demonstrated, based on field measurements, that the energy flux over the breaking waves doubles. This fact can lead to a significant altering of the wind-input energy rates in the case of a substantial amount of breaking in a wave field, and the parameterisation of the respective wave-growth rates is described.

9
Role of wave breaking in the air–sea interaction

Wave breaking, apart from serving as the main dissipation source in the wave field, plays a number of other roles both on the ocean surface, and away from the surface. Chapter 8 was dedicated to such breaking-related other-than-dissipation features in the surface-wave system itself, and this penultimate chapter of the book, prior to Conclusions will extend the topic into a description of the roles wave breaking performs in air–sea interactions in a broader sense.

Section 9.1 and subsections are dedicated to the breaking-related effects in the atmospheric boundary layer. These are not limited to the wind–wave energy/momentum exchanges which have been discussed throughout the book. The breaking alters the sea drag, which then affects the near-surface air flow in general. The breaking is a source of spray, and the presence of the spray has a most essential influence on the boundary layer again. And finally, all these features have their peculiarities or even new behaviours, to a great extent unknown at this stage, in extreme conditions.

Section 9.2 also includes a number of subsections on important roles which the breaking plays in the upper-ocean dynamics and mixing. These cover descriptions of interface momentum and gas transfer, of which the breaking is part. In particular, two subsections are dedicated to the physical phenomena by means of which breaking facilitates such air–sea exchanges and mixing, i.e. the wave-induced turbulence and injection of the air bubbles under the water surface.

9.1 Atmospheric boundary layer

Like with any turbulent fluid flow over a boundary, the dynamic balance near the boundary is very different from that of the free flow (e.g. Holton, 1962; Komen *et al.*, 1994). In the boundary layer, the flow starts feeling the surface, and its dynamics changes essentially, even if there are no interactions other than friction against the surface.

In the case of geophysical atmospheric flow, the boundary layer can be further subdivided into sublayers, with different forces and their balances being important. Very close to the ocean surface, where the friction dominates and other forces can be neglected, the system of equations for the horizontal momentum reduces to

$$\frac{\partial \tau}{\partial z} = 0 \tag{9.1}$$

where τ is the wind-stress vector (3.7).

This defines the so-called constant-flux layer, that is the layer where the solution of (9.1) produced flux τ which does not depend on the distance to the surface. The height of this layer is of the order of 10 m (e.g. Komen *et al.*, 1994), and it may be less than that for light winds or extend a few tens of metres high for stronger winds.

It is within this sublayer that the wind–wave interactions happen. The main difference of the boundary layer over the ocean, compared to the atmospheric flow over the land and fluid flow near the wall in general, is that the ocean surface is mobile and ever changing. That is, the surface roughness is not constant, it evolves in response to the wind. In addition, the roughness elements are not stationary or even steady, they move and their motion accelerates as the system evolves (see e.g. Kitaigorodskii, 1973).

Indeed, the wind pumps energy and passes momentum to the waves, which grow and move faster as their spectrum peak downshifts. In turn, the waves can also alter the wind (see, for example, Sections 7.4 and 7.5). Most importantly, the waves, and breaking waves in particular, can alter the properties of this lower boundary layer itself. Thus, wind impacts the waves, which then impact the wind and the boundary layer, the effects of which then again impact the waves in this coupled small-scale air–sea system where such feedback influences can hardly be unambiguously separated.

The constant-flux-layer dynamics can be further subdivided into subdynamics, and the very constant flux itself consists of different contributions whose relative magnitude depends on the wind forcing and actually depends on distance to the surface. In (7.36), such contributions are subdivided into Reynolds turbulence stress τ_{turb}, wave-coherent-turbulence stress τ_{wave} and viscous stress τ_{ν}. The former vanishes at the surface, but dominates away from the surface, the latter dominates in light winds and becomes negligible in strong winds (e.g. Kudryavtsev *et al.*, 2001).

The wave-induced turbulence represents 100% of the turbulent stress on the air-side of the interface and rapidly decays upwards, becoming negligible at the height of the order of $\lambda/2$ (e.g. Donelan, 1999), which layer we will call the wave boundary layer (WBL). Here, τ_{wave} can be further subdivided into the contributions of regular waves and enhancement due to breaking (see Section 8.3). The breaking-induced stress constitutes up to 50% of the wave stress at moderate-to-strong winds (e.g. Kudryavtsev *et al.*, 2001). This fact highlights the role of the wave breaking as the most essential in the air–sea interactions.

At extreme conditions, the breaking takes on yet another role in the boundary layer. Makin (2005), for example, describes a situation when the sea drops saturate the near-surface air to such an extent that one can talk about a thin (about 10% of significant wave height) sublayer attached to the surface, with a yet different dynamics.

Subsections of this Section 9.1 outline these roles of the breaking in the dynamics and description of WBL. Section 9.1.1 will highlight the potential of wave breaking to alter the sea drag in parameterisations of the constant-stress layer. The generation of spray due to breaking is a very large topic, and it will be briefly reviewed in Section 9.1.2. The last subsection, 9.1.3, will indicate changes to the roles of breaking in extreme wind-forcing situations.

9.1.1 Sea-drag dependence on wave breaking

Coupling between the atmospheric boundary layer and the ocean surface is often parameterised in terms of the drag coefficient C_D (3.8), see Section 3.1. The very definition of C_D relies on the idea of a constant-flux layer outlined above. This concept has proved very consistent in general fluid mechanics when constant-speed flows over solid walls are considered. In the case of ocean waves, evolving simultaneously at multiple time scales from very long and continuous (slow growth due to wind input and nonlinear interactions) through to very short and intermittent (wave breaking), with their very complex physics and multiple mechanism for imparting feedback on the atmospheric flow, deviations from the assumed simple friction forcing can be expected, particularly as the winds are ever changing and gusty too, with a continuous spectrum of temporal and spatial inhomogeneities which may disrupt the very concept of the layer with a constant flux of momentum. Besides, at low wind speeds, the height of this layer can be less than 10 m, and (3.8) would not be valid. When applicable, however, knowledge of C_D enables a simple determination of the wind stress or the flux of momentum from the wind to the waves τ (3.7), if U_{10} is specified. In many models of the air–sea interaction, in fact, and particularly in large-scale models such as climate models, the sea-drag coefficient is the only property which defines this momentum exchange between the atmosphere and the ocean. Here, we will follow Babanin & Makin (2008) to introduce this property in a broader context of the air–sea-interaction properties.

Accurate evaluation of C_D has proven to be a major challenge experimentally, since it requires precise field measurements of fine turbulent fluctuations in the atmospheric boundary layer close to the wavy surface. The available field data have resulted in a number of quite different parameterisations. Guan & Xie (2004), for example, counted as many as 25 of them produced over a period of 50 years. Routinely, C_D is parameterised as a function of mean wind speed U_{10}, but the scatter of experimental data around such parametric dependences is very significant and has not improved noticeably over some 40 years, in spite of all the advances in the instrumentation, measurement techniques and data-logging systems. This scatter imposes a serious limitation on forecasts and predictions that make use of the sea-surface-drag parameterisations.

Also, since measuring in extreme wind and wave conditions is logistically particularly difficult, the majority of the data has been obtained during light to moderate winds. This, in addition to the scatter, further limits the applicability of the available parameterisations because their extrapolation into extreme conditions is questionable. Different physics is expected to drive the air–sea interaction in very strong wind-forcing conditions as has been shown in an escalating series of recent studies (see Section 9.1.3). This limitation is particularly important for modelling extreme events, such as tropical cyclones, and also for long-term climate prognosis, which predicts an increasing frequency of such events. It is felt, however, even in moderate-wind conditions as some parts of the continuous wave spectrum are always subject to strong forcing and thus further contribute to uncertainties and scatter of C_D estimates (Donelan *et al.*, 2006; Kudryavtsev & Makin, 2007; Tsagareli *et al.*, 2010).

Parameterising C_D in terms of mean wind speed U_{10} bears further deficiencies. The mean wind speed U_{10} does not define the wave properties unambiguously. For example, the mean or dominant wave height and length, the saturation level of the wave frequency spectrum and its directional spread, all are contributors to the sea drag and all can vary greatly for the same wind speed as a function of other properties and phenomena in the air–sea system, even in ideal wave-development situations. Depending on the duration of the wind action and on the wave fetch, the waves will evolve from short young seas into much longer old seas. Young waves are on average much steeper compared to the old ones, break more frequently and most of the other wave characteristics evolve too. This is known as the sea-state dependence, with U_{10}/c_p usually being the sea-state (or inverse wave age) parameter. A sea-state dependence in C_D has long been foreshadowed (e.g. Stewart, 1974), but only relatively recently has it been observed in field measurements. Some support has been found in a number of data sets (e.g. Smith *et al.*, 1992; Donelan *et al.*, 1993; Oost *et al.*, 2001; Drennan *et al.*, 2003), but notably not in others (e.g. Yelland *et al.*, 1998). Another effort at reconciling this fundamental issue has been on the basis of the dominant wave steepness (e.g. Oost *et al.*, 2001; Taylor & Yelland, 2001). The dominant sea waves, however, are known to play a relatively small direct role in determining the wind stress, except possibly for very young wind seas. Unless the waves are young, dominant waves are fast and their interaction with the wind is weak, but the effect of the dominant waves on the sea drag may be indirect – by means of modulating the shorter waves (Kudryavtsev & Makin, 2002; Hara & Belcher, 2002) or due to air-flow separation from breaking or non-breaking dominant waves (Makin & Kudryavtsev, 2002; Donelan *et al.*, 2006; Babanin *et al.*, 2007b). This highlights the need to understand more completely the basic physics of the sea-surface wind stress, including the wave-breaking effects, in order to parameterise it reliably in the form of a drag coefficient.

Many other effects can contribute significantly to the wind stress. The gustiness of the wind, which is always a feature of real wind fields, is accommodated in a number of theories (Janssen, 1986a; Miles & Ierley, 1998) and may result in either reduction of the stress or its enhancement, but certainly in increasing the scatter of C_D dependences if not accounted for (Babanin & Makin, 2008). The effects of the gustiness, however, are very difficult to take into account, even if parameterisation of C_D for these effects were available. Indeed, most of the measurements and the majority of models produce wind-speed variations at temporal scales which average the gustiness out, and therefore it remains unknown. Babanin & Makin (2008) suggested a way to bypass this difficulty. They estimated the minimal relative gustiness, which is approximately constant, and produced an experimental dependence for the maximal gustiness as a function of mean wind speed. That is, at each value of U_{10}, the maximal and minimal magnitudes of gustiness are known, and therefore the limits of its contribution can be estimated and parameterised depending on the wind speed.

The winds and waves are also non-stationary, which has been shown to have a major effect on estimating the wind input. Uz *et al.* (2002) concluded that the wind stress tends to be higher in decreasing winds than with increasing winds at a given wind speed, mainly due to the delayed response of the short waves to varying wind forcing (see Donelan & Plant, 2009). Skafel & Donelan (1997) demonstrated modulation of the wind stress by the passage of wave groups. Makin (1988) and Agnon *et al.* (2005) found wind-input oscillations due to nonlinear wind–wave interactions.

Another uncertain source of potentially significant stress variations is the swell present on the ocean surface. Dobson *et al.* (1994) did not find noticeable influence of the swell on the sea drag, whereas Donelan *et al.* (1997), Drennan *et al.* (1999) and Guo-Larsen *et al.* (2003) revealed significantly enhanced drag coefficients for cross-wind and, particularly, for adverse-to-the-wind swell. These features of swell were theoretically explained by Kudryavtsev & Makin (2004). Drennan *et al.* (1999), Smedman *et al.* (1999) and Grachev *et al.* (2003) observed negative stress – momentum flux from the waves to the wind – for swell following the wind. Potentially, swell can influence the dominant and short wind–wave spectra through hydrodynamic interactions, and through the interaction of the changed spectra with the wind – the sea drag (see also Babanin & Makin, 2008). Even in the absence of swell, the directional spread of wind waves themselves can alter the sea drag by up to 60% (Ting *et al.*, 2010).

This brief review of the long-standing problem emphasises the need for a complex approach to account for multiple phenomena that may simultaneously affect the sea drag. Babanin & Makin (2008) used the Lake George field data (see Section 3.5, Young *et al.*, 2005) in combination with the wind-over-waves coupled (WOWC) approach (Makin *et al.*, 1995; Makin & Kudryavtsev, 1999, 2002; Kudryavtsev *et al.*, 1999; Kudryavtsev & Makin, 2001) to address the complex processes, or rather some of them, associated with small-scale air–sea interactions, and therefore the sea drag.

Babanin & Makin (2008) outlined a list of physical properties and phenomena whose effect on sea drag should be investigated and incorporated in the final parameterisation in order to reduce the scatter. It includes, among other possibilities,

(1) mean wind speed;
(2) sea-state dependence;
(3) wave steepness;
(4) full flow separation for strongly forced wind–waves;
(5) enhancement of sea drag due to wave breaking;
(6) rising and falling winds;
(7) gustiness of the wind;
(8) temperature stratification in the atmospheric boundary layer;
(9) swell;
(10) nonlinear wind–wave interactions;
(11) wave horizontal skewness and vertical asymmetry;

(12) variation of the wavy-surface properties at wave-group and wavelength scales;
(13) wave directionality;
(14) wave short-crestedness;
(15) coupled effects in the air/sea boundary layers.

The 16th and separate item would be that due to peculiarities of air–sea interaction in extreme wind-forcing conditions which include an entire set of new features irrelevant in moderate winds (see Section 9.1.3 below). Here, we would also include the sea surface temperature explicitly, in addition to the stratification item (8) above (see argument in Bortkovskii, 1997). Even then, this list is far from exhaustive, and does not mention, for example, properties and processes which breach the validity of the constant-flux-layer approximation, as in such circumstances the notion of the drag coefficient (3.8) becomes uncertain. Since a significant number of large-scale processes in the atmosphere disrupt the constant-flux physics, parameterisations for the drag coefficient are bound to have some residual scatter.

This scatter, therefore, cannot be helped by the technological advances as it is not necessarily due to insufficient accuracy of the measurements. The bulk of this scatter is apparently due to the fact that, at the same wind speed, C_D can vary due to influences other than the wind as such. These influences are very many and the variations are very substantial.

To support this perception, Babanin & Makin (2008) used the approach which combined an experimental study with theoretical investigations conducted by means of the WOWC model. Experiment, although an ultimate truth, is hardly able to separate the effects of the multiple influences acting simultaneously. This can be done within the WOWC model, by switching on and off different physical mechanisms. If, for particular conditions, the experiment and the model produce identical or close results, it can be assumed that the physics included in the model is adequate for the relevant field circumstances. If, on the contrary, there are essential discrepancies between the measurement and the model, such cases should be scrutinised to find the cause.

Out of the many properties listed, Babanin & Makin (2008) considered the behaviour of C_D in terms of wind speed U_{10} and sea state U_{10}/c_p and, in addition to these traditional properties, looked at the effects that the rising and falling winds (6) and gustiness (7) (from the list above) have on the C_d parameterisations. The latter effects were found to be the major source of disagreement between the experiment and the model.

The Lake George finite-depth field experiment is well-documented in the literature and is also described in this book in Section 3.5. Here, we will only mention that an integrated set of instruments was deployed in all four relevant environments: the atmospheric boundary layer, the water surface, the water column and the bottom boundary layer. The waves were recorded with a stationary directional eight-probe wave array and by a set of mobile one-probe arrays which were used to record the short-scale spatial variability of wave trains. Detection of breaking events was also carried out by multiple means (Babanin *et al.*, 2001, see also Sections 3.5 and 8.3).

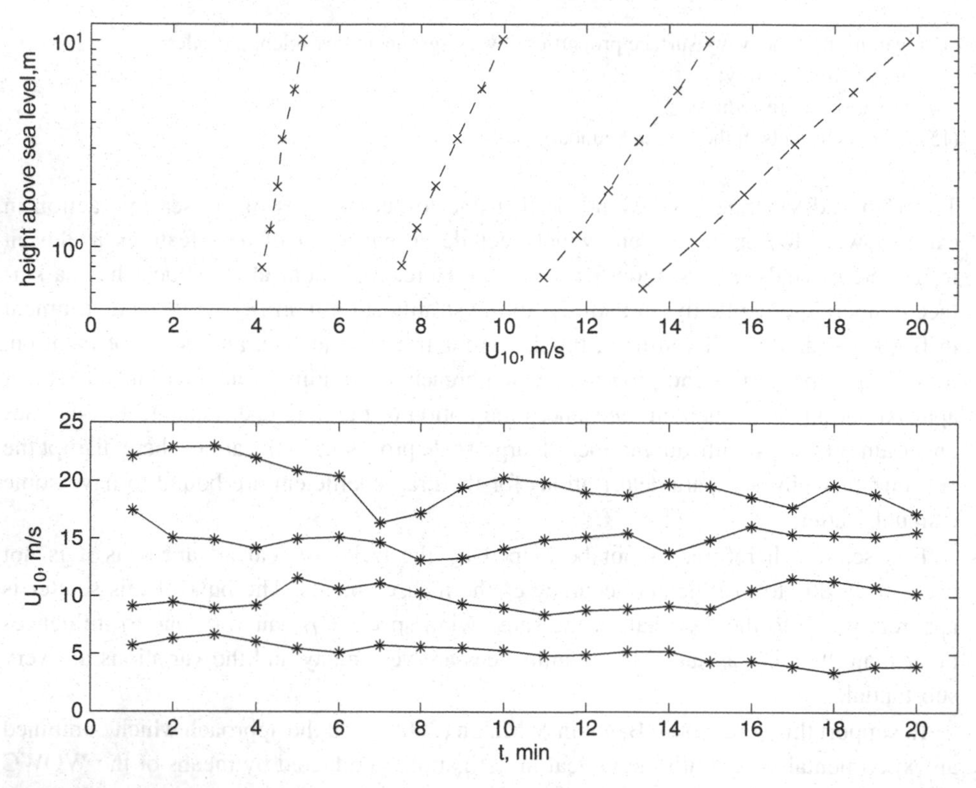

Figure 9.1 (top) 20 min-average measured wind profiles for wind speeds of U_{10} = 5.1, 10.0, 15.0 and 19.8 m/s, respectively. (bottom panel) One-minute-average wind speeds at the 10 m height for these records. Figure is reproduced from Babanin & Makin (2008) by permission of American Geophysical Union

Here, we are mainly interested in the air-side boundary-layer observations. The wind profile was obtained by means of the anemometer mast with six cup anemometers logarithmically-spaced from 10 m height down to 22 cm above the mean sea level. The wind directions were measured at 10 and 0.89 m heights. The wind probes were Aanderaa Instruments Wind Speed Sensors 2740 and Wind Direction Sensors 3590. The speed sensor provided 1 min-average wind speeds and gusts. The accuracy of the wind-speed measurements is 12% or 0.2 m/s whichever is greater. Additionally, for redundancy in the wind-speed and momentum-flux estimates, a Gill Instruments ultrasonic anemometer was also deployed on the mast and sampled the three-dimensional air velocity at 21 Hz rate.

The capability of the Lake George anemometer-mast measurements is demonstrated in Figure 9.1. In the top panel, profiles of the near-surface winds are shown for records of mean wind conveniently separated by 5 m/s: 5.1, 10.0, 15.0 and 19.8 m/s. No significant or systematic deviations from the logarithmic profile (3.19) are evident. Overall, correlation coefficients for the logarithmic profiles were above 99%. Therefore, stratification

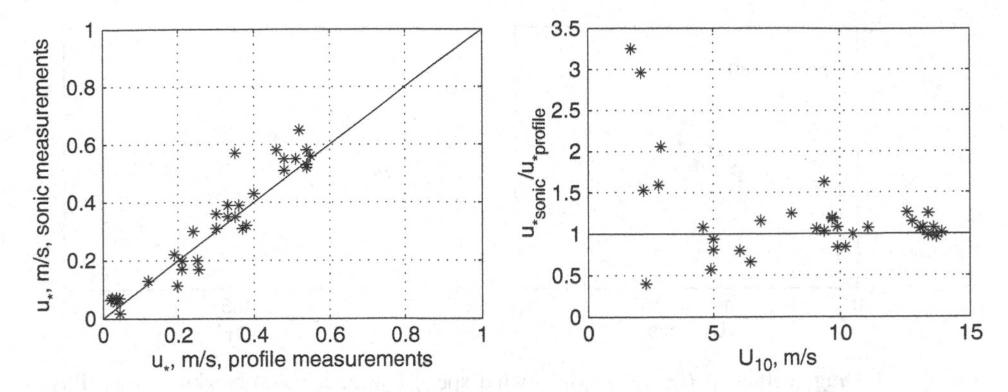

Figure 9.2 Comparison of friction velocity u_* obtained by means of stress measurements (sonic anemometer) and measurements of the mean wind profile in the boundary layer (vertical array of cup anemometers). (left panel) $u_{*\text{sonic}}$ versus $u_{*\text{profile}}$. (right panel) Ratio of $u_{*\text{sonic}}$ to $u_{*\text{profile}}$ versus U_{10}. Figure is reproduced from Babanin & Makin (2008) by permission of American Geophysical Union

effects were small and the Monin–Obukhov-scale function ψ can be neglected in (3.19). In any scenario, the stratification is expected to be insignificant as air–water temperature differences in the 1 m-deep Lake George were usually small, within a few degrees, with water temperature of the shallow lake fast-tracking any atmospheric changes.

In the bottom panel, one-minute-average wind speeds at the 10 m height for the four records are shown which exhibit both trends of the mean and the gustiness. Gustiness is the largest for the lightest wind and smallest for the strongest winds, as is generally the case (see Babanin & Makin, 2008, for details).

The selection of records in Figure 9.1, therefore, illustrates a variety of small-scale wind unsteadiness. This does not appear to cause deviations from the logarithmic-boundary-layer wind profile (3.19), but has the potential to affect sea drag in a serious way as will be outlined below.

The friction velocity u_* obtained from the wind profiles (3.19) can also be measured using the sonic anemometer:

$$u_* = \frac{\tau}{\rho_a} = \overline{-U'w'} \tag{9.2}$$

where U' and w' are oscillations of the horizontal (i.e. length of vector sum of the down-wind and cross-wind components) and vertical velocities, correspondingly. Thus, the constancy of τ and adequacy of the wind profile measurements, can be verified.

In Figure 9.2, u_* measurements by the sonic anemometer ($u_{*\text{sonic}}$) and by the anemometer mast ($u_{*\text{profile}}$) are compared (for the data, see Table 2 in Babanin & Makin (2008)). As seen in the left panel of $u_{*\text{sonic}}$ versus $u_{*\text{profile}}$, the scatter is significant, but overall matching in terms of absolute values of u_* is quite good, with a correlation coefficient of 95% and a sampling standard deviation of 0.06 m/s.

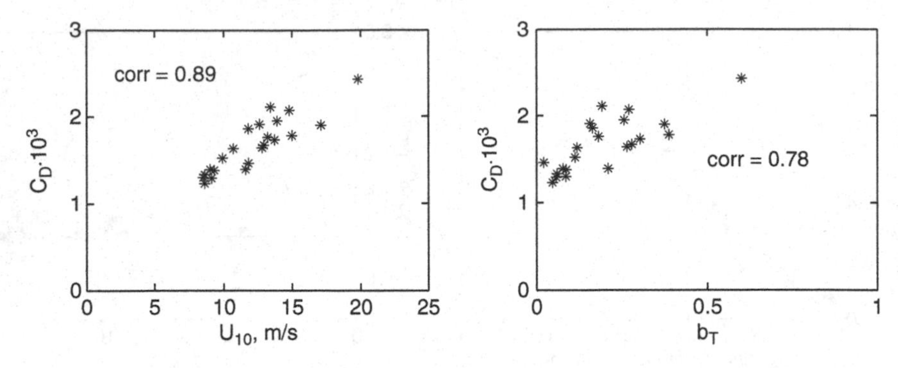

Figure 9.3 Drag coefficient C_D versus (left) wind speed U_{10} and (right) breaking probability of dominant waves b_T (2.3) – for wave fields with $b_T \geq 0.02$

The right panel, however, identifies a potential problem. In this subplot, the ratio of u_{*sonic} to $u_{*profile}$ is plotted versus U_{10}. At the low U_{10} wind speeds, the ratio can be as large as 3. This considerable scatter of the relative values of u_* shows that the physics of the constant-flux layer could have been affected by background processes in the atmosphere and at $U_{10} < 4\,\mathrm{m/s}$ the concept of the constant flux may not be valid over the ocean as mentioned in Komen *et al.* (1994). This can be due to the non-stationarity and non-homogeneity of the mean wind, which is very likely for fields of light winds (see, for example, Figure 8 in Babanin & Makin, 2008), or perhaps the height of the constant-flux layer can be less than 10 m. In any case, the light-wind records were excluded from the analysis below, as well as cases with very low breaking rates when plotting C_D as a function of breaking occurrence in Figure 9.3.

One of the findings of Babanin & Makin (2008), based on the Lake George data, were the limiting dependences of sea drag. These dependences envelope the Lake George data from below:

$$C_D = 1.92 \cdot 10^{-7} U_{10}^3 + 0.00096 \tag{9.3}$$

and

$$C_D = 9.33 \cdot 10^{-7} \left(\frac{U_{10}}{c_p} \right)^4 + 0.00096, \tag{9.4}$$

whereas in open-sea observations sea-drag values below these limiters are a regular occurrence. Dependences (9.3) and (9.4) provide some Lake George "ideal" relationships for C_D. At this stage, it is not obvious what physical properties constitute the ideal conditions. But almost any deviation from such conditions, at a given wind speed U_{10}, causes the drag at Lake George to increase. Babanin & Makin (2008) suggested that decrease of the drag with respect to the ideal conditions, which exhibits itself in a number of known open-ocean data sets, would be caused by a momentum flux back from the waves to the wind due to long waves outrunning the wind. In Lake George, swell and such waves do not exist.

Furthermore, based on comparisons of the model and experiment, Babanin & Makin (2008) revealed that the most distant outliers in their dependences of C_D on U_{10} and U_{10}/c_p were field cases with high wind gustiness. Once these obvious outliers were removed, some crude dependences emerged. They are in approximate agreement with some other known field parameterisations. The remaining scatter, however, appeared to be brought about by causes other than the wind instabilities only and does not asymptote to cases of the zero trend and minimal gustiness.

One of such causes can be wave breaking. As discussed throughout the book, breaking is only indirectly related to the wind, unless the wind is very strong (see Section 9.1.3), and for the same wind speed there can be different breaking rates (see, for example Figure 5.20). If breaking happens, however, it does affect the momentum and energy fluxes (Kudryavtsev & Makin, 2002; Babanin *et al.*, 2007b, Section 8.3). Therefore, correlation of the drag coefficient C_D with wave breaking can be expected.

In Figure 9.3, C_D dependences are plotted for records used by Babanin & Makin (2008), but only those in their Table 1 for which the dominant breaking rates b_T (2.3) were known and they were $b_T \geq 0.02$. As in Section 5.3.2 (see Figure 5.30), this is done to avoid bias and scatter due to zero-breaking contributions, which are not meant to belong to these dependences anyway, when the rates are low.

The left panel is a regular C_D versus U_{10} plot. It is instructive to note that there are no wind-speed values less than 7.5 m/s. This, again, points to the change in the wave-system dynamics at such wind speed, due to the absence of breaking, at least at Lake George. In Section 5.3.4, this change was discussed with respect to the dissipation in the water column beneath the waves (see parameterisation (5.74)), and now it is apparent with respect to sea drag and the air-side boundary layer.

The right panel of Figure 9.3 is a plot of C_D versus b_T. The correlation of 0.8 is lower than that of 0.9 for the wind, but it is still quite high in every regard and therefore the dependence is significant. This is particularly essential if we notice that the breaking as such is not an independent player in WBL, it only alters the drag imposed by the wind-generated waves when those are breaking. Such a large correlation of C_D with the breaking rates of dominant waves points out that these alterations, which underlie any existing drag-versus-wind dependence, need attention, examination and parameterisation to reduce the scatter.

Indeed, as we know very well, wave breaking is distributed across all wave scales, and regardless of whether they produce whitecapping or not the breaking events are likely to introduce the flow separations at respective scales and associated changes to the momentum fluxes. Therefore, at the very least, the link between sea drag and breaking is not as straightforward as a correlation between C_D and dominant breaking rates b_T. Whitecapping, when present, certainly changes the roughness of the surface locally, and its patches would affect the sea drag too, particularly if the whitecap coverage area is large. And the extent of the whitecapping plumes relates to the breaking severity which is not directly identified by the breaking-probability parameter in the right subplot of Figure 9.3 either (see Section 3.1). Also, in the turbulent wake of breaking waves, there is an active dissipation of short waves (see Section 7.3.4). Since those are the main contributors to the sea drag, then, depending on the frequency of breaking occurrences and their strength, this aspect of the breaking can also prove important for the C_D dependences and their scatter. The first feature would lead to enhancement of the drag, the last one should result in its reduction, and the whitecap coverage can potentially go either way, and these ambiguous roles of the breaking only add to the complexity and the importance of the topic.

Thus, it has to be recognised that a new approach to the problem of sea drag is needed. C_D does increase, on average, once the wind goes up, but it is not a simple function of mean wind speed and, therefore, attempts to parameterise it in terms of U_{10} or sea state U_{10}/c_p are bound to have a great scatter, no matter how extensive and how precise are the measurements of wind and waves.

Babanin & Makin (2008) suggest a complex approach to tackle the problem. In short, it is based on recognition of the complex nature of air–sea interaction at small scales, where multiple mechanisms affect and alter the sea drag simultaneously, sometimes in opposite directions. Wave breaking is one of such mechanisms, and the breaking alone can cause both increase and decrease of the drag depending on the relative importance of its multiple consequences.

Seventeen features are listed above which can influence the drag, and this list is not exhaustive. Babanin & Makin (2008) demonstrated that most significant exaggeration of the drag occurs due to wind trends and, particularly, due to wind gustiness. The breaking in Figure 9.3 also has a very high correlation with C_D, as has the directional spread of the wave spectra (Ting *et al.*, 2010). We believe that this complexity is an underlying reason for the observed large scatter of C_D data. Adopting this perspective, we should eventually be able to significantly reduce the scatter of C_D parameterisations. The contributing mechanisms, including the breaking, need to be singled out, studied separately, evaluated and then reunited in a joint parameterisation for the sea drag. An analytical approach should be applied to the mechanisms, wherever it is possible. This will enhance our understanding of their physics to create a complete picture of the complex phenomenon and to produce a general parameterisation.

9.1.2 Generation of spray

Sea spray is a large topic in its own right, and even at the present stage can easily constitute a subject for a separate book. And the present state is still characterised by the lack of experimental observations and knowledge, and by a selection of theories and numerical simulations, largely based on semi-empirical approaches, which, at the very least, are not necessarily compatible, and at the other end are even contradictory.

In the meantime, out of all topics in the field of air–sea interactions, this was one of the very first attended to both by experimentalists and theoreticians. Here, we will quote Andreas (2004) on the most significant role of sea spray in the dynamics of extreme wind–wave conditions and on the origins of its research:

"When sea spray droplets are thrown into the air, they accelerate almost immediately to the local wind speed. This process extracts momentum from the near-surface wind and therefore slows it.

When these droplets then crash back into the sea, they transfer this momentum to the sea surface as a surface stress. Thus, conceptually at least, sea spray has the potential to alter the near-surface distributions of momentum and stress on both the air and the water sides of the air–sea interface. Munk (1955) may have been the first to identify this process."

In the broader context of the effects of particles suspended in an incompressible gas on its dynamics, this physical problem had been scrutinised even earlier (Barenblatt, 1953, 1955; Kolmogorov, 1954). This general-fluid-mechanics approach has been successfully employed across a very wide range of geophysical applications such as dust storms, snow avalanches, and suspended sediment both in the ocean and rivers (e.g. Barenblatt & Golitsyn, 1974; Bridge & Dominic, 1984; Bintanja, 2000). With respect to wave boundary layer (WBL) in general and high-wind conditions in particular, these old papers have now been actively referenced too, and their theories are revisited and further advanced (e.g. Barenblatt, 1996; Makin, 2005; Barenblatt *et al.*, 2005; Kudryavtsev, 2006; Kudryavtsev & Makin, 2007, 2009, 2010; Rastigejev *et al.*, 2011). Other mechanisms of the sea-spray influence on the dynamics of WBL have also been explored over the years (e.g. Bortkovskii, 1973; Borisenkov, 1974; Ling & Kao, 1976; Korolev *et al.*, 1990; Lighthill, 1999, among others).

In this context, that is the influence of the suspended particles on the turbulent gas (air) flow, the dynamic impact of the particles (droplets) has two effects: redistribution of the momentum between the droplets and air–water as in Andreas (2004) quoted above, and damping the turbulence in the air. Kudryavtsev (2006) and Kudryavtsev & Makin (2009, 2010) point out that the second effect is stronger.

In general terms, this effect

"leads to suppression of the turbulent mixing and, as a consequence, to the acceleration of the wind velocity and suppression of the sea-surface drag"

(Kudryavtsev & Makin, 2010).

For modelling this phenomenon, more detailed knowledge on spume generation is needed in terms of the size of the droplets, the rate of their production and the altitude of their emission, and the shape of the spume cloud.

The role of the spray emitted into, suspended in and then taken from the air is, however, not limited to its contributions to the mechanics of air–sea interactions. Around the same time as the early 'mechanical' papers mentioned above, Riehl (1954) suggested that the spray plays a part in providing the heat necessary for the generation and development of tropical storms.

Different theories have been proposed to explain such heat fluxes over the years. Andreas & Emanuel (2001), for example, suggested a mechanism based on a combination of sensible and latent heat exchanges between the suspended droplets and surrounding air. That is, their spray particles cool faster than they evaporate, i.e. they give away their sensible heat while airborne, but are deposited back into the sea before extracting the latent heat from the air.

Thus, both mechanical and thermodynamic processes in the air filled by sea droplets are most important. Some of them can be considered separately, some cannot, as the mechanical–thermodynamic effects are often coupled. Vakhguelt (2007) and Chai & Vakhguelt (2009), for example, suggest a model of pressure-wave propagating through gas with liquid particles, and describe dissipation of the wave due to thermodynamic interactions. While this is set up as a more general problem, compared to the water droplets suspended in WBL where wave-induced pressure pulsations play the main role in the wave growth by the wind, physically the problem is the same.

Generally speaking, in this regard oceanographers can learn a lot from the fields of physics, technology and engineering where applications related to liquid particles suspended in gas are abundant (e.g. Borisov & Vakhguelt, 1981). For example, such an engineering problem as rocket combustion engines deals with liquid-fuel droplets in the gas chamber, and this problem has at least a few decades' head start in theoretical and practical terms, compared to detailed studies of tropical-cyclone physics which came into the spotlight lately.

A range of geophysical applications, which oceanographers can relate and refer to, have also been mentioned above in the context of suspended particles influencing the air/water turbulence. For example, Kudryavtsev (2006) elaborated his stratification approach for WBL, having started in general terms from the model of Barenblatt & Golitsyn (1974) for dust storms over the land (see also Goroch *et al.*, 1981). With respect to the other side of the mechanical influence of spray, its momentum-balance impact on the surface stress, analogies can be extended, for instance, to the effect of the rain in WBL (e.g. Caldwell & Elliott, 1971, 1972; Andreas, 2004).

Basic theory of tropical cyclones directly connects the mechanical and thermodynamic processes, including those roles played by the spray. Emanuel (1986) showed that the intensity of the cyclones depends both on the sea-drag coefficient C_D and on the enthalpy-transfer coefficient C_k. Curiously enough, the large-scale modelling foreshadowed saturation of C_D magnitude at high wind speeds well before this topic exploded in the field of wind-generated waves and small-scale air–sea interactions (see Section 9.1.3). That is, according to Emanuel (1986), the maximal wind speed should be inversely proportional to $\sqrt{C_D}$. Emanuel (1995) then concluded that, if the sea-drag parameterisations measured in moderate winds were extrapolated into high winds (where they had not been measured), the hurricanes would hardly be possible, certainly not beyond their lowest category.

This indirect-argument discussion of the sea-drag saturation is not free of controversy, and the investigations and discussions still continue. Bister & Emanuel (1998) and Emanuel (2003) modified the previous approaches to stress that the enthalpy and drag coefficients should become independent of the wind speed in extreme conditions. Makarieva *et al.* (2010), however, argued that previous derivations of the dissipative heating terms contradict the fundamental thermodynamic laws.

The main problem here, as in most of the other theories and models related to hurricanes, is the lack of direct measurements and observation-based estimates. Zhang *et al.* (2008) pointed out that, with respect to the ratio of C_k/C_D, no such measurements have been available for wind speeds in excess of 28 m/s. In this regard, the new laboratory study by Haus *et al.* (2010) should be mentioned, which extends the measurement range of this property up to 38 m/s wind speeds, scaled as environmental winds.

In this section, we obviously do not intend to review the full topic, that is the physics of spray, its measurements, theories and applications. Rather, we will just outline it and direct the reader to some references where he/she can find more details and further references from this research field.

Spray must be mentioned in a book dedicated to the effects of wave breaking, of course, as most of its existence is either due to wave breaking or is part of the wave breaking as such. Like almost anything related to the breaking, when details are looked into, a phenomenon splits into further subgroups and subphenomena, where different physical arguments and different approaches apply, and with different relative importance. In this regard, spray produced by breaking waves can be broadly classified into three groups (e.g. Andreas, 1998; O'Dowd & de Leeuw, 2007; Rastigejev *et al.*, 2011; Soloviev & Lukas, 2010); but note the fourth mechanism of droplet generation as a result of modulational instability of capillary waves (Shats *et al.*, 2010).

The first group (smallest size) and the last group (largest size) of droplets are routinely produced in benign wave conditions by the breakers. The tiny submicrometre- to several-μm-diameter droplets result from bursting of bubbles in regular whitecaps. These are sometimes called film droplets, but they can be further subdivided into two groups, one of which (the submicrometer sub-group) is indeed due to instability of the bubble-produced films, and the other (the supermicrometer sub-group) is caused by the jet formed in the centre of the bursting bubble.

At the other end of the scale are a few orders of magnitude larger droplets from plunging jets and spilling splashes. Neither of the two groups play an essential part in the momentum exchange and surface stresses. The first kind of droplets are too small, and the second kind are too few to affect the dynamics of the lower boundary layer and the interface. The film spray, however, is very important in producing aerosols (see e.g. Monahan *et al.*, 1986), with all the meteorological, applied and other consequences attached.

Their absolute-mass production may be increasing in high-wind conditions, but then their relative amount gives way to the second type of spray, the spume (Koga, 1981). These are particles of water resulting from direct tearing of wave crests by the wind. Their size is from tens to hundreds of μm, and in strong wind they fill the air within the lower WBL, from wave troughs up to a vertical distance above the wave crests of the order of wave height.

Quantitative account of this phenomenon, however, is not well defined. While there seems to be a general agreement that this effect is small in moderate winds up to some 20 m/s (e.g. Monahan, 1966; Wu, 1973), estimates for high-wind conditions differ. Pielk & Lee (1991), for example, found consequences of the spray significant in winds of the order of 40 m/s, whereas Fairall *et al.* (1994) concluded that it is negligible (i.e. of the order of 10%) up to 50 m/s winds. In a more recent model, Andreas (2004) suggests, on the contrary, that the same 10% of the spray's contribution to the total stress is achieved for a 30 m/s wind, and at 60 m/s it supports all of the stress.

Direct measurements in this field in general, however, are rare, particularly in the ocean, and particularly so are the measurements of the spray-production function and spray's effects on the stress. Therefore, the experimental knowledge is not always convincing and conclusive (see also Section 9.1.3 on further discussions of boundary-layer research in extreme conditions). Models, on the other hand, do help to reveal and analyse the physics, but without firm experimental guidance they have to rely on assumptions and indirect evidence and sometimes lead to conflicting conclusions or even contradictions.

Indeed, the measurements available were either conducted in the laboratory or, if in the field, primarily in surf zones (see O'Dowd & de Leeuw, 2007, for a recent review). Here, we would like to point out measurements by Bortkovskii (1977, 1980) which escaped this review and are particularly valuable as they were conducted in the open sea. Bortkovskii (1980) concluded that the structure of the two-phase (air–water) whitecapping medium defines both the spectrum of the bubbles in the water and of the droplets in the air which result from collapse of the bubbles. In any case, the typical range of the wind speeds involved in similar measurements is less than 10 m/s, with few exemptions up to 20 m/s (e.g. Geever *et al.*, 2005). These, therefore, mostly deal with film droplets as the spume production at such wind speeds is marginal if any.

Measurements at much stronger wind speeds are hampered by apparent practical difficulties. These are due to both the need for specialised and sensitive equipment and for that equipment to be enduring enough to withstand the harsh conditions, and due to the actual ability to conduct experiments in such conditions.

The equipment and measurement techniques for field observations of spray production are being developed and becoming more robust. These include both indirect methods, such as the whitecap method, refractory measurements and surf-zone profiles (see O'Dowd & de Leeuw, 2007), and direct measurement of the spray (e.g. Koga, 1981; Koga & Toba, 1981; Nilsson *et al.*, 2001; Geever *et al.*, 2005). The latter estimates the fluxes through the eddy covariance by using the sonic anemometer and a condensation particle counter combined.

Consistency of the measurements and their reliability, however, is still an issue. A number of experimental data on spume generation are available (e.g. Lai & Shemdin, 1974; Monahan & MacNiocaill, 1986; De Leeuw, 1993; Smith *et al.*, 1993; Wu, 1993; Andreas, 1998), but review of the spray-production function by Andreas (2002) (see also Schultz *et al.*, 2004) exhibited a variation by approximately six orders of magnitude. Using such estimates for testing and calibration of the analytical models is clearly a problem.

The discrepancies are not necessarily due to inaccurate measurements as such. As in the case of the sea drag parameterisations discussed in Section 9.1.1 above, the fact that spray production can be affected by many processes other than just the wind speed, would certainly contribute to the scatter, and the scatter will not be possible to reconcile unless these additional factors and features are taken into account. O'Dowd & de Leeuw (2007) argued that, at least for the film spray, sea-surface temperature is important too (see also Bortkovskii, 1997). Apart from the physical sources of spray, chemical source functions should be taken into consideration as well. Correspondingly, O'Dowd & de Leeuw (2007) report reduction of the discrepancy down to a factor of two.

These are measurements and estimates of the micrometer-scale spray. Knowledge of this spray is essential in meteorology where such spray can be responsible for condensation, in practical applications which rely on transparency of the atmosphere. This spray occurs even in more or less benign wind–wave conditions.

From the point of view of the dynamics of the boundary layer, the spume is more important, which occurs at higher winds, and the measurements become much more complicated in such circumstances (e.g. Anguelova *et al.*, 1999). The obvious retreat for measuring spume and WBL at extreme winds would be a laboratory wind–wave tank capable of producing hurricane-like winds. Such tanks are available and the dedicated experiments have been conducted.

Donelan *et al.* (2004), Fairall *et al.* (2009) and Mueller & Veron (2009) carried out tests in such tanks, with the centre-line wind speed in the flume varying in the range 0 to 30 m/s in the first experiment, and being 16.7 m/s and 17.8 m/s in the second. Obviously, the full boundary layer is not reproduced in such tanks, with the layer of air above the waves being of the order of 1 m under the tank's lid, but if the wind speed is doubled as an approximate estimate of the corresponding U_{10} value, which produces hurricane-like winds in both cases, definitely in the range well above $U_{10} = 20$ m/s, spume starts to be actively generated.

Given the challenge of the problem of direct physical modelling of extreme winds, both experiments are quite outstanding, but their direct applicability to real hurricanes is in question. Donelan *et al.* (2004), for example, did obtain the saturation of the surface drag at wind speeds greater than 30 m/s, as was observed in the field (Powell *et al.*, 2003; Jarosz *et al.*, 2007, see discussions in Section 9.1.3). Kudryavtsev & Makin (2007), however, (see also Kukulka *et al.*, 2007) argued that in the laboratory this saturation was due to reasons other than in the ocean. According to them, at the very short fetches the spray does not play a major role in the dynamics of WBL, and the observed drag reduction was due to air-flow separation over continuously breaking crests. This is not the enhancement effect described in Section 8.3, but an opposite effect as far as the momentum/energy fluxes are concerned – because of the small-scale waves, which contribute most to the drag, being under the separated bubble.

For the spume, similarly to the conclusion of O'Dowd & de Leeuw (2007) for the film droplets mentioned above, Fairall *et al.* (2009) suggest that u_* is not the only scaling parameter to describe the spray-production strength. According to them, particulars of the wave breaking can be important, such as whitecap fraction, breaking severity, length of the breaking crests. Speaking of the ocean conditions, we can add the sea-water temperature as another factor here (e.g. Exton *et al.*, 1986; Blanchard, 1989; Bortkovskii, 1997). These all may be different and bring about different contributions to the spray generation, but undistinguishable if scaled by the surface stress or u_*, particularly as the experiment was only conducted for two wind speeds. As the authors say:

> "we do not claim that the spray function we measure in the laboratory is the same as over the ocean. Rather, we assert that the wind stress interactions with breaking waves in the laboratory are reasonably similar to oceanic waves."

Thus, direct measurements in the field are necessary, but obviously difficult. While spray measurements from research vessels in hurricane conditions are hardly feasible, ocean sensing can be conducted by instruments mounted on offshore platforms (e.g. Forristall, 2000). In this regard, the spray-estimate method by means of laser wave probes suggested by Toffoli *et al.* (2010b) shows significant promise.

Down-looking lasers are often used these days to measure sea surface elevations. Typically, they are associated with high-precision non-intrusive sensing of the surface in the laboratory (e.g. Donelan *et al.*, 2004; Babanin *et al.*, 2007a), but they have also been operational for wave measurements from industrial oil rigs at least since 1984 (see e.g. Krogstad *et al.*, 2006).

These devices are based on the fact that the distance between the instrument and the sea surface can be measured as a function of the time-of-flight of a short pulse of infrared laser radiation, which returns to the instrument of origin after being reflected from the water surface. Typically, the received signal is interpreted in an on/off way, that is only used to measure the distance it travelled, when the return signal is received.

If the laser pulse crosses clouds of spray droplets on its way above the water surface, however, its intensity is attenuated (e.g. Hauser *et al.*, 2005; Dysthe *et al.*, 2008). Thus, it is reasonable to assume that an estimate of the sea-spray concentration can be extracted from a direct measurement of the intensity of the laser radiation. If properly calibrated, these laser sensors deployed on *in situ* platforms can be used for estimates of the total amount of spray. Since the industrial rigs are not evacuated and the instrumentation can operate autonomously, such field measurements can potentially be extended into the most extreme oceanic conditions.

Toffoli *et al.* (2010b) attempted to verify this conjecture through a number of laboratory tests in the ASIST wind–wave tank (see Section 5.1 about this facility). The effect of the sea spray on the intensity of a laser signal was tested, by means of an Optec laser, in a wide range of wind speeds, with corresponding magnitude of U_{10} from moderately strong 20 m/s to ultimately extreme 60 m/s.

For the sea-spray study, the facility was equipped with a high-definition LEG camera for estimating the sea spray volume and with a nearby down-looking laser. Results confirmed that attenuation of the laser intensity becomes more prominent with the increase of wind speed and hence of the amount of spray droplets above the water surface. This attenuation was calibrated against the direct measurement of the spray, and a simple regression model was suggested, to extract the sea-spray volume from the returned laser intensity.

In Figure 9.4, the sea-spray volume is presented as a function of U_{10}. In the experiment, only the total volume of suspended water was measured, and in order to be able to compare the measurements with data available in the literature, the size of the droplets was assumed to be 100 μm on average. Then, parameterisations of Andreas (1998) and Gong (2003) as well as the ASIST laboratory observations are shown. Overall, the concentration of spray droplets grows monotonically with the increase of the wind speed. For $U_{10} \leq 30$ m/s, the parameterisations and the data are quite close (given the great uncertainty of the parameterisations themselves as mentioned above). For even higher wind

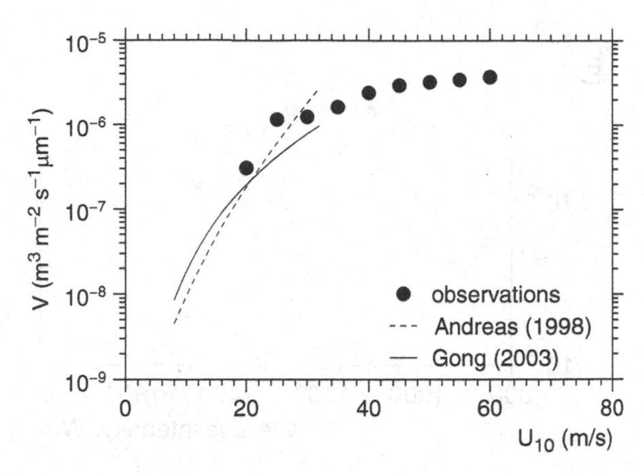

Figure 9.4 Sea-spray volume as a function of U_{10} wind speed. Dots – laboratory observations of Toffoli *et al.* (2010b); lines – parameterisations of Andreas (1998) (dashed) and Gong (2003) (solid)

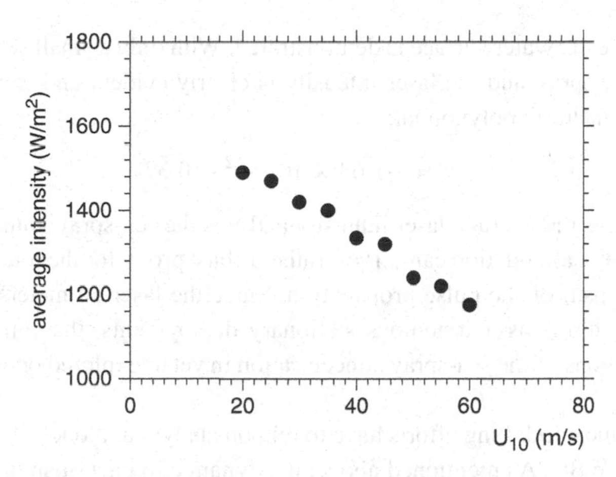

Figure 9.5 Average intensity of the laser signal as a function of the reference wind speed U_{10}

speeds, the field dependences and laboratory data are not expected to agree, as discussed above. These wind speeds were simply employed to generate larger spray concentrations in order to extend the laser calibration.

The attenuation of the laser intensity due to the droplet concentration is shown in Figure 9.5. Here, the average intensity is presented as a function of the wind speed U_{10}. With an increase of wind speed, and hence of sea-spray concentration (see Figure 9.4), the laser beam clearly decays more strongly. This confirms that there is a well-defined relationship between the magnitude of the returned laser radiation and the concentration of spume droplets.

Dependence between the laser intensity and the sea-spray volume is shown in Figure 9.6. Here, a regression model for an estimate of the amount of spray volume that remains in

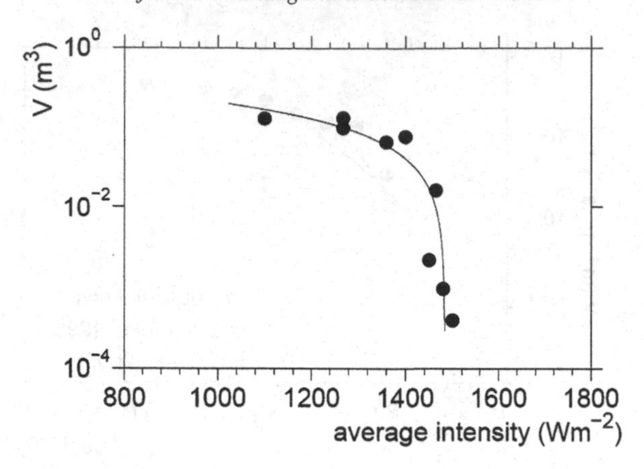

Figure 9.6 Sea-spray volume versus average intensity. Dots signify the experimental data, solid line is the polynomial fit (9.5)

suspension above the water surface is demonstrated. With only a small scatter, the relationship between the spray and the laser intensity is clearly evident and can be conveniently described by a quadratic polynomial:

$$y = -1.68 \times 10^{-7} x^2 + 0.37 \tag{9.5}$$

where x represents the average laser intensity and y is the sea-spray volume.

Thus, laser-pulse attenuation can serve a rather robust proxy for the total amount of spray in the air in the path of the pulse propagation. Since the laser altimeters can be operated during tropical storms as autonomous stationary deployments, the calibration (9.5) can provide observations of the sea-spray concentration in yet unexplored open-ocean extreme-wind conditions.

In the meantime, modelling efforts have to rely on analytical models of spray production and spray-filled WBL. As mentioned above, the dynamic impact of spume can be roughly divided into two parts: the momentum-damping effect and the turbulence-damping effect.

The momentum-balance effect of spray is most apparent, and we start this section from quoting its illustration: when water droplets enter the air flow they accelerate and, correspondingly, slow the flow down. To accommodate this effect in their model, Andreas & Emanuel (2001) split the total stress into two parts: the turbulent part (3.7) which they call 'interfacial surface stress' and spray contribution τ_{sp}. With respect to the latter, Andreas (2004) later argued that a misconception occurred in Andreas & Emanuel (2001) where the two stresses were just added up:

"Because the spray system is closed, however, this assumption violates the conservation of momentum."

Still, as in Andreas & Emanuel (2001), it was argued that

$$\tau_{sp} \sim u_*^4. \tag{9.6}$$

Andreas (2004) explained that within this approach the spray's role is to redistribute the momentum. As a result, the total stress does not change at a given wind speed and keeps growing towards high winds, but the role of the turbulent stress (3.7), proportional to u_*^2, diminishes by comparison with (9.6) towards winds at around 60 m/s as mentioned above.

Still, according to Andreas (2004), spray can affect the air–sea momentum coupling by providing a negative feedback on the wind–wave growth in response to increasing wind forcing. That is, as more spray is generated, which then falls back into the sea and knocks down the short waves, the relative stress is reduced because those short waves would support an extra stress should they persist.

Most of the research on the 'dynamic spray impact' lately has concentrated on the role of the suspended droplets in altering the physics of WBL directly. Makin (2005) started such studies by revisiting the old theory of suspended particles of Barenblatt (1955) and suggested a model where a thin (of the order of wave height) sublayer, saturated with 'light water droplets', appears very near the surface. Surface roughness seen by the outer sublayer changes, and so does the sea drag, with the effect of C_D being reduced at U_{10} in excess of 30 m/s as was observed in the field by Powell *et al.* (2003) and Jarosz *et al.* (2007). The dimensionless 'weight' of the droplets is determined by a combination of the terminal fall velocity u_{term} and frictional velocity u_*

$$\omega_w = \frac{u_{\text{term}}}{\kappa u_*}, \tag{9.7}$$

with $\omega_w < 1$ being light. Later, Kudryavtsev & Makin (2010) pointed out that the existence of such a 'light-water-particle' sublayer is lacking experimental evidence at this stage.

Barenblatt *et al.* (2005), Bye & Jenkins (2006), Kudryavtsev (2006), Kudryavtsev & Makin (2010) and Rastigejev *et al.* (2011) presumed that the particles were 'heavy' ($\omega_w \gg 1$) and damp the turbulence in the lower WBL. Estimates of Barenblatt *et al.* (2005) led to almost an order of magnitude increase of the wind speed as a result of this. When Rastigejev *et al.* (2011) attempted to reproduce these results, however, they did not find such a large growth of the wind. Their acceleration effect significantly depends on a choice of the maximal-spray-concentration function. From those available, it ranged from 7% for the slowest production term (Andreas, 1998) to 32% for the fastest (Wu, 1993).

An essential new outcome of the model by Rastigejev *et al.* (2011) is a critical value of the wind speed, at which the effects start to occur. For the fastest spray-production function of Wu (1993), it would be $U_{10_{\text{critical}}} \approx 30$–$40$ m/s, and for the slowest term of Andreas (1998) $U_{10_{\text{critical}}} \approx 50$ m/s. If so, then the faster model of Wu (1993) is more realistic since it leads to reduction of the sea drag at those wind speeds where such a reduction is actually observed.

Kudryavtsev (2006) suggested another way of interpreting and modelling the impact of sea droplets on turbulence in WBL, by means of the Monin–Obukhov approach to the temperature-stratified boundary layer (see Section 3.1). He argued that both the temperature stratification and the distribution of suspended spume affect the turbulence through the buoyancy force, and thus the Monin–Obukhov similarity theory, for the stable stratification, can be adopted in this situation.

Kudryavtsev (2006) found that the spume, if injected into the air at the level of breaking-wave crests, can slow down the turbulence mixing, and thus lead to acceleration of the wind speed and reduction of the sea drag as observed in the field. Outcomes of the model depended essentially on the assumed size of the droplets, and that needed further reanalysis. Also reanalysis was needed for situations when the density of the droplet–air mixture was becoming large and therefore a non-Boussinesq approximation of the momentum balance for such flow would have to be employed.

These revisions were done by Kudryavtsev & Makin (2007, 2009, 2010). One essential outcome of this series of research studies was a conclusion that the second dynamic effect of the spume, that is the attenuation of the turbulence due to density stratification produced by the suspended droplets, dominates. The momentum effect is still present, but in terms of its influence it can be disregarded by comparison with the other.

The governing equations for conservation of mass and momentum in Kudryavtsev & Makin (2010) are written as

$$\frac{\partial}{\partial z}(\overline{s'w'} - a\overline{s}) = V_s \tag{9.8}$$

and

$$\frac{\partial}{\partial z}(\overline{\rho}\overline{u'w'}) = 0. \tag{9.9}$$

The model is two-dimensional, and u' here is the x-component of the horizontal-velocity oscillations. The Reynolds terms in (9.8) and (9.9) are the vertical turbulent kinematic fluxes of the droplets and momentum, respectively; $\overline{s}$ and $\overline{\rho}$ are the mean droplet concentration and two-phase-fluid density, correspondingly, and V_s is the volume source of the spume – the total volume of spume created per unit of time and per unit of volume of air.

Equation (9.9) is the constant-momentum flux equation (9.1), but for WBL with density $\overline{\rho}$, and it turns into (3.7) far enough from the surface layer filled with spray:

$$\rho_a\overline{u'w'} + \Delta\rho\overline{s}(z)\overline{u'w'} = \rho_a u_*^2. \tag{9.10}$$

Here, $\Delta\rho = \rho_w - \rho_a$, and closer to the surface the difference with (3.7) is due to the second term on the left-hand side which describes the contribution of the droplets into the momentum balance.

(9.10) can be rewritten in terms of what Kudryavtsev & Makin (2010) call 'local friction velocity' v_*:

$$v_*^2 = \frac{\rho_a}{\rho}u_*^2 = \frac{u_*}{1 + (\Delta\rho/\rho_a)\overline{s}}. \tag{9.11}$$

The ratio at the bottom of (9.11) is of the order $\Delta\rho/\rho_a \sim 10^3$, and if $\overline{s} \sim 10^{-3}$, which it can be, the stress is halved. That is the spray–stress impact may be very significant.

Thus, to conclude and briefly summarise this section, we can say that there is a general consensus that the effects of spray are important across a broad range of applications and processes in the atmosphere, from being a condensation seed in meteorological phenomena to altering the dynamics of the lower boundary layer in the air–sea-interaction physics.

Depending on the classification, three to four different types of spray can be mentioned, all of them due to wave breaking. (Note that in the case of Shats *et al.*, (2010), this is the breaking of capillary waves.) Thus, in this regard the breaking takes on a different role in the general system of air–sea interactions.

The spume's contribution to the dynamics of the wave boundary layer is most significant at wind speeds in excess of 20 m/s. At winds greater than 30 m/s this leads (perhaps in conjunction with some other, e.g. aerodynamic effects), to the reduction of the sea drag.

As far as the spume in such conditions is concerned, models of its impacts have been developing intensively, and at the present stage appear quite feasible in describing the dynamics of the droplet-filled WBL. The persisting problem here is calibration of such models and spray impacts. They rely on spray-production terms which are diverse and not consistent in their estimates and predictions. Those would need experimental guidance, but field data are largely unavailable, and the laboratory measurements do not seem to be able to fill the gap.

9.1.3 Boundary layer at extreme breaking

The term 'extreme conditions' which we have loosely used in this chapter, requires definition. Obviously, what is extreme from the point of view of one set of applications can be treated as moderate or even ordinary in other regards. This book is dedicated to wave breaking, and from this perspective we will call the circumstances extreme when the U_{10} wind exceeds 20 m/s.

Indeed, as discussed in Section 9.1.2, it is at this wind speed that spume starts being torn from the top of wave crests to fill the lower boundary layer. This reduces the wave height and takes energy away. Thus, by definition, this is the wave breaking, but this is a new kind of breaking not seen at wind speeds below, when the breaking behaved differently and mostly affected the ocean-surface features and dynamics of the upper ocean rather than the atmospheric boundary layer. Even though the breaking in non-extreme winds does influence the wind input and therefore the sea drag (Section 8.3), at low breaking rates the overall effect is small.

Even visually, this threshold signifies a transition to different surface coverage by breaking and its consequences. Sporadic whitecaps at strong wind forcing in Figure 1.1 are replaced by a surface where the foam extends from crest to crest and the interface becomes hazy in Figures 1.5 and 1.9.

New roles of the breaking in such extreme conditions are yet to be fully understood. Their traditional roles, those parameterised as dependences for frequency of breaking occurrence, for breaking strength, for whitecapping production and coverage, also need to be re-evaluated and perhaps re-quantified.

As a result of the breaking and its consequences, as well as other physical processes close to the air–sea interface, which are either altered or only appear in extreme conditions, the sea drag saturates at wind speeds in excess of 30 m/s and potentially even goes down at even higher winds. This effect was long anticipated based on larger-scale air–sea-interaction dynamics (e.g. Emanuel, 1995, see also discussion in Section 9.1.2), and lately it was confirmed experimentally (Powell *et al.*, 2003; Jarosz *et al.*, 2007) where the critical wind speed was found as

$$U_{10} = 32\text{–}33\,\mathrm{m/s}. \tag{9.12}$$

Powell (2007) reanalysed his earlier data and detailed the previous conclusions. He now provided distribution of the possible values for sea drag rather than a unique dependence on the wind speed. The range of wind speeds analysed was $U_{10} = 20\text{–}79\,\mathrm{m/s}$. It was now concluded that the drag coefficient increases linearly with the wind speed until it reaches the maximum of

$$C_D = 2 \cdot 10^{-3}, \tag{9.13}$$

at $U_{10} = 41\,\mathrm{m/s}$. It then starts decreasing down to the minimum of

$$C_D = 0.6 \cdot 10^{-3}, \tag{9.14}$$

at $U_{10} = 61\,\mathrm{m/s}$. Values (9.13) and (9.14) are quite low in absolute rather than relative terms, and can be exceeded even in benign wind-forcing conditions of $U_{10} \sim 10\,\mathrm{m/s}$ (e.g. Babanin & Makin, 2008). Therefore, one can wonder whether the high winds indeed quench the sea drag so much or if there is some systematic measurement bias here.

Qualitatively, however, the account provided by Powell (2007) for the non-uniform distribution of the sea drag within the ocean surface covered by a tropical cyclone is most interesting. Within a 30 km-distance from the hurricane centre, the measurements showed low values of $C_D \approx 1 \cdot 10^{-3}$, with no dependence on the actual wind speed in the hurricane. The (9.13)–(9.14) dependence is therefore for the drag in the outer zone of the tropical cyclone. In this outer zone, the 'increase-then-decrease behaviour was found to be confined to the front left sector'. As Powell (2007) points out this is the region where the wave field is characterised by swell presence (see also Young, 2006). Dependence of the sea drag on the wind, for the wind-sea crossed by swell, is very complicated in general (e.g. Dobson *et al.*, 1994; Donelan *et al.*, 1997; Drennan *et al.*, 1999; Smedman *et al.*, 1999; Grachev & Fairall, 2001; Grachev *et al.*, 2003; Guo-Larsen *et al.*, 2003; Kudryavtsev & Makin, 2004, see also Section 9.1.1), and apparently this complexity is greatly enhanced in tropical cyclones (see Young, 2006).

It should be commented here that it is not clear how the peculiar wind-dependent increase-then-decrease behaviour, confined to one front sector in the outer zone of the hurricane, that is to a relatively small area with respect to the total area of the tropical cyclone, can define the behaviour of sea drag over the entire hurricane. Whether this is because C_D dependence on the wind in the other sectors is marginal (as in the 30 km inner hurricane zone) or due to something else, is still to be clarified.

The observation of drag saturation in hurricanes was followed by a number of theoretical, modelling and experimental efforts which offered explanations, simulations and summary reviews of this effect (Andreas, 2004; Donelan *et al.*, 2004, 2006; Barenblatt *et al.*, 2005; Makin, 2005; Bye & Jenkins, 2006; Kudryavtsev, 2006; Kudryavtsev & Makin,

2007, 2009, 2010; Black *et al.*, 2007; Stiassnie *et al.*, 2007; Vakhguelt, 2007; Troitskaya & Rybushkina, 2008; Rastigejev *et al.*, 2011; Soloviev & Lukas, 2010, among others, see also Section 7.3.5 for a relevant discussion). Most often, these studies have been interpreting dynamic and thermodynamic influences of the suspended spray. As discussed in Section 9.1.2, the spray models in this regard can be approximately subdivided into four groups in terms of the physics which they imply. That is, the effect of damping the near-surface atmospheric turbulence due to suspended spume (Barenblatt *et al.*, 2005; Makin, 2005; Kudryavtsev & Makin, 2010; Rastigejev *et al.*, 2011), the momentum-balance effect of the spume acceleration (Andreas & Emanuel, 2001; Andreas, 2004), the rain-like effect of damping the short waves by the droplets falling back down on the wavy surface (Andreas, 2004), and the thermodynamic effects of wave-induced-pressure damping by suspended droplets (Vakhguelt, 2007).

In addition to the spume models, it has been shown that aerodynamic effects can also cause relative or even absolute reduction of the sea drag. One of such effects is the full flow separation of strong winds over steep waves (e.g. Donelan *et al.*, 2004, 2006; Kudryavtsev & Makin, 2007). Such separation causes relative reduction of the wave-induced pressure which is then accompanied by a decreased wind-energy input to the waves and therefore results in the wind effectively experiencing a smaller sea drag. Such a reduction depends on a combination of high winds and steep waves and can occur at moderate wind speeds as low as $U_{10} \sim 12\,\mathrm{m/s}$ (Donelan *et al.*, 2006), progressively increasing towards higher winds. To avoid confusion, it should be mentioned again that in benign conditions the full-flow separation and the separation due to wave breaking lead to opposite effects in terms of the wind input and sea drag (Donelan *et al.*, 2006; Babanin *et al.*, 2007b, and Section 8.3).

Additionally, in the case of the full-flow separation, a part of the wave profile will now be residing under the separated flow and thus will not contribute to the total sea drag. This may be essential if a proportion of short waves, riding this wave, is excluded from the wind–wave interaction or eliminated this way, because the short waves support most of the sea drag (Kudryavtsev & Makin, 2007). Qualitatively, this effect is similar to that of the rain-like droplets mentioned above (Andreas, 2004).

Another type of aerodynamic model was suggested by Troitskaya & Rybushkina (2008). This is a quasi-linear model based on Reynolds equations closed through the eddy viscosity which takes into account the viscous sublayer. Wave-induced disturbances in the air are treated in the linear context, whereas the mean wind profile is derived by considering nonlinear wind stresses. The model is similar to that of Jenkins (1992, 1993), but is extended to directly incorporate the contribution of the short-wave spectrum tail, rather than to parameterise it.

This is the only model of the sea-drag saturation which does not rely on wave breaking to bring about features which then lead to the reduction of the drag. This original theoretical approach required neither sea spray nor flow separation.

While being conceptually different, however, in terms of the physics which eventually leads to drag reduction and saturation, this model is similar to the sea-spray model of the type which relys on the turbulence damping by suspended droplets. Troitskaya & Rybushkina (2008) attribute this suppression of the turbulence to the role of the strong wind which stimulates transfer of the air-turbulence energy near the water surface into the energy of pressure waves in the air.

An additional factor which leads to reduction of the aerodynamic drag in the Troitskaya & Rybushkina (2008) model is smoothing of the sea surface under extreme wind forcing, i.e. removing or reducing the magnitude of short waves which support a significant part of the drag. In this regard, the qualitative consequences of this model are similar to those in the rain-like model of Andreas (2004) or the full-flow separation models of Donelan *et al.* (2004, 2006) and Kudryavtsev & Makin (2007).

Another model was proposed by Soloviev & Lukas (2010). This most original approach can hardly be classified into any of the above-mentioned schemes, as it allows for and relies on the existence of almost all the features of the other models mentioned above, that is the two-phase layer near the surface filled with spray and bubbles, as well as on the air-flow separation due to wave breaking. This breaking, however, is a micro-breaking of capillary parasitic waves riding the front face of dominant waves (see e.g. Cox, 1958; Crapper, 1970; Ebuchi *et al.*, 1987; Longuet-Higgins, 1995, about such parasitic modes), rather than the breaking of gravitational waves mostly discussed in this book, and it is caused by Kelvin–Helmholtz (KH) instability.

Following Koga (1981), Soloviev & Lukas (2010) assume that KH instability happens and disrupts the water surface when the wind-drift current velocity U_s exceeds the minimal phase speed of capillary waves $c_{\min} = 0.232\,\mathrm{m/s}$. This condition is characterised by what they call the 'Koga Number' K:

$$K = \frac{U_s}{c_{\min}} = \frac{u_*}{(g\sigma\rho_w/\rho_a^2)^{1/4}}. \tag{9.15}$$

KH instability occurs if

$$K > K_{\text{critical}} \tag{9.16}$$

where the critical Koga Number ranges from $K_{\text{critical}} = 1$ to $K_{\text{critical}} = 0.26$, which values correspond to $U_{10} \approx 10$–30.

That is, sporadic flow separation due to KH microbreaking can happen at wind speeds as low as $U_{10} = 10\,\mathrm{m/s}$. Note that this is approximately the wind speed where the sporadic full flow separation was found in field experiments of Donelan *et al.* (2006). These latter authors, however, provided a different explanation of their observation: i.e. centrifugal acceleration of the flow over the crest of steep waves was exceeding that due to the pressure gradient which forces the flow to stay attached to the surface.

One way or another, once the flow separation is a permanent feature at wind speeds in excess of $U_{10} = 30\,\mathrm{m/s}$, it leads to what Soloviev & Lukas (2010) call a 'transition layer' filled with spray and bubbles. Once created, this layer experiences a shear stress which leads to another type of KH instability. This instability leads to entraining the water and air parcels and to thickening the transition layer. The thickness reaches some equilibrium level when this mixing is balanced by the gravitational force which acts on the droplets and bubbles and tends to shrink the transition layer. Soloviev & Lukas (2010) call this condition 'marginal stability'.

The condition of marginal stability determines the lower limit for the sea-drag coefficient. The upper limit is defined by the customary Charnock wave resistance. Soloviev & Lukas (2010) verify these limits against available data obtained at extreme winds in the field and laboratory and argue that all such data are found between the two limits identified in their model.

Thus, in this section we have briefly discussed the consequences of wave breaking in extreme conditions. Direct observations and accounts of such breaking are rare, but multiple models point to a number of peculiarities of this kind of breaking which distinguish it from the breaking in benign environmental situations. These peculiarities include full flow separation, disruption of the surface due to microbreaking caused by Kelvin–Helmholtz instability, and tearing the wave crests by the high winds.

Speaking of breaking in extreme conditions, it is worth reminding that at approximately the same wind speed at which the saturation of the sea drag occurs (9.12), surface-wave asymmetry (1.3) also saturates, at least in the laboratory experiments of Leikin *et al.* (1995) (see detailed discussion of these observations in Section 3.3). This fact can be interpreted in two ways. Firstly, this may mean that breaking probability saturates at such a wind speed (see Section 7.3.3 and Eqs. (7.18)–(7.19)). Alternatively (or additionally), this fact can mean that the breaking regime is changed, that is the waves no longer break because of the modulational instability, which leads to the negative average asymmetry, but due to, for example, direct wind forcing. One way or another, the onset of the hurricane wind speeds appears to signify essential changes for the wave-breaking process as well, which may in fact have an impact on the sea drag also.

Tearing the crests leads to production of spume in the atmospheric boundary layer, and this spume along with other types of spray are discussed in Section 9.1.2. All kinds of spray result from the breaking and its consequences, such as whitecapping, and play a variety of roles in the dynamics and thermodynamics of the boundary layer, air–sea exchanges, including genesis of tropical cyclones, in producing aerosols, transferring moisture, serving as condensation matter in meteorological processes, among many others.

In Section 9.1.1, we touched on the topic of sea drag and the influences of wave breaking in this regard. Thus, the current section, 9.1, briefly outlines and discusses issues of the physics and dynamics of the atmospheric boundary layer where knowledge and understanding of the wave breaking and its consequences can be important or helpful.

9.2 Upper-ocean mixing

Surface waves in general and wave breaking in particular are phenomena which owe their existence to the presence of an air–water interface. The wave oscillations, however, also involve orbital motions of the water particles through the water depths at wavelength scale, and at a similar scale wave-induced fluctuations of velocity and pressure are felt in the atmospheric boundary layer.

Breaking waves disrupt the continuous surface-wave trains, and these events do affect locally the momentum and energy fluxes in the atmospheric boundary layer (Section 8.3). In terms of the influence of the breaking on the wave orbital motion, however, this effect seems to be small and thus the disruption of the orbital velocity is confined to the near-surface area (Section 7.3.3, see Figure 7.5 and its discussion).

Similarly, the direct effect of turbulence generation due to wave breaking in the water column appears to be limited to the depths of the order of wave height beneath the surface as discussed in Section 9.2.2. This subsection is preceded by Section 9.2.1 which provides a necessary brief overview of the general topic of momentum and energy transfer from the wind to the ocean, including the waves and breaking waves as a mediator.

In the context of upper-ocean influences of the wave breaking, Section 9.2.3 will discuss injection of the bubbles under the ocean surface. Like production of the spray (Section 9.1.2), generation of the bubbles is primarily due to the breaking and therefore is a relevant issue to be mentioned. This is an extensive research topic in its own right, as such bubbles play a number of important roles in the air–sea interactions, from dynamic dissipative consequences through acoustic noise to the heat and gas exchanges across the surface which will be briefly outlined and will conclude Section 9.2.

9.2.1 Transfer of energy and momentum from the wind to the ocean

Upper-ocean mixing is an outcome of a complicated chain of momentum and energy transformations which connect energy and momentum fluxes from the outer atmosphere through the boundary layer including WBL to the waves, currents and subsurface turbulence. As a result, the upper tens of metres of the ocean are typically well-mixed. Under further external forcing, this layer deepens or stratifies depending on the wind forcing and buoyancy flux.

The rate of deepening depends on the production of turbulent kinetic energy (e.g. Richman & Garrett, 1977). Under unstable stratification, that is in autumn and the winter months when the air is colder than the water, deepening of the mixed layer is mostly determined by the vertical convection, which is stronger than the wind-caused turbulence production. Such convection is apparently not induced by the waves and their breaking and will not be discussed here.

The wind, as a forcing source, does not produce the turbulence in the water directly. It creates the mean currents and waves which do that on its behalf.

The shear instability of such mean flow generates turbulence through the water column and potentially at the base of the mixed layer if MLD is not very deep and the vertical velocity gradients are still large enough. Such a mean current can be either caused directly by the wind surface drift through the tangential viscous stress τ_ν in (7.36) or indirectly through the waves' Stokes drift.

The wave orbital velocities are typically at least an order of magnitude larger than those of the mean current, and with the exponentially decaying velocity profile are able to produce substantial shear stresses and turbulence (see Sections 7.5 and 9.2.2). Waves can also inject turbulence directly when breaking.

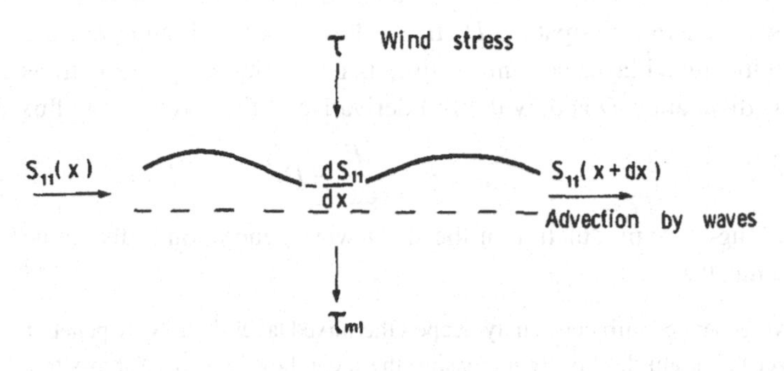

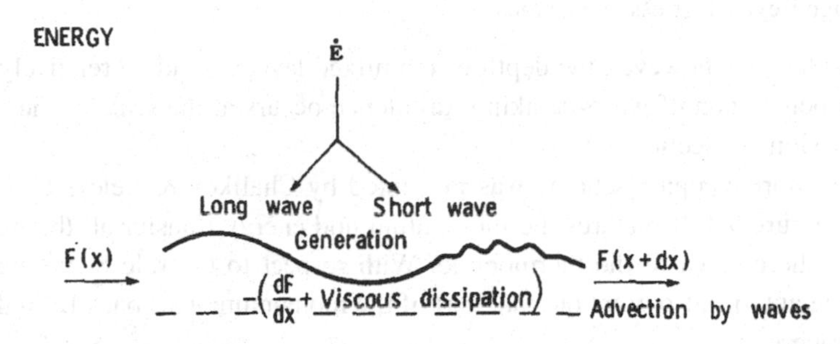

Figure 9.7 Schematic models of the momentum (top) and energy (bottom) transfer from the wind to the mixed layer through the surface waves. Figure is reproduced from Richman & Garrett (1977) © American Meteorological Society. Reprinted with permission

These roles of waves will be discussed and compared in Section 9.2.2. In the current section, we would like to outline the more general pattern of the momentum and energy transfer from the atmosphere to the upper ocean and point out the place of the waves and their breaking in this pattern.

Literature on this topic is considerable and we will not be attempting to provide its review in this book. Rather, we will mention a selection of such transfer schemes suggested over the years and will refer the reader to the respective publications and references in these papers and books (see e.g. Thorpe, 1995, 2005, for substantial reviews).

Figure 9.7 reproduces the scheme of Richman & Garrett (1977). Here, only the role of the momentum and energy fluxes supported by the waves is depicted.

In the momentum-transfer part of the scheme, the tangential stress τ_{ν} and corresponding drift currents are disregarded, and the entire wind stress τ is transferred to the momentum flux τ_{ml} from the waves to the mixed layer, less the temporal growth of wave momentum and less the momentum carried away by the surface waves:

$$\tau_{ml} = \tau - \frac{dS_{11}}{dx}. \tag{9.17}$$

Here, $\mathbf{S}$ is the momentum-flux tensor.

The energy transfer is only a little bit more complicated, as it additionally includes the viscous wave-energy dissipation D. In the lower part of Figure 9.7, the energy $\dot{E}_{ml}$ transferred to the mixed layer per unit of time is less than the energy $\dot{E}$ transferred to the waves, by this dissipation D and by the full derivative of the wave energy flux $\mathbf{F}$:

$$\dot{E}_{ml} = \dot{E} - \frac{dF}{dx} - D. \tag{9.18}$$

In the figure, long-wave modulation of the short-wave generation is also indicated.

In this scheme, the

"breaking-wave-generated turbulence may deepen the mixed layer directly by penetrating to the bottom of the layer to entrain fluid or by increasing the mean flow and shear across the bottom of the layer through Reynolds stress interactions."

For that to happen, however, the depth of the mixed layer should be relatively shallow as the direct penetration of wave-breaking turbulence occurs at the scale of the wave height (see discussions in Section 9.2.2).

A much more complex scheme was presented by Chalikov & Belevich (1993) and is shown in Figure 9.8. It pictures the momentum and energy transfer all the way from the free atmosphere down to the thermocline. With respect to the role of the waves in this transfer, the authors also point out that part of the momentum goes back from the waves to the atmosphere.

While the wall-turbulence analogy is often assumed for descriptions of WBL, and for a good reason – since the logarithmic constant-flux layer (9.1) and (3.19) is indeed observed, the difference between the WBL and solid-wall boundary layer is in fact substantial. The near-water turbulent momentum flux is dominated by the wave-induced fluctuations of pressure, velocity and turbulent stress (see Section 7.4 and (7.36)).

Because of these oscillations, the total flux (7.36) is constant only on average. Dynamically, the constant-flux idealised regime is not maintained where the wave-caused oscillations become significant. As a result, deviations of the wind profile from the logarithmic occur inside WBL and cause changes to the profile outside this boundary layer. Chalikov & Belevich (1993) suggest that it is convenient, within the quasi-logarithmic boundary layer, to parameterise these effects by still using the logarithmic profile but adjusting the total roughness in such a way that it takes into account contributions of both the dominant waves and spectrum tail.

Figure 9.8 demonstrates the general scheme of dynamic interactions in the system of wave boundary layer and wave-mixed layer. A lower boundary condition for the momentum balance equation in WBL is applied at some height $z = z_r$ (see Chalikov & Belevich, 1993, for details):

$$\nu_t \frac{\partial \mathbf{u}}{\partial z} = \mathbf{T}_r = C_r |\mathbf{u}_r| \mathbf{u}_r \tag{9.19}$$

where ν_t is the coefficient of turbulent diffusion, $\mathbf{u}$ is a vector of wind speed, and $\mathbf{T}_r$ and C_r are local momentum-flux vector and drag coefficient, respectively. $\mathbf{T}_r$ defines viscous tangential stress and direct momentum exchange between WBL and mixed layer currents.

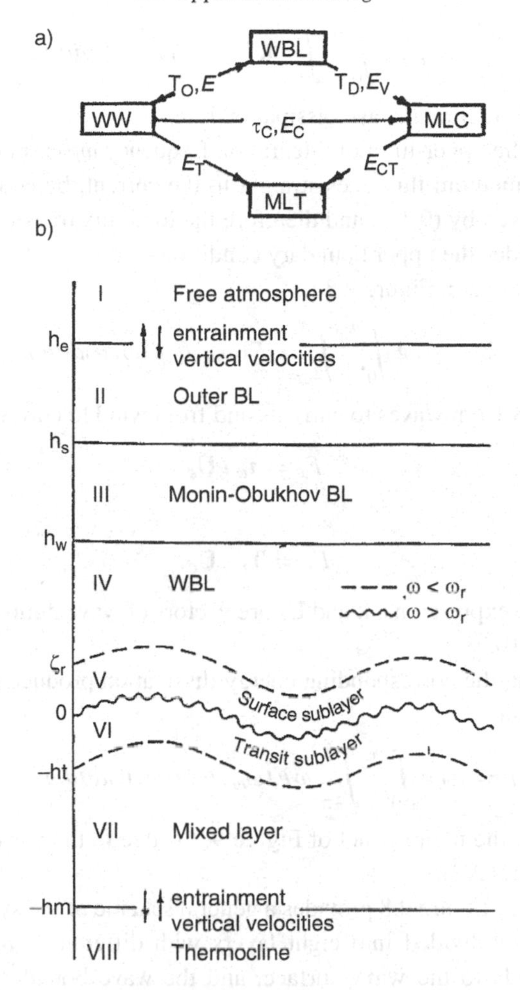

Figure 9.8 a) Scheme of energy and momentum exchange between Wave Boundary Layer (WBL), Wind Waves (WW), Mixed Layer Currents (MLC) and Mixed Layer Turbulence (MLT). Here, τ_0 is the momentum exchange between WBL and WW (9.20), τ_c between WW and MLC (9.21), and T_r between WBL and MLC (9.19). E is the energy exchange between WBL and WW (9.22), E_c between WW and MLC (9.23), E_v between WBL and MLC (9.24), E_T between WW and MLT (9.25), and E_{CT} between MLC and MLT. b) General scheme for the atmosphere–ocean interactions. Figure is reproduced from Chalikov & Belevich (1993) with kind permission from Springer Science+Business Media

The momentum flux from wind to waves is

$$\tau_0 = \rho_w g \int_0^{\omega_r} \int_{-\pi}^{\pi} \mathbf{k} F(\omega, \theta) \beta(\omega, \theta) d\theta d\omega \tag{9.20}$$

where $\beta(\omega, \theta)$ is the wave fractional-growth coefficient (see e.g. Donelan *et al.*, 2006, for detailed definitions of the wave-growth coefficients). The momentum transferred from waves to currents because of wave breaking is equal to

$$\tau_c = \rho_w g \int_0^{\omega_r} \int_{-\pi}^{\pi} \mathbf{k} F(\omega, \theta) \gamma(\omega, \theta) d\theta d\omega \tag{9.21}$$

where $\gamma(\omega, \theta)$ is the fractional wave dissipation rate.

In (9.20)–(9.21), the upper-limit of integration frequency ω_r is such that at the spectrum tail $\omega > \omega_r$ the momentum flux goes straight to the current, bypassing the wave-growth stage. This flux is given by (9.19), and therefore the total flux transferred to the currents is $\tau_c + \mathbf{T}_r$ which provides the upper boundary condition.

Total energy exchange in Figure 9.8 is

$$E = \rho_w g \int_0^{\omega_r} \int_{-\pi}^{\pi} \omega F(\omega_m, \theta) \beta(\omega, \theta) d\theta d\omega. \tag{9.22}$$

Thus, the energy flux from waves to currents and from wind to currents is

$$E_c = \tau_c \cdot \mathbf{U}_c \tag{9.23}$$

and

$$E_v = \mathbf{T}_r \cdot \mathbf{U}_c, \tag{9.24}$$

respectively. In these expressions, $\mathbf{k}$ and $\mathbf{U}_c$ are vectors of wavenumber and surface current (see also Section 2.10).

Wave breaking and the corresponding energy dissipation produce turbulent energy flux E_T to the upper ocean:

$$E_T = \rho_w g \int_0^{\omega_r} \int_{-\pi}^{\pi} \omega F(\omega_m, \theta) \gamma(\omega, \theta) d\theta d\omega - E_c. \tag{9.25}$$

The last term E_{cT} in the upper panel of Figure 9.8 is due to the wave-induced turbulence (see Sections 7.5 and 9.2.2).

The bottom panel of Figure 9.8 provides a general scheme of the system of atmosphere–ocean interactions, subdivided into eight layers with different dynamics. Zero level in the figure corresponds to the wavy surface, and the wave boundary layer in the atmosphere includes regions *IV* and *V*. The outer WBL extends to the height $h_w \sim 0.1\lambda_p$. It is here that the dominant-wave-induced pressure oscillations and therefore the wave-induced momentum flux are felt. The 'dominant wave' scale in this scenario is determined by $\omega < \omega_r$.

The meaning of the ω_r scale now becomes clearer as frequencies $\omega > \omega_r$ play a role in the internal WBL sublayer V. These short waves form the local roughness parameter and the tangential stress (9.19). The height of this layer ζ_t is linked to the oscillating surface rather than to the mean water level.

Layer *III* is the part of WBL where stratification effects become important as described by the Monin–Obukhov theory (see (3.19) and Section 3.1). In the case of neutral stratification, the profile here is logarithmic, and the roughness parameter is defined by the waves and the friction drag. Since h_w is usually much smaller than the Monin–Obukhov scale L, stratification effects can be neglected in layers *I* and *II*. At the upper-boundary height h_s of the Monin-Obukhov layer *III*, Coriolis force has to be introduced.

Layers *I* and *II* depict the free atmosphere, which 'does not know about the ocean surface', and the outer (planetary) boundary layer, respectively. At the boundary height h_e between the two layers, interactions start to occur through the entrainments due to turbulent vertical velocities. In the case of the full development (Pierson & Moskowitz, 1964), this height can be estimated (Benilov *et al.*, 1978). At such height, total kinetic energy should be equal to the energy of the fully-developed sea which leads to

$$h_e \approx 10\frac{u^2}{g} \tag{9.26}$$

where u is the wind-velocity scale. Thus, the height of the boundary layer can reach several hundred metres.

Chalikov & Belevich (1993) argue that the velocities and entrainments at h_e may be due to variations of the stress field over the ocean caused by the non-uniform and non-stationary wavy surface, and therefore 'it is quite possible that wind–waves affect the weather and climate'.

Below the interface, the mixed layer *VII* is located above the thermocline *VIII* which starts at depth $-h_m$. This mixed layer obtains momentum and energy through the transition zone *VI* whose depth $-h_t$ is linked to the wavy surface from below. The total amount of the momentum and energy are described by expressions (9.20) and (9.22). Because of the turbulent vertical velocities at the thermocline border, mixing through the thermocline due to the waves is possible (e.g. Martin, 1985). As mentioned above, direct impact of the breaking waves at this border is hardly feasible unless depth $-h_m$ is small (that is comparable to wave height).

Thus, the surface waves play an important role in the dynamic regime of both the atmospheric boundary layer and the upper ocean. In the ocean-circulation models, it is quite typical to disregard the waves (although some trends in this regard are noticeable lately, see also Section 9.2.2). In reality, however, the waves support a major part of the momentum flux from the atmosphere to the ocean, and they do not release it immediately upon receiving. It is redistributed across the spectrum by means of nonlinear interactions S_{nl} in (2.61), and can go both back to the atmosphere and to the current through the wave breaking and non-breaking dissipation. The transition-delay depends on the temporal and spatial scales of the wave-energy dissipation, and at this scale the waves are an independent player rather than a mere momentum-transfer means. Additionally, both breaking and non-breaking waves are a source of turbulence in the mixed ocean layer.

As the role of the waves in the lower-atmosphere and upper-ocean dynamics gains recognition, multiple implementations of this role are being discussed which correspond to different physical mechanisms. Some of them are depicted in a recent scheme reproduced from Ardhuin *et al.* (2005) (Figure 9.9). This scheme also shows the body of the ocean below thermocline all the way to the bottom boundary layers.

At the top, we see the momentum fluxes already described above in the Figure 9.8 scheme of Chalikov & Belevich (1993), i.e. total stress τ^a, momentum flux to waves

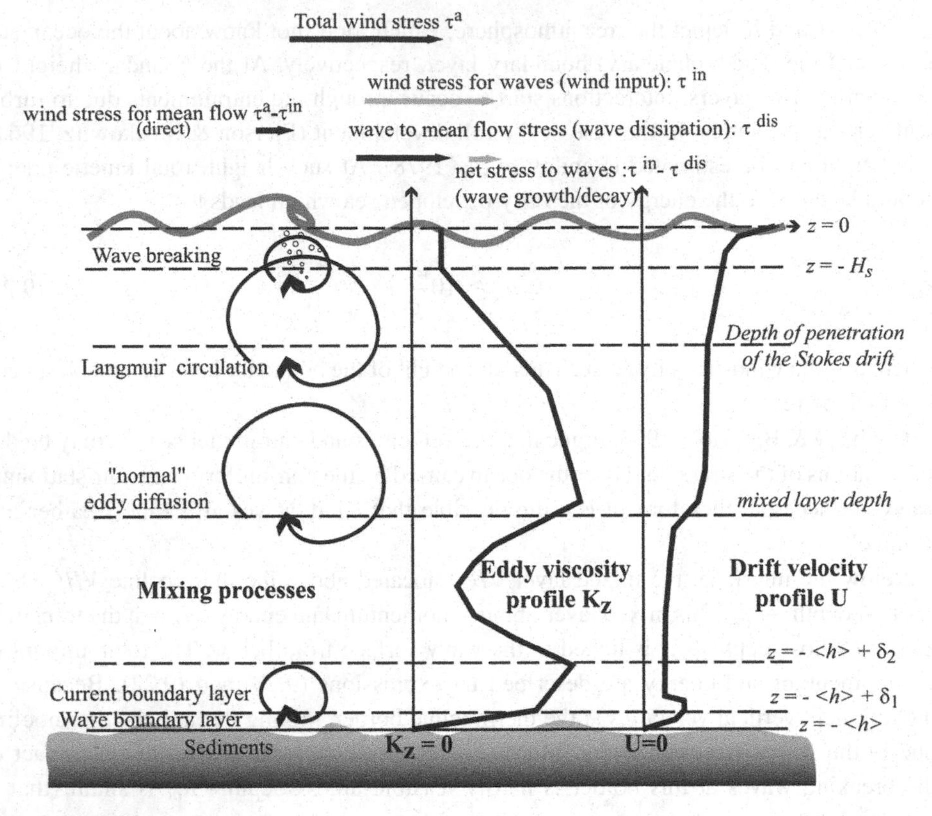

Figure 9.9 Scheme of momentum fluxes and mixing processes which couple waves and currents. Processes are indicated for horizontally uniform conditions, and possible profiles of eddy viscosity K_Z and drift velocity U are shown. Figure is reproduced from Ardhuin *et al.* (2005) by permission of American Geophysical Union

$\tau^{in} = \tau_0$, direct flux to the mean flow $\tau^a - \tau^{in} = \mathbf{T}_r$, and the flux from the waves to the mean flow $\tau^{dis} = \tau_c$. Net stress to the waves is also demonstrated as $\tilde{\tau}_w = \tau^{in} - \tau^{dis}$ which goes into wave growth (or wave decay for that matter, if the wave-breaking stress is greater than the input stress, see (9.27) below).

More details of the wave influence are shown below the surface. Here, the wave-breaking sublayer is highlighted where the momentum fluxes are dominated by the direct wave-breaking action. Its vertical scale is of the order of significant wave height H_s. Other vertical scales indicated are the characteristic depth of penetration of Stokes drift and MLD. Near the bottom at depth h, two other vertical scales can be mentioned which correspond to the wave bottom-boundary layer δ_1 (provided that the ocean is shallow enough to still experience the wave action at the bottom) and to the thicker current-bottom boundary layer δ_2. Vertical solid lines demonstrate profiles of eddy viscosity K_Z and drift velocity U.

On the left, turbulent mixing mechanisms are indicated. Wave-breaking turbulence is located near the water surface, Langmuir circulation penetrates deeper. This circulation is a consequence of an instability resulting from the waves interacting with both the Stokes and wind drifts (e.g. Langmuir, 1938; Craik & Leibovich, 1976; Phillips, 1998). Thus, this is still a wave-related mechanism.

Below lies what is labelled as 'normal' eddy diffusion, that is turbulent diffusion unrelated to the waves. This is turbulence produced by the mean shear flow. Bottom boundary-layer wave and current turbulence concludes the scheme. It should be mentioned in passing that internal waves at the thermocline and throughout the stratified ocean, their breaking and interactions with other circulation and turbulent processes, are another source of momentum fluxes in the ocean, which will not be, however, considered here.

In this context, we will also point out that the overall pattern of the upper-ocean mixing, even in the purely wind–sea case, is much more complex than the general schemes outlined here. For example, Ardhuin *et al.* (2009b) further subdivided the Stokes drift and quasi-Eulerian currents, likely related to the wave breaking and/or Langmuir circulation, and concluded that such decomposition is important for the estimation of energy fluxes to the upper ocean.

We will be discussing the concept of the different turbulence layers and turbulence mechanisms in more detail in Section 9.2.2 below. In this section, dedicated to the general scheme of momentum and energy fluxes across the interface and in the upper ocean, we would like to additionally mention two mechanisms of turbulence generation, that is Langmuir turbulence and non-breaking wave-induced turbulence. In the schemes presented above they are indicated, but one needs to be particularly perceptive to notice that.

Langmuir turbulence is essentially the same physical mechanism as Langmuir circulation (e.g. McWilliams *et al.*, 1997; Sullivan & McWilliams, 2010). It deserves a specific highlight, however, both in phenomenological terms and because of its potentially special role in the upper-ocean mixing.

Langmuir circulation is typically pictured as rolls rotating perpendicularly to the main wave direction and displaying themselves as streaks on the ocean surface, parallel to the wave propagation (e.g. Langmuir, 1938; Smith, 1992; Phillips, 2001, 2002, 2005). Their surface spacing can grow from millimetres to kilometres (Phillips, 2001; Thorpe, 2004), and their depth is more or less restricted by the depth of the mixed layer. Their rate of rotation is relatively slow (of the order of cm/s); nevertheless, Langmuir circulations provide a means to advect turbulence kinetic energy vertically (see Babanin *et al.*, 2009b, for a discussion).

What McWilliams *et al.* (1997) called Langmuir turbulence is a continuous spectrum of Langmuir circulations, generated by the waves interacting with background drift currents through the water column. They would provide superposition of different spatial and velocity rotational scales and not necessarily extend and exhibit themselves at the water surface. Effectively, this is a turbulence distributed in the upper ocean where the instabilities of wave–drift interactions can persist. As the turbulence is distributed vertically, it can play an obvious role in the upper-ocean mixing and the schemes mentioned in this section.

The other turbulent-production wave-related mechanism is that due to shear stresses imposed by the orbital wave motion and the vortex instability of potential surface waves (Benilov *et al.*, 1993; Babanin, 2006). In Chalikov & Belevich (1993), this is E_{cT}, which, for some reason is shown as turbulent energy flux from the mean layer currents rather than from wind–waves in their scheme reproduced here in Figure 9.8. This turbulence was considered in Section 7.5 in the context of wave-energy dissipation and will be further discussed in the next section, 9.2.2, specifically with respect to the upper-ocean mixing.

Thus, in this section, we have discussed general schemes of transfer of the energy and momentum from the wind to the ocean. These include descriptions of fluxes all the way from the free atmosphere down to the ocean bottom. In such an overall pattern, the role of the waves and their breaking may not be most determining, but it is essential, and it has been highlighted and its relative place is shown in this general context. The next section will specifically consider generation of ocean turbulence by the waves and due to wave breaking.

9.2.2 Generation of turbulence

The most comprehensive and recent update on the effects of surface wave breaking in the upper ocean is given by Benilov & Ly (2002), and we refer the reader to this paper for details and numerous further references. Here, we will briefly summarise this wave-breaking role. It should be pointed out from the very beginning that although, as mentioned in Section 9.2.1, the breaking is unlikely to directly facilitate the mixing through the thermocline unless it is very shallow, its influence is felt by the turbulence production profiled all the way through the mixed layer.

Benilov & Ly (2002) start from emphasising that, in contrast to the wall-layer turbulent flows or geophysical flows over the land, turbulence in the upper ocean is governed not just by the mean shear, but also by surface waves. (To an extent, it is also true for the atmospheric flow over the ocean (see Chalikov & Belevich, 1993, and discussions in Section 9.2.1). Thus, the transport of momentum, energy, heat, moisture and gases across the ocean interface are all affected or even controlled by the waves (see also Section 9.1.2 on the role of spray and Section 9.2.3 on the role of bubbles).

If approached analytically, turbulence in the ocean is a very interesting phenomenon of fluid mechanics rather than just geophysics: turbulent motion in a liquid whose free surface is subject to wind forcing which makes the surface vertically unstable and as a result covered by growing and propagating oscillations (waves). These waves receive a considerable amount of momentum and energy from the wind, which they do not accumulate, but pass over to currents and ocean turbulence. When doing this, however, they redistribute this momentum and energy in time and space by means of breaking.

Benilov & Ly (2002) write:

"Breaking waves create a highly turbulent environment within the top few meters of the ocean. Wave breaking provides a mechanism for injection of both momentum and turbulent kinetic energy from the surface winds to the water... Hence, the turbulence of the upper ocean is nourished by the energy supplied in the waves. Consequently, the turbulence characteristics should depend on the state of the ocean surface."

In a retrospective, starting from historical works by Dobroklonskiy (1947) and Bowden (1950) who first attempted to take into account the wave influence on the ocean turbulence theoretically, credit has to be given to Kitaigorodskii & Mitropolskiy (1968), Longuet-Higgins (1969a), Korotaev *et al.* (1971), Yefimov & Khristoforov (1971), Benilov (1973), Benilov & Lozovatskiy (1977), Kitaigorodskii & Lumley (1983), Benilov *et al.* (1993), Craig & Banner (1994) and Craig (1996), among many others. Benilov & Ly (2002), in terms of the momentum τ_0 (9.20), describe the wave-breaking stress as

$$\tau_w = \tilde{\tau}_w + \tau_c \tag{9.27}$$

where $\tilde{\tau}_w$ is spent on the wave growth and is carried away by the waves, and τ_c is the momentum (9.21) flux produced by the wave breaking (it finally goes to the mean current as discussed in Section 9.2.1).

Benilov & Ly (2002) suggested a three-layer model of the mixed-layer turbulence, the upper part being controlled by this wave breaking. They used a $k-\epsilon$ model for the evolution of turbulent kinetic energy (TKE), with the energy budget having an extra term for the turbulent diffusion of wave kinetic energy.

In the upper layer, the potential wave field possesses the best part of the kinetic energy, and production of the turbulence by breaking significantly exceeds the mean-shear effect. The turbulent diffusion of the wave kinetic energy dominates over the diffusion of TKE at these depths of the order of wave height.

Below this layer, as the wave kinetic energy rapidly drops, diffusion of turbulent kinetic energy exceeds the wave effect in the energy budget, and direct wave impact rapidly diminishes. Indirectly, however, the wave-breaking turbulence affects the dynamics of the entire Ekman layer below it.

Thus, immediately below the wave-breaking diffusive layer there is a transitional turbulent diffusive layer. Here, the turbulent diffusion still exceeds the mean shear contribution to the TKE budget, that is the main source of turbulence here is the TKE flux from the breaking-controlled layer above.

And it is only below that we can see the logarithmic mean-velocity profile, which is an analogy to the wall-law turbulence. In this turbulent sublayer, the mean-shear production of TKE dominates. As in the case of the atmospheric boundary layer discussed in Section 9.2.1, however, this logarithmic layer 'knows' about the surface waves and their breaking. The roughness scale of this profile, the water-side friction velocity u_{*w}, the volumetric dissipation rate ϵ_{dis} of the constant-flux layer, and the vertical extent of this logarithmic layer – they all are defined by the wave breaking and the dynamics of the wave-breaking and transitional sublayers above.

Thus, the mixing scheme of Benilov & Ly (2002) provides a full and convincing picture of the turbulence profile in the water column with the free wind-forced surface. This profile is quite different to the turbulence near the solid boundary. There is additional turbulence production due to wind-caused shear at the mobile interface, but the main difference is due to the powerful surface-located additional source of turbulence caused by the wave breaking.

It is instructive and relevant to notice here, however, that for consistency the picture lacks the wave-induced turbulence caused by the wave orbital motion, unrelated to wave breaking. The intensity of this turbulence production may be small near the surface, if compared with direct turbulence injection by the breaking, but the source of such turbulence is distributed through the water column, rather than being localised at the surface (Babanin, 2006).

It is interesting to note that Benilov & Ly (2002) mention this turbulence, but do not elaborate or include it into the turbulence budget (9.27), even if implicitly. It is even more intriguing if we notice that Benilov *et al.* (1993) are authors of a theory of such turbulence and that Benilov & Ly (2002) actually explicitly point to this turbulence source:

> "Another mechanism of the upper layer turbulence generation may be associated with the vortex instability of potential surface waves (Benilov *et al.*, 1993)."

Indeed, Benilov *et al.* (1993) suggested and described physically and mathematically a mechanism of production of turbulence by potential waves. In this paper, they asked and answered a question of

> "what follows from hydrodynamic equations about vortex disturbances in the potential flow?"

They considered a problem of stability of such disturbances in the potential velocity field of linear waves. The problem is essentially three-dimensional (even for two-dimensional waves), as the instability develops in the dimension perpendicular to the plane of two-dimensional waves. A second-order solution (with respect to depth-filtered steepness $\epsilon = ak \cdot \exp(-kz)$ as the perturbation parameter) is always unstable.

The obvious question of the origin of the vortex disturbances in the potential waves was not approached by Benilov *et al.* (1993), but the answer is actually quite apparent. First of all, the ocean is always turbulent and in any scenario perturbations of the velocity field, even if the background mean wave motion is potential, are inevitable. Secondly, as was discussed in Section 7.5 (see also Kinsman, 1965; Babanin & Haus, 2009), the potential wave theory is an approximate approach to begin with, based on the Euler equation as a boundary condition which neglects the water viscosity. Since the water is a viscous fluid, even though the viscosity is small, the shear stresses should be possible in the orbital motion with strong vertical velocity gradients, and may serve as the source of the vorticity. This latter issue, however, is subject to some controversy which we cannot solve here and we address the reader to the literature (Benilov & Lozovatskiy, 1977; Kitaigorodskii & Lumley, 1983; Benilov *et al.*, 1993).

Thus, the theory of Benilov *et al.* (1993) is a consistent physical mechanism which can explain generation of turbulence by waves whose mean orbital motion may be well approximated by the potential and even linear wave theory. Following this approach, Chalikov

(2010, personal communication) developed a numerical model of turbulence in potential waves. The model consists of three parts: fully nonlinear potential model of two-dimensional (i.e. long-crested) waves (Chalikov & Sheinin, 2005), LES three-dimensional model based on full Reynolds equations with subgrid turbulence, and three-dimensional model of evolution of subgrid turbulence. The last two modules are new and written in cylindrical conformal coordinates. Small pertubations of the potential motion are introduced and then allowed to develop as dictated by the theory. As such, this model does not have tuning parameters and can be used for qualitative and even quantitative investigation of the phenomenon.

In the absence of breaking (that is for waves with small steepness and transitional Reynolds numbers – see discussion in Section 7.5), the turbulence in the model is strongly intermittent because the time scale for dissipation is very short. It concentrates at the rear face of the waves. This is what was also observed in laboratory experiments of Babanin & Haus (2009) with such turbulence. Comparison of the volumetric dissipation rates produced by the model and those measured by Babanin & Haus (2009) are given in Figure 9.10.

Such dissipation rates *max*(*Diss*) are plotted as a function of wave amplitude a, like ϵ_{dis} in Figure 7.24. It is, however, not exactly the same property as ϵ_{dis} in Figure 7.24 and in Babanin & Haus (2009). Those were measured below the troughs of the highest waves, i.e. always at some distance below the surface and even below the mean water level.

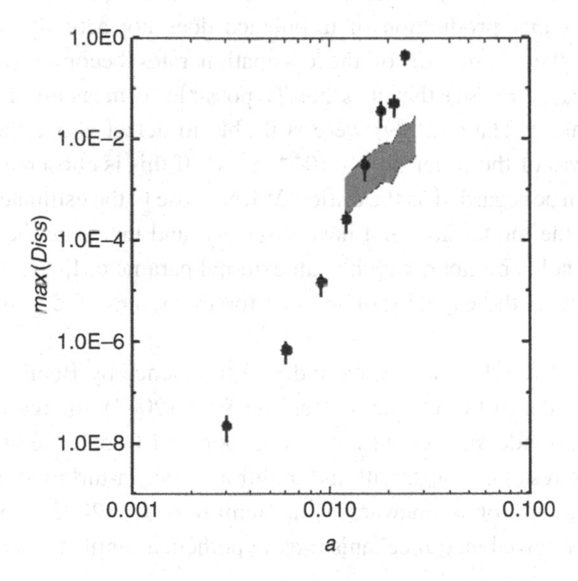

Figure 9.10 Maximal volumetric dissipation rates *max*(*Diss*) versus wave amplitude a (Chalikov, 2010, personal communication, numerical model of non-breaking wave-induced turbulence in potential waves). The shaded area corresponds to the range of measurement and scatter of the observational data of Babanin & Haus (2009)

The *max(Diss)* in Figure 9.10 indicates the maximum dissipation in the wave-induced-turbulence dissipation profile. In practice, this is an estimate of the volumetric dissipation rate near the surface and above the mean water level. Since most of the wave-induced turbulent energy is known to concentrate within wave crests (Gemmrich, 2010), it is expected that such a *max(Diss)* should be greater than ϵ_{dis} of Babanin & Haus (2009). In the model, this happens because the generation of turbulence depends on the gradient of orbital velocity $(du_{\text{orb}}/dz)^2$ which has a maximum at the surface.

In the figure, the shaded area corresponds to the range of measurements and scatter of the observational data of Babanin & Haus (2009). There is quantitative agreement for wave amplitudes of $\sim$1.5 cm (wavelength in the model, as in the experiment, corresponds to the frequency 1.5 Hz), and for the higher waves *max(Diss)* within the crests is greater than that measured below the troughs. The theoretical growth of the maximal dissipation rate, as a function of steepness, is faster than that of the background dissipation ϵ_{dis} in (7.76) and Figure 7.24.

In Figure 9.11, vertical profiles of the the volumetric dissipation ϵ_{dis} (denoted as *Diss*, units of m^2/s^3) are plotted as a function of depth $Z(m)$, averaged over horizontal planes in Cartesian coordinates. The thin horizontal line corresponds to the depth below troughs. The model allows us to obtain these values also above this line within segments occupied by the water, so there are positive depths present too. The solid line is the average dissipation, and dashed lines indicate the standard deviation.

Different subplots show profiles for different wave amplitudes a and corresponding wave Reynolds numbers (7.70) denoted as *RE*, as indicated within these different panels. It is quite obvious that production of turbulence does not actually stop at low amplitudes/WRNs, but the magnitude of the dissipation rates becomes so marginal ($\epsilon_{\text{dis}} \sim 10^{-8}\,\text{m}^2/\text{s}^3$ at $\text{Re}_{\text{wave}} = 84$) that it is hardly possible to measure. The lowest dissipation which Babanin & Haus (2009) were still able to detect above the noise level with the PIV method was of the order $\epsilon_{\text{dis}} \sim 10^{-4}\,\text{m}^2/\text{s}^3$. If this is chosen as a reference, then $\text{Re}_{\text{wave}} \approx 2000$ can be regarded as the critical WRN, close to the estimate (7.67) of Babanin (2006). Note that the model does not have viscosity, and therefore the physical meaning of the critical Reynolds number, which is an external parameter linked to the wave amplitude and frequency, is the capacity of viscous forces to damp the growth of these small instabilities.

Returning back to the theory of wave-induced turbulence by Benilov *et al.* (1993) and the upper-ocean model of turbulence by Benilov & Ly (2002), the reason why Benilov & Ly (2002) do not provide an account for such a source of turbulence in their upper-ocean turbulence scheme rests, perhaps, with the tradition in ocean-turbulence modelling which neglects this turbulence or is unaware of it. Benilov & Ly (2002), apparently, treat the non-breaking wave-turbulence mechanism as hypothetical until it is explicitly confirmed experimentally.

Ocean-circulation modellers have been facing an opposite dilemma. While having no formal theoretical reason to introduce such turbulence, they often felt a need for an additional source of turbulence, definitely connected with the surface waves and not necessarily with

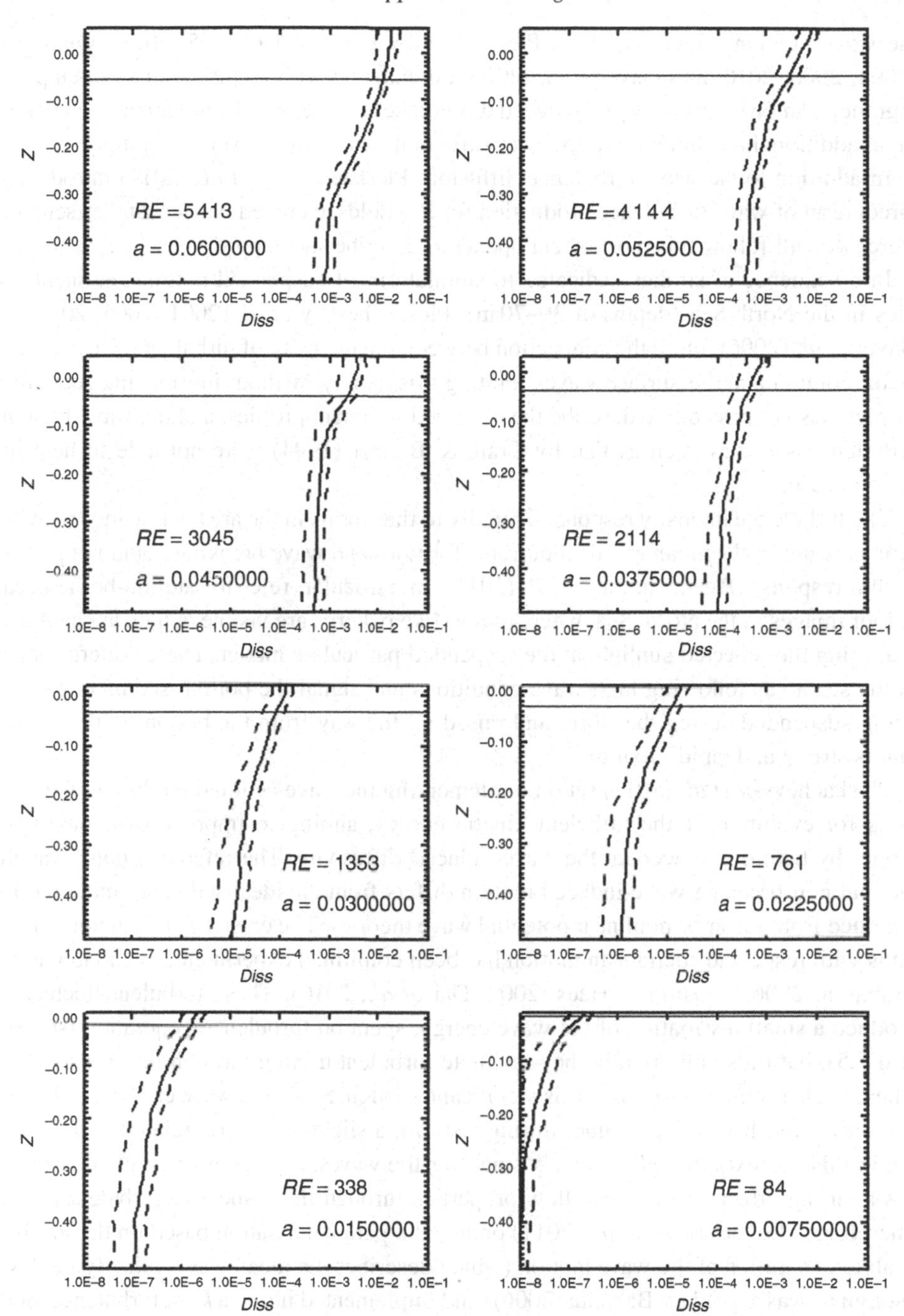

Figure 9.11 Vertical profiles of the volumetric dissipation $Diss(m^2/s^3)$ as a function of depth $Z(m)$ (Chalikov, 2010, personal communication, numerical model of non-breaking wave-induced turbulence in potential waves). Different subplots show profiles for different wave amplitudes a and corresponding wave Reynolds numbers (7.70) denoted as RE, as indicated within the panels

the wave breaking. Jacobs (1978), Pleskachevsky *et al.* (2001, 2005, 2011), Qiao *et al.* (2004, 2008, 2010) and Gayer *et al.* (2006) all came up with empirical or semi-empirical arguments in favour of the wave-induced turbulence and suggested similar parameterisation of an additional coefficient of turbulence diffusion, scaled by the wave amplitude.

In addition to the wave-turbulence diffusion, Pleskachevsky *et al.* (2011) introduced a direct term of wave-turbulence production for Reynolds averaged Navier–Stokes schemes. Here, we will follow this most recent update to describe such approaches.

In a sequence of studies dedicated to simulations of suspended bottom-sediment profiles in the North Sea (depths of 30–70 m), Pleskachevsky *et al.* (2001, 2005, 2011) and Gayer *et al.* (2006) found the connection between the intensity of turbulent diffusion in the water column and the surface waves. Putting this simply, without introducing such diffusion, it was not possible to describe the observed sediment profiles, and the wave-breaking turbulence schemes such as that by Craig & Banner (1994) were not able to help in a general case.

The turbulence intensity responded rapidly to the storms in the area, while models which took account of the mean and drift currents for tides and wave breaking could not produce such a response. Pleskachevsky *et al.* (2011), in particular, refer to satellite-borne ocean-colour images of the North Sea, where plume-like patterns are visible, which are caused by scattering the reflected sunlight at the suspended particulate matter. These patterns appear in the sea areas following high-wave conditions and signal the bottom-sediment particles being suspended during the storm and raised all the way from the bottom to the surface, that is strong and rapid mixing.

Pleskachevsky *et al.* (2011) set on implementing the wave action directly into the equations for evolution of the turbulent kinetic energy, aiming to improve their circulation model by taking into account the wave-induced diffusivity. The reference point was that the real non-breaking wave-induced motion differs from the idealised mean-motion orbits obtained from linear or nonlinear potential wave theories. The existence of random fluctuations with respect to such mean motion has been confirmed experimentally in laboratories (Babanin, 2006; Babanin & Haus, 2009; Dai *et al.*, 2010). These turbulent fluctuations produce a small dissipation of the wave energy, spent on turbulence generation (see Section 7.5), but most importantly they facilitate turbulent mixing through the water column where such a source of turbulence is significant enough. Since the wave energy (and orbital velocities) reach very high values during a storm, a slightest loss (unrelated to the breaking in this context), though relatively small for the waves, but being vertically distributed, has a strong effect on the mean-flow properties through the respective turbulent mixing. Therefore, Pleskachevsky *et al.* (2011) obtained a parameterisation based on the idealised analytical solution of the wave motion (which describes the mean wave-velocity field well enough as was argued by Babanin (2006)) and implemented it into a $k-\epsilon$ turbulence model (GOTM, Burchard *et al.*, 1999).

For clarity of the argument, Pleskachevsky *et al.* (2011) introduce what they call 'symmetrical' and 'asymmetrical' wave motion. The so-called symmetrical oscillations of the individual waves relate to applications where the wave dissipation is not relevant or is not taken into account (that is the wave stays symmetric over its period/length). Such are the various approaches which led, for example, to the concept of the wave-caused turbulent diffusion (Jacobs, 1978; Qiao *et al.*, 2004, 2010; Gayer *et al.*, 2006).

The asymmetry of the wave-induced water-particle motion, as a residual effect of the gradual (i.e. non-breaking) dissipation, can be included in the mean flow, and this possibility has also been considered before (e.g. Ardhuin & Jenkins (2006) obtained it by using interaction of the Stokes drift and wind-induced currents). Pleskachevsky *et al.* (2011) argued that both symmetric and asymmetric effects are important and must be taken into account. In the models, the overall effect on turbulent diffusivity can be obtained based on an idealised solution for symmetric wave profiles (e.g. linear wave theory), if the dissipation of the wave is known and can be taken into account independently (e.g. from measurements or wave modelling which includes the dissipation term).

In fluid flows, the turbulence is described by the random fluctuations of velocity u_i (i.e. see (9.2) above) for $i = x, y, z$ directions. The instantaneous velocity u_i is generally (also due to wave motion) presented as

$$u_i = \overline{u_i} + u_i' \tag{9.28}$$

where $\overline{u_i}$ is the mean fluid velocity (for wave motion, the integration increment should be $dt \ll T$ in order to capture variations of u_i). By definition, the time-averaged fluctuations u_i' are zero, and therefore at the time scale of wave period T this turbulence is 'invisible', but it does influence turbulent viscosity ν_t and turbulent diffusion.

If the turbulence intensity (that is ratio $u_i'/\overline{u_i}$) is less than 1% of the mean flow, this is usually classified as a low-level turbulence, 1–5% corresponds to a medium-level turbulence intensity and greater than 5% is a high-level turbulence case. Since wave orbital velocities can exceed 3–5 m/s, even the low-turbulence condition will signify mean-orbit fluctuations of the order of several cm/s and can produce turbulent mixing orders of magnitude higher than the mixing due to molecular viscosity or due to background circulation currents (for depths of 30–50 m, these currents are about 0–0.2 m/s in the lower water layers of the North Sea).

The mean orbital velocity $\overline{u_{\text{orb},i}}$ is a solution of linear theory, a function of H_s and T and is a particular case of $\overline{u_i}$ in (9.28). Input of H_s and T into $\overline{u_{\text{orb},i}}$ (which is how the wave information is transferred in the circulation models) presents static values for the wave height, period and length, whereas these properties change dynamically: such input does not include information about wave energy dissipation of any kind. Therefore, turbulent fluctuations and damping effects due to them (loss of energy) are not presented in the idealised $\overline{u_{\text{orb},i}}$ which is typically used for the technical applications.

In reality, even at laminar motion the idealised $\overline{u_{\text{orb},i}}$ will be slowing down due to molecular water viscosity ν. The difference can be denoted as u_{Ti}:

$$u_{Ti} = \overline{u_{\text{orb},i}} - \overline{u_i} \tag{9.29}$$

and characterises deviation of the real mean-orbital wave motion in viscous fluid from ideal non-viscid fluid. While WRN (7.70) is below critical value (7.67), the wave-induced motion remains laminar and the turbulent viscosity $\nu_t = 0$ ($u_i' = 0$). If the damping component u_{Ti} overcomes a certain limit, $\mathrm{Re}_{\mathrm{wave}}$ exceeds the critical value and the wave motion becomes turbulent as known from experiments (Babanin, 2006; Babanin & Haus, 2009; Dai *et al.*, 2010). Then, turbulent stresses

$$\tau = u'w' = \nu_t \frac{du_{\mathrm{orb}}}{dz} \tag{9.30}$$

will significantly and rapidly enhance the effective total viscosity which is now a combination of the eddy and molecular viscosity:

$$\nu_{\mathrm{tot}} = \nu_t + \nu. \tag{9.31}$$

Now, it is possible to define the turbulent viscosity ν_t produced by the resultant mean wave velocity $\overline{u_i}$ in (9.29). Note that this mean velocity is different from the ideal mean orbital velocity $\overline{u_{\mathrm{orb},i}}$ as it takes into account the fact that the turbulent diffusion feeds back to the wave motion and therefore contributes to a wave-energy damping.

As noticed above, the wave motion can be sufficiently represented by the linear-theory outcomes. Keeping in mind that the existence of u_{Ti} is physically and mathematically responsible for the appearance of turbulence in the first place (in a non-viscous fluid $u_{Ti} = 0$), it was further assumed that the magnitude of u_{Ti} is negligible compared to $\overline{u_{\mathrm{orb},i}}$, i.e. $\overline{u_i} \approx \overline{u_{\mathrm{orb},i}}$. This way, it is possible to obtain ν_t by using the shear frequency M

$$M^2 = M_{x,z}^2 = \left(\frac{d\overline{u}_x}{dz}\right)^2 + \left(\frac{d\overline{u}_z}{dx}\right)^2, \tag{9.32}$$

based on the idealised motion $\overline{u_{\mathrm{orb},i}}$, which will be subsequently returned to the mean-flow model with a time scale coarser than dt. Using linear wave theory (e.g. Young, 1999), the mean square value (integrated over wavelength) of the shear frequency for the wave motion can be expressed as

$$\overline{M_{\mathrm{wave}}^2}(z) = \left(\frac{\pi k H_s}{T} \frac{\sinh(k(z+d))}{\sinh(kd)}\right)^2. \tag{9.33}$$

Circulation models simulate the mean flow, and turbulence models interconnect these mean currents $\mathbf{U}$ and their fluctuations $\mathbf{U}'$, through the turbulent kinetic energy equation for the evolution of TKE (denoted as K here, see Burchard *et al.*, 1999):

$$\frac{\partial K}{\partial T_{mean}} + U_i \frac{\partial K}{\partial X_i} = D_K + P_S + G - E_K \tag{9.34}$$

where U_i is the mean-current component for i coordinate, G is the production of TKE by buoyancy, D_K is the turbulent and viscous transport term and E_K is the dissipation term. The temporal and spatial resolution dT_{mean} and X_i (capital letters), correspondingly, indicate an equation for the mean flow (which simulates the process on scales that are coarser than the scales dt, dx_i needed to simulate an individual wave). P_S, in the default formulation of TKE equation (9.34), signifies TKE production by mean currents:

$$P_S = P_{\mathrm{current}} = \nu_t M_{\mathrm{current}}^2, \tag{9.35}$$

that is M_{current} is the shear frequency due to the respective mean-current component U_i. The eddy viscosity due to the mean-flow currents is usually defined as

$$\nu_t = \nu_{\text{current}} = c_\mu \sqrt{K l}, \tag{9.36}$$

with stability function c_μ and length scale l.

Now, to take the wave motion into account in the circulation model, Pleskachevsky *et al.* (2011) decomposes the overall wave motion into two parts and parameterises them separately. The sub-processes occur on a $dt < T$ micro-time scale and the mean-flow processes occur on the coarser dT_{mean} mean-flow scale. Following the definitions above, the sub-processes will be referred to as symmetric motion, and mean-flow processes as asymmetric motion.

Symmetric motion (SM), if integrated, does not contribute to the mean currents (the mean horizontal velocity, for example, due to such motion is zero; hence the motion is symmetric), but during time dT_{mean}, on the dt-scale, a strong influence of waves on turbulence is possible. The impact of SM on the flow can be parameterised using shear

$$M_{\text{wave}}^{\text{SM}} = \overline{M_{\text{wave}}}, \tag{9.37}$$

from (9.33). Note that $\overline{M_{\text{wave}}}$ uses wave parameters H, T and k. These integral parameters are obtained from the wave spectra, estimated by the wave model with the dissipation accounted for. Various implementations of this effect have been conducted before (e.g. Jacobs, 1978; Pleskachevsky *et al.*, 2001, 2005; Qiao *et al.*, 2004, 2010; Gayer *et al.*, 2006).

Asymmetric motion (AM) represents the dissipation of the primary wave motion: the orbital tracks are no longer closed due to the damping, and the mean flow gains a weak residual-current effect. The asymmetric component can be parameterised by using shear $\overline{M_{\text{wave}}}$ and by employing ratio $k_{\text{wave}}^{\text{AM}}$ between wave energy dissipated and total wave energy:

$$M_{\text{wave}}^{\text{AM}} = k_{\text{wave}}^{\text{AM}} \overline{M_{\text{wave}}}. \tag{9.38}$$

Coefficient $k_{\text{wave}}^{\text{AM}}$ here signifies the degree of efficiency of wave-energy dissipation:

$$k_{\text{wave}}^{\text{AM}} = \frac{E_{\text{diss}}}{E} \tag{9.39}$$

where E_{diss} means the depth-integrated dissipated wave energy per unit of surface, and E is the wave energy in (2.23) before the dissipation is taken into account by the wave model.

Now, we can summarise the wave effect, both SM and AM parts, in (9.34). First, the TKE production has to be increased by the additional wave source:

$$P_{\text{wave}} = \nu_t M_{\text{wave}}^{\text{AM}^2}. \tag{9.40}$$

This is the asymmetric part of the wave motion which influences the mean flow $\mathbf{U}$ directly:

$$P_S = P_{\text{current}} + P_{\text{wave}} = \nu_t \left(M_{\text{current}}^2 + M_{\text{wave}}^{\text{AM}^2} \right). \tag{9.41}$$

In order to account for the influence of the symmetric component, the eddy viscosity has to be modified by viscosity from the wave action ν_{wave}:

$$\nu_t = \nu_{\text{current}} + \nu_{\text{wave}}. \tag{9.42}$$

These are sub-scale effects in the circulation models, and the standard closure hypothesis is adopted (e.g. Kapitza, 2002):

$$\nu_{\text{wave}} = l^2_{\text{wave}} M^{\text{SM}}_{\text{wave}} \tag{9.43}$$

where l_{wave} is the length scale for the wave-induced turbulence. Following the argument of Qiao *et al.* (2004) and Babanin (2006), l_{wave} is assumed to be proportional to the wave amplitude (radius of the wave orbit) which characterises the motion scale of physical particles in the waves:

$$l_{\text{wave}}(z) = \alpha \cdot a(z) = \alpha \frac{H}{2} \frac{\cosh(k(z+d))}{\sinh(kd)}. \tag{9.44}$$

The proportionality coefficient was set to $\alpha = 1$ in Qiao *et al.* (2004), and Dai *et al.* (2010) and Pleskachevsky *et al.* (2011), but there are indications that it can be smaller than that.

Finally, the TKE production term P_S can be summarised by taking into account the wave effects of both SM and AM components:

$$P_S = (\nu_{\text{current}} + \nu_{\text{wave}})\left(M^2_{\text{current}} + M^{\text{AM}^2}_{\text{wave}}\right). \tag{9.45}$$

This version of the production term describes the mean flow and includes the wave effects. The symmetric part of the wave motion modifies the turbulent viscous term (influences the mean velocity **U** through the eddy viscosity). The asymmetric part of the wave motion, due to dissipation of the wave energy, imparts the mean current directly and appears explicitly in the term for shear frequency.

To complete the wave-mixing scheme, the transition from laminar to turbulent wave motion has to be parameterised too. This is achieved by using the critical Wave Reynolds Number (7.67) suggested by Babanin (2006).

Pleskachevsky *et al.* (2011) continued on to successfully simulate the laboratory experiment with an explosion-like transition from laminar to turbulent wave motion (i.e. the experiment with dissolution of dye in Babanin, 2006, see also Section 7.5 and Figure 7.21) and the North Sea observations of suspended particular matter. As mentioned above, similar schemes were used by Qiao *et al.* (2004, 2010) on a global scale, where a wave model was explicitly coupled with ocean-circulation models.

Results of Qiao *et al.* (2010) are particularly impressive. The influence of the wave-diffusion term ν_{wave} was tested using the Princeton Ocean Model (POM) and compared with the standard Mellor–Yamada scheme which does not include the wave effects. With the wave-induced (non-breaking) turbulent diffusion, the performance of the model in predicting sea surface temperature, the vertical thermal structure in the upper ocean and depth of the mixed layer – all improved dramatically at latitudes where the wave activity is significant (outside the tropical and polar areas). For example, the correlation coefficient between simulations and data (World Ocean Atlas, Levitus, 1982), for globally averaged values of SST at 35° N, increased from 68% to 93%.

In this non-breaking wave-mixing context, the laboratory experiment of Dai *et al.* (2010) deserves a special mention. While the scale of the small wave tank is incomparable to the open-sea and global-ocean simulations described above, the idea of the experiment is very convincing and highlights the wave-induced mixing most prominently.

The experiments were carried out in a wave tank at the First Institute of Oceanography, Qingdao, China. The wave tank had refrigeration tubes installed under the tank's bottom in order to cool the water from below and create a stable stratification. A vertical array of temperature sensors was deployed to record the water-temperature variations automatically. Nine sensors were vertically spaced 1.0 cm apart, from 4 cm through 12 cm below the water surface. The temperature difference between the upper-most and lower-most sensors was set to 5°C.

Then, the refrigeration was switched off and the molecular diffusion was allowed to take its course. The start time for observations and numerical simulations was chosen as the moment when the ice at the bottom just disappeared. The slow destratification process took about 20.5 hours for the temperature difference to reduce from 5.0°C to 0.6°C.

In the next set of experiments, gentle non-breaking surface waves were generated mechanically once the ice melted. Measurements were conducted at the centre of the tank, away from the boundary layers, in the absence of wind, internal waves, shear currents and other potential sources of turbulence. In all cases the destratification time decreased by some two orders of magnitude. For example, for waves with amplitude $a = 1$ cm and wavelength $\lambda = 75$ cm, i.e. steepness of $\epsilon = ak = 0.08$, it took only 25 minutes to reach the same temperature difference which the molecular viscosity achieved over 20 hours. A one-dimensional diffusion model, with and without wave-turbulence diffusion ν_{wave}, was able to describe both the molecular-diffusion observation and wave-mixing outcomes.

The effect of the wave-induced turbulence is apparent in the experiment of Dai *et al.* (2010), but interesting further questions arise with respect to the role of stratification in such mixing. The stable stratification is known to suppress the turbulence, at least to some extent (see, for example, discussions in Section 9.1.2 with respect to the atmospheric turbulence). This role appears to be small in the described experiment, where numerical simulations were able to reproduce the experiment by accounting for ν_{wave} and without considering the stratification effects, but such a role needs further examination as it has significant implications for the wave-induced mixing through the thermocline. If the wave orbital motions are such that corresponding wave Reynolds numbers (7.70) are above the critical value (7.67) within the thermocline, will the turbulence be generated and effectively facilitate the mixed-layer deepening? The answer is essential for modelling upper-ocean mixing and air-sea interactions at all temporal scales, from hours (e.g. hurricanes) to years (i.e. climate). For example, for hurricanes with their asymmetric structure of the wave fields, asymmetric mixing has been indicated both in measurements and in modelling (e.g. Fedorov *et al.*, 1979; Pudov *et al.*, 1979; Price, 1981).

To finalise this section, we will say that the importance of wave-induced turbulence, both due to the breaking and due to the mean orbital motion, i.e. unrelated to the breaking, is briefly reviewed. The presence of this turbulence poses a principal difference to the wall-layer turbulence schemes, which are often employed to describe the upper-ocean mixing. Near the ocean surface, production of turbulence by breaking dominates over the mean-shear effects, and the wall-layer approach is inapplicable. Immediately below, there is a transitional layer, followed further down by the layer where the shear stresses are dominant and the wall-law logarithmic profile is observed. Characteristics of this profile, however, still 'know' about the wave-breaking turbulence above and to some extent are defined by the breaking at the surface.

A substantial part of this subsection is dedicated to a discussion of the non-breaking wave-induced turbulence. We believe this discussion is relevant here since most of the present-day turbulence-production and ocean-mixing schemes miss this source of turbulence generation and turbulence diffusion, and without it the picture of the upper-ocean dynamics is not complete. Out of the wave-turbulence interaction mechanisms on the water side of the interface, the non-breaking turbulence is weak by comparison with the breaking turbulence only near the surface. While the breaking is the surface-concentrated source of the turbulence, the non-breaking wave-induced turbulence production is distributed through the water column and potentially can be responsible even for the mixing through the thermocline (see also Section 7.5). Needless to say that in the absence of breaking (i.e. swell case), this is the main source of turbulence production by the waves.

In this section, the theoretical background for such non-breaking wave-induced turbulence is outlined, direct numerical simulations of this turbulence are presented, and laboratory experiments and field observations of the mixing due to such turbulence are described. The wave-turbulence diffusion coefficient and production term for Reynolds averaged Navier–Stokes schemes are discussed in detail, and improvements due to their inclusion into water-mixing models, ranging through applications from laboratory wave tanks through sediment transport in a sea to global ocean-mixing applications, are highlighted.

9.2.3 Injecting the bubbles; gas exchange across the surface

The topic of the bubbles on and under the water interface and of their role in gas exchange across the ocean surface is so vast that there are books dedicated specifically to this issue (e.g. Bortkovskii, 1987a; Brennen, 1995) and there are series of international conferences on this particular subject (e.g. International Symposiums on Gas Transfer at Water Surfaces, the 6th of which took place in Kyoto, Japan in May, 2010). Since in the ocean the bubble production and injections are primarily due to wave breaking, which is the topic of this book, we should outline such a connection in this subsection. Note that other natural sources of bubbles also exist, e.g. Langmuir vortices, but at a much lesser extent than those due to the breaking. A review of bubble physics, however, and its relevance for the wave-energy dissipation, ocean dynamics and air–sea exchanges cannot be attempted here. Rather, we will only indicate the problem and point the interested reader to some relevant references.

All but micro-breaking events of ocean waves produce whitecapping which is a bubble cloud. The significance of whitecaps in general and individual bubbles which form the clouds in particular, for remote sensing, for underwater acoustics, and for spray and aerosol production has been discussed extensively throughout the book (e.g. Sections 3.1, 3.5, 3.6 and 9.1.2).

Surface whitecap signatures of the breaking are important in remote sensing, for good or bad reasons, i.e. they either provide a useful signal for measurements of, for example, wave dissipation and waves as a source of spray, for estimates of surface winds, or alternatively an undesirable optical noise when dealing with information based on the ocean colour. In the latter case, knowledge and understanding of the whitecap-bubble generation is still necessary in order to filter their contribution out.

Similarly, the acoustic underwater noise, produced by the breaking waves when the bubbles are formed and by the passive whitecaps when the bubbles coalesce or collapse, can be a useful or unwanted signal. When dealing with the breaking, such noise allows to identify the breaking events, study their statistics, the speed of propagation of breakers and other relevant breaking properties. Acoustic signals emitted by individual bubble-formation events can be used to classify the spectral distribution of the breaking probability and even to measure the severity of the wave breaking and ultimately the wave-energy dissipation due to breaking, which are the most difficult properties to estimate in field observations of ocean waves. At the other end of the underwater acoustic applications, the breaking-bubble-produced sound is a distraction if other sources of underwater bubble production, whether natural or artificial, or other sources of underwater sound are sought.

Bubble bursting at the ocean surface is a primary mechanism of generation of small-scale spray in moderate weather conditions. It plays a pivotal role in negotiating moisture fluxes to the atmosphere and providing aerosols which are then carried up and around by the atmospheric convection and winds, even across continents.

Most essential is the role of the bubbles in the wave-energy dissipation physics. Laboratory measurements estimate from 14% to 50% of the energy lost in a breaking event to be spent on work against the buoyancy forces when injecting the air bubbles into the water in the course of the breaking (see e.g. Section 3.4). If so, the bubble dynamics has to be given a proper account when modelling the wave-breaking dissipation, along with the physics of the turbulence generation by breaking waves.

Since all these features of the wave-breaking bubbles have already been discussed or at least outlined in the sections mentioned above, in this section we will concentrate on the process of injecting the bubbles as such, and on its consequences in terms of the bringing to and dissolving of atmospheric gases in the ocean. As usual, there are a number of physical mechanisms facilitating such exchanges, even if all of them are by means of the bubbles, and there are differences in those mechanisms in benign, moderate and extreme weather circumstances. We will try to indicate this variety too.

Before discussing the injection and exchange issues, however, it is helpful to outline the bubble-field structure. This includes both the external and internal structure of the whitecapping, such as the total whitecap coverage, geometry and spatial distribution of whitecaps, their thickness, as well as bubble size and their concentration, rate of production, and depends both on the breaking frequency of occurrence, i.e. how often the waves break (see e.g. Sharkov, 2007) and the surface distribution of the bubbling patches due to breakings of different scales (e.g. Zaslavskii & Sharkov, 1987), and on breaking severity, whose absolute strength defines the size of the bubbles produced (Garrett *et al.*, 2000; Manasseh *et al.*, 2006) and perhaps the thickness of the foam (e.g. Reul & Chapron, 2003).

The external structure of the whitecapping was broadly discussed in Chapters 3 and 5. The internal structure, before the onset of the remote-sensing (i.e. acoustic) methods of bubble identification, was mainly studied either in the laboratory (e.g. Longuet-Higgins & Turner, 1974) or in the surf zone (e.g. Blanchard & Woodcock, 1957; Raizer & Sharkov, 1980). Field measurements, however, even though extremely difficult since in those days they had to rely on high-resolution photography of the surface and underwater distributions of the bubbles, were also available and produced reliable quantitative estimates (e.g. Kolobaev, 1975; Johnson & Cooke, 1979; Bortkovskii & Timanovskii, 1982; Bezzabotnov *et al.*, 1986).

Bezzabotnov *et al.* (1986) observed the bubble structure in large enclosed seas, i.e. the Baltic Sea and the Caspian Sea, and in the Pacific Ocean, at wind speeds from 9 to 20 m/s. The main observed difference between the seas and the ocean, where the salinity is some three times greater, is the volume concentration N_V of air in the bubbles which was much lower in the Baltic and the Caspian.

In Bezzabotnov *et al.* (1986), bubble structure was statistically and dynamically analysed separately for active and passive whitecaps, as well as for surface and underwater (10 cm below the surface) bubbles. For active breaking fronts, histograms of the bubbles sizes are quite broad and asymmetric, whereas they are very narrow for the bubbles remaining behind the breaking wave. The mean values of these histograms are also very different for the active whitecaps this was about 1 mm, and for the passive foam some 5.5 mm. The speed of horizontal bubble propagation reached up to 1 m/s in the breaking waves, which apparently corresponds to the orbital velocity of the water particles at the propagating wave crest, and in the residual foam the bubbles remained more or less stationary.

Bezzabotnov *et al.* (1986) explain the observations through the difference in lifetime of small and larger bubbles. For the small surface bubbles it is of the order of 0.1 s (McIntyre, 1972; Bortkovskii, 1987a), and by the time the larger bubbles surface those smaller ones disappear. The authors also noticed a negative correlation, which reaches values of −0.6 to −0.8, between bubble size and their horizontal propagation speed. In a way, this is counter-intuitive since this means that slower-propagating breakers produce large bubbles, at least among those which stay on the surface without being injected under the water first.

Below the surface, the structure of the two-phase medium, both the bubble-size distribution and void fraction, depends on the breaking severity. The latter, according to Bezzabotnov *et al.* (1986), can be characterised by the area of the whitecap and orbital velocities in the active breaker. Statistically, in their study the whitecap area depends on the wind (in this regard, see review of the dependences of the whitecap coverage W on the wind speed in Section 3.1).

The weak breakers are characterised by a broader distribution of the bubble sizes below the surface, with the distribution having a relatively larger fraction of bigger bubbles if compared with the stronger breakers, which observation is consistent with the above-mentioned negative correlation of the breaker speed and surface bubble-sizes. In time, bubble-size distributions narrow down and their median value shifts toward smaller bubble radii. This is due to smaller bubbles being injected deeper and surfacing slower because of weaker buoyancy forces. Ultimately, the mean bubble size becomes that of the minimal bubbles originally generated. The rate of decreasing of such mean bubble sizes is greater for weaker breakers, apparently due to the fact that they produce fewer large bubbles and inject them down to a smaller depth (see, however, model of Garrett *et al.*, 2000, in this regard).

According to Bezzabotnov *et al.* (1986), the typical behaviour for all the observed underwater bubble distributions is a rapid and nonlinear reduction of the void fraction N_V in time: at 10 cm below the surface, this air-volume concentration drops 4–6 times 1 s after the breaking and falls to a mere 1–5% of its original magnitude during the following 2 s for weak breakers, and over a further 5 s for moderate-strength and 8 s for strong breaking events. Bezzabotnov *et al.* (1986) classified the breaking strength not in terms of the wave properties, but in terms of the wind speed U_{10} which classification produces some ambiguity as to what actual breaking severities are compared: $U_{10} = 10$–$13\,\mathrm{m/s}$ corresponds to weak, 14–16 m/s to moderate and 17–19 m/s to severe breaking. In principle, some relation between wind speeds and the properties of the waves observed in field experiments of Bezzabotnov *et al.* (1986) can be figured out by using their parameterised dependence between orbital velocities in the breakers and U_{10}:

$$u_{\mathrm{orb}} = 3.7 \cdot 10^{-4} U_{10}^{2.67} + 0.13 \tag{9.46}$$

where all the velocities are in m/s.

For the weak breaking, quantitative time dependence is suggested:

$$N_V(t) = N_{V_0} \exp(-t/t_0) \tag{9.47}$$

where N_{V_0} is the initial air-volume concentration and $t_0 \approx 0.7$ s is the damping constant. Overall, if all the data of Bezzabotnov *et al.* (1986), corresponding to the different winds and various water bodies, are put together, the dependence of N_V on wind speed can be approximated as follows:

$$N_V(t) = 0.007 U_{10} - 0.042 \tag{9.48}$$

where N_V is dimensionless (cm^3 of the bubble air per cm^3 of the total water-bubble volume) and U_{10} is in m/s.

An interesting observation was that at wind speeds $U_{10} > 17$ m/s, at the measurement level of 10 cm below the surface, small bubbles of diameter less than 0.5 mm are always present. While quantitatively the volume concentration of this background void fraction, $N_V \sim 0.1\%$ of the concentration in the active whitecapping, is quite small, qualitatively this wind speed indicates a transition to a new phase of air–sea interaction in extreme-wind conditions, when the bubble and gas exchange through the interface becomes a continuously distributed feature rather than sporadic and intermittent.

Johnson (1986), Hwang *et al.* (1991), Thorpe *et al.* (1992) and Bortkovskii (1997) pointed out another environmental dependence for bubble formation, entrainment and collapse – on the water temperature. This apparently happens through temperature dependence of the air–water fluid properties such as cavitation stability, surface tension, air-to-water diffusivity and water viscosity. The latter, for example, can change more than two times under realistic temperature variations of the water (7.77).

The onset of the acoustic methods of bubble detection and measurements (Thorpe, 1982) signalled, in a way, a new era in this kind of investigation. The theory and practice of these methods was discussed in Section 3.5, and here we will only outline applications relevant to the present section.

Such methods do have their limitations when dealing with individual bubble events, under high void-fraction concentrations, because of the peculiarities of the acoustic-wave propagation through the two-phase medium with essentially different densities and phase speeds in the air and the water. Overall, however, the capabilities of the active acoustic sensing by sonars and even of passive listening to the bubble-produced noise by hydrophones are much greater by comparison to the photographing of the bubbles in the hostile and non-stationary environment, particularly as far as the underwater investigations are concerned.

These methods were reviewed and discussed in Section 3.5 with respect to detection and quantifying the wave-breaking events as such. In short, they either make bubbles ring by exciting them artificially by means of active acoustic radiation emitted by sonars, or utilise the acoustic signatures of bubbles naturally produced when the bubbles are created, coalesce or collapse.

As an example of a modern application of acoustic methods to researching the bubble-cloud structure, we refer to the recent study by Gemmrich & Farmer (2004). They used free-floating underwater acoustic resonators to determine the size-distribution of microbubbles with radii of 20–200 μm, at a sampling rate of 2.2 Hz. As mentioned above, the method is not suited for high levels of void fraction, but for $10^{-8} \leq N_V \leq 10^{-4}$ it produced data hardly possible to achieve otherwise. Note that these magnitudes of air fraction are much lower than those in (9.47), but Gemmrich & Farmer (2004) measured this fraction some 10 times deeper than Bezzabotnov *et al.* (1986) and dealt with microbubbles which would be impossible to detect by the photographic method of the latter study in principle.

In particular, Gemmrich & Farmer (2004) analysed in detail the underwater signatures of a breaking wave with period $T = 5$ s, recorded at the depth of approximately 1 m below the surface. These are noticeable both in the velocity field (turbulence) and void fraction (bubbles).

Turbulence due to the breaking causes the standard deviation of fluctuations of the vertical component of the orbital velocity to rise six times. The peak of this rise almost coincides with the passage of the crest of the breaker above (slightly lags) and the overall duration of this response lasts for a quarter of the wave period. The volumetric turbulence-dissipation rate ϵ_{dis} increases by more than three orders of magnitude. It is interesting that the generation of the turbulence precedes the arrival of the whitecapping crest, and 0.7 s before this arrival ϵ_{dis} has already grown some 100 times. High levels of the turbulence dissipation rate, i.e. an order of magnitude higher than the background dissipation, persist for at least ten wave periods.

Bubble production and the corresponding air fraction also respond to the breaking passage most sensitively. The background radius of the microbubbles at 1 m depth is less than 100 μm, which radius rises to greater than 300 μm following the air entrainment due to the breaking. Estimates show that the increased vertical velocities are sufficient to counteract the buoyancy of 200 μm bubbles, and this agrees with the dynamics of the bubble-size distributions observed. These larger bubbles leave the sensor depth some 10 s (two wave periods) later, but the mean bubble radius does not drop to the pre-breaking value for an extended period of time. Due to the higher levels of the turbulence mentioned above, the size of the largest bubbles remaining in suspension, although decreasing gradually, stays greater than 100 μm, up to 120 μm.

The background air fraction was $N_V = 5 \cdot 10^{-6}$. One second after the passage of the whitecap it starts rapidly growing and 2 s later reaches the saturation level $N_V \sim 10^{-4}$ of the resonator located at 0.85 m depth. It remains at this high level for two wave periods, which correspond to the larger-bubbles being kept in suspension as described above. Afterwards, N_V starts to decline gradually, but ten periods after the breaking it is still twice the background level, apparently due to the buoyancy of those larger-than-background microbubbles still being counteracted by the enhanced turbulence.

Using the sonars and resonators requires active acoustic probing, and it has been shown that passive acoustic methods are capable of achieving similar accuracy and have potential advantages in many regards. These are simply techniques tuned to listening to the subsurface sounds by means of an underwater microphone (hydrophone) and of more or less complicated pre- or post-processing of the acoustic data. If the data are pre-processed, then requirements for the data storage can be remarkably low, for example, just the time of occurrence of a bubble-formation event and the bubble size rather than acoustic records read at a kHz-sampling rate. Being a passive device, the hydrophone energy consumption is also very low by comparison with sonars and therefore such instrumentation can be operated on batteries for extended periods of time or even for routine observations. In addition, hydrophones are not necessarily limited to a particular area or volume of sensing (see e.g. Manasseh *et al.*, 2006, and Section 3.5 for relevant discussions).

The capability of the passive acoustic method of Manasseh *et al.* (2006) in resolving sizes of individual bubbles in the course of breaking, in obtaining the distribution of breaking probability across different scales of waves with a continuous spectrum and even in measuring the severity of the breaking both in the field and laboratory are demonstrated and discussed in Sections 3.5 and 6.2. In particular, it was shown that in a laboratory experiment with monochromatic 2.8 m-long breaking waves, the radius of bubbles produced ranged from 1.1 mm to 1.9 mm depending on the breaking strength (Figure 3.10). In the field, for waves with a continuous spectrum with peak frequencies in the range from 0.3 to 0.5 Hz and significant wave height from 0.2 to 0.4 m, the size of the bubbles generated was also in the range from 1 mm to 2 mm (Figure 6.6). This is, however, not a random scatter, and these passive-acoustic observations reveal interesting dependences for the bubble size of breaking occurring in a variety of environmental conditions, with wind speeds from 10 m/s to 20 m/s and across the spectral band ranging from $0.8 f_p$ below the spectral peak f_p up to double peak frequency $2 f_p$.

Now, moving to the process of injection of the bubbles by the breaking waves, this process can be approximately subdivided into two essentially different phases: the bubble-submersion stage and the bubble-rise stage (e.g. Bortkovskii, 1998). Because of these two dynamic phases, the structure of active and passive whitecaps is also different as described above. Dynamically, in the course of active breaking, the structure is more or less permanent, but it rapidly changes following the breaker passage – towards the quasi-stationary residual bubble foams with the mean bubble size several time larger. At strong wind speeds, this transition occurs in the space of 0.5–2.5 s.

Bortkovskii (1998) assumes that descending water motion is formed beneath the breaking crest which transports the bubbles created at the surface downward, and that the velocity of this motion gradually decreases with depth. Entrainment of the bubbles at the surface followed by their vertical submersion can happen in a number of ways. For spilling breakers, for example, Koga & Toba (1981) demonstrated in laboratory experiments that at the front edge of the bulge sliding down the front face of a breaking wave, diffusive penetration of the bubbles into the water happens, having significant vertical velocity components. For plunging breakers, the entrainment is direct due to the impact of the plunging jet. The vertical velocity of the core of the plunging water volume u_{pl} decreases exponentially away from the surface, and at the wind speed $U_{10} = 15$ m/s in field observations, for example, Bortkovskii (1998) measured the following dependence:

$$u_{pl} = 70 \exp(-0.011 z) \tag{9.49}$$

where this velocity is expressed in cm/s. Accordingly, close to the surface, measured bubble velocities are of the order of 60–100 cm/s (Bezzabotnov *et al.*, 1986). It is interesting to note that these values are actually close to estimates of the vertical component of the orbital velocity at wave crests with the limiting inclination (2.56), corresponding to the breaking onset, which should be of the order of 35–60 cm/s (Longuet-Higgins & Turner, 1974).

Estimates show that the bubbles reach their lowest position in several seconds (Bortkovskii, 1998). Production of the bubbles of all sizes stops simultaneously, upon the passage of the whitecap, and the entrainment rate of bubbles of all sizes by the downward water motion is the same, at least in the plunging jet. The buoyancy of the bubbles of different sizes, however, is different and therefore their rates of surfacing are not the same. As a result, depth distribution of the bubble sizes is not uniform: smaller bubbles penetrate deeper. Again, temperature plays a role here. The lifetime of the bubbles in cold water is longer (3.18), and therefore the mass loss is slower, buoyancy is higher and depth penetration is smaller (Hwang *et al.*, 1991; Bortkovskii, 1998).

Transfer of gases from the atmosphere to the ocean due to bubbles is a well-known phenomenon (e.g. Thorpe, 1982, 1992; Merlivat & Memery, 1983; Melville, 1996; Woolf, 1997; Donelan, 1998; Bortkovskii, 2003; McNeil & D'Asaro, 2007, among many others), although the exact mechanisms are perhaps not that well understood and described. At lower wind speeds, turbulent and molecular diffusion of gases is essential. For higher wind speeds in excess of 10–12 m/s, conclusions have been made that the bubbles may dominate the air–sea gas exchange (Thorpe, 1982; Merlivat & Memery, 1983). In any scenario, the diffusive gas exchange weakens as the gas-concentration difference across the interface decreases at strong winds (e.g. Merlivat & Memery, 1983; Oost *et al.*, 1995), but the wave breaking, which significantly enhances the gas transfer through the bubble injection (e.g. Jahne, 1990), is on the rise.

In the bubbles, the pressure is higher than in the surrounding waters and therefore the gas transfer into the water continues even if it is already saturated and even over-saturated under the given pressure (e.g. Bortkovskii, 2003). Therefore, bubbles enhance the gas flux from the atmosphere to the ocean. A model capable of describing such fluxes, which agrees with observations of bubble production at strong winds up to 20 m/s, was suggested by Bortkovskii (2003).

McNeil & D'Asaro (2007) tried to explicitly parameterise the air–sea gas fluxes at winds even stronger than that, those fluxes measured during Hurricane Frances in 2004 at U_{10} winds up to 55 m/s, by means of air-deployed floats. In particular they intended to answer questions regarding whether there are existing parameterisations not suitable at such winds, what functional dependences could possibly describe such fluxes across the entire range of wind speeds and, most importantly, what are the actual physical processes responsible for the air–sea gas exchange in hurricanes?

The models to be excluded are those which predict cubic dependence on the wind speed (Wannikhoff & McGillis, 1999). According to McNeil & D'Asaro (2007), at hurricane wind speeds the fluxes still grow rapidly, but in relative terms at a much lower rate by comparison with the moderately strong winds.

A change of regime of the air–sea gas transfer is observed at

$$U_{10} > 35\,\mathrm{m/s}. \tag{9.50}$$

This is remarkably close to the wind speed which characterises saturation of the sea drag coefficient (9.12) and of the surface-wave asymmetry (3.26). The drag coefficient C_D describes the momentum flux from the atmosphere to the ocean, and the asymmetry signifies the breaking probability and/or the breaking mechanism on the surface itself, and such coincidence of the regime change for dynamic and gas-transfer properties of air–sea interactions can hardly be random. Some fundamental physics should be responsible for the alteration of all exchanges at the ocean interface at the same time, even though sea drag and bubble-injection rates seem to be weakly related phenomena. As usual, such physics is necessarily complex and hardly limited to a single phenomenon, but the wave breaking may be one of the dynamics involved as it relates to both the surface roughness and bubble production.

In this high-wind regime of gas transfer, bubbles of 1 mm radius and smaller are transported down by vertical currents with speeds of 20–30 cm/s and mostly dissolve due to hydrostatic compression before they reach depths of some 20 m. This additional mechanism of air injection deeper into the upper-ocean layer leads to supersaturation of the sea water with the gases. McNeil & D'Asaro (2007) make a conclusion about the dominant role of the bubble injection in gas transfer in such extreme conditions.

As intended, McNeil & D'Asaro (2007) proposed a parameterisation which spans the entire range of natural wind speeds including hurricanes. This accounts for both under-critical mechanisms of gas transfer, that is the turbulent diffusion of microbubbles through the interface and direct ventilation by means of wave breaking, and the super-critical supersaturating dissolution of the subducted bubbles.

Thus, to conclude this last section of Chapter 9 we should say that we have briefly reviewed a very large topic of bubble production and air-to-sea gas transfer due to the breaking. Three main issues were outlined, the structure of the bubble clouds, the dynamics of bubble submersion and surfacing, and the role of the bubbles in gas transport to the upper ocean.

Overall, Chapter 9 is dedicated to the various roles the breaking plays in the broader system of air–sea interactions. Section 9.1 describes these roles with respect to the atmospheric boundary layer. These include the dependence of sea drag on the breaking, the production of spray by the breaking, and discussion of the boundary-layer physics in extreme weather conditions. While the effects of the breaking on the drag coefficient are secondary, that is the breaking alters the sea drag which is due to turbulent fluxes from the wind to non-breaking waves, the wave-breaking part in the generation of spray is primary. Most of the spray is produced by the breakers. In extreme conditions, the physics of this generation and the structure of the droplet distributions change and that is linked to other changes in the boundary-layer physics. The significance of spray and aerosol production for other fields of atmospheric sciences and applications is also discussed.

The upper-ocean section, 9.2, apart from the topic of the present subsection, discusses in some detail schemes of the momentum and energy transfer from the wind to the ocean and of the water-turbulence generation. This is done in a more general context than just the influence of the breaking, because without accounting for other effects of the dynamic transfer and turbulence behaviour such a description would be incomplete. The roles of breaking, however, are most essential in all these processes, and in the respective subsections they are identified and highlighted.

10

Conclusions. What else do we need to know about wave breaking?

Wave breaking represents one of the most interesting and most challenging problems for both fluid mechanics and physical oceanography. It is an intermittent random process, very fast by comparison with other processes in the wave system. The distribution of wave breaking on the water surface is not continuous, but its role in maintaining the energy balance within the continuous wind–wave field is critical.

The challenges thus outlined make understanding of such wave breaking and even an ability to describe its onset very difficult, and as a result knowledge of the physics of the breaking, and even practical parameterisations of the phenonemon have been hindered for decades. Recently, knowledge of the breaking phenomenon has significantly advanced, and this book is an attempt to summarise the facts into a consistent, even if still incomplete, picture of the phenomenon.

If this picture were to be formulated into a few paragraphs of these conclusions, we would like to say the following. The waves break because the water surface reaches some limiting steepness. Apparently, the fluid interface has to have a limit beyond which it will collapse. In the system of nonlinear water-surface waves subject to a variety of external forcings and internal instabilities, there are a number of physical processes which can lead to such a steepness. They are the modulational instability, linear (dispersive and directional) and nonlinear (amplitude-dispersion) focusing, modulation of steepness of shorter waves by longer waves, direct forcing by the wind (if very strong) or by the current, and wave-bottom interactions, among others. It is not, however, these processes, for example, not the modulational instability or the focusing which make the waves break; their importance is only in bringing the wave steepness to the limit. In nonlinear wave fields with a continuous frequency-directional spectrum, both the nonlinear evolution and the focusing are always active, but if the result of their activity does not lead to the limiting steepness, the wave subjected to such action will not break. The wave will only collapse if it is steep beyond the limit. Breaking is a surface phenomenon and therefore subsurface and other accompanying features related to the breaking are simply indicators that the steepness limit is being reached.

While the breaking as such is triggered by the limiting steepness regardless of the mechanism which led to this steepness, in terms of the dynamics of the wave-energy dissipation due to the breaking the wave appears to 'remember' what made it that steep. This dynamics and the resulting spectral-dissipation signature of the modulational-instability breaking is different to that of the linear-focusing breaking. The former leads to loss of energy by the primary waves, accompanied by the cumulative effect at higher frequencies, whereas the latter mainly results in the dissipation of smaller-scale waves. There are indications that breakings brought about by other physical mechanisms, e.g. amplitude focusing, are characterised by yet different spectral signatures.

Thus, the potential reasons for the waves reaching the breaking limiting steepness may be different, but there are multiple indications that in moderate-wind deep-water conditions modulational instability is the most likely cause of dominant breaking. The answer to the question of whether this kind of instability is active in directional wave fields with typically observed mean steepness, emerges as positive. This conclusion does not deny the breaking due to, for example, linear superposition, but the probability of the latter is low since the limiting steepness would require a superposition of a large number of ocean waves of the typical background steepness. Breaking of waves shorter than dominant scales is induced by the underlying longer waves, particularly of those further away from the peak in Fourier space.

Apart from dissipation of the wave energy, breaking plays other important and sometimes pivotal roles across many processes at the ocean interface, as well as above it and below. Among those most essential in the system of surface waves are the downshifting of the wave energy, defining the level of the wave spectrum, enhancement of the energy and momentum input from the wind to the waves. The last feature illustrates the connection of wave breaking with the dynamics of the atmospheric boundary layer where the breaking alters the sea drag and ejects spray, which effect is particularly significant in extreme weather conditions. Below the ocean surface, the breaking is a major player in the upper-ocean mixing schemes, in the production of turbulence and generation of bubbles which facilitate the air-to-sea gas transfer.

In this Conclusion, we will avoid using references unless the relevant papers have not been mentioned before – the references are very detailed in the chapters above. A wide spectrum of achievements in the research on wave breaking, and research related to wave breaking, have been attended to through the nine chapters. The first six chapters largely follow the review of Babanin (2009). This review was dedicated to studies of the breaking phenomenon as such, but respective chapters in this book had to be updated or even essentially rewritten. Progress in the field is rapid, and in two years since publishing the review the advances have to be noted, and some pending questions have already been answered, notably the question about modulational instability in directional wave fields.

Introductory Chapter 1 presented the topic and brought in the concept of wave breaking. This concept is initiated through its main physical consequence – dissipation of wave energy.

Chapter 2 formulated definitions pertinent to wave breaking, to be used throughout the book. One of the advantages of presenting the wave-breaking phenomenon is the fact that, unlike many other physical processes, including those oceanographic, this phenomenon is apparent even to non-specialists. Everybody has a perception of wave breaking on the ocean surface. In a way, however, this is a disadvantage too, when it comes to accurate descriptions of the process and its properties. Phases of the breaking progress and its characteristics are not necessarily commonly defined because they are left to obvious intuitive understanding.

Thus, for example, wave-breaking onset is often investigated by measuring properties of the breakers which have already produced whitecapping and related acoustic, optical or other distinct signatures. It is argued in the book that, if whitecapping is present, then the wave is already collapsing, that is, the breaking is in progress, which is characterised by physics different to that which led to the breaking onset. And it is hard, if not impossible to understand the causes of the collapse by investigating its progress and consequences rather than pre-history. In the case of wave breaking, this is even more complicated due to the fact that the dynamics of the breaking progress differ essentially depending on the physics which led to the breaking onset as mentioned above.

Therefore, Chapter 2 suggested a definition of the breaking onset and classification of different phases of the wave breaking. It defined the breaking probability and breaking severity which is broadly employed throughout the book, and a combination of which determines the wave energy dissipation. Types of breakers were outlined too, and breaking criteria described in some detail. Finally, some unrelated to breaking general knowledge of wave features and terminology, frequently used in the book, such as dispersion relation and radiative transfer equation, were introduced for completeness.

Chapter 3 described methods of detecting and measuring wave breaking. There is a great variety of these. We started from traditional means of observing the whitecaps, which have regained importance in recent years in the light of advances of remote-sensing environmental applications. Other methods include those visual and instrumental, experimental and theoretical. Experimental techniques were divided into contact and remote-sensing measurements, laboratory experiments and field observations. These were further subdivided, laboratory methodologies into those dealing with deterministic and random wave trains and fields, remote-sensing methods into acoustic, optical, infrared and radar applications, the acoustic techniques into active probing and passive listening. Theoretical approaches were separated into two sections, dedicated to analytical and statistical methods.

Contact measurements are direct and obviously most accurate, as they serve as calibration means for remote-sensing techniques and theoretical approaches to begin with. It should be emphasised that measuring the surface elevations (wave height) in very steep and nonlinear waves, such as those at the breaking onset or in the process of breaking, is not always sufficient to estimate the respective wave energies. Relationships between the potential and kinetic energies in strongly nonlinear waves are not known, and therefore measurements of the wave-motion velocity field are necessary to obtain the kinetic and thus the total energy. The good news, however, is that the surface-wave nonlinearities appear to decay rapidly away from the surface. Therefore, at least in deep water, we would expect the deviations from linear wave orbits to be essential only very close to the interface, perhaps even within the crest-trough volume.

The contact measurements are apparently more difficult, and often impossible in the field, given the sporadic, powerful and destructive nature of the breaking events to be measured. Many elaborative remote-sensing methods have been developed to help this problem. The infrared techniques based on disruption of the surface skin layer, for example, have been shown not only to detect the breaking occurrences, but also to indicate the breaking severity.

Most promising in this regard, in our view, are the passive acoustic methods. In particular, the bubble-detection method allows us to identify the wave which is breaking and producing bubbles, and therefore to obtain the distribution of breaking occurrences and of rates of such occurrences along the spectral frequency and wavenumber scales. The size of the produced bubbles can also be estimated which has a connection with the wave-energy loss, and thus information on the breaking severity is also available within this method. So, the technique has a demonstrated potential to measure all the elements of the spectral distribution of the wave-energy dissipation, and to do that by using a hydrophone. The hydrophones are simple and cheap sensors, and can be deployed below the surface and thus avoid the violent power of the wave breaking as such. Their energy consumption is small and they can be operated on batteries for long-term observations. Information on the time of bubble appearance and on its size is basically just two numbers, and therefore memory requirements for such devices are very modest too.

Analytical methods, also described in Chapter 3, allow us to point out the presence of breaking events in wave records where the breaking was not registered by any experimental means. Since there are large quantities of wave records available, such methods can be very helpful in enhancing their value in the case where some information on breaking occurrence and breaking statistics is necessary, even if approximate, but was not recorded directly.

Also in the absence of direct counting and measuring the wave breaking, a set of statistical approaches outlined interprets the probability functions in order to infer information on the dissipation due to wave breaking. If compared with other theoretical methods intended for this purpose, this method has the least limiting underlying assumptions. Basically, all that is needed are the valid maximal and minimal experimentally confirmed values of wave steepness of individual waves achieved in the course of breaking events.

Chapter 4 has been partially re-written compared to the review (Babanin, 2009). The first section of this chapter remained. It is dedicated to numerical simulations of the evolution of nonlinear waves leading to the breaking onset, and of the imminent-breaking stage itself, by means of analytical approaches. The fully nonlinear Chalikov–Sheinin model was used, which was also coupled with a model for the atmospheric boundary layer, and allowed investigations of the wind influences on breaking onset. Many new and interesting findings of this section were further verified, investigated and researched into the stages beyond the breaking onset in experimental Chapters 5 and 6.

The former second section of Chapter 4 on the kinematics of the wave-breaking onset has now been removed: this was a scientific criticism of one of the analytical approaches and was important to point out in the journal review, but does not have an independent value for the book. Instead, a new section on Lagrangian fully nonlinear methods of wave-breaking numerical simulations was added.

Chapters 5 and 6 are the core of the book, discussing wave-breaking probability and severity. Here, knowledge of the first four chapters is used, that is, understanding of the breaking concept, of the definitions, of the breaking measurements and of fully-nonlinear wave behaviour. The structure of Chapter 5 is essentially altered compared to the earlier review.

The breaking probability was discussed for the monochromatic (or rather quasi-monochromatic, since the unstable sidebands will grow from the background noise if they were not imposed initially) wave trains in Section 5.1 and for waves with a continuous spectrum in Section 5.3. Section 5.2 is dedicated to the important feature of the wave-breaking threshold in terms of the background mean wave steepness. It is not to be confused with the limiting steepness of individual waves which signifies the breaking onset.

The threshold behaviour means that the breaking in the wave train/field does not happen unless the mean steepness is above some certain level. Once it is above this level, then the breaking rates/probabilities are determined by the magnitude of excess of mean background steepness over the level. In the case of a continuous-spectrum environment, there is a threshold value for the spectral density at each spectral scale.

The significance of the threshold aspect of wave breaking, which is a certain observational and experimental feature, should not be underestimated. At present, for example, none of the operational wave-forecast models account for the threshold behaviour in their dissipation functions, although dedicated developments in this regard are already under way. This means that the models implicitly employ the physics which is not adequate in some circumstances and will predict an active whitecapping dissipation in environments where the waves do not actually break (for example, swell).

It should be emphasised that wave trains/fields falling below the breaking threshold (i.e. such conditions where the breaking does not happen) do not, however, represent some different kind of nonlinear behaviour compared to those waves above the threshold. All the nonlinear features of the below-threshold wave fields should be the same, just of a lesser magnitude of course, with allowance for the smaller steepness/nonlinearity. What is different is that individual waves in the course of such impaired nonlinear evolution do not reach a limiting steepness beyond which the water surface collapses.

This limiting steepness of $Hk/2 \approx 0.44$, which follows both from the fully nonlinear numerical simulations of the breaking onset as a transient wave condition, and from the experiments, is remarkably close to the well-known 150-year old Stokes limit. The waves at the breaking onset, however, as they are simulated and observed, do not look like the stationary point-crested Stokes waves and rather exhibit the characteristic features of the crest instabilities (that is the crest is leaning forward while the wave is still symmetric) and other transient dynamic properties. In this regard, it is not obvious whether the agreement is coincidental or based on fundamental limitations for the water surface in the gravity field.

In the context of this limiting breaking-onset steepness, modulational instability and background mean wave-train steepness, the issue of freak waves is discussed. It is concluded that the modulational instability can produce the freak waves if the background mean steepness is close to $ak = 0.12$–0.13.

The breaking threshold and the limiting steepness should be equally important to all the breaking waves, but it appears that in the spectral fields these arguments become of secondary importance for waves relatively short with respect to dominant waves. The amount of short waves breaking due to inherent reasons, i.e. due to the same reasons as the long waves, is small compared to their induced breaking. The latter can be triggered by the dominant breaking, or caused by modulation of the short waves with the carrying longer waves which modifies the steepness of short waves, or both.

In any case, the experimental evidence shows that the induced breaking essentially changes the breaking rates of shorter waves, and at spectral scales greater than some $3f_p$ the induced breaking completely dominates over the inherent breaking, to such an extent that the breaking probabilities at those frequencies should be disconnected from the local spectral density and be determined by the integral of the spectrum over frequencies below, including first of all the spectral peak. In passing, we will mention that this frequency should also identify the transition from the ω^{-4} to ω^{-5} behaviour at the spectral equilibrium interval, as the latter associates with the purely breaking-defined spectrum tail (this is discussed in detail in a different chapter in Section 8.2).

The induced breaking brings about the so-called cumulative effect, that is, at each higher frequency the breaking is determined by all the waves with frequencies below. As with the breaking-threshold feature mentioned above, the cumulative effect is a definite observational physical property which is missing in the present dissipation functions. That is, all the operational models and the majority of the research spectral wave models currently treat whitecapping dissipation at smaller scales in an unrealistic way.

Following the logic of the fully-nonlinear analytical and numerical investigations of the breaking in quasi-monochromatic two-dimensional wave trains in Chapter 4, Section 5.1 investigated the evolution of nonlinear waves to breaking, breaking onset and wind influences for such wave trains, and provides direct measurements of a great variety of properties of an imminent breaker. The wave-breaking probability is parameterised as the distance to the breaking dependent on initially monochromatic steepness of the nonlinear trains.

Section 5.3 described the breaking of spectral waves, and its structure is necessarily different to that of Section 5.1. Some analogies between monochromatic wave trains and spectral wave fields are applicable to the dominant waves in the spectrum which correspond to the narrow-banded spectral peak and therefore exhibit the characteristic features of modulated wave groups. For these waves, parameterisation in terms of background mean wave steepness is possible, in a way similar to that above based on the IMS.

The analogies clearly do not work at smaller scales away from the spectral peak. Here, depending on how far the scale is from the peak, breaking of short waves is induced by the longer waves one way or another. Wind-forcing effects on the spectral distribution of the breaking are also discussed.

Particularly challenging is understanding the directional distribution of the breaking probability. In the context of directionality, Section 5.3.3 also contains a new important result on the efficiency of modulational instability in the directional wave fields. Here, the directional modulation index was introduced, by analogy with the Benjamin–Feir index for two-dimensional wave trains, which describes the balance between steepness and directional spread. It is shown that for the magnitudes of this index typical for the oceanic waves, the modulational instability remains active.

Section 5.3.4 discussed another breaking threshold, the wind speed below which the wave breaking does not happen. Estimates of this range from 3 m/s to 8 m/s and the discrepancy may be due to natural rather than observational-accuracy reasons. Transition from no-breaking to breaking is important across the entire range of the air–sea features, that is in the context of wave evolution, of the near-surface underwater turbulence and of the sea drag (see also Section 9.1.1).

If the results of Chapters 4 and 5 are combined, the most important conclusions are the breaking threshold in terms of the background statistical wave steepness, which is applicable both to the wave trains and wave fields, the limiting value of individual wave steepness, which identifies a breaking onset for this wave, and the induced-breaking effect for short waves modulated by longer ones. The directional behaviour of the breaking probability is not trivial and very intriguing, but is also very unclear at the present stage. Wind-forcing impacts, both for the breaking probability in wave trains and in the fields, are marginal and even negligible unless the winds are very strong.

Due to rather incomplete knowledge on the wave-breaking severity, Chapter 6 dedicated to this topic was relatively short. It incorporated two sections which describe the breaking strength of waves in modulated wave trains and in waves with a continuous spectrum, respectively. Again, as with the many topics mentioned above and below, important updates to this subject have happened recently which were introduced in the chapter, in addition to the material of the respective sections in the review.

The breaking severity appears to vary in a very broad range, virtually between 0% and 100% energy loss in an individual breaking event. The most important finding in this regard is the role of the wind which appears to control the rate of development of the modulational instability. While wind influence has always been mentioned as marginal as far as the breaking onset and breaking probability are concerned, it emerges as being of primary significance for the breaking strength.

As shown in laboratory experiments with modulated wave trains, the larger wind forcing increases the breaking probability, but decreases the breaking severity, to the point of practically cancelling the breaking as such. It appears, however, that the wind achieves this effect not directly, but rather by means of gradually affecting the instability growth in nonlinear wave groups. Slowing down of this instability, because of the wind, is indicated by the depth of the wave-group modulations in the course of wave-train evolution. As a result, the modulation depth of wind-forced trains is much smaller compared to those unforced, and this appears to lead to the different energetics of the breaking once it happens.

In Section 6.1, a laboratory experiment was described, where dependence on the depth of the modulation was isolated: that is variation of this depth was achieved by purely hydrodynamic means, without involving the wind. The dependence indicates a robust trend between the depth of the modulation and breaking severity. Significant scatter of the data points shows the importance of other mechanisms superimposed over the observed trend and the complexity of the problem, even in the absence of the wind. The influence of the wind on the depth of the modulation, if added, further contributes to this complexity as it is most essential.

Since spectral models of wave forecast are one of the main consumers of dissipation-function knowledge, it should be mentioned that the modulation depth will be a difficult property to account for in such models where information on the wave-group properties is basically lost, let alone on the depth of the modulation and development of instability modes. Such relevant properties, e.g. rate of instability growth, will have to be parameterised through the characteristics of the wind forcing and available wave properties, for example, the steepness of appropriately bandpassed waves.

The last three chapters of the book are new compared to the review. These chapters almost doubled the volume of the original review by considering breaking-related topics in addition to the research on the wave breaking itself in the previous chapters. Their investigations and significance are in many regards equivalent to the wave-breaking research as such, and the separation of the wave-breaking studies and wave-breaking-related studies is in fact quite superficial. For example, topics of whitecapping dissipation of surface-wave energy, spray generation in the atmospheric boundary layer, and bubble production in the upper ocean are inseparable from the topic of wave breaking itself.

Indeed, research on the whitecapping dissipation routinely stands as almost a synonym for wave-breaking studies. The wave-breaking chapters looked in detail at the breaking probability and breaking severity, and these two features combined basically define the dissipation. As it happens, however, the dissipation research is not a straightforward amalgamation of the two topics, but has rather taken on its own direction of investigations. A variety of innovative analytical, statistical, probabilistic, numerical, experimental and observational approaches have been developed to deal specifically with the whitecapping dissipation. Effectively, this is a research area in its own right which constitutes a large volume of knowledge, and had to be reviewed separately in Chapter 7, particularly given its great applied importance with respect to wave-forecast modelling.

Chapter 7 started with a brief review of analytical theories which have tried to explicitly investigate the wave-energy dissipation (Section 7.1). These interpret pre-breaking and post-breaking features of the wave fields, bypassing the stage of individual waves as such. Here, we have added a new type of such analytical models, the kinematic-dynamic model, to the classifications available in previous reviews of these modelling issues.

Explicit modelling of the breaking in progress is a complex problem of fluid mechanics, where a two-phase turbulent medium has to be simulated based on first principles, with the strongly nonlinear wavy air–water interface, filled with air bubbles and cavities at and below the water surface, and with spray droplets in the air. Such a task is demanding analytically and expensive computationally. It is, however, possible nowadays and has been a subject of active development over the past decade, predominantly with respect to engineering applications ranging from the impact of breaking on structures to simulation of breaking and its consequences in the surf zone. A variety of such models are available these days and some of them have attempted to investigate different aspects of the wave-energy dissipation directly during the breaking in progress. This topic was discussed in Section 7.2.

The main difficulty with respect to the use of the phase-resolvent models in systematic investigation and parameterising the dissipation due to wave breaking is their heavy demand in terms of computational resources. It is possible to simulate a few periods of the breaking in progress, but it is not feasible to simulate the evolution of nonlinear wave trains leading to the breaking onset over tens of wave periods. Here, an obvious compromise can be employed. Such evolution is described very well by the fully nonlinear potential models, which are less computationally expensive, as described in Chapter 4. As discussed in this chapter, such potential models, however, cannot go past the breaking onset, mostly because their physics becomes inadequate: that is, they are not designed to describe turbulence, bubbles and spray. This is what the two-phase models are designed to do, and therefore the apparent solution would be to couple the output of the fully nonlinear potential models with the input of the two-phase models – at the point of the breaking onset.

In Section 7.3, direct and indirect measurements of the dissipation were discussed. These start with a description of dedicated laboratory experiments followed by a subsection which concentrates on and summarises the difference between signatures of the modulational-instability and linear-focusing breaking. Measurement of the dissipation in the field waves with a continuous spectrum are described, again followed by a subsection with the most important summary, which is mainly dedicated to the cumulative effect imposed by the large waves on the small-scale breakers.

A subsection on the dissipation at extreme wind forcing followed. This demonstrates that at very strong winds the dissipation apparently balances the excessive wind input, as the spectrum level does not change significantly in response to such extreme forcing. Data on the forcing of pre-existing waves by a hovering helicopter are presented to support the argument. Such dissipation behaviour is quite different to the benign conditions, where it is a hydrodynamic phenomenon that is determined by the wave properties (and the wave spectrum) rather than by the wind.

The directional distribution of the dissipation (like the directional distribution of the wind input in this regard) is usually subject to a speculative or residual-approach argument and has hardly been measured. In the last subsection of Section 7.3, experimental evidence of the bimodal distribution of the dissipation function was presented.

The bimodal distribution means that the breaking dissipation is weakest in the main wave-propagation direction and therefore the breaking tends to make the directional spreading narrower. In Section 6.2, it was noted that the frequency distribution of the breaking severity is also smallest at the spectral peak. That is, the breaking appears to narrow down the spectral peak both in the frequency and in the directional domains.

A large Section 7.4 was dedicated to actual implementation of the dissipation terms in spectral wave models. For many years, these terms were treated as tuning knobs to close the wave-growth balances. Following the recent advances, this section reviewed the physical features of the dissipation functions which have actually been observed in field experiments, and techniques which allow us to calibrate the dissipation function independently, before substituting it into the model where many source terms are operational simultaneously and their contributions are not possible to separate. One of the main constraints in the wind–wave system is the total wind stress, whose parameterisations are available and have to be matched by the wind-input integral. This integral allows independent calibration of the input, and the integrand of the dissipation spectrum has to be a fraction of it, whose magnitude and its variations are again known independently from field observations. Testing of a new observation-based dissipation function by such constraint-based means and its implementation in a research spectral model are also described.

In the context of the dissipation function in spectral modelling, non-breaking dissipation of wave energy has to be clearly outlined and defined too, and studied in conjunction with the breaking dissipation. Such dissipation is often forgotten about, but not by the wave forecasters. Not all the wave-energy dissipation comes through the breaking, and this fact has to be realised and clearly stated. This will help confusions in forecasting, for example, the swell propagation. Non-breaking loss of wave momentum/energy, passed to the ocean, facilitates the upper-ocean mixing, and that passed back to the atmosphere is another potentially significant physical element at the atmosphere–ocean interface. Both of these may claim importance in the case of large-scale applications including climate modelling.

Thus, the topic of wave-energy dissipation is incomplete without discussing non-breaking dissipation and this is done in Section 7.5 of the book. Two mechanisms of non-breaking dissipation, which appear to be of the same order of magnitude, are discussed and compared to measurements of the swell propagation across the oceans. These are wave-induced non-breaking turbulence and tangential turbulent stresses imposed by the waves against the air above the interface.

While small, by comparison, in the presence of breaking, the non-breaking loss of energy spent on the generation of turbulence is only insignificant as far as the wave attenuation is concerned. There is an emerging argument that the role of non-breaking wave-induced turbulence in the upper ocean may be more essential than that of the breaking-produced turbulence. This argument is further discussed in the book in Chapter 9, which is dedicated to the multiple roles that the breaking plays in the system of air–sea interactions.

The consequences of the wave-breaking physics are many, and these are not necessarily related to the energy dissipation. Non-dissipative impacts of the breaking are multiple and they deserve special attention and revision in Chapter 8. Among others, they include the breaking-caused spectral peak downshift, the role of wave breaking in maintaining the level of the spectrum tail, and wind-input enhancement due to wave breaking. The first of the three features mentioned, the downshift, is most universally attributed to the weak nonlinear interactions, and this is how it is treated in spectral wave models with very few exceptions. In the meantime, there is compelling experimental evidence supported to an extent by a theoretical argument that the breaking does move the spectral peak to lower frequencies/wavenumbers. And under appropriate conditions, the rate of this peak-frequency change is much faster compared to the downshifting brought about by the weak nonlinear resonance. This topic is outlined in Section 8.1 where a model of such downshifting due to breaking is also presented.

Next Section 8.2 is dedicated to the role of breaking in maintaining the level of the high-frequency wind–wave spectrum tail. Mention of this level, and in particular of the transition from ω^{-4} to ω^{-5} behaviour, are scattered around the book, and the section concentrates on this topic and highlights part of the breaking in this regard. It is argued that the ω^{-5} subinterval is always present, and it is the breaking of the short waves which is responsible for this. For all but mature stages of wave development, the level of the ω^{-5} interval is constant on average, and must be supported by a combined reaction of the severity and probability of the small breakers in response to changing wind forcing.

An important outcome of Section 8.2 is the dependence of the transitional frequency ω^{-4} to ω^{-5} as a function of a combination of the wind speed and peak frequency, or, in a broader analytical sense, of the energy flux from the wind to the waves. This dependence is a result of discussion in this section of two approaches to describing the physics which forms the spectrum tail, those due to the wind action and due to nonlinear fluxes within the wave system. It is argued that these approaches do not contradict each other, as they have routinely been indicated to, but in fact are complementary and both explain the same spectrum shape of ω^{-4}, that is one through the energy input and the other through redistribution of this input. The wave breaking competes with this mechanism for control over the spectrum shape, and where it starts to dominate the transition to ω^{-5} occurs. It is feasible that this transition also separates bands of predominantly-inherent and predominantly-induced breakings in the wave spectrum.

The last section of Chapter 8 discusses the breaking-caused wind-input enhancement. It approximately doubles over the breaking waves and can prove a significant increase if the breaking rates are high. Therefore, while taking the energy and momentum from the wave system and passing them on to the ocean turbulence and mean currents, the breaking at the same time promotes the wind-energy/momentum intakes and in this way facilitates the atmosphere–ocean exchanges.

Thus, Section 8.3 is further linked to a wider topic of the role of wave breaking in the small-scale air–sea interactions in general which is the subject of the last chapter, Chapter 9. These interactions affect the physics of the atmospheric bottom-boundary layer as discussed in Section 9.1 and of the upper ocean in Section 9.2.

It has been shown, for example, that the wave-breaking characteristics alter the sea-drag dependences in this layer. While such dependences bear a profound importance in the large-scale air–sea interaction modelling, because the sea-drag coefficient is often the parameter which solely defines the momentum flux between the atmosphere and the ocean in such models, the influences of the breaking on the variations of this coefficient have received very scant attention so far. This is the topic of Section 9.1.1.

Another contribution of breaking to the air–sea interactions is generation of spray, which makes the air close to the interface a two-phase fluid, as discussed in Section 9.1.2. Such a fluid has a different density since the water droplets are much heavier compared to the surrounding air, and as a result the spray absorbs some momentum of the existing air flow. The spume also creates density stratification and damps the near-surface turbulence. In these and other ways, the suspended droplets may alter the behaviour of the very bottom of the boundary layer, where the interactions as such occur, in a significant manner. Such effects can potentially modify the air–sea interactions beyond recognition, if the amount of suspended water matter is essential, as it is under extreme wind-forcing/wave-breaking conditions such as bora winds or hurricanes. Measurements of spume in these conditions is particularly difficult and a new experimental technique able to handle such *in situ* estimates is discussed.

The important role of the spray has been the topic of an escalating number of papers published lately. The main models of the spray influence and spray production are discussed or outlined in the section. Spray classification is presented, and both its mechanic and thermodynamic capacities are highlighted. In this regard, models of the spray in the atmospheric boundary layer can be related to a gross number of applications, ranging from rocket combustion engines, with fuel droplets suspended in a turbulent gas, through dust storms, where the two-phase medium consists of the air and solid particles, to tropical cyclones where the sea droplets can serve as a source of sensible heat.

Section 9.1.3 is dedicated to the change in the dynamic regime of air–sea interactions at hurricane-like wind forcing. The role of the breaking is most important here as the majority of models relate this regime change to the breaking directly or indirectly.

Saturation of the sea drag at high wind speeds was foreshadowed by large-scale modellers of air–sea interaction a while ago, but only over the last decade was it actually measured. This experimental finding stimulated a boom in theoretical models striving to reproduce and explain the effect. In Section 9.1.3 the models are briefly reviewed by being subdivided into two broad classes: those due to the spray effects and due to changes in the ocean-surface aerodynamics.

The spume models can be conditionally further subdivided into four types. These are those which rely on the effect of damping the near-surface atmospheric turbulence due to suspended spume, on the momentum-balance effect of the spume acceleration, on the rain-like effect of smoothing the short waves by the droplets falling back down on the wavy surface, and on the thermodynamic effects of wave-induced-pressure reduction by suspended droplets.

The aerodynamic effects are basically two: full-flow separation and smoothing the short waves. The latter, apart from the rain-like effect of spume mentioned above, can be achieved directly by a very strong wind without flow separation. The separation, at the same time, can be caused in a variety of ways, that is due to centrifugal acceleration of strong wind cornering the steep crests, due to wave breaking, and due to Kelvin–Helmholtz instability of capillary waves leading to their microbreaking.

The wave breaking itself in extreme conditions also experiences a number of unusual behaviours. Direct observations and accounts of such breaking are rare, but multiple models point to a number of peculiarities of this kind of breaking which distinguish it from the breaking in benign environmental situations. These include the full flow separation, disruption of the surface due to microbreaking caused by Kelvin–Helmholtz instability, and tearing the wave crests by high winds. Most importantly, there appears to be a change of regime for the wave breaking at approximately the same wind speed as the saturation of the sea drag, as concluded in Section 9.1.3. This is either due to saturation of the breaking probability, or because the direct wind forcing replaces hydrodynamics mechanisms as a cause of wave breaking, or both.

More byproducts of the wave-breaking studies can be found in the adjacent field of upper-ocean mixing (Section 9.2). Such mixing, at a seasonal scale, for example, regulates the dynamic balance between the air and the water bodies, and ultimately negotiates the weather and climate conditions. In this regard, the breaking is responsible for a large proportion of the transfer of the momentum from the wind to the ocean, for generation of turbulence, and for injecting the bubbles and thus facilitating the gas exchange across the interface.

While being the dominant source as far as bubble generation is concerned, the role of the breaking in mixing the upper layer is important, but not pivotal. The breaking-injected turbulence is a surface energy source (that is occurs at depths comparable with the wave height), and it does provide the boundary condition which then imparts on the turbulence profiles throughout the layer. Schemes of the momentum exchange through the ocean surface and the place of breaking in these schemes are described in Section 9.2.1.

More specifically, the generation of turbulence by waves, including the breaking waves, is discussed in Section 9.2.2. Here, however, the non-breaking mean orbital motion can serve as a distributed source of turbulence, that is this motion can produce such turbulence directly at the depths scaled with the wave length. This turbulence source has been generally overlooked, and in this section it is discussed in some detail in the context of the upper-ocean mixing schemes and turbulence generation, in order to provide a complete picture of such mixing.

Section 9.2.3 of the book discusses the physics and dynamics of the bubbles. The majority of the bubbles at the ocean interface and in the upper ocean are due to wave breaking, so the connection of the book theme to this issue is apparent. The topic, however, is very large and could only have been outlined here. Three main issues are discussed: the structure of the bubble clouds, the dynamics of the bubble submersion and surfacing, and the role of the bubbles in the gas transfer from the air to the upper ocean.

As there is a change in the dynamic regime of the atmospheric boundary layer mentioned above at wind in excess of 30 m/s (9.12), and of the surface-wave breaking at approximately the same wind speed (3.26), so there is a change in the gas-transport regime at similar threshold winds (9.50). This poses an interesting question of whether such a simultaneous change in the physics of the low-atmosphere, ocean-surface and upper-ocean properties can be incidental. The answer is most likely negative, and the question certainly needs further attention in the light of its apparent significance across the range of research and applied topics and even fields. In any case, as far as the bubbles are concerned, at such winds and perhaps even at the lighter wind speeds, the role of the bubbles and therefore of the breaking is dominant in the air–sea gas exchange.

Many more breaking-related topics can be mentioned, but were not included in this book. One set of such applications is the breaking in finite water depths and shallow waters. These topics are depth-limited wave development, i.e. forecast of waves in finite-depth environments, surf-zone breaking (we mention it here, separately from the finite-depth conditions, for obvious reasons), and wave–bottom interactions, including sediment suspension and transport, among many others. While the ultimate criteria for wave breaking, i.e. the conditions (steepness) of the water surface when it loses stability and starts collapsing, should be similar in deep and finite waters, the physics of the processes leading to the breaking onset and of the outcomes of the breaking event can be very different. In many regards, the wave breaking and certainly the dissipation in finite and shallow depths is another research field rather than an extension of the deep-water topic into a new environment (e.g. Thornton & Guza, 1983; Herbers *et al.*, 2000, 2003; Lowe *et al.*, 2005, 2007; Rogers & Holland, 2009). In the present book, the issue of the limiting steepness and breaking probability in finite depths is discussed in Section 5.3, but the overall topic is left to other books.

Among many more wave-breaking-related research areas, which are not addressed in this book, or not addressed in any detail, we can mention wave breaking in the presence of currents, remote sensing applications of wave breaking, and engineering tasks. The latter set deals with a variety of problems, whose solutions have to be different to those research products sought after in this book and whose significance covers the needs of marine transport, including underwater pipelines, navigation, ship design, coastal and offshore industries, maritime safety issues, pollution mitigation, naval exercises, fishery, recreational activities at sea, and many others. Typical engineering problems are the impacts of the breaking waves on structures or on the bottom in shallow areas, and suspension of the sediment and generating currents to carry out long-shore and cross-shore sediment transport, among many others.

With this, we finish the summary of what we know about the breaking and open the discussion. The topic of the discussion is the sub-title of the current Chapter 10: "What else do we need to know about wave breaking?"

In terms of wave breaking as such, much essential progress has been made lately in understanding what is the breaking onset and in describing the breaking probability. That is we can, with reasonable confidence, answer the question of why waves break. Hand in hand with that, the breaking probability is sufficiently well described and even quantitatively parameterised both for monochromatic two-dimensional wave trains and for oceanic directional wind-forced waves with a continuous spectrum, even for those in finite depths.

In this regard, it is worth mentioning again that field investigations of the breaking onset and of micro-breaking, which seem quite difficult because these are features where whitecapping-related properties cannot be employed, can be approached by radar monitoring of sea spikes. As discussed in Section 3.6, the radar measurements do detect the incipient breakers rather than breakers in progress and are one of the oldest experimental means of wave-breaking investigations, but for some reason they are essentially under-used in wave-breaking research. This is particularly surprising as they can even estimate orbital velocities in breaking field waves, the elusive property which serves as a breaking criterion for many theoretical approaches.

The issue of the breaking onset in finite depths, and particularly in shallow waters is quite a different one (e.g. Miche, 1944). While the waves should certainly collapse if they reach the limiting steepness, how much could such a limit be modified when the effect of the bottom is crucial? The transient wave is a dynamic phenomenon and the wave may collapse because the velocity field cannot be sustained (for example, the orbital velocity is higher than the phase speed), and whether this occurs at the same limiting steepness in, for example, very shallow environments is not that obvious.

Breaking probability in such very shallow environments also deserves special consideration. In Section 5.3, adjustment of the deep-water breaking-probability dependence was done for finite-depth conditions and it worked at the Lake George field site, but the analysis showed that, while affected by the bottom proximity, the majority of waves were still breaking due to inherent reasons. How should the breaking probability be described in shallow non-dispersive waters, where deep-water nonlinear mechanisms, including the modulational instability, are deactivated? Would the dependences be the same for flat and sloped bottoms, or for different slope angles?

A very separate issue is waves on currents with horizontal velocity gradients both following and adverse currents. In the latter circumstance the waves can be blocked and breaking occurs at the blocking point (e.g. Chawla & Kirby, 2002). Will such an external forcing modify the breaking onset? What is the rate of penetration and reflection of energy of nonlinear wave trains in such circumstances? This needs to be understood in order to estimate the wave dissipation correctly.

A particularly interesting question is the end of the wave breaking (see also Section 2.2 about the subsiding stage of wave breaking). We now understand why the breaking starts, but why does it stop? The severity of the breaking can be anything, from almost 0% to 100%, so the fraction of energy lost does not signal the end of breaking. An interesting insight was produced by two-phase modelling, discussed in Section 7.2. At least for some spilling breakers, the wave breaks until the energy is reduced back to the steepness level which defined the no-breaking threshold in the wave system. This is not so for plunging breakers which do not stop breaking at such a threshold. And this issue is very important for understanding, describing and parameterising the wave-energy dissipation. Since a wave can completely disappear in the course of the wave breaking, this means that there is no lower-steepness limit which signifies the end of breaking, and the stop-trigger here is perhaps some dynamic property.

Since phase-resolving wave modelling is mentioned at this point of the discussion, the modelling issue raised above in the descriptive part of the Conclusions has to be reiterated. Two-phase models can explicitly simulate the breaking process, and they are well developed analytically and numerically, and broadly used in engineering applications. The main problem of their transfer into physical-oceanography studies in general, and wave breaking in particular, is their high computational cost. Such a cost makes long-term simulation of the nonlinear wave evolution by two-phase models unfeasible, but this can be overcome by coupling the two-phase models with fully nonlinear potential models. The latter cannot simulate the breaking in progress, but do exceptionally well in modelling the long-term evolution to the breaking and are much less demanding in terms of computing resources.

From the point of view of experimental research on the dissipation, one of the major problems is the breaking severity. Combined with the breaking-probability dependences, the breaking-severity parameterisations would provide experimental estimates of the dissipation and its behaviour. Chapter 6, dedicated to the breaking strength, however, offers more questions for future studies on the breaking severity than answers that are available. It is relevant to reiterate them in the Conclusions.

What does the severity depend on? Will it be constant, even if on average, in different wave fields with the same background steepness, for instance? Does the breaking strength change across the spectrum, i.e. is the relative energy loss different for the dominant waves and the shorter waves away from the peak? For the latter waves, does their severity depend on whether they break due to inherent reasons or their breaking is induced by larger waves? Is the severity dependent on the wind? How do the wave directionality and shortcrestedness impact the breaking strength, if they do? Also, for example, how much of the energy loss is spent on work against the buoyancy forces while entraining the bubbles into the water, and what fraction is passed on to the water turbulence? These and other questions should be answered and parameterised for the practical applications, and this topic represents a very interesting physics too.

Certainly, the severity is more difficult to measure and approach, compared to the frequency of breaking occurrence, particularly by contact *in situ* means, but it is not impossible, and a variety of remote-sensing techniques capable of quantifying the breaking strength are available, like those acoustic, infrared and others described in Chapter 3. Therefore, the problem of answering the many questions raised should be a matter of dedicated effort rather than of principal difficulty.

One difficulty, however, has to be specifically mentioned and highlighted: the directional dependence of the breaking strength. Since, as discussed in the book, there are indications that the directional distribution of breaking probability is more or less uniform, and there is evidence that the dissipation is not, this implies that breaking strength can be a function of an angle between the propagation direction of wave groups, where the breaking occurs, and the main direction of the waves/wind. What it is exactly and how this can be explained in the context of the physics of the air–sea interactions, is an interesting question to keep in mind. Dependences and parameterisations for the breaking severity may turn out to be more complex than it seems at first sight.

Directional behaviours related to the breaking, not only to the breaking strength, but in general, are among the most significant outstanding problems of wave-breaking studies. Understandably, a lot of the research was conducted in two dimensions, in theoretical, numerical and laboratory approaches, but as we can see the directional properties can be most essential.

In this regard, the word 'directional' artificially unifies quite different and even unrelated physics. For example, the directional distributions of the breaking probability, breaking severity, and whitecapping dissipation discussed above is one set of problems. Breaking due to linear directional focusing is another issue, whereas the impact of wave directionality on the modulational instability is a completely different topic. Yet another topic is the shortcrestedness of waves, particularly of nonlinear waves progressing to breaking. All these problems are due to an additional level of difficulty and complexity when studying a three-dimensional as opposed to a two-dimensional phenomenon, whether these are theoretical or experimental investigations. This additional level, however, should be addressed as the directionality obviously provides a principal impact on the physics of the wave breaking rather than a mere set of corrections to the existing knowledge.

Some problems, such as the role of the modulational instability in the breaking of realistic oceanic waves, cannot be answered by means of two-dimensional studies in principle. Such instability is a well-established and understood phenomenon for wave trains, but its very existence in the directional fields has been questioned lately. In Section 5.3.3, however, it is argued that breaking due to modulational instability is active in directional wave fields with typical magnitudes of wave steepness and directional spread.

If so, the relative role of wave focusing and of the modulational instability in the breaking of realistic waves observed in field conditions can now be compared. Indeed, as has been discussed throughout the book, it appears that in order to break all a wave needs is to reach some limiting steepness. It should not really matter what process leads it to this steepness, and potentially such processes are many. Broadly, however, it is the focusing and the instability of nonlinear wave groups which have been the main competitors for leading the waves to the no-return steepness in benign and moderate deep-water conditions. As already mentioned above, the first mechanism, wave focusing, can be further subdivided into frequency focusing, amplitude focusing and directional focusing.

The frequency focusing and amplitude focusing both exploit variation in the phase speeds of waves: either because of different frequencies or due to different amplitudes respectively. The frequency dispersion is quite large if the frequencies differ significantly and seems like a good candidate to often provide superpositions of various waves in the field with a continuous wave spectrum. The problem, however, is the fact that the drop in spectral density (and the wave height) away from the peak is much more rapid than changes in the phase speed, as a function of frequency. That is, the waves with close frequencies/heights propagate with close speeds and are not likely to superpose, and the waves which are likely to superpose have such different wave heights that addition of the smaller ones to the primary wave does not alter the height of the latter appreciably. In a wave field with a typical primary steepness of $ak = 0.1$, superposition of very many such waves at a single point would be needed in order to reach the limiting steepness of $ak = 0.44$ which becomes an event with quite a low probability.

The frequency dispersion has been one of the standard techniques of making waves break in the laboratory, and correspondingly features of the breaking achieved this way have been well investigated. Some of these features are similar to those of the modulational type of breaking, some are not, as will be discussed below. What is surprising, however, is the apparent lack, if not absence, of studies which would provide a parameterisation or even a reasonable conclusion on how frequent breaking due to frequency focusing is in a wave field with a typical background wave steepness and typical wave spectrum.

The amplitude dispersion intuitively seems a less likely event compared to the frequency focusing because differences in wave phase speeds due to the amplitude variation are only of the second order of magnitude. This type of focusing, however, can bring together waves of the same frequency and therefore of close wave heights, and thus can make the breaking of a primary wave more probable. Again, parameterisation of the breaking rates due to this type of focusing, or in combination with the other types is not available.

The directional focusing does not rely on the dispersion, but on the contrary on waves propagating with close phase speeds, at an oblique angle with respect to each other. As with the amplitude focusing above, the advantage of directional focusing is the fact that waves of a similar frequency/height are brought together, and therefore fewer waves are required in order to reach the steepness limit, but the disadvantage is that the steepness of the waves in the converging trains, if these are two trains, has to be close to half of the limiting steepness which is too high for typical wave fields. This, again, makes such a type of focusing a rare event, as shown in Section 5.3.3.

As a result, none of the focusing mechanisms lends themselves as obvious likely candidates to provide the reasonably high breaking rates observed in oceanic wave fields. Quantitative comparisons of the focusing-breaking probability with the probability due to modulational instability are impossible because of the lack of respective experimental or theoretical dependences for the former.

Parameterisations of the wave breaking brought about by the instability of quasi-two-dimensional wave trains are available and the corresponding breaking rates are quite consistent with those observed in the field (Section 5.1.4). The fact that the modulational instability is active in the directional fields makes it a likely cause for the majority of the dominant breaking in the ocean (breaking of the shorter waves is mostly induced, unless the coherent-wavelet structure is present at those scales, as discussed below).

In this regard, it is worth briefly revisiting a number of features of nonlinear wave behaviour leading to the modulational-instability breaking, which were revealed both in the two-dimensional simulations/measurements and in field observations, as mentioned throughout the book. The main such feature is the wave-breaking threshold in terms of the mean background steepness of field waves. This is the typical behaviour of modulational instability, leading to the breaking in two-dimensional wave trains also, and even the steepness thresholds are consistent in the 2D trains and directional fields as discussed in Section 5.1.4. The focusing, whether this is linear or nonlinear, cannot explain such a threshold feature. If the focusing was the main reason for wave breaking, it should lead to the gradual vanishing of breaking as the steepness asymptotes zero, rather than to a cessation of the breaking at some threshold value of steepness.

Other features observed both in the wave flumes and in the field, and characteristic of the modulational instability, are the upshifting of spectral energy prior to breaking and the oscillations of the skewness/asymmetry. Loss of energy by the primary breaking component is a certain feature of the wave-instability breaking, which in this way is similar to the breaking due to amplitude focusing but is opposite to the frequency-focusing dissipation outcomes. The fact that doubling the wind input brings about modulational wave breaking four times as fast is consistent with field experiments on wind input.

In experiments described in Section 5.3.3, the directional wave fields were created by two-dimensional waves coming from different directions. In the resulting three-dimensional pattern, the waves are often called short-crested, because of their appearance, but in fact they are a superposition of long-crested waves. This is the typical Fourier approach, explicitly or implicitly prevailing in ocean wave studies of any kind.

The term short-crestedness is often used as a counterpart for directionality, but this causes great confusion. Wave trains can be perfectly two-dimensional (unidirectional), but still apparently short-crested – for example, steep waves in wave flumes (see also discussion in Section 3.4).

As a result, the same Fourier spectrum can potentially be measured in very different directional wave fields: those created by the superposition of long-crested waves, and those due to quasi-two-dimensional short-crested short-lived wave groups (wavelets) coming from different directions. Those groups can be coming to the measurement point one at a time, but would produce a directional spectrum if the measurements are conducted long enough. If such groups remain coherent at the time scale of a few dozens of periods of dominant waves, then the basic two-dimensional Benjamin–Feir-like modulational-instability mechanism will be active. In the wave fields with a typical background steepness $ak \geq 0.1$, this will lead to the breaking without a need for any further superpositions or explanations like those offered in Section 5.3.3 in terms of the directional modulational index.

Evidence for such coherent wave groups is available among remote-sensing studies of surface ocean waves. If such a structure of the wave fields is confirmed, this could prove an exciting novel development for the entire set of analytical and statistical approaches in wave dynamics, not just for wave breaking.

The answer to the question of whether it is the modulational instability or something else that causes the dominant waves to break is not just a subject of scientific curiosity. This answer has the most serious implications for the topic of wave-energy dissipation, as the spectral signatures of the dissipation brought about by different physical mechanisms are different. In this regard, if the short-crested short-lived wavelet structure is a reality, then such instability may even be applicable to the short-scale wave trains. In the present continuous-wave-train Fourier perception of wave fields, such waves at the spectrum tail do not have a characteristic bandwidth and therefore are not expected to have a group structure leading to such instabilities and breaking.

Hand-in-hand with the question about the role of the modulational instability comes the non-dissipative question of the downshifting of the wave energy due to wave breaking. For decades, such downshifting was attributed to the resonant nonlinear interactions, and if the modulational-breaking does bring it about in oceanic wave fields, many features of the wave evolution may need to be reconsidered, as well as those of the wave modelling.

Another issue, related, at least partially, to the breaking is the question of the aerodynamic surface roughness due to small-scale waves and of the sea drag. Progress in describing these most important properties of the air–sea interactions has stagnated for decades, unable to improve the large scatter of experimental data and parameterisations, and it appears that the sensible way to progress here is to take into account multiple influences which form such a drag, wave breaking among them.

Speaking generally, a proper account of the wave-breaking effects both in the dynamic models of the atmospheric boundary layer and in mixing schemes of the upper ocean should also be on the agenda. In this regard, not only the wave breaking, but the waves in general, are an important player near the ocean surface, above and below, and negotiate all the exchanges through the surface. This includes the large-scale air–sea interactions too, all the way to the climate scale. The wave effects and influences are important, but largely overlooked or underestimated by large-scale models, and coupling of them with the wave models, which is already under way as mentioned throughout the book, is the reasonable and sensible way to advance modelling physics and improve the respective forecasts.

Below the surface, the topic of bubbles generated by the breaking is also in need of attention. Such bubbles are a major player both in the dynamics of the subsurface water layer and in the air–sea gas and moisture exchange. On the technical side, they appear to be a proxy for wave-energy dissipation, the most challenging property due to wave breaking to measure and investigate.

In principle, the problems discussed so far should not take too long to solve. As discussed in Chapter 3 and throughout, all the instrumentation, measurement techniques and theoretical understanding are available, and it only requires attention and a concentrated effort by the wave-breaking research community.

It is not so, however, with respect to wave breaking and breaking-related issues in extreme weather conditions such as tropical cyclones. Here, change in the regime of the air–sea exchanges is observed in every regard, with the wave breaking in the centre of it. In the air, it generates spume which may slow down the momentum fluxes, affect the thermodynamics of the boundary layer and thus allow the cyclones to grow. Below the surface, the new mechanisms of bubble submersion are produced which can enhance the air-to-sea gas transfer. The wave breaking itself, with the wind directly tearing wave crests rather than gradually leading to the water surface collapse, takes on another regime.

Three layers of difficulties seem to prevent solution of the extreme-weather problems in the immediate future. Firstly, there is an apparent lack of instruments and techniques which can operate, conduct measurements and collect data in such circumstances. As a result, the observations are few and cannot support conclusive theoretical research. Secondly, there is no clarity in what exactly should be measured, that is, what are the most essential properties of wave breaking and its products which are needed for adequate physical modelling. In some way, this field is now in the position in which the general wave-breaking field was some 20–30 or even more years ago, that is we measure what we can measure and are guided by these measurements. The choice of available analytical theories of extreme air–sea interactions has been abundant in recent years, but it has not seemed to help to guide the experiment consistently. This is the third difficulty: there is no leading recognised theory or even set of theories for waves and wave breaking under extreme wind forcing, and for air–sea interactions in such situations, which means there is no clear understanding of the physics.

In the meantime, this physics must be something very special and must be strongly coupled across all the ocean-interface environments. As highlighted in this book, through discussions in Sections 3.3, 9.1.3, 9.2.3 and in this conclusion, at wind speeds $U_{10} \sim$ 30–35 m/s a change in the dynamic regimes occurs in the atmospheric boundary layer (saturation of the sea drag), on the surface (saturation of wave asymmetry and alteration of the breaking mechanism) and below the surface (dynamics of bubble submersion). The latter also leads to a change in the regime of gas transfer. It is time for a combined effort/approach across the different air–sea-interaction sciences to address this issue which apparently needs a complex coupled insight.

Another change in the regime observed in this book should also be pointed out. This may not seem as exciting as the new set of air–sea interactions in hurricanes just mentioned, but it may signify a new set of physical relationships in the wave system in benign and moderate conditions, which are much more typical compared to the tropical cyclones. This is the transition of the dynamic behaviour within the wave spectrum, at wave scales which can be characterised by forcing $U_{10/C} \approx 1.45$, as discussed in Sections 5.3.4 and 8.2. At these scales, alteration of the spectrum-tail behaviour happens, from the tail-level being constant on average to becoming wave-age-dependent. At such wind forcing, the highest dimensional transitional frequency between the f^{-4} and f^{-5} subintervals occurs. And at the same dimensionless frequency, the transition from the inherent-breaking to induced-breaking dissipation is observed. Each of the three features has importance and needs attention in its own right, but the fact that the respective transitions all occur at the same dynamic wind-forcing and frequency range can hardly be incidental and implies some interesting coupled physics which is still to be understood.

Overall, in spite of the difficulties and gaps in understanding, we would like to conclude on an optimistic note. The wave-breaking research and knowledge is much better structured today compared to even ten years ago. The physical picture of wave breaking is still not complete, but a consistent image of this interesting and challenging phenomenon is emerging.

References

Abadie, S., Caltragirone, J.-P. & Waltremez, P. 1998 Mecanisme de generation du jet secondaire ascendant dans un deferlement plongeant. *C. R. Mec.* **326**, 553–559 (in French)

Abe, T., Ono, T. & Kishino, N. 1963 A fundamental study on the prevention of the salty damages due to the foaming of sea water (preliminary report). *J. Oceanogr. Soc. Japan* **18**, 185–192

Agnon, Y., Babanin, A. V., Chalikov, D. *et al.* 2005 Fine scale inhomogeneity of wind–wave energy input, skewness and asymmetry. *Geophys. Res. Lett.* **32**, L12603, doi:10.1029/2005GL022701, 4p

Agrawal, Y. C., Terray, E. A., Donelan, M. A., *et al.* 1992 Enhanced dissipation of kinetic energy beneath surface waves. *Nature* **359**, 219–220

Alber, I. E. 1978 The effects of randomness on the stability of two-dimensional wavetrains. *Proc. R. Soc. Lond.* **A363**, 525–546

Alves, J. H. G. M. & Banner, M. L. 2003 Performance of a saturation-based dissipation-rate source term in modeling the fetch-limited evolution of wind waves. *J. Phys. Oceanogr.* **33**, 1274–1298

Alpers, W., Ross, D. B. & Rufenach, C. L. 1981 On the detectability of ocean surface wave by real and synthetic aperture radar. *J. Geophys. Res.* **86**, 6481–6498

Andreas, E. L. 1998 A new spray generation function for wind speeds up to 32 m/s. *J. Phys. Oceanogr.* **28**, 2175–2184

Andreas, E. A. & Emanuel, K. A. 2001 Effects of sea spray on tropical cyclone intensity. *J. Atmos. Sci.* **58**, 3741–3751

Andreas, E. L. 2002 A review of sea spray generation function for the open ocean. *Atmosphere–Ocean Interactions* **1**, Ed. W. A. Perrie, Southampton, UK: WIT Press, 1–46

Andreas, E. L. 2004 Spray stress revisited. *J. Phys. Oceanogr.* **34**, 1429–1440

Anguelova, M. D., Barber Jr., R. P. & Wu, J. 1999 Spume drops produced by the wind tearing of wave crests. *J. Phys. Oceanogr.* **29**, 1156–1165

Anguelova, M. D. & Webster, F. 2006 Whitecap coverage from satellite measurements: A first step toward modeling the variability of oceanic whitecaps. *J. Geophys. Res.* **C111**, C0307, doi:10.1029/2005JC003158

Annenkov, S. Y. & Shrira, V. I. 2006 Role of non-resonant interactions in evolution of nonlinear random water wave fields. *J. Fluid Mech.* **561**, 181–207

Ardhuin, F. & Jenkins, A. D. 2005 On the effect of wind and turbulence on ocean swell. *Proc. 15th Intl. Offshore and Polar Eng. Conf.* **3**, International Society of Offshore and Polar Engineers (ISOPE), Seoul, Korea, 429–434

Ardhuin, F., Jenkins, A. D., Hauser, D., Reniers, A. & Chapron, B. 2005 Waves and operational oceanography: toward a coherent description of the upper ocean. *EOS* **86**, 37–44

Ardhuin, F. & Jenkins, A. D. 2006 On the interaction of surface waves and upper ocean turbulence. *J. Phys. Oceanogr.* **36**, 551–557

Ardhuin, F., Herbers, T. H. C., van Vledder, G. P., *et al.* 2007 Swell and slanting-fetch effects on wind wave growth. *J. Phys. Oceanogr.* **37**, 908–931

Ardhuin, F., Chapron, B. & Collard, F. 2009a Observation of swell dissipation across oceans. *Geophys. Res. Lett.* **36**, L06607, doi:10.1029/2008GL037030, 4p

Ardhuin, F., Marie, L., Rascle, N., Forget, P. & Roland, A. 2009b Observations and estimation of Lagrangian, Stokes and Eulerian currents induced by wind and waves at the sea surface. *J. Phys. Oceanogr.* **39**, 2820–2838

Ardhuin, F., Rogers, W. E., Babanin, A. V., *et al.* 2010 Semi-empirical dissipation source functions for ocean waves. Part I: definitions, calibration and validations. *J. Phys. Oceanogr.*, **40**, 1917–1941.

Arsenyev, S. A., Dobroklonsky, S. V., Mamedov, R. M. & Shelkovnikov, N. K. 1975 Direct measurements of some characteristics of fine-scale turbulence from a stationary platform in the open sea. *Izvestiya, Atmospheric and Oceanic Physics* **11**, 530–533

Babanin, A. V. & Soloviev, Y. P. 1987 Parameterization of width of directional energy distributions of wind-generated waves at limited fetches. *Izvestiya, Atmospheric and Oceanic Physics* **23**, 645–651

Babanin, A. V. 1988 Connection of parameters of wind surface current with the wind in the North-West part of the Black Sea. *Morskoi Gidrofizicheskii Zhurnal* **4**, 55–58 (in Russian, English abstract)

Babanin, A. V., Verkeev, P. P., Krivinskii, B. B. & Proshchenko, V. G. 1993 Measurements of wind waves by means of a buoy accelerometer wave gauge. *Phys. Oceanogr.* **4**, 387–393

Babanin, A. V. 1995 Field and laboratory observations of wind wave breaking. In *The Second Intl. Conf. on the Mediterranean Coastal Environment, Tarragona, Spain, October 24–27, 1995* **3**, Ed. E.Ozhan, Autoritat Potuaria de Tarragona, Spain, 1919–1928

Babanin, A. V. & Polnikov, V. G. 1995 On the non-Gaussian nature of wind waves. *Phys. Oceanogr.* **6**, 241–245

Babanin, A. V. & Soloviev, Y. P. 1998a Field investigation of transformation of the wind wave frequency spectrum with fetch and the stage of development. *J. Phys. Oceanogr.* **28**, 563–576

Babanin, A. V. & Soloviev, Y. P. 1998b Variability of directional spectra of wind-generated waves, studied by means of wave staff arrays. *Marine & Freshwater Res.* **49**, 89–101

Babanin, A. V., Young, I. R. & Banner, M. L. 2001 Breaking probabilities for dominant surface waves on water of finite constant depth. *J. Geophys. Res.* **C106**, 11659–11676

Babanin, A. V. & Young, I. R. 2005 Two-phase behaviour of the spectral dissipation of wind waves. *Proc. Ocean Waves Measurement and Analysis, Fifth Intern. Symposium WAVES2005, 3–7 July, 2005, Madrid, Spain*, Eds. B. Edge, J. C. Santas, paper no. 51, 11p

Babanin, A. V., Young, I. R. & Mirfenderesk, H. 2005 Field and laboratory measurements of wave-bottom interaction. *Proc. 17th Australasian Coastal and Ocean Engineering*

Conf. and 10th Australasian Port and Harbour Engineering Conf., 20–23 September 2005, Adelaide, South Australia, Eds. M. Townsend, D. Walker, The Institution of Engineers, Canberra, Australia, 293–298

Babanin, A. V. 2006 On a wave-induced turbulence and a wave-mixed upper ocean layer. *Geophys. Res. Lett.* **33**, L20605, doi:10.1029/2006GL027308, 6p

Babanin, A. V. & Young, I. R. 2006 Experimental source terms for spectral wave forecast models. *Proc. Third Chinese-German Symp. Coast. and Ocean Eng., 8–16 November 2006*, Eds. S.-H. Ou, C. C. Kao, T.-W. Hsu, National Cheng Kung University, Tainan, Taiwan, 49–60

Babanin, A. V., Chalikov, D., Young, I. R. & Savelyev, I. 2007a Predicting the breaking onset of surface water waves. *Geophys. Res. Lett.* **34**, L07605, doi:10.1029/2006GL029135, 6p

Babanin, A. V., Banner, M. L., Young, I. R. & Donelan, M. A. 2007b Wave follower measurements of the wind input spectral function. Part 3. Parameterization of the wind input enhancement due to wave breaking. *J. Phys. Oceanogr.* **37**, 2764–2775

Babanin, A. V., Young, I. R., Manasseh, R. & Schultz, E. 2007c Spectral dissipation term for wave forecast models, experimental study. *Proc. 10th Intl. Workshop on Wave Hindcasting and Forecasting and Coastal Hazards, Oahu, Hawaii, November, 11–16, 2007*, WMO/IOC Joint Technical Commission for Oceanography and Marine Meteorology (JCOMM), 19p

Babanin, A. V., Tsagareli, K. N., Young, I. R. & Walker, D. 2007d Implementation of new experimental input/dissipation terms for modeling spectral evolution of wind waves. *Proc. 10th Intl. Workshop on Wave Hindcasting and Forecasting and Coastal Hazards, Oahu, Hawaii, November, 11–16, 2007*, WMO/IOC Joint Technical Commission for Oceanography and Marine Meteorology (JCOMM), 12p

Babanin, A. V., Alves, J.-H., Ardhuin, F., *et al.* 2007e Spectral dissipation in deep water. *Progr. Oceanogr.* **75**, 622–632 (In and The WISE Group (2007))

Babanin, A. V. & Makin, V. K. 2008 Effects of wind trend and gustiness on the sea drag: Lake George study. *J. Geophys. Res.* **C113**, C02015, doi:10.1029/2007JC004233, 18p

Babanin, A. V. & van der Westhuysen, A. J. 2008 Physics of "saturation-based" dissipation functions proposed for wave forecast models. *J. Phys. Oceanogr.* **38**, 1831–1841

Babanin, A. V. 2009 Breaking of ocean surface waves. *Acta Physica Slovaca* **59**, 305–535

Babanin, A. V. & Haus, B. K. 2009 On the existence of water turbulence induced by non-breaking surface waves. *J. Phys. Oceanogr.* **39**, 2675–2679

Babanin, A. V., Chalikov, D. & Young, I. R. 2009a Breaking of two-dimensional waves in deep water. In *Intl. Conf. in Ocean Eng., ICOE 2009 IIT Madras, Chennai, India, 1–5 Feb. 2009*, Eds. V. A. Anantha Subramanian, S. Nallayarasu, S. A. Sannasiraj, 386–395

Babanin, A. V., Ganopolski, A. & Phillips, W. R. C. 2009b Wave-induced upper-ocean mixing in a climate modelling of intermediate complexity. *Ocean Model.* **29**, 189–197

Babanin, A. V. 2010 Wind input, nonlinear interactions and wave breaking at the spectrum tail of wind-generated waves; transition from f^{-4} to f^{-5} behaviour. In *Ecological Safety of Coastal and Shelf Zones and Comprehensive Use of Shelf Resources*, Marine Hydrophysical Inst., Sevastopol, **21**, 173–187

Babanin, A. V., Chalikov, D., Young, I. R. & Savelyev, I. 2010a Numerical and laboratory investigation of breaking of steep two-dimensional waves in deep water. *J. Fluid Mech.* **644**, 433–463

Babanin, A. V., Waseda, T., Kinoshita, T. & Toffoli, A. 2011a Wave breaking in directional fields. *J. Phys. Oceanogr.*, **41**, 145–156

Babanin, A. V., Tsagareli, K. N., Young, I. R. & Walker, D. J. 2010c Numerical investigation of spectral evolution of wind waves. Part 2. Dissipation function and evolution tests. *J. Phys. Oceanogr.* **44**, 667–683

Babanin, A. V., Hsu, T.-W., Roland, A., *et al.* 2010d Spectral modelling of Typhoon Krosa. *Nat. Hazards Earth Syst. Sci.*, in press

Babanin, A. V., Waseda, T., Shugan, I. & Hwung, H.-H. 2011b Modulational instability in directional wave fields, and extreme wave events. *Proc. 30th Intl. Conf. on Ocean, Offshore and Arctic Eng. OMAE 2011*, July 19–24, 2011, Rotterdam, The Netherlands, submitted

Badulin, S. I., Pushkarev, A. N., Resio, D. & Zakharov, V. E. 2005 Self-similarity of wind-driven seas. *Nonlin. Processes Geophys.* **12**, 891–945

Balmorth, N. J. 1999 Shear instability in shallow water. *J. Fluid Mech.* **387**, 97–127

Bandou, T., Mitsuyasu, H. & Kusaba, T. 1986 An experimental study of wind waves and low frequency oscillations of the water surface. *Report of Res. Inst. for Appl. Mech.* **XXXIII**, 13–32

Banner, M. L. & Melville, W. K. 1976 On the separation of air flow above water waves. *J. Fluid Mech.* **77**, 825–842

Banner, M. L. 1990 The influence of wave breaking on the surface pressure distribution in wind–wave interactions. *J. Fluid Mech.* **211**, 463–495

Banner, M. L. & Tian, I. R. 1994 Modeling spectral dissipation in the evolution of wind waves. Part I. Assessment of existing model performance. *J. Phys. Oceanogr.* **24**, 1550–1571

Banner, M. L., Jones, I. S. F. & Trinder, J. C. 1989 Wavenumber spectra of short gravity waves. *J. Fluid Mech.* **198**, 321–344

Banner, M. L. & Peirson, W. L. 1998 Tangential stress beneath wind-driven air–water interfaces. *J. Fluid Mech.* **364**, 115–145

Banner, M. L. & Tian, X. 1998 On the determination of the onset of breaking for modulating surface gravity waves. *J. Fluid Mech.* **367**, 107–137

Banner, M. L., Babanin, A. V. & Young, I. R. 2000 Breaking probability for dominant waves on the sea surface. *J. Phys. Oceanogr.* **30**, 3145–3160

Banner, M. L., Gemmrich, J. R. & Farmer, D. M. 2002 Multi-scale measurements of ocean wave breaking probability. *J. Phys. Oceanogr.* **32**, 3364–3375

Banner, M. L. & Peirson, W. L. 2007 Wave breaking onset and strength for two-dimensional and deep-water wave groups. *J. Fluid Mech.* **585**, 93–115

Barenblatt, G. I. 1953 On the motion of suspended particles in a turbulent flow. *Prikl. Mat. Mekh.* **17**, 261–274 (in Russian)

Barenblatt, G. I. 1955 On the motion of suspended particles in a turbulent flow in a half space or a plane open channel in finite depth. *Prikl. Mat. Mekh.* **19**, 61–88 (in Russian)

Barenblatt, G. I. & Golitsyn, G. S. 1974 Local structure of mature dust storms. *J. Atmos. Sci.* **31**, 1917–1933

Barenblatt, G. I. 1996 *Scaling, Self-Similarity, and Intermediate Asymptotics*, Cambridge University Press, 386p

Barenblatt, G. I., Chorin, A. J. & Prostokishin, V. M. 2005 A note concerning the Lighthill "sandwich model" of tropical cyclones. *Proc. Nation. Acad. Sci.* **102**, 1148–1115

Bass, S. J. & Hey, A. E. 1997 Ambient noise in the natural surf zone: wave-breaking frequencies. *IEEE J. Oceanic Eng.* **22**, 411–424

Belberov, Z. K., Zhurbas, V. M., Zaslavskiy, M. M. & Lobisheva, L. G. 1983 Integral characteristics of frequency spectra of wind-generated waves. In *Interaction of Atmosphere, Hydrosphere and Lithosphere in Sea Coastal Zone*, Sofia, Bulgarian Academy of Sciences Press, 143–154

Belcher, S. E. & Hunt, J. C. R. 1993 Turbulent shear flow over slowly moving waves. *J. Fluid Mech.* **251**, 109–148

Belcher, S. E. & Hunt, J. C. R. 1998 Turbulent flow over hills and waves. *Ann. Rev. Fluid Mech.* **30**, 507–538

Benilov, A. Y. 1973 The turbulence generation in the ocean by surface waves. *Izvestiya, Atmospheric and Oceanic Physics* **9**, 160–164

Benilov, A. Y., Gumbatov, A., Zaslavsky, M. & Kitaigorodskii, S. A. 1978 Non-steady model of development of turbulent boundary layer above sea under generating of surface waves. *Izvestiya, Atmospheric and Oceanic Physics* **14**, 1177

Benilov, A. Y. & Ly, L. N. 1977 Semi-empirical methods of the ocean turbulence description. In *Turbulence and Impurities Diffusion in the Sea, The Co-ordination Center of the COMECON (SEV), Information Bulletin* **5**, Moscow, 89–97 (in Russian)

Benilov, A. Y., McKee, T. G. & Safray, A. S. 1993 On the vortex instability of the linear surface wave. *Numerical Methods in Laminar and Turbulent Flow* **VIII**, Part 2, Ed. C. Taylor, Pineridge Press, UK, 1323–1334

Benilov, A. Y. & Ly, L. N. 2002 Modelling of surface waves breaking effects in the ocean upper layer. *Mathematical and Computer Modelling* **35**, 191–213

Benjamin, T. B. & Feir, J. E. 1967 The disintegration of wave trains in deep water. Part 1. Theory. *J. Fluid Mech.* **27**, 417–430

Benney, D. J. & Newell, A. C. 1967 The propagation of nonlinear wave envelopes. *J. Math. Phys.* **46**, 133–139

Bezzabotnov, V. S., Bortkovskii, R. S. & Timanovskii, D. F. 1986 On structure of two-phase medium created by breaking wind waves. *Izvestiya Akademii Nauk SSSR. Fizika Atmosferi i Okeana* **22**, 1186–1193 (in Russian, English abstract)

Bidlot, J., Janssen, P. A. E. M. & Abdalla, S. 2007 A revised formulation for ocean wave dissipation and its model impact. *Technical Memorandum*, 509, ECMWF, Reading, U.K., 27p

Bintanja, R. 2000 Snowdrift suspension and atmospheric turbulence. Part 1: Theoretical background and model description. *Boundary-Layer Meteorol.* **95**, 343–368

Bister, M. & Emanuel, K. 1998 Dissipative heating and hurricane intensity. *Meteorol. Atmos. Phys.* **65**, 223–240

Black, P. G., D'Asaro, E. A., Drennan, W. M., *et al.* 2007 Air–sea exchange in hurricanes: synthesis of observations from the coupled boundary layer air-sea transfer experiment. *Bull. Am. Meteorol. Soc.* **88**, 357–374

Blanchard, D. C. & Woodcock, A. H. 1957 Bubble formation and modification in the sea and its meteorological significance. *Tellus* **9**, 145–158

Blanchard, D. C. 1963 The electrification of the atmosphere by particles from bubbles in the sea. *Progr. Oceanogr.* **1**, 71–202

Blanchard, D. C. 1989 The size and height to which jet drops are ejected from bursting bubbles in sea water. *J. Geophys. Res.* **C94**, 10999–11002

Blenkinsopp, C. E. & Chaplin, J. R. 2007 Void fraction measurements in breaking waves. *Proc. R. Soc. Lond.* **A463**, 3151–3170

Bondur, V. G. & Sharkov, E. A. 1982 Statistical properties of whitecaps on a rough sea. *Oceanology* **22**, 274–279

Bonmarin, P. 1989 Geometric properties of deep-water breaking waves. *J. Fluid Mech.* **209**, 405–433

Booij, N., Ris, R. C. & Holthuijsen, L. H. 1999 A third-generation wave model for coastal regions. Part I. Model description and validation. *J. Geophys. Res.* **C104**, 7649–7666

Borisenkov, E. P. 1974 Some mechanisms of atmosphere–ocean interaction under stormy weather conditions. *Fluid Mechanics – Soviet Research* **43–44**, 72–83

Borisov, A. A. & Vakhguelt, A. F. 1981 Wave processes in solid-laden gas flows. *Fluid Mechanics – Soviet Research* **10**, 70–80

Bortkovskii, R. S. 1973 On the mechanism of interaction between the ocean and the atmosphere during a storm. *Fluid Mechanics – Soviet Research* **2**, 87–94

Bortkovskii, R. S. 1977 Experimental investigation of sea spray over wind waves. *Trudy GGO* **398**, 34–40 (in Russian)

Bortkovskii, R. S. & Kuznetsov, M. A. 1977 Some results of investigations of the state of sea surface. In *Typhoon-75* **1**, Gidrometeoizdat, Leningrad, 90–104 (in Russian)

Bortkovskii, R. S. 1980 On determining intensity of spray production over the sea surface. *Trudy GGO* **444**, 17–22 (in Russian)

Bortkovskii, R. S. & Timanovskii, D. F. 1982 On microstructure of breaking crests of wind waves. *Izvestiya Akademii Nauk SSSR. Fizika Atmosferi i Okeana* **18**, 327–329 (in Russian, English abstract)

Bortkovskii, R. S. 1987a *Air–Sea Exchange of Heat and Moisture during Storms*, D. Riedel, 194p

Bortkovskii, R. S. 1987b Spatio-temporal characteristics of whitecaps and foam patches produced by the breaking wind waves. *Meteorologiya i Gidrologiya* **5**, 68–75

Bortkovskii, R. S. & Novak, V. A. 1993 Statistical dependencies of sea state characteristics on water temperature and wind–wave age. *J. Mar. Sys.* **4**, 161–169

Bortkovskii, R. S. 1997 On the influence of water temperature on the ocean-surface state and transfer processes. *Izvestiya Akademii Nauk SSSR. Fizika Atmosferi i Okeana* **33**, 266–273 (in Russian, English abstract)

Bortkovskii, R. S. 1998 Aeration of the ocean layer due to bubbles generated in whitecaps. In *Remote Sensing of the Pacific Ocean by Satellites*, Ed. R. A. Brown, 354–362

Bortkovskii, R. S. 2003 Gas transfer across the ocean surface under a strong wind and its contribution to the mean gas exchange. *Izvestiya Akademii Nauk SSSR. Fizika Atmosferi i Okeana* **39**, 809–816 (in Russian, English abstract)

Bos, W. J. T., Clark, T. T. & Rubinstein, R. 2007 Small scale response and modeling of periodically forced turbulence. *Phys. Fluids* **19**, 055107, 11p

Bowden, K. F. 1950 The effect of eddy viscosity on ocean waves. *Phil. Mag.* **41**, 320

Boyd, J. W. R. & Varley, J. 2001 The uses of passive measurement of acoustic emissions from chemical engineering processes. *Chem. Eng. Sci.* **56**, 1749–1767

Brennen, C. E. 1995 *Cavitation and Bubble Dynamics*, Oxford Univ. Press, 304p

Bretschneider, C. L. 1958 Revisions in wave forecasting. *Tech. Mem. Beach Erosion Bd.*, Washington

Bridge, J. S. & Dominic, D. F. 1984 Bed load velocities and sediment transport rates. *Water Resour. Res.* **20**, 476–490

Brown, M. G. & Jensen, A. 2001 Experiments in focusing unidirectional water waves. *J. Geophys. Res.* **C106**, 16917–16928

Brown, R. A. 1986 On satellite scatterometer capabilities in air–sea interaction. *J. Geophys. Res.* **C91**, 2263–2282

Burchard, H., Bolding, K. & Villarreal, M. R. 1999 GOTM – a general ocean turbulence model. Theory, applications and test cases. *Tech. Rep. EUR 18745* European Commission, 103p

Burling, R. W. 1955 Wind generation of waves on water. *PhD Dissertation*, Imperial College, University of London, 71–202

Bye, J. A. T. & Jenkins, A. D. 2006 Drag coefficient reduction at very high wind speeds. *J. Geophys. Res.* **C111**, C03024, doi:10.1029/2005JC003114

Bye, J. A. T. & Babanin, A. V. 2009 Wave generation by wind. *Encyclopedia of Ocean Sciences* **6**, Academic Press, Elsevier, 304–309

Calabresea, M., Buccinoa, M. & Pasanisi, F. 2008 Wave breaking macrofeatures on a submerged rubble mound breakwater. *J. Hydro-Environ. Res.* **1**, 216–225

Caldwell, D. R. & Elliott, W. P. 1971 Surface stresses produced by rainfall. *J. Phys. Oceanogr.* **1**, 145–148

Caldwell, D. R. & Elliott, W. P. 1972 The effect of the rainfall on the wind in the surface layer. *Boundary-Layer Meteorol.* **3**, 146–151

Capon, J. 1969 High-resolution frequency-wavenumber spectrum analysis. *Proc. IEEE* **57**, 1408–1418

Cardone, V. J. 1969 Specification of the wind distribution in the marine boundary layer for wave forecasting. *Tech. Rept. GSL-69-1* New York University, 131p

Carey, W. M. & Bradley, M. P. 1985 Low-frequency ocean surface noise sources. *J. Acoust. Soc. Am.* **78**, S1–S2

Cartmill, J. W. & Su, M. Y. 1993 Bubble size distribution under saltwater and freshwater breaking waves. *Dyn. Atmos. Oceans* **20**, 25–31

Caulliez, G. 2002 Self-similarity of near-breaking short gravity wind waves. *Phys. Fluids* **14**, 2917–2920

Cavaleri, L., Ewing, J. A. & Smith, N. D. 1978 Measurement of the pressure and velocity field below surface waves. In *Turbulent Fluxes Through the Sea Surface, Wave Dynamics and Prediction*, Eds. A. Favre and K. Hasselmann, Plenum, New York, 257–272

Cavaleri, L. 1979 Resistance wave staff – accuracy of the measurements. *L'Energia Elettrica* **6**, 299–306

Cavaleri, L. & Zecchetto, S. 1987 Reynolds stresses under wind waves. *J. Geophys. Res.* **C92**, 3894–3904

Cavaleri, L. 2009 Wave modeling – missing the peaks. *J. Phys. Oceanogr.* **39**, 2757–2778

Cengel, Y. A. & Cimbala, J. M. 2006 *Fluid Mechanics. Fundamentals and applications*, Mc Graw Hill, Higher Education, 956p

Chai, A. & Vakhguelt, A. 2009 Computational simulation of wave propagation energy dissipation in liquid-gas medium. *Int. Conf. CODE 2009*, Seoul, Korea, paper s224-5, 157–160.

Chalikov, D. & Belevich, M. 1993 One-dimensional theory of the wave boundary layer. *Boundary-Layer Meteorol.* **63**, 65–96

Chalikov, D. & Sheinin, D. 1998 Direct modeling of one-dimensional nonlinear potential waves. In *Nonlinear Ocean Waves*, Ed. W. Perrie, Advances in Fluid Mechanics **17**, 207–258

Chalikov, D. 2005 Statistical properties of nonlinear one-dimensional wave fields. *Nonlin. Processes Geophys.* **12**, 1–19

Chalikov, D. & Sheinin, D. 2005 Modeling extreme waves based on equations of potential flow with a free surface. *J. Comp. Phys.* **210**, 247–273

Chalikov, D. 2007 Simulation of Benjamin-Feir instability and its consequences. *Phys. Fluids.* **19**, 016602, doi:10.1063/1.2432303

Chalikov, D. 2009 Freak waves: their occurrence and probability. *Phys. Fluids.* **21**, 016602, doi:10.1063/1.3155713

Chalikov, D. 2011 Transformation of harmonic waves. *Fundamental and Applied Hydrophysics*, in press (in Russian)

Chalikov, D. & Rainchik, S. 2011 Coupled numerical modelling of wind and waves and theory of the wave boundary layer. *Boundary-Layer Meteorol.*, **138**, 1–41

Challenor, P. G. & Srokosz, M. A. 1984 Extraction of wave period from altimeter data. *Proc. Workshop on ERS-1 Radar Altimeter Data Products*, Frascati, Italy, (ESA, SP-221, Aug. 1984), 121–124

Chaudry, A. H. & Moore, R. K. 1984 Tower-based backscatter measurements of the sea. *IEEE J. Ocean. Eng.* **9**, 309–316

Chawla, A. & Moore, J. T. 2002 Monochromatic and random wave breaking at blocking points. *J. Geophys. Res.* **C107**, 10.1029/2001JC001042

Chen, G., Kharif, C., Zaleski, S. & Li, J. 1999 Three-dimensional Navier-Stokes simulation of breaking waves. *Phys. Fluids* **11**, 121–133

Chen, L., Manasseh, R., Nikolovksa, A. & Norwood, A. 2003 Noise generation by an underwater gas jet. *Proc. 8th Western Pacific Acoustics Conf.*, Melbourne, Australia, 7–9 April, WC21, 1–4

Cherneva, Z., Tayfun, M. A. & Soares, C. G. 2009 Statistics of nonlinear waves generated in an offshore wave basin. *J. Geophys. Res.* **C114**, C08005, doi:10.1029/2009JC005332, 8p

Cheung, T. K. & Street, R. L. 1988 The turbulent layer in the water at an air–water interface. *IEEE J. Fluid Mech.* **194**, 133–151

Chitre M. A., Potter, J. R. & Ong, S. H. 2006 Optimal and near-optimal signal detection in snapping shrimp dominated ambient noise. *IEEE J. Oceanic Eng.* **31**, 497–503

Collard F., Ardhuin, F. & Chapron, B. 2009 Monitoring and analysis of ocean swell fields from space: New methods for routine observations. *J. Geophys. Res.* **C114**, doi:10.1029/2008JC005215, 15p

Commander, K. W. & Prosperetti, A. 1989 Linear pressure waves in bubbly liquids: comparison between theory and experiments. *J. Acoust. Soc. Am.* **85**, 732–746

Cox, C. S. 1958 Measurements of slopes of high-frequency wind waves. *J. Mar. Res.* **16**, 199–225

Craig, P. D. & Sulem, M. L. 1994 Modeling wave-enhanced turbulence in the ocean surface layer. *J. Comp. Phys.* **108**, 73–83

Craig, P. D. 1996 Velocity profiles and surface roughness under breaking waves. *J. Geophys. Res.* **101**, 1265–1277

Craig, W. & Sulem, C. 1993 Numerical simulation of gravity waves. *J. Comp. Phys.* **108**, 73–83

Craik, A. D. D. & Leibovich, S. 1976 A rational model for Langmuir circulations. *J. Fluid Mech.* **73**, 401–426

Crapper, G. D. 1957 An exact solution for progressive capillary waves of arbitrary amplitude. *J. Fluid Mech.* **2**, 532–540

Crapper, G. D. 1970 Non-linear capillary waves generated by steep gravity waves. *J. Fluid Mech.* **40**, 149–159

Christensen, E. D. & Deigaard, R. 2001 Large eddy simulation of breaking waves. *Coast. Eng.* **42**, 53–86

Cummins, W. 1962 The impulse-response function and ship motion. *Schiffstechnik Forshugsh. Schiffbau Schiffsmaschineubau* **9**, 101–109

Dai, D., Qiao, F., Sulisz, W., Han, L. & Babanin, A. V. 2010 An experiment on the non-breaking surface-wave-induced vertical mixing. *J. Phys. Oceanogr.* **40**, 2180–2188

Dalrymple, R. A. & Rogers, B. D. 2006 Numerical modeling of water waves with the SPH method. *Coast. Eng.* **96**, 141–147

Dao, M. H., Xu, H., Tkalich, P. & Chan, E. S. 2010 Modeling of extreme wave breaking. *Proc. 17th Congr. of the Asia and Pacific Division of the Intl. Assoc. Hydro-Environment Eng. and Res., incorporating 7th Intl. Urban Watershed Management Conf.*, 9p

Davis, M. & Zarnik, E. 1964 Testing ship models in transient waves. *Proc. Fifth Symp. on Naval Hydrodynamics*, Washington, D.C., Office of Naval Research, 509–540

De Leeuw, G. 1993 Aerosol near the air–sea interface. *Trends in Geophysical Research* **2**, 55–70

Demchenko, P. F. 1993 Integral model of atmospheric planetary boundary layer with non-stationary equations for turbulent kinetic energy and its dissipation rate. *Izvestiya Akademii Nauk SSSR. Fizika Atmosferi i Okeana* **21**, 315–320 (in Russian, English abstract)

Dias, F., Pushkarev, A. & Zakharov, V. E. 2004 One-dimensional wave turbulence. *Physics Reports* **398**, 1–65

Didenkulov, I. N. 1992 The influence of wave-breaking bubbles on low-frequency underwater ambient noise formation. In *Breaking Waves. IUTAM Symposium, Sydney, Australia, 1991*, Eds. M. L. Banner, R. H. J Grimshaw, Springer-Verlag, Berlin, Heidelberg, 181–186

Dillon, T. M., Richman, J. C., Hansen, C. G. & Pearson, M. D. 1981 Near-surface turbulence measurements in a lake. *Nature* **290**, 390–392

Ding, L. & Farmer, D. M. 1994 Observations of breaking surface wave statistics. *J. Phys. Oceanogr.* **24**, 1368–1387

Dobroklonskiy, S. V. 1947 Eddy viscosity in the surface layer of the ocean, and waves. *Dokl. Acad. Nauk SSSR (Transactions of USSR Academy of Sciences – English Translation)* **58**, 1157–1159

Dobson, F. W., Smith, S. D. & Anderson, R. J. 1994 Measuring the relationship between wind stress and sea state in the open ocean in the presence of swell. *Atmosphere-Ocean* **32**, 237–256

Doering, J. C. & Donelan, M. A. 1997 Acoustic measurements of the velocity field beneath shoaling and breaking waves. *Coast. Eng.* **32**, 321–330

Dold, J. W. & Peregrine, D. H. 1986 Water-wave modulation. *Proc. 20th Intl. Conf. Coastal Eng.*, Taipei, ASCE, 163–175

Dold, J. W. 1992 An efficient surface-integral algorithm applied to unsteady gravity waves. *J. Comp. Phys.* **103**, 90–115

Dommermuth, D. G., Yue, D. K. P., Lin, W. M. & Rapp, R. J. 1988 Deep-water plunging breakers: a comparison between potential theory and experiments. *J. Fluid Mech.* **189**, 423–442

Donelan, M. A., Longuet-Higgins, M. S. & Turner, J. S. 1972 Periodicity in whitecaps. *Nature* **239**, 449–451

Donelan, M. A. 1978 Whitecaps and momentum transfer. In *Turbulent Fluxes Through the Sea Surface, Wave Dynamics and Prediction*, Favre, NATO Conf. Series, Ed. K. Hasselmann, NY: Plenum Press, 74–94

Donelan, M. A., Hamilton, J. & Hui, W. H. 1985 Directional spectra of wind-generated waves. *Phil. Trans. R. Soc. Lond.* **A315**, 509–562

Donelan, M. A. & Pierson, W. J. 1987 Radar scattering and equilibrium ranges in wind-generated waves – with application to scatterometry. *J. Geophys. Res.* **C92**, 4971–5029

Donelan, M. A. 1990 Air–sea interaction. In *The Sea: Ocean Eng.* **9**, Eds. B. LeMehaute, D. M. Hanes, 239–292

Donelan, M. A., Dobson, F. W., Smith, S. D. & Anderson, R. J. 1993 On the dependence of sea roughness on wave development. *J. Phys. Oceanogr.* **23**, 2143–2149

Donelan, M. A. & Yuan, Y. 1994 Wave dissipation by surface processes. In Komen *et al.* (1994), 143–155

Donelan, M. A., Drennan, W. M. & Magnusson, A. K. 1996 Nonstationary analysis of the directional properties of propagating waves. *J. Phys. Oceanogr.* **26**, 1901–1914

Donelan, M. A., Drennan, W. M. & Katsaros, K. B. 1997 The air–sea momentum flux in conditions of wind sea and swell. *J. Phys. Oceanogr.* **27**, 2087–2099

Donelan, M. A. 1998 Air–water exchange processes. In *Physical Processes in Lakes and Oceans. Coastal and Estuarine Studies* **54**, Ed. J. Imberger, 19–36

Donelan, M. A. 1999 Wind-induced growth and attenuation of laboratory waves. In *Wind-over-Wave Couplings. Perspective and Prospects*, Eds. S. G. Sajadi, N. H. Thomas, J. C. R. Hunt, Clarendon Press, Oxford, 183–194

Donelan, M. A. 2001 A nonlinear dissipation function due to wave breaking. In *ECMWF Workshop on Ocean Wave Forecasting, 2–4 July, 2001*, Series ECMWF Proceedings, 87–94

Donelan, M. A., Haus, B. K., Reul, N., *et al.* 2004 On the limiting aerodynamic roughness of the ocean in very strong winds. *Geophys. Res. Lett.* **31**, L18306, doi:10.1029/2004GL019460

Donelan, M. A., Babanin, A. V., Young, I. R., Banner, M. L. & McCormick, C. 2005 Wave follower field measurements of the wind input spectral function. Part I. Measurements and calibrations. *J. Atmos. Oceanic Tech.* **22**, 799–813

Donelan, M. A., Babanin, A. V., Young, I. R. & Banner, M. L. 2006 Wave follower field measurements of the wind input spectral function. Part II. Parameterization of the wind input. *J. Phys. Oceanogr.* **36**, 1672–1688

Donelan, M. A. & Plant, W. J. 2009 A threshold for wind–wave growth. *J. Geophys. Res.* **C114**, C07012, doi:10.1029/2008JC005238

Donelan, M. A., Haus, B. K., Plant, W. J. & Troianowski, O. 2010 The modulation of short wind waves by long waves. *J. Geophys. Res.* **C115**, 1672–1688

Doong, D.-J., Kao, C. C., Liu, P. C. & Chen, H. S. 2008 An observed extreme large wave. *Proc. of 2008 Taiwan-Polish Joint Seminar on Coastal Protection*, C39–C48

Dore, B. D. 1978 Some effects of the air–water interface on gravity waves. *Geophys. Astrophys. Fluid Dyn.* **10**, 215–230

Drazen, D., Melville, W. K. & Lenain, L. 2008 Inertial scaling of dissipation in unsteady breaking waves. *J. Fluid Mech.* **611**, 307–332

Drennan, W. M., Donelan, M. A., Terray, E. A. & Katsaros, K. B. 1996 Oceanic turbulence dissipation measurements in SWADE. *J. Phys. Oceanogr.* **26**, 808–815

Drennan, W. M., Donelan, M. A., Terray, E. A. & Katsaros, K. B. 1997 On waves, oceanic turbulence, and their interaction. *Geophysica* **33**, 17–27

Drennan, W. M., Kahma, K. K. & Donelan, M. A. 1999 On momentum flux and velocity spectra over waves. *Boundary-Layer Meteorol.* **92**, 489–515

Drennan, W. M., Donelan, M. A., Terray, E. A. & Katsaros, K. B. 2003 On the wave age dependence of wind stress over pure wind seas. *J. Geophys. Res.* **C108**, doi:10.1029/2000JC000715

Ducrozet, G., Bonnefoy, F., Le Touse, D. & Ferrant, P. 2010 A modified high-order spectral method for wavemaker modeling in a numerical wave tank. *J. Fluid Mech.*, submitted

Duncan, J. H. 1981 An experimental investigation of breaking waves produced by a towed hydrofoil. *Proc. R. Soc. Lond.* **A377**, 331–348

Dulov, V. A., Kudryavtsev, V. N. & Bol'shakov, A. N. 2002 A field study of whitecap coverage and its modulations by energy containing surface waves. *Geophysical Monograph* **127**, 187–192

Duraiswami, R., Prabhukumar, S. & Chahine, G. L. 1998 Bubble counting using an inverse scattering method. *J. Acoust. Soc. Am.* **104**, 2699–2717

Dyachenko, A. I., Newell, A. C., Pushkarev, A. & Zakharov, V. E. 1992 Optical turbulence: weak turbulence, condensates, and collapsing filaments in the nonlinear Schrödinger equation. *Physica D* **57**, 96–160

Dyachenko, A. I., Kuznetsov, E. A., Spector, M. D. & Zakharov, V. E. 1996 Analytical description of the free surface dynamics of an ideal fluid (canonical formalism and conformal mapping). *Physics Lett. A* **221**, 73–79

Dyachenko, A. I. & Zakharov, V. E. 2005 Modulation instability of Stokes wave – freak wave. *Pis'ma v ZhETF* **81**, 318–322

Dysthe, K. B. 1979 Note on a modification to the nonlinear Schrödinger equation for application to deep water waves. *Proc. R. Soc. Lond.* **A369**, 105–114

Dysthe, K., Krogstad, H. E. & Muller, P. 2008 Oceanic rogue waves. *Ann. Rev. Fluid Mech.* **40**, 287–310

Ebuchi, N., Kawamura, H. & Toba, Y. 1987 Fine structure of laboratory wind-wave surfaces studied using an optical method. *Boundary-Layer Meteorol.* **39**, 133–151

Emanuel, K. A. 1986 An air–sea interaction theory for tropical cyclones. Part I: Steady-state maintenance. *J. Atmos. Scie.* **43**, 585–604

Emanuel, K. A. 1995 Sensitivity of tropical cyclones to surface exchange coefficients and a revised steady-state model incorporating eye dynamics. *J. Atmos. Sci.* **52**, 3969–3976

Emanuel, K. A. 2003 A similarity hypothesis for air–sea exchange at extreme wind speeds. *J. Atmos. Sci.* **60**, 1420–1428

Ester, L. & Arnone, R. 1994 Effect of whitecaps on determination of chlorophyll concentration from satellite data. *Remote Sens. Environ.* **50**, 328–334

Evans, K. C. & Kibblewhite, A. C. 1990 An examination of fetch-limited wave growth off the west coast of New Zealand by a comparison with the JONSWAP results. *J. Phys. Oceanogr.* **20**, 1278–1296

Exton, H. J., Latham, J., Park, P. M., Smith, M. H. & Allan, R. R. 1986 The production and dispersal of marine aerosol. In *Oceanic Whitecaps and their Role in Air–Sea Exchange Processes*, Riedel Publ. H., Dortrecht, 175–193

Fairall, C. W., Kepert, J. D. & Holland, G. J. 1994 The effect of sea spray on surface energy transport over the ocean. *Global Atmos. Ocean Syst.* **2**, 121–142

Fairall, C. W., Banner, M. L., Peirson, W. L., Asher, W. & Morison, R. P. 2009 Investigation of the physical scaling of sea spray spume droplet production. *J. Geophys. Res.* **C114**, C10001, doi:10.1029/2008JC004918

Farge, M. 1992 Wavelet transforms and their applications to turbulence. *Ann. Rev. Fluid Mech.* **24**, 395–457

Farmer, D. M. & Vagle, S. 1988 On the determination of breaking surface wave distribution. *J. Geophys. Res.* **C93**, 3591–3600

Farmer, D. M., Vagle, S. & Booth, A. D. 1998 A free-flooding acoustic resonator for measurement of bubble size distribution. *J. Atmos. Oceanic Technol.* **15**, 1132–1146

Fedorov, A. V., Melville, W. K. & Rozenberg, A. 1998 An experimental and numerical study of parasitic capillary waves. *Phys. Fluids* **10**, doi:10.1063/1.869657

Fedorov, K. N., Varfolomeev, A. A., Ginzburg, A. I., *et al.* 1979 Thermal reaction of the ocean on the passage of the hurricane Ella. *Okeanologiya* **19**, 992–1001

Feir, J. E. 1967 Discussion: Some results from wave pulse experiments. *Proc. R. Soc. Lond.* **A299**, 54

Felizardo, F. C. & Melville, W. K. 1995 Correlation between ambient noise and the ocean surface wave field. *J. Phys. Oceanogr.* **25**, 513–532

Filipot, J.-F., Ardhuin, F., Babanin, A. V. & Magne, R. 2008 Parametrage du deferlement des vagues dans les modeles spectraux: approches semi-empirique et physique. *Journees Nationales Genie Cotier – Genie Civil, 14–16 octobre 2008, Sophia Antipolis*, Eds. D. Levacher, P. Gaufres, 335–344 (in French, English abstract)

Filipot, J.-F., Ardhuin, F. & Babanin, A. V. 2010 A unified deep-to-shallow-water spectral wave-breaking dissipation formulation. Part 1. Breaking probability. *J. Geophys. Res.*, **115**, C04022, doi:10.1029/2009JC005448

Fochesato, C., Grilli, S. & Dias, F. 2007 Numerical modeling of extreme rogue waves generated by directional energy focusing. *Wave Motion* **26**, 395–416

Forristall, G. Z. 1981 Wave crest distributions: Observations and second-order theory. *J. Phys. Oceanogr.* **30**, 1931–1943

Forristall, G. Z. 2000 Measurements of a saturation range in ocean wave spectra. *J. Geophys. Res.* **86**, 8075–8084

Fouques, S., Krogstad, H. E. & Myrhaug, D. 2006 A second-order Lagrangian model for irregular ocean waves. *J. Offshore Mech. Artic Eng.* **128**, 177–183

Fougues, S. & Stansberg, C. T. 2009 A modified linear Lagrangian model for irregular long-crested waves. *Proc. 28th Intl. Conf. on Ocean, Offshore and Artic Eng. OMAE 2009, May 31–June 5, 2009, Honolulu, USA*, 8p

Galchenko, A., Babanin, A. V., Chalikov, D., Young, I. R. & Hsu, T.-W. 2010 Modulation depth and breaking strength for deep-water wave groups. *J. Phys. Oceanogr.*, **40**, 2313–2324

Gallego, G., Benetazzo, A., Yezzi, A. & Fedele, F. 2008 Wave statistics and spectra via a variational wave acquisition stereo system. *Proc. 27th Intl. Conf. on Offshore Mechanics and Artic Eng. OMAE 2008, June 15–20, 2008, Estoril, Portugal*, 8p

Gargett, A. E. & Smith, B. A. 1972 On the interaction of surface and internal waves. *J. Fluid Mech.* **52**, 179–191

Garratt, J. R. 1977 Review of drag coefficients over oceans and continents. *Mon. Wea. Rev.* **105**, 915–927

Garrett, C. & Smith, J. 1976 On the interaction between long and short surface waves. *J. Phys. Oceanogr.* **6**, 925–930

Garrett, C., Li, M., & Farmer, D. 2000 The connection between bubble size spectra and energy dissipation rates in the upper ocean. *J. Phys. Oceanogr.* **30**, 62–77

Garrett, W. D. 1967 The influence of surface-active material on the properties of air bubbles at the air/sea interface. *Naval Research Lab. Rept. 6545*, Washington, D.C., 14p

Gathman, S. & Trent, E. M. 1968 Space charge over the open ocean. *J. Atmos. Sci.* **25**, 1075–1079

Gayer, G., Dick, S., Pleskachevsky, A., & Rosenthal, W. 2006 Numerical modeling of suspended matter transport in the North Sea. *Ocean Dynamics* **56**, 62–77

Geever, M., O'Dowd, C., van Ekeren, S., *et al.* 2005 Sub-micron sea-spray fluxes. *Geophys. Res. Lett.* **32**, L15810, doi:10/1029/2003GB002079

Gemmrich, J. R. & Farmer, D. M. 1999 Observations of the scale and occurrence of breaking surface waves. *J. Phys. Oceanogr.* **29**, 2595–2606

Gemmrich, J. R. & Farmer, D. M. 2004 Near-surface turbulence in the presence of breaking waves. *J. Phys. Oceanogr.* **34**, 1067–1086

Gemmrich, J. R. 2005 On the occurrence of wave breaking. *Proc. 14th 'Aha Huliko'a Hawaiian Winter Workshop, January 2005*, 123–130

Gemmrich, J. R. 2006 The spectral scale of surface wave breaking. *Proc. 9th Intl. Workshop on Wave Hindcasting and Forecasting, Victoria, B.C., Canada, September 24–29, 2006*, 7p

Gemmrich, J. R., Banner, M. L. & Garrett, C. 2008 Spectrally resolved energy dissipation rate and momentum flux of breaking waves. *J. Phys. Oceanogr.* **38**, 1296–1312

Gemmrich, J. R. 2010 Strong turbulence in the wave crest region. *J. Phys. Oceanogr.* **40**, 583–595

Giovanangeli, J. P., Reul, N., Garat, M. H. & Branger, H. 1999 Some aspects of wind–wave coupling at high winds: an experimental study. In *Wind-Over-Wave Couplings*, Ed. S. G. Sajjadi, Oxford University Press, 81–90

Gordon, S. L. 2003 A parametrization of sea-salt aerosol source function for sub- and super-micron particles. *Global Biogeochem. Cycles* **17**, 1097, doi:10.1029/2003GB002079

Gordon, H. R. 1997 Atmospheric correction of ocean color imagery in the Earth Observing System Era. *J. Geophys. Res.* **102**, 17081–17106

Gordon, J. R. 1986 Theory of the soliton self-frequency shift. *Optics Lett.* **11**, 662–664

Goroch, A., Burk, S. & Davidson, K. L. 1981 Stability effects on aerosol size and height distribution. *Tellus* **32**, 245–250

Grachev, A. A. & Fairall, C. W. 2001 Upward momentum transfer in the marine boundary layer. *J. Fluid Mech.* **478**, 325–343

Grachev, A. A., Fairall, C. W, Hare, J. E., Edson, J. B. & Miller, C. D. 2003 Wind stress vector over ocean waves. *J. Phys. Oceanogr.* **33**, 2408–2429

Gramstad, O., Agnon, Y. & Stiassnie, M. 2010 Localized Zakharov equation; derivation and validation. *Europ. J. Mech.-B/Fluids*, in press

Greenslade, D. J. M. 2001 A wave modelling study of the 1998 Sydney to Hobart yacht race. *Aust. Met. Mag.* **50**, 53–63

Griffin, O. M. 1984 The breaking of ocean surface waves. *NRL Memorandum Rep.* **5337**, 76p

Griffin, O. M., Peltzer, R. D. & Wang, H. T. 1996 Kinematic and dynamic evolution of deep water breaking waves. *J. Geophys. Res.* **C101**, 16515–16531

Grilli, S. T., Guyenne, P. & Dias, F. 2001 A fully non-linear model for three-dimensional overturning waves over an arbitrary bottom. *Int. J. Numer. Meth. Fluids* **35**, 829–867

Groeneweg, J. & Battjes, J. A. 2003 Three-dimensional wave effects on a steady current. *J. Fluid Mech.* **478**, 325–343

Grue, J. & Jensen, A. 2006 Experimental velocities and accelerations in very steep wave events in deep water. *Eur. J. Mech.-B/Fluids* **25**, 554–564

Guan, C. & Xie, L. 2004 On the linear parameterisation of drag coefficient over sea surface. *J. Phys. Oceanogr.* **32**, 2847–2851

Guan, C., Hu, W., Sun, J. & Li, R. 2007 The whitecap coverage model from breaking dissipation parameterizations of wind waves. *J. Geophys. Res.* **C112**, doe:10.1029/2006JC003714, 9p

Guignard, S., Marcer, R., Rey, V., Kharif, C. & Fraunie, P. 2001 Solitary wave breaking on sloping beaches: 2-D two-phase flow numerical simulation by SL-VOF method. *Eur. J. Mech. B – Fluids.* **20**, 57–74

Gunther, H., Hasselmann, S. & Janssen, P. A. E. M. 1992 The WAM Model Cycle-4.0, User Manual. *Deutsche Klimarechenzentrum, Technical Report* **4**, 102

Guo-Larsen, X., Makin, V. K. & Smedman, A. S. 2003 Impact of waves on the sea drag: measurements in the Baltic Sea and a model interpretation. *Glob. Atm. Oc. System.* **9**, 97–120

Haines, M. A. & Johnson, B. D. 1995 Injected bubble populations in seawater and fresh water measured by a photographic method. *J. Geophys. Res.* **100**, 7057–7068

Hanley, K. E., Belcher, S. E. & Sallivan, P. P. 2010 A global climatology of wind–wave interaction. *J. Phys. Oceanogr.* **40**, 1263–1282

Hara, T. & Johnson, S. E. 2002 Wind forcing in the equilibrium range of wind–wave spectra. *J. Fluid Mech.* **470**, 223–245

Harris, D. L. 1966 The wave-driven wind. *J. Atmos. Sci.* **23**, 628–693

Hasselmann, D. E., Dunckel, M. & Ewing, J. A. 1980 Directional spectra observed during JONSWAP 1973. *J. Phys. Oceanogr.* **10**, 1264–1280

Hasselmann, K. 1960 Grundgleichungen der Seegangsvoraussage. *Schiffstechnik* **7**, 191–195

Hasselmann, K. 1962 On the non-linear energy transfer in a gravity-wave spectrum. Part I. General theory. *J. Fluid Mech.* **12**, 481–500

Hasselmann, K., Barnett, T. P., Bouws, E., *et al.* 1973 Measurements of wind–wave growth and swell decay during the Joint North Sea Wave Project (JONSWAP). *Dtsch. Hydrogh. Z. Suppl.* **A8**(12), 1–95

Hasselmann, K. 1971 On the mass and momentum transfer between short gravity waves and larger-scale motions. *J. Fluid Mech.* **50**, 189–205

Hasselmann, K. 1974 On the spectral dissipation of ocean waves due to white capping. *Boundary-Layer Meteorol.* **6**, 107–127

Hasselmann, K., Janssen, P. A. E. M. & Komen, G. J. 1994 Wave–wave interaction. In *Dynamics and Modelling of Ocean Waves*, G. J. Komen, L. Cavaleri, M. Donelan, K. Hasselmann, S. Hasselmann, P. A. E. M. Janssen, Cambridge University Press, 113–143

Haus, B. K., Dahai, J., Donelan, M. A., Zhang, J. A. & Savelyev, I. 2010 The relative rates of sea-air heat transfer and frictional drag in very high winds. *Geophys. Res. Lett.* **37**, L07802, doi:10.1029/2009GL042206.

Hauser, D., Kahma, K. K., Krogstad, H. E., Lehner, S., Monbaliu, J. & Wyatt, L. W. 2005 *Measuring and analysing the directional spectrum of ocean waves*, Cost Office, Brussels, 465p

Herbers, T. H. C., Russnogle, N. R. & Elgar, S. 2000 Spectral energy balance of breaking waves within the surf zone. *J. Phys. Oceanogr.* **30**, 2723–2737

Herbers, T. H. C., Orzech, M., Elgar, S. & Guza, R. T. 2003 Shoaling transformation of wave frequency-directional spectra. *J. Geophys. Res.* **C108**, doi:10.1029/2001JC001304

Hieu, P. D., Katsutoshi, T. & Ca, V. T. 2004 Numerical simulation of breaking waves using a two-phase flow model. *Appl. Math. Model.* **288**, 983–1005

Holthuijsen, L. H. & Herbers, T. H. C. 1986 Statistics of breaking waves observed as whitecaps in the open sea. *J. Phys. Oceanogr.* **16**, 290–297

Holthuijsen, L. H. 2007 *Waves in Oceanic and Coastal Waters*, Cambridge University Press, 387p

Holton, J. E. 1962 *An Introduction to Dynamic Meteorology*, Academic Press, New York 511p

Houmb, O. G. & Overvik, T. 1976 Parameterization of wave spectra and long term joint distribution of wave height and period. *Proc. Symp. Behavior of Offshore Structures* **1**, 144–169

Hristov, T., Friehe, C. & Miller, S. 1998 Wave-coherent fields in air flow over ocean waves: identification of cooperative behavior buried in turbulence. *Phys. Rev. Lett.* **81**, 5245–5248

Hsiao, S. V. & Shemdin, O. H. 1983 Measurements of wind velocity and pressure with wave follower during MARSEN. *J. Geophys. Res.* **88**, 9841–9849

Hsu, T.-W., Ou, S.-H. & Liau, J.-M. 2005 Hindcasting nearshore wind waves using a FEM code for SWAN. *Coast. Eng.* **52**, 177–195

Hua, F. & Yuan, Y. 1992 Theoretical study of breaking wave spectrum and its application. In *Breaking Waves. IUTAM Symposium, Sydney, Australia, 1991*, Eds. M. L. Banner, R. H. J. Grimshaw, Springer-Verlag, Berlin, Heidelberg, 177–282

Huang, N. E., Long, S. R. & Bliven, L. F. 1983 A non-Gaussian statistical model for surface elevation of nonlinear random wave fields. *J. Geophys. Res.* **C88**, 7597–7606

Huang, N. E., Long, S. R., Bliven, L. F. & Tung, C. C. 1984 The non-Gaussian joint probability density function of slope and elevation for a non-linear gravity wave field. *J. Geophys. Res.* **C89**, 1961–1972

Huang, N. E. 1986 An estimate of the influence of breaking waves on the dynamics of the upper ocean. In *Wave Dynamics and Radio Probing of the Sea Surface*, Eds. O. M. Phillips, K. Hasselmann, New York: Plenum, 295–313

Huang, N. E., Shen, Z., Long, S. R., *et al.* 1998 The empirical mode decomposition and the Hilbert spectrum for nonlinear and non-stationary time series analysis. *Proc. R. Soc. Lond.* **A454**, 903–995

Hwang, P. A., Xu, D. & Wu, J. 1989 Breaking of wind-generated waves: breaking and characteristics. *J. Fluid Mech.* **202**, 177–200

Hwang, P. A., Poon, Y.-K. & Wu, J. 1991 Temperature effects on generation and entrainment of bubbles induced by a water jet. *J. Phys. Oceanogr.* **21**, 1602–1605

Hwang, P. A. & Wang, D. W. 2004 An empirical investigation of source term balance of small scale surface waves. *Geophys. Res. Lett.* **31**, L15301, doi:10.1029/2004GL020080

Hwang, P. A. 2007 Spectral signature of wave breaking in surface wave components of intermediate-length scale. *J. Mar. Sys.* **66**, 28–37

Hwung, H.-H., Chiang, W.-S., Liu, P. L. F. & Linett, P. 2004 Sideband evolution of initially uniform deep water wave. *Proc. 29th Intl. Conf. Coast. Eng. (Lisbon)*, 157–168

Hwung, H.-H. & Chiang, W.-S. 2005 Measurements of wave modulation and breaking. *Meas. Sci. Technol.* **16**, 1921–1928

Hwung, H.-H., Huang, L.-H., Lee, J.-F. & Tsai, C.-P. 2005 The characteristics of nonlinear wave transformaion on sloping bottoms. *Overall Performance Report* **91-E-FA09-7-3**, National Cheng Kung University, 123p

Hwung, H.-H., Chiang, W.-S. & Hsao, S.-C. 2007 Observations of the evolution of wave modulation. *Proc. R. Soc.* **A463**, 85–112

Hwung, H.-H., Yang, R.-Y., Shugan, I. V. & Chiang, W.-S. 2009 Evolution of the Stokes wave side-band instability along a super tank: experiments and threshold modification of Tulin NLS model. *Proc. 19th Intl. Offshore and Polar Eng. Conf.*, International Society of Offshore and Polar Engineers (ISOPE), Osaka, Japan, 854–859

Hwung, H.-H., Chiang, W.-S., Yang, R.-Y. & Shugan, I. V. 2011 Threshold model on the evolution of Stokes wave side-band instability. *Europ. J. Mech.-B/Fluids*, in press

Hyun, B. S., Balachandar, K. Y. & Patel, V. C. 2003 Assessment of PIV to measure mean velocity and turbulence in open-ocean flow. *Experiments in Fluids* **35**, 262–267

Iafrati, A. & Campana, E. F. 2005 Free-surface fluctuations behind microbreakers: space-time behaviour and subsurface flow field. *J. Fluid Mech.* **529**, 311–347

Iafrati, A. 2009 Numerical study of the effects of the breaking intensity on wave breaking flows. *J. Fluid Mech.* **622**, 371–411

Iafrati, A. 2011 Energy dissipation mechanisms in wave breaking processes: spilling and highly aerated plunging breaking events. *J. Geophys. Res.*, submitted

Irisov, V. & Voronovich, A. 2011 Numerical simulation of wave breaking. *J. Phys. Oceanogr.*, in press

Jacobs, C. A. 1978 Numerical simulations of the natural variability in water temperature during BOMEX using alternative forms of the vertical eddy exchange coefficients. *J. Phys. Oceanogr.* **8**, 119–141

Jahne, B. 1990 New experimental results on the parameters influencing air–sea gas exchange. *Proc. Air–Water Mass Trans. 2nd Intl. Symp.*, Eds. S. C. Wilhelms, J. S. Gulliver, 582–592

Janssen, P. A. E. M. 1986a On the effects of gustiness on wave growth. *KNMI Afdeling Oceanografisch Onderzoek memo* **00-86-18**, DeBilt, 17p

Janssen, P. A. E. M. 1986b Laboratory observations of the kinematics in the aerated region of breaking waves. *Coast. Eng.* **9**, 453–477

Janssen, P. A. E. M. 1991 Quasi-linear theory of wave generation applied to wave forecasting. *J. Phys. Oceanogr.* **21**, 1631–1642

Janssen, P. A. E. M. 1994 Wave growth by wind. In *Dynamics and Modelling of Ocean Waves*, G. J. Komen, L. Cavaleri, M. Donelan, *et al.* Cambridge University Press, 71–112

Janssen, P. A. E. M. 2003 Nonlinear four-wave interaction and freak waves. *J. Phys. Oceanogr.* **33**, 863–884

Janssen, P. A. E. M. 2004 *The Interaction of Ocean Waves and Wind*, Cambridge University Press, 308p

Janssen, C. & Krafczyk, M. 2010 A lattice Boltzmann approach for free-surface-flow simulations on non-uniform block-structured grids. *Computers and Mathematics with Applications* **59**, 2216–2235

Jarosz, E., Mitchell, D. A., Wang, D. W. & Teague, W. J. 2007 Bottom-up determination of air–sea momentum exchange under a major tropical cyclone. *Science* **315**, 1707–1709

Jeffreys, H. 1924 On the formation of waves by wind. *Proc. Roy. Soc.* **107A**, 189–206

Jeffreys, H. 1925 On the formation of waves by wind. II. *Proc. Roy. Soc.* **110A**, 341–347

Jenkins, A. D. 1992 Quasi-linear eddy-viscosity model for the flux of energy and momentum to wind waves using conservation-law equations in a curvilinear coordinate system. *J. Phys. Oceanogr.* **22**, 843–858

Jenkins, A. D. 1993 Simplified quasi-linear model for wave generation and air–sea momentum flux. *J. Phys. Oceanogr.* **23**, 2001–2018

Jensen, B. L., Sumer, B. M. & Fredsoe, J. 1989 Turbulent oscillatory boundary layers at high Reynolds numbers. *J. Fluid Mech.* **206**, 265–297

Jessup, A. T., Keller, W. C. & Melville, W. K. 1990 Measurements of sea spikes in microwave backscatter at moderate incidence. *J. Geophys. Res.* **95**, 9679–9688

Jessup, A. T., Zappa, C. J. & Yeh, H. 1997a Defining and quantifying microscale wave breaking with infrared imagery. *J. Geophys. Res.* **C102**, 23145–23153

Jessup, A. T., Zappa, C. J. Loewen, M. R. & Hesany, V. 1997b Infrared remote sensing of breaking waves. *Nature* **385**, 52–55

Jessup, A. T. & Phadnis, K. 2005 Measurement of the geometric and kinematic properties of microscale breaking waves from infrared imagery using a PIV algorithm. *Measur. Sci. Tech.* **16**, 1961–1969

Jiang, J. Y., Street, R. L., & Klotz, S. P. 1990 A study of wave-turbulence interaction by use of a nonlinear water wave decomposition technique. *J. Geophys. Res.* **95**, 16037–16054

Johnson, B. D. & Cooke, R. C. 1979 Bubble population and spectra in coastal waters: a photographic approach. *J. Geophys. Res.* **84**, 3761–3766

Johnson, B. D. 1986 Bubble population: background and breaking waves. In *Oceanic Whitecaps and their Role in Air–Sea Exchange Processes*, Riedel Publ. H., Dordrecht, 69–73

Jones, I. S. F. 1985 Turbulence below wind waves. In *The Ocean Surface – Wave Breaking, Turbulent Mixing and Radio Probing*, Eds. Y. Toba, H. Mitsuyasu, Reidel, Dordrecht, 437–442

Kahma, K. K. & Calkoen, C. J. 1992 Reconciling discrepancies in the observed growth of wind-generated waves. *J. Phys. Oceanogr.* **22**, 1389–1405

Kapitza, H. 2002 *TRIM documentation manual*, GKSS-Research Center, Geesthacht

Kara, A. B., Wallcraft, A. J. & Hurlburt, H. E. 2005 How does solar attenuation depth affect the ocean mixed layer? Water turbidity and atmosphere forcing impacts on the simulation of seasonal mixed layer variability in the turbid Black Sea. *J. Climate* **18**, 389–409

Katsaros, K. B. 1980 The aqueous thermal boundary layer. *Boundary Layer Meteorol.* **18**, 107–127

Katsaros, K. B. & Atakturk, S. S. 1992 Dependence of wave-breaking statistics on wind stress and wave development. In *Breaking Waves. IUTAM Symposium, Sydney, Australia, 1991*, Eds. M. L. Banner, R. H. J. Grimshaw, Springer-Verlag, Berlin, Heidelberg, 119–132

Keller, W. C., Plant, W. J. & Valenzuela, G. R. 1986 Observations of breaking ocean waves with coherent microwave radar. In *Radio Probing of the Ocean Surface*, Eds. O. M. Phillips, K. Hasselmann, Plenum, New York, 285–293

Kennedy R. M. & Snyder, R. L. 1983 On the formation of whitecaps by a threshold mechanism. Part I: Monte-Carlo experiments. *J. Phys. Oceanogr.* **13**, 1493–1504

Kerman, B. R. 1988 *Sea Surface Sound: Natural Mechanisms of Surface Generated Noise in the Ocean*, Kluwer Academic, 639p

Kerman, B. R. 1992 *Natural Physical Sources of Underwater Sound: Sea Surface Sound*, Kluwer Academic, 749p

Kinsman, B. 1965 *Wind Waves: Their Generation and Propagation on the Ocean Surface*, Englewood Cliffs, N.J.: Prentice-Hall, 676p

Kitaigorodskii, S. A. & Mitropolsky, Y. Z. 1968 Dissipation of turbulent energy in the surface layer of the ocean. *Izvestia, Atmospheric and Oceanic Physics* **4**, 647–659.

Kitaigorodskii, S. A. 1973 *The Physics of Air–Sea Interaction*, Israel Program for Scientific Translations, Jerusalem, 237p

Kitaigorodskii, S. A., Donelan, M. A., Lumley, J. L. & Terray, E. A. 1983 Wave–turbulence interactions in the upper ocean. Part 2: statistical characteristics of wave and turbulent components of the random velocity field in the marine surface layer. *J. Phys. Oceanogr.* **13**, 1988–1999

Kitaigorodskii, S. A. & Lumley, J. L. 1983 Wave–turbulence interactions in the upper ocean. Part I: The energy balance of the interacting fields of surface wind waves and wind-induced three-dimensional turbulence. *J. Phys. Oceanogr.* **13**, 1977–1987

Kjeldsen, S. P. & Myrhaug, D. 1978 Kinematics and dynamics of breaking waves. Report A78100, *Ships in Rough Seas*, Part 4, Norwegian Hydrodynamic Laboratory

Kjeldsen, S. P. & Myrhaug, D. 1980 Wave–wave interactions, current–wave interactions and resulting extreme waves and breaking waves. *Proc. 17th Coastal Eng. Conf.*, 2277–2303

Kjeldsen, S. P. 1990 Breaking waves. In *Water Wave Kinematics*, Eds. A. Torum, O. T. Gudmestad, Kluwer Academic Publishers, Netherlands, 453–473

Kleiss, J. M. & Melville, W. K. 2010 Observation of wave breaking kinematics in fetch-limited seas. *J. Phys. Oceanogr.*, **40**, 2575–2604

Kleiss, J. M. & Melville, W. K. 2011 The analysis of sea surface imagery for white cap kinematics. *J. Atmos. Ocean. Technol.*, submitted

Knudsen, V., Alford, R. S. & Emling, J. W. 1948 Underwater ambient noise. *J. Marine Res.* **7**, 410–429

Koga, M. 1981 Direct production of droplets from breaking wind-waves – its observation by a multi-colored overlapping exposure technique. *Tellus* **3**, 552–563

Koga, M. & Toba, Y. 1981 Droplet distribution and dispersion processes in breaking wind waves. *Tohoku Geophys. J.* **28**, Sci. Rep. Tohoku Univ., Ser. 5, 1–25

Kolmogorov, A. N. 1954 New variant of the gravitational theory of motion of suspended sediment. *Vestink MGU* **3**, 41–45 (in Russian)

Kolobaev, P. A. 1975 Investigation of concentration and statistical distribution of bubbles created by the wind in the near-surface ocean layer. *Okeanologiya* **15**, 1013–1017 (in Russian)

Komen, G. I., Hasselmann, K. & Hasselmann, S. 1984 On the existence of a fully-developed wind sea spectrum. *J. Phys. Oceanogr.* **14**, 1271–1285

Komen, G. I., Cavaleri, L., Donelan, M., *et al.* 1994 *Dynamics and Modelling of Ocean Waves.* Cambridge University Press, UK, 554p

Korolev, V. S., Petrichenko, S. A. & Pudov, V. D. 1990 Heat and moisture exchange between the ocean and atmosphere in tropical storms Tess and Skip. *Sov. Meteorol. Hydrol.* **3**, 92–94

Korotaev, G. K., Kuftarkov, Y. M. & Fel'zenbaum, A. I. 1971 Non-linear theory of the Ekman boundary layer of the deep sea. *Morskiye Gidrofizicheskiye Issledovaniya* **55**, MGI, Sevastopol, 17–28

Kraan, C., Oost, W. A. & Janssen, P. A. E. M. 1996 Wave energy dissipation by whitecaps. *J. Atmos. Ocean. Technol.* **13**, 262–267

Krasitskii, V. P. 1994 On reduced equations in the Hamiltonian theory of weakly nonlinear surface waves. *J. Fluid Mech.* **272**, 1–20

Krogstad, H. E., Donelan, M. A. & Magnuson, A. K. 2006 Wavelet and local directional analysis of ocean waves. *Int. J. Offshore Polar Eng.* **16**, 97–103

Kubo, H. & Sunamura, T. 2001 Large-scale turbulence to facilitate sediment motion under spilling waves. *Proc. Coastal Dynamics '01*, ASCE, 212–221

Kudryavtsev, V. N., Makin, V. K. & Chapron, B. 1999 Coupled sea surface atmosphere model. 2. Spectrum of short wind waves. *J. Geophys. Res.* **104**, 7625–7639
Kudryavtsev, V. N. & Makin, V. K. 2001 The impact of air-flow separation on the drag of the sea surface. *Boundary-Layer Meteorol.* **98**, 155–171
Kudryavtsev, V. N., Makin, V. K. & Meirink, J. F. 2001 Simplified model of the air flow above the waves. *Boundary-Layer Meteorol.* **98**, 155–171
Kudryavtsev, V. N. & Makin, V. K. 2002 Coupled dynamics of short wind waves and the air flow over long surface waves. *J. Geophys. Res.* **C107**, doi:10.1029/2001JC001251, 13p
Kudryavtsev, V. N. 2004 Impact of swell on the marine atmospheric boundary layer. *J. Phys. Oceanogr.* **34**, 934–949
Kudryavtsev, V. N. & Makin, V. K. 2004 Impact of swell on the marine atmospheric boundary layer. *J. Phys. Oceanogr.* **34**, 934–949
Kudryavtsev, V. N. 2006 On the effect of sea drops on the atmospheric boundary layer. *J. Geophys. Res.* **C111**, C07020, doi:10.1029/2005JC002970
Kudryavtsev, V. N. & Makin, V. K. 2007 Aerodynamic roughness of the sea surface at high winds. *Boundary-Layer Meteorol.* **125**, 289–303
Kudryavtsev, V. N. & Makin, V. K. 2009 Model of the spume sea spray generation. *Geophys. Res. Lett.* **36**, L06801, doi:10.1029/2008GL036871
Kudryavtsev, V. N. & Makin, V. K. 2010 Impact of ocean spray on the dynamics of the marine atmospheric boundary layer. *Boundary-Layer Meteorol.*, in press
Kukulka, T., Hara, T. & Belcher, S. E. 2007 A model of the air-sea momentum flux and breaking-wave distribution for strongly forced wind waves. *J. Phys. Oceanogr.* **37**, 1811–1828
Kwoh, D. S. & Lake, B. M. 1984 A deterministic, coherent, and dual-polarized laboratory study of microwave backscattering from water waves, Part I: Short gravity waves without wind. *IEEE J. Ocean. Eng.* **9**, 291–308
Lai, R. J. & Shemdin, O. H. 1974 Laboratory study of the generation of spray over water. *J. Geophys. Res.* **79**, 3055–3063
Laing, A. K. 1986 Nonlinear properties of random gravity waves in water of final depth. *J. Phys. Oceanogr.* **16**, 2013–2030
Lake, B. M., Yuen, H. C., Rungaldier, H. & Rheem, W. E. 1977 Nonlinear deep-water waves: theory and experiment. Part 2. Evolution of a continuous wave train. *J. Fluid Mech.* **83**, 75–81
Lake B. M. & Yuen H. C. 1978 A new model for nonlinear wind waves. Part 1. Physical model and experimental evidence. *J. Fluid Mech.* **88**, 33–66
Lakehal, D. & Liovic, P. 2011 LEIS of turbulent structure and interaction with breaking wave surfaces. *J. Fluid Mech.*, in press, doi:10.1017/jfm.2011.3
Lammarre, E. & Melville, W. K. 1991 Air entrainment and dissipation in breaking waves. *Nature* **351**, 469–472
Lammarre, E. & Melville, W. K. 1992 Void-fraction measurements and sound-speed fields in bubble flumes generated by breaking waves. *J. Acoust. Soc. Amer.* **95**, 1317–1328
Lamb, H. 1932 *Hydrodynamics*, Dover, New York, 738p
Lamonth-Smith, T., Waseda, T. & Rheem, C.-K. 2007 Measurements of the Doppler spectra of breaking waves. *IET Radar Sonar Navig.* **1**, 149–157
Landau, L. D. & Lifshitz, E. M. 1987 *Fluid Mechanics*, Butterworth-Heinemann, 539p
Landrini, M., Oshri, O., Waseda, T. & Tulin, M. P. 1998 Long time evolution of gravity wave systems. *Proc. 13th Intl. Workshop on Water Waves and Floating Bodies (Delft)*, Ed. A. J. Hermans, 75–78

Langmuir, I. 1938 Surface motion of water induced by the wind. *Science* **87**, 119–123

Large, W. G. & Pond, S. 1981 Open ocean momentum flux measurements in moderate to strong winds. *J. Phys. Oceanogr.* **11**, 324–336

Lavrenov, I. V. 2003 *Wind-Waves in Oceans: Dynamics and Numerical Simulations*, Springer, 377p

Lavrenov, I. V. 2004 Weak turbulent fluxes estimation in surface water wave spectrum. *Proc. 8th Intern. Workshop on Wave Hindcasting and Forecasting and Coastal Hazards, Oahu, Hawaii, November, 14–19, 2004*, 14p

Leighton, T. G. 1994 *The Acoustic Bubble*. Academic Press, London, 613p

Leikin, I. A., Donelan, M. A., Mellen, R. H. & McLaughlin, D. J. 1995 Asymmetry of wind generated waves studied in a laboratory tank. *Nonlin. Processes Geophys.* **2**, 280–289

Lenau, C. W. 1966 The solitary wave of maximal amplitude. *J. Fluid Mech.* **26**, 309–320

Levitus, S. 1982 Climatological atlas of the world ocean. *NOAA Prof. Paper* **No. 13**, US Government Printing Office, 173p

Levitus, S. & Boyer, T. P. 1994 *World Ocean Atlas 1994. Temperature* **4**, NOAA Atlas NESDIS 4, U.S. Govt. Print. Off., Washington, D.C., 117p

Levitus, S., Burgett, R. & Boyer, T. P. 1994 *World Ocean Atlas 1994. Salinity* **3**, NOAA Atlas NESDIS 3, U.S. Govt. Print. Off., Washington, D.C., 99p

Lewis, B. L. & Olin, I. D. 1980 Experimental study and theoretical model of high-resolution radar backscatter from the sea. *Radio Sci.* **15**, 815–828

Li, Y. & Raichlen, F. 2003 Energy balance model for breaking solitary wave runup. *J. Waterway, Port, Coastal, Ocean Eng.* **129**, 47–59

Lighthill, J. 1999 Ocean spray and thermodynamics of tropical cyclones. *J. Eng. Math.* **35**, 11–49

Lighthill, M. J. 1965 Contributions to the theory of waves in non-linear dispersive systems. *J. Inst. Math. Appl.* **1**, 269–306

Ling, S. C. & Kao, T. W. 1976 Parameterization of the moisture and heat transfer process over the ocean under whitecap sea states. *J. Phys. Oceanogr.* **6**, 306–315

Liovic, P. & Lakehal, D. 2007 Multi-phase treatment in the vicinity of arbitrary deformable gas-liquid interfaces. *J. Comput. Phys.* **222**, 504–535

Liu, P. C. 1993 Estimating breaking statistics from wind–wave time series data. *Ann. Geophysicae* **11**, 970–972

Liu, P. C. 2000 Wavelet transform and new perspective on coastal and ocean engineering data analysis. In *Advances in Coastal and Ocean Engineering* **6**, Ed. P. L. F. Liu, World Scientific, 57–101

Liu, P. C., Schwab, D. J. & McLaughlin, R. E. 2002 Has wind-wave modeling reached its limit? *Ocean Eng.* **29**, 81–98

Liu, P. C. & Babanin, A. V. 2004 Using wavelet spectrum analysis to resolve breaking events in the wind wave time series. *Ann. Geophysicae* **22**, 3335–3345

Liu, P. C., Chen, H. S., Doong, D.-J., Kao, C. C. & Hsu, Y.-J. G. 2007 Monstrous ocean waves during typhoon Krosa. *Ann. Geophysicae* **26**, 1327–1329

Liu, P. C., Chen, H. S., Doong, D.-J., Kao, C. C. & Hsu, Y.-J. G. 2009 Freak waves during typhoon Krosa. *Ann. Geophysicae* **27**, 2633–2642

Long, M. W. 1974 On a two-scatterer theory of sea echo. *IEEE Trans. Antennas Propag.* **AP-22**, 667–672

Longuet-Higgins, M. S. & Stewart, R. W. 1960 Changes in the form of short gravity waves on long waves and tidal currents. *J. Fluid Mech.* **8**, 564–585

Longuet-Higgins, M. S., Cartwright, D. E. & Smith, N. D. 1963 Observations of the directional spectrum of sea waves using the motions of the floating buoy. *Proc. Conf. on Ocean Spectra*, Prentice-Hall, Englewood Cliffs, NJ, 111–136

Longuet-Higgins, M. S. 1969a On wave breaking and the equilibrium spectrum of wind-generated waves. *Proc. R. Soc. Lond.* **A310**, 151–159

Longuet-Higgins, M. S. 1969b A nonlinear mechanism for generation of sea waves. *Proc. R. Soc. Lond.* **A311**, 371–389

Longuet-Higgins, M. S. 1974 Breaking waves in deep or shallow water. *Proc. 10th Conf. Naval Hydrodynamics*, M.I.T., 597–605

Longuet-Higgins, M. S. & Turner, J. S. 1974 An "entrainment plume" model of a spilling breaker. *J. Fluid Mech.* **63**, 1–20

Longuet-Higgins, M. S. 1975a Integral properties of periodic waves of finite amplitude. *Proc. R. Soc. Lond.* **A342**, 157–174

Longuet-Higgins, M. S. 1975b On the joint distribution of the periods and amplitudes of sea waves. *J. Geophys. Res.* **C80**, 2688–2694

Longuet-Higgins, M. S. & Cokelet, E. D. 1976 The deformation of steep surface waves on water. I. A numerical method of computation. *Proc. R. Soc. Lond.* **A350**, 1–26

Longuet-Higgins, M. S. & Fox, M. G. H. 1977 Theory of the almost highest wave: The inner solution. *J. Fluid Mech.* **80**, 721–741

Longuet-Higgins, M. S. & Cokelet, E. D. 1978 The deformation of steep surface waves on water. II. Growth of normal-mode instabilities. *Proc. R. Soc. Lond.* **A364**, 1–28

Longuet-Higgins, M. S. 1983 On the joint distribution of wave periods and amplitudes in a random wave field. *Proc. R. Soc. London* **A389**, 241–258

Longuet-Higgins, M. S. & Smith, N. D. 1983 Measurement of breaking waves by a surface jump meter. *J. Geophys. Res.* **88**, 9823–9831

Longuet-Higgins, M. S. 1984 Statistical properties of wave groups in a random sea state. *Phil. Trans. R. Soc. Lond.* **312A**, 219–250

Longuet-Higgins, M. S. 1985 Accelerations in steep gravity waves. *J. Phys. Oceanogr.* **15**, 1570–1579

Longuet-Higgins, M. S. 1987 The propagation of short surface waves on longer gravity waves. *J. Fluid Mech.* **177**, 293–306

Longuet-Higgins, M. S. 1989 Monopole emission of sound by asymmetric bubble oscillations. Part 1. Normal modes. *J. Fluid Mech.* **201**, 525–541

Longuet-Higgins, M. S. 1995 Parasitic capillary waves: a direct calculation. *J. Fluid Mech.* **301**, 79–107

Longuet-Higgins, M. S. & Dommermuth, D. G. 1997 Crest instabilities of gravity waves. Part 3. Nonlinear development and breaking. *J. Fluid Mech.* **336**, 51–68

Longuet-Higgins, M. S. & Tanaka, M. 1997 On the crest instabilities of steep surface waves. *J. Fluid Mech.* **336**, 33–50

Lowe, R. J., Falter, J. L., Bandet, M. D., *et al.* 2005 Spectral wave dissipation over a barrier reef. *IEEE J. Geophys. Res.* **C110**, C04001, doi:10.1029/2004JC002711

Lowe, R. J., Falter, J. L., Koseff, J. R., Monismith, S. G. & Atkinson, M. J. 2007 Spectral wave flow attenuation within submerged canopies: Implications for wave energy dissipation. *IEEE J. Geophys. Res.* **C112**, C05018, doi:10.1029/2006JC003605

Lowen, M. R. & Melville, W. K. 1991a Microwave backscatter and acoustic radiation from breaking waves. *J. Fluid Mech.* **224**, 601–623

Lowen, M. R. & Melville, W. K. 1991b A model of the sound generated by breaking waves. *J. Acoust. Soc. Am.* **90**, 2075–2080

Lowen, M. R. & Siddiqui, M. H. K. 2006 Detecting microscale breaking waves. *Meas. Sci. Technol.* **17**, 771–780

Lu, N. Q., Prosperetti, A. & Yoon, S. W. 1990 Underwater noise emissions from bubble clouds. *IEEE J. Oceanic Eng.* **15**, 275–281

Lubin, P., Vincent, S., Abadie, S. & Caltagirone, J.-P. 2006 Three-dimensional Large Eddy Simulation of air entrainment under plunging breaking waves. *Coast. Eng.* **53**, 631–655

Lucy, L. B. 1977 A numerical approach to the testing of fission hypothesis. *Astronomical Journal* **82**, 1013–1024

Lumley, J. L. & Terray, E. A. 1983 Frequency spectra of frozen turbulence in a random wave field. *J. Phys. Oceanogr.* **13**, 2000–2007

Lyzenda, D. R., Maffett, A. L. & Shuchman, R. A. 1983 The contribution of wedge scattering to the radar cross section of the ocean surface. *IEEE Trans. Geosci. Remote Sens.* **GE-21**, 502–505

Maat, N. & Makin, V. K. 1992 Numerical simulation of air flow over breaking waves. *Boundary-Layer Meteorol.* **60**, 77–93

Makarieva, A., Gorshkov, V., Li, B. & Nobre, A. 2010 A crtique of some modern applications of the Carnot heat engine concept: the dissipative heat engine cannot exist. *Proc. R. Soc. Lond.* **A466**, 1893–1902.

Makin, V. K. & Chalikov, D. V. 1980 Calculation of the energy flux to real waves. *Dokl. Acad. Nauk SSSR (Transactions of USSR Academy of Sciences – English Translation)* **253**, 1458–1462

Makin, V. K. 1988 Effect of wave interactions in the seawater-adjoining atmospheric boundary layer on flow of energy to wind waves. *Soviet Physics. A Translation of the Physics Section of the Proceedings of the Academy of Sciences of the USSR* **299**, 276–279

Makin, V. K., Kudryavtsev, V. N. & Mastenbroek, C. 1995 Drag of the sea surface. *Boundary-Layer Meteorol.* **73**, 159–182

Makin, V. K. & Kudryavtsev, V. N. 1999 Dynamical coupling of surface waves with the atmosphere. In *Air–Sea Exchange: Physics, Chemistry and Dynamics*, Ed. G. L. Geernaert, Kluwer Acad. Publ., 73–126

Makin, V. K. & Kudryavtsev, V. N. 2002 Impact of dominant waves on sea drag. *Boundary-Layer Meteorol.* **103**, 83–99

Makin, V. K. 2005 A note on drag of the sea surface at hurricane winds. *Boundary Layer Meteorol.* **115**, 169–176

Manasseh, R. 1997 Acoustic sizing of bubbles at moderate to high bubbling rates. In *Experimental Heat Transfer, Fluid Mechanics and Thermodynamics*, Eds. M. Giot, F. Mayinger, G. P. Celata, Edizoni ETS, 943–947

Manasseh, R., Yoshida, S. & Rudman, M. 1998 Bubble formation processes and bubble acoustic signals. *Proc. Third Intl. Conf. Multiphase Flow*, Lyon, France, 8–12 June, 1

Manasseh, R., LaFontaine, R. F., Davy, J., Shepherd, I. C. & Zhu, Y. 2001 Passive acoustic bubble sizing in sparged systems. *Exp. Fluids* **30**, 672–682

Manasseh, R., Nikolovska, A., Ooi, A. & Yoshida, S. 2004 Anisotropy in the sound field generated by a bubble chain. *J. Sound Vibration* **278**, 807–823

Manasseh, R., Babanin, A. V., Forbes, C., Rickards, K., Bobevski, I. & Ooi, A. 2006 Passive acoustic determination of wave-breaking events and their severity across the spectrum. *J. Atmos. Ocean. Technol.* **23**, 599–618

Manasseh, R., Riboux, G. & Risso, F. 2008 Sound generation on bubble coalescence following detachment. *Intl. J. Multiphase Flows* **34**, 938–949

McCowan, J. 1894 On the highest wave of permanent type. *Phil. Mag.* **38**, 351–358
McIntyre, F. 1972 Flow patterns in breaking bubbles. *J. Geophys. Res.* **77**, 5211–5228
McLean, J. W. 1982 Instabilities of finite-amplitude water waves. *J. Fluid Mech.* **114**, 315–330
McLean, P. J. 1985 Simulation of the mixed layer at OWS November and Papa with several models. *J. Geophys. Res.* **C90**, 903–916
McNeil, C. & D'Asaro, E. 2007 Parameterization of air–sea gas fluxes at extreme wind speeds. *J. Mar. Sys.* **66**, 110–121
McWilliams, J. C., Sullivan, P. P. & Moeng, C.-H. 1997 Langmuir turbulence in the ocean. *J. Fluid Mech.* **334**, 1–30
Medwin, H. 1989 Bubble sources of the Knudsen sea noise spectra. *J. Acoust. Soc. Am.* **86**, 1124–1130
Medwin, H. & Daniel, A. C. 1990 Acoustic measurements of bubble production by spilling breakers. *J. Acoust. Soc. Am.* **88**, 408–412
Melville, W. K. 1982 Instability and breaking of deep-water waves. *J. Fluid Mech.* **115**, 165–185
Melville, W. K., Loewen, M., Felizardo, F., Jessup, A. & Buckingham, M. 1988 Acoustic and microwave signatures of breaking waves. *Nature* **336**, 54–56
Melville, W. K., Loewen, M. R. & Lamarre, E. 1992 Sound production and air entrainment by breaking waves: a review of recent laboratory experiments. In *Breaking Waves. IUTAM Symposium, Sydney, Australia, 1991*, Eds. M. L. Banner, R. H. J. Grimshaw, Springer-Verlag, Berlin, Heidelberg, 139–146
Melville, W. K. 1994 Energy dissipation by breaking waves. *J. Phys. Oceanogr.* **24**, 2041–2049
Melville, W. K. 1996 The role of surface wave breaking in air–sea interaction. *Ann. Rev. Fluid Mech.* **28**, 279–321
Melville, W. K., Shear, R. & Veron, F. 1998 Laboratory measurements of the generation and evolution of Langmuir circulations. *J. Fluid Mech.* **364**, 31–58
Melville, W. K. & Matusov, P. 2002 Distribution of breaking waves at the ocean surface. *Nature* **417**, 58–63
Melville, W. K., Veron, F. & White, J. 2002 The velocity field under breaking waves: coherent structures and turbulence. *J. Fluid Mech.* **454**, 202–233
Merlivat, L. & Memery, L. 1983 Gas exchange across an air–water interface: experimental results and modeling of bubble contribution to transfer. *J. Geophys. Res.* **88**, 707–724
Meyer, Y. 1989 Wavelets and operators. In *Analysis at Urbana I. London Math. Soc. Lecture Notes*, **137**, Cambridge University Press, 256–365
Meza, E., Zhang, J. & Seymour, R. J. 2000 Free-wave energy dissipation in experimental breaking waves. *J. Phys. Oceanogr.* **30**, 2404–2418
Miche, A. 1944 Mouvements ondulatoires de la mer en profondeur croissante ou decroissante. Forme limite de la houle lors de son deferlement. Application aux digues maritimes. Troisieme partie. Forme et proprietes des houles limites lors du deferlement. Croissance des vitesses vers la rive. *Annales des Ponts et Chaussees* **114**, 369–406 (in French)
Miles, J. W. 1957 On the generation of surface waves by shear flows. *J. Fluid Mech.* **3**, 185–204
Miles, J. W. 1959 On the generation of surface waves by shear flows. Part 2. *J. Fluid Mech.* **6**, 568–582
Miles, J. W. 1960 On the generation of surface waves by turbulent shear flows. *J. Fluid Mech.* **7**, 469–478

Miles, J. & Ierley, G. 1998 Surface-wave generation by gusty wind. *J. Fluid Mech.* **357**, 21–28

Minnaert, M. 1933 On musical air bubbles and the sound of running water. *Phil. Mag.* **16**, 235–248

Mironov, A. S. & Dulov, V. A. 2008 Detection of wave breaking using sea surface video records. *Meas. Sci. Technol.* **19**, doi:1018/0957-0233/19/1/015405, 10p

Mitsuyasu, H., Tasai, F., Suhara, T., Mizuno, S., Ohkuzo, M., Honda, I. & Rikiishi, K. 1975 Observations of the directional spectrum of ocean waves using a cloverleaf buoy. *J. Phys. Oceanogr.* **5**, 750–758

Miyake, Y. & Abe, T. 1948 A study on the foaming of the water. Part 1. *J. Mar. Res.* **7**, 67–73

Monaghan, J. J. 1977 Smoothed particle hydrodynamics. *Ann. Rev. Astronomy and Astrophysics* **30**, 543–574

Monahan, E. C. 1966 Sea spray and its relationship to low elevation wind speed. *PhD dissertation* **26**, Massachusetts Institute of Technology, 147p

Monahan, E. C. 1969 Fresh water whitecaps. *J. Atmos. Sci.* **26**, 1026–1029

Monahan, E. C. & Zietlow, C. R. 1969 Laboratory comparisons of fresh-water and salt-water whitecaps. *J. Geophys. Res.* **74**, 6961–6966

Monahan, E. C. 1971 Oceanic whitecaps. *J. Phys. Oceanogr.* **1**, 139–144

Monahan, E. C., O'Muircheartaigh, I. G. & Fitzgerald, M. P. 1981 Determination of surface wind speed from remotely measured whitecap coverage: a feasibility assessment. *SP-167*, Eur. Space Agency, Paris, 103–109

Monahan, E. C. & MacNiocaill, G., Eds. 1986 *Oceanic Whitecaps and Their Role in Air–Sea Exchange Processes*, D. Riedel, 214p

Monahan, E. C. & O'Muircheartaigh, I. G. 1986 Whitecaps and passive remote sensing of the ocean surface. *Intl. J. Remote Sens.* **7**, 627–642

Monahan, E. C., Spiel, D. E. & Davidson, K. L. 1986 A model of marine aerosol generation via whitecaps and wave disruption. In *Oceanic Whitecaps and Their Role in Air-Sea Exchange Processes*, Eds. E. C. Monahan, G. MacNiocaill, D. Riedel, 167–174

Monahan, E. C. & Lu, M. 1990 Acoustically relevant bubble assemblages and their dependence on meteorological parameters. *IEEE J. Oceanic Eng.* **15**, 340–349

Monahan, E. C. 1993 Occurrence and evolution of acoustically relevant subsurface bubble plumes and their associated, remotely monitorable surface whitecaps. In *Natural Physical Sources of Underwater Sound*, Ed. B.R. Kerman, Kluwer Acad., Norwell, Mass., 503–517

Monin, A. S. & Yaglom, A. M. 1971 *Statistical fluid mechanics: The mechanics of turbulence* **1**, M. I. T. Press, 769p

Moore, K. D., Voss, K. J. & Gordon, H. 1997 Spectral reflectance of whitecaps: Their contribution to water leaving radiance. *J. Geophys. Res.* **105**, 6493–6499

Mori, N. & Yasuda, T. 1994 Orthonormal wavelet analysis for deep-water breaking waves. *Proc. 24th Int. Conf. Coast. Eng.* **31**, Koba, Japan, 412–426

Mueller, J. A. & Veron, F. 2009 A sea-state dependent spume generation function. *J. Phys. Oceanogr.* **39**, 2363–2372

Munk, W. H. 1947 A critical wind speed for air–sea boundary processes. *J. Marine Res.* **6**, 203–218

Munk, W. H. 1955 Wind stress over water: An hypothesis. *Quart. J. Roy. Meteorol. Soc.* **81**, 320–332

Mutsuda, H. & Yasuda, T. 2000 Numerical simulation of turbulent air–water mixing layer within surf-zone. *Proc. 27th Int. Conf. Coast. Eng.*, 755–768

Myrhaug, D. & Kjeldsen, S. P. 1986 Steepness and asymmetry of extreme waves and the highest waves in deep water. *Ocean Eng.* **13**, 549–568

Nadaoka, K., Hino, M. & Kayano, Y. 1989 Structure of the turbulent flow field under breaking waves in the surf zone. *J. Fluid Mech.* **204**, 359–387

Nath, J. H. & Ramsey, F. L. 1976 Probability distributions of breaking wave heights emphasizing the utilization of the JONSWAP spectrum. *J. Phys. Oceanogr.* **6**, 316–323

Nepf, H. M., Wu, C. H. & Chan, E. S. 1998 A comparison of two- and three-dimensional wave breaking. *J. Phys. Oceanogr.* **28**, 1496–1510

Nepf, H. M., Wu, C. H. & Chan, E. S. 2001 Turbulent aerosol fluxes over the Arctic Ocean. 2. Wind-driven sources from the sea. *J. Geophys. Res.* **106**, 32139–32154

Neumann, J. 1954 Zur Charakteristik des Seeganges. *Arch. Meteorol. Geophys. Biokl.* **A7**, 352

Oakey, N. S. & Elliott, J. A. 1982 Dissipation within surface mixed layer. *J. Phys. Oceanogr.* **12**, 171–185

Ochi, M. K. & Tsai, C.-H. 1983 Prediction of occurrence of breaking waves in deep water. *J. Phys. Oceanogr.* **13**, 2008–2019

O'Dowd, C. D. & de Leeuw, G. 2007 Marine aerosol production: a review of the current knowledge. *Phil. Trans. R. Soc. Lond.* **A365**, 1753–1774

Oh, S.-H., Mizutani, N. & Suh, K.-D. 2008 Laboratory observation of coherent structures beneath microscale and large-scale breaking waves under wind action. *Experimental Thermal and Fluid Sci.* **32**, 1232–1247

Okamura, M. 1986 Maximum wave steepness and instabilities of finite-amplitude standing gravity waves. *Fluid Dynamics Res.* **1**, 201–214

Onorato, M., Osborne, A. R., Serio, M. & Bertone, S. 2001 Freak wave in random oceanic sea states. *Phys. Rev. Lett.* **86**, 5831–5834

Onorato, M., Osborne, A. R. & Serio, M. 2002 Extreme wave events in directional, random oceanic sea states. *Phys. Fluids* **14**, 25–28

Onorato, M., Cavaleri, L., Fouques, S., *et al.* 2009a Statistical properties of mechanically generated surface gravity waves: a laboratory experiment in a three-dimensional wave basin. *J. Fluid Mech.* **637**, 235–257

Onorato, M., Waseda, T., Toffoli, A., *et al.* 2009b On the statistical properties of directional ocean waves: the role of the modulational instability in the formation of extreme events. *Phys. Rev. Lett.* **102**, doi:10.1103/PhysRevLett.102.114502, 4p

Oost, W. A., Kohsiek, W., de Leeuw, G., *et al.* 1995 On the discrepancies between CO_2 flux measurement methods. *Select. Pap. Third Intl. Symp. Air–Water Gas Transf.*, Eds. B. Jahne, E. C. Monahan, Heidelberg, 723–733

Oost, W. A., Komen, C. J., Jacobs, C. M. J. & van Oort, C. 2001 New evidence for a relation between wind stress and wave age from measurements during ASGAMAGE. *Preprint 2001–05 Koninklijk Ned. Met. Inst. (KNMI)*, De Bilt, Netherlands, 30p

Packham, B. 1952 The theory of symmetrical gravity waves of finite amplitude. 2. The solitary wave. *Proc. Roy. Soc.* **213A**, 238–249

Papadimitrakis, Y. A. & Huang, N. E. 1988 An estimate of wave breaking probability for deep water waves. *Sea Surface Sound*, Ed. B. R. Kerman, 71–83

Papadimitrakis, Y. A. 2005 On the probability of wave breaking in deep water. *Deep-Sea Res. II* **52**, 1246–1269

Peirson, W. & Garcia, A. 2008 On the wind-induced growth of slow water waves of finite steepness. *J. Fluid. Mech.* **608**, 243–274

Penney, W. G. & Price, A. T. 1952 Finite periodic stationary gravity waves in a perfect liquid. *Phil. Trans. R. Soc. Lond.* **A244**, 254

Perlin, M., Lin, H. & Ting, C.-L. 1993 On parasitic capillary waves generated by steep gravity waves: an experimental investigation with spatial and temporal measurements. *J. Fluid Mech.* **255**, 597–620

Phelps, A. D., Ramble, D. G. & Leighton, T. G. 1996 The use of a combination frequency technique to measure the surf zone bubble population. *J. Acoust. Soc. Am.* **101**, 1981–1989

Phillips, O. M. 1957 On the generation of waves by turbulent wind. *J. Fluid Mech.* **2**, 417–445

Phillips, O. M. 1958 The equilibrium range in the spectrum of wind generated waves. *J. Fluid Mech.* **4**, 426–434

Phillips, O. M. 1963 On the attenuation of long gravity waves by short breaking waves. *J. Fluid Mech.* **16**, 321–332

Phillips, O. M. 1977 *The Dynamics of the Upper Ocean*, Cambridge University Press, 336p

Phillips, O. M. 1984 On the response of short ocean wave components at a fixed number to ocean current variations. *J. Phys. Oceanogr.* **14**, 1425–1433

Phillips, O. M. 1985 Spectral and statistical properties of the equilibrium range of wind-generated gravity waves. *J. Fluid Mech.* **156**, 505–531

Phillips, O. M., Posner, F. L. & Hansen, J. P. 2001 High resolution radar measurements of the speed distribution of breaking events in wind-generated ocean waves: surface impulse and wave energy dissipation rates. *J. Phys. Oceanogr.* **31**, 450–460

Phillips, W. R. C. 1998 Finite amplitude rotational waves in viscous shear flows. *Stud. Appl. Math.*, **101**, 23–47

Phillips, W. R. C. 2001 On an instability to Langmuir circulation and the role of Prandtl and Richardson numbers. *J. Fluid Mech.* **442**, 335–358

Phillips, W. R. C. 2002 Langmuir circulation beneath growing or decaying surface waves. *J. Fluid Mech.* **469**, 317–342

Phillips, W. R. C. 2003 Langmuir circulation. In *Wind Over Waves II: Forecasting and Fundamentals of Applications,* Eds. S. G. Sajjadi, J. C. R. Hunt, Horwood Publishing, Chichester, UK, 157–167

Phillips, W. R. C. 2005 On the spacing of Langmuir circulation in strong shear. *J. Fluid Mech.* **525**, 215–236

Pielk, R. A. & Lee, T. J. 1991 Influence of sea spray and rainfall on the surface wind profile during conditions of strong winds. *Boundary-Layer Meteorol.* **55**, 305–308

Pierson, W. J., Jr. & Moskowitz, L. 1964 A proposed spectral form for fully developed wind seas based on the similarity theory of S. A. Kitaigordskii. *J. Geophys. Res.* **69**, 5181–5190

Pierson, W. J., Silvester, W. B. & Donelan, M. A. 1986 Aspects of determination of winds by means of scatterometry and on the utilization of vector wind data for the meteorological forecasts. *J. Geophys. Res.* **C91**, 2263–2282

Pierson, W. J., Donelan, M. A. & Hui, W. H. 1992 Linear and nonlinear propagation of water wave groups. *J. Geophys. Res.* **C97**, 5607–5621

Pigeon, V. W. 1968 Doppler dependence of radar sea return. *J. Geophys. Res.* **73**, 1333–1341

Pleskachevsky, A., Horstmann, J. & Rosental, W. 2001 Modeling of sediment transport in synergy with ocean color data. *Proc. 4th Berlin Workshop on Ocean Remote Sensing*, Wissenschaft und Tchnick Verlag, Berlin, 30 May–01 June 2002, 177–182

Pleskachevsky, A., Gayer, G., Horstmann, J. & Rosenthal, W. 2005 Synergy of satellite remote sensing and numerical modeling for monitoring of suspended particulate matter. *Ocean Dynamic* **55**, 2–9

Pleskachevsky, A., Dobrynin, M., Babanin, A. V., Gunther, H. & Stanev, E. 2011 Turbulent diffusion due to ocean surface waves indicated by suspended particulate matter. Implementation of satellite data into numerical modelling. *J. Phys. Oceanogr.*, in press, doi:10.1175/2010JPO4328.1

Polnikov, V. G. 1991 A third generation spectral model for wind waves. *Izvestia, Atmospheric and Oceanic Physics* **27**, 615–623

Polnikov, V. G. 1993 On a description of a wind-wave energy dissipation function. *The Air–Sea interface. Radio and Acoustic Sensing, Turbulence and Wave Dynamics*, Eds. M. A. Donelan, W. J. Hui, W. J. Plant, Rosenstiel School of Marine and Atmospheric Science, University of Miami, Miami, FL, 277–282

Powell, M. D., Vickery, P. J. & Reinhold, T. A. 2003 Reduced drag coefficient for high wind speeds in tropical cyclones. *Nature* **422**, 279–283

Powell, M. D. 2007 Drag coefficient distribution and wind speed dependence in tropical cyclone. *Final Report to the National Oceanic & Atmospheric Administration.* Joint Hurricane Testbed Program, 26p

Price, J. F. 1981 Upper ocean response to a hurricane. *J. Phys. Oceanogr.* **11**, 153–175

Price, R. K. 1970 Detailed structure of the breaking wave. *J. Geophys. Res.* **75**, 5276

Price, R. K. 1971 The breaking of water waves. *J. Geophys. Res.* **76**, 1576–1581

Prosperetti, A. 1985 Bubble-related ambient noise in the ocean. *J. Acoust. Soc. Am.* **78**, S2

Prosperetti, A. 1988 Bubble-related ambient noise in the ocean. *J. Acoust. Soc. Am.* **84**, 1042–1054

Pudov, V. D., Varfolomeev, A. A. & Fedorov, K. N. 1979 Vertical structure of the wake of a typhoon in the upper ocean. *Okeanologiya* **21**, 142–146

Pushkarev, A., Resio, D. & Zakharov, V. E. 2003 Weak turbulent approach to the wind-generated gravity sea waves. *Physica D* **184**, 29–63

Qiao, F., Yeli, Y., Yang, Y., *et al.* 2004 Wave-induced mixing in the upper ocean: Distribution and application to a global ocean circulation model. *Geophys. Res. Lett.* **31**, L11303, doi:10.1029/2004GL019824

Qiao, F., Ma, J., Xia, C., Yang, Y. & Yuan, Y. 2006 Influences of the surface wave-induced mixing and tidal mixing on the vertical temperature structure of the Yellow and East China Seas in summer. *Progr. in Natural Sci.* **16**, 739–746

Qiao, F., Yang, Y., Xia, C. & Yuan, Y. 2008 The role of surface waves in the ocean mixed layer. *Acta Oceanologica Sinica* **27**, 30–37

Qiao, F., Yuan, Y., Ezer, T., *et al.* 2010 A three-dimensional surface wave–ocean circulation coupled model and its initial testing. *Ocean Dyn.*, **60**, 1339–1355

Qiao, H. & Duncan, J. H. 2001 Gentle spilling breakers: crest flow-field evolution. *J. Fluid Mech.* **439**, 57–85

Raizer, V. Y. & Sharkov, E. A. 1980 On scaling structure of the sea whitecapping. *Izvestiya Akademii Nauk SSSR. Fizika Atmosferi i Okeana* **16**, 773–776 (in Russian, English abstract)

Ramberg, S. E. & Bartholomew, C. L. 1982 Computer-based measurements of incipient wave breaking. In *Computational Methods and Experimental Measurements*, Eds. G. A. Keramidas, C. A. Brebbia, Springer-Verlag, Berlin, 102–115

Ramberg, S. E., Barber, M. E. & Griffin, O. M. 1985 Laboratory study of steep and breaking deep water waves in a convergent channel. *NRL Memorandum Rep.* **5610**, 42p

Ramberg, S. E. & Griffin, O. M. 1987 Laboratory study of steep and breaking deep water waves. *J. Waterway, Port, Coastal, Ocean Eng.* **113**, 493–507

Rapp, R. J. & Melville, W. K. 1990 Laboratory measurements of deep-water breaking waves. *Phil. Trans. R. Soc. Lond.* **A311**, 735–800

Rastigejev, Y., Suslov, S. & Lin, Y.-L. 2011 Effect of ocean spray on vertical momentum transport under high-wind conditions. *Boundary-Layer Meteorol.*, in press

Rayleigh, Lord 1917 On the pressure developed in a liquid during the collapse of a spherical cavity. *Philos. Mag.* **34**, 94–98

Reid, J. S. 1992 The sideband instability and the onset of wave breaking. In *Breaking waves. IUTAM Symposium, Sydney, Australia, 1991*, Eds. M. L. Banner, R. H. J Grimshaw, Springer-Verlag, Berlin, Heidelberg, 155–160

Resio, D. T., Long, C. E. & Vincent, C. L. 2004 Equilibrium-range constant in wind-generated wave spectra. *J. Geophys. Res.* **109**, C01018, doi:10.1029/2003JC001788

Reul, N. 1998 Etude experimentale de la structure de l'ecoulement d'air au-dessus de vagues courtes d'eferlantes. *PhD Thesis*, Universite de la Mediterranee, Aix-Marseille II, 334p

Reul, N. & Chapron, B. 2003 A model of sea-foam thickness distribution for passive microwave remote sensing applications. *J. Geophys. Res.* **C108**, C103321, doi:10.1029/2003JC001887

Reynolds, O. 1883 On the experimental investigation of the circumstances which determine whether the motion of water shall be direct or sinuous, and the law of resistance in parallel channels. *Phil. Trans. R. Soc. Lond.* **174**, 935–982

Rice, S. O. 1954 Mathematical analysis of random noise. In *Noise and Stochastic Processes*, Ed. N. Wax, Dover, New York, 133–294

Richman, J. & Garrett, C. 1977 The transfer of energy and momentum by the wind to the surface mixed layer. *J. Phys. Oceanogr.* **7**, 876–881

Riehk, S. O. 1954 *Tropical Meteorology*, McGraw-Hill, 392p

Robinson, I. S., Wells, N. C. & Charnock, H. 1984 The sea surface thermal boundary layer and its relevance to the measurement of sea surface temperature by airborne and spaceborne radiometers. *Intl. J. Remote Sens.* **5**, 19–45

Rogers, W. E., Hwang, P. A. & Wang, D. W. 2003 Investigation of wave growth and decay in the SWAN model: Three regional-scale applications. *J. Phys. Oceanogr.* **33**, 366–389

Rogers, W. E. & Holland, K. T. 2009 A study of dissipation of wind–waves by viscous mud at Cassino Beach, Brazil: prediction and inversion. *Cont. Shelf Res.* **29**, 676–690

Rogers, W. E., Babanin, G. V. & Wang, D. W. 2011 Observation-based input and whitecapping dissipation in a model for wind-generated surface waves: description and simple calculations. *J. Phys. Oceanogr.*, submitted

Roland, A. 2009 Development of WWMII-Spectral wave modeling on unstructured meshes. PhD thesis, Inst. for Hydraulics and Water Resources Eng., Univ. of Technology Darmstadt, 221p

Schultz, M., de Leeuw, G. & Balkanski, Y. 2004 Sea-salt aerosol source functions and emissions. In *Emissions of Atmospheric Trace Compounds*, Eds. C. Granier, P. Artaxo, C. Reeves, Dordecht, The Netherlands: Kluwer, 333–359

Schneggenburger, C., Gunther, H. & Rosenthal, W. 2000 Spectral wave modeling with non-linear dissipation: validation and applicaions in a coastal tidal environment. *Coast. Eng.* **41**, 201–235

Schulz, E. W. 2009 The Riding Wave Removal Technique: recent developments. *J. Atmos. Ocean. Technol.* **26**, 135–144

Schwartz, L. W. & Fenton, J. D. 1982 Strongly nonlinear waves. *Ann. Rev. Fluid Mech.* **14**, 39–60

Segur, H., Fenderson, D., Carter, J., *et al.* 2005 Stabilizing the Benjamin-Feir instability. *J. Fluid Mech.* **539**, 229–271

Sharkov, E. A. 2007 *Breaking Ocean Waves. Geometry, Structure and Remote Sensing.* Springer, 278p

Shats, M., Punzmann, H. & Xia, H. 2010 Capillary rogue waves, *Phys. Rev. Lett.*, **104**, 104503

Shemer, L., Kit, E. & Jiao, H.-Y. 2002 An experimental and numerical study of the spatial evolution of unidirectional nonlinear water-wave groups. *Phys. Fluids* **14**, 3380–3390

Shemer, L., Goulitski K. & Kit, E. 2007 Evolution of wide-spectrum unidirectional wave groups in a tank: an experimental and numerical study. *Europ. J. Mech.-B/Fluids* **26**, 193–219

Shemer, L. & Dorfman, B. 2008 Spatial vs. temporal evolution of nonlinear wave groups – experiments and modeling based on the Dysthe equation. *Proc. 27th Intl. Conf. on Offshore Mechanics and Arctic Eng. OMAE 2008, June 15–20, 2008, Estoril, Portugal*, 10p

Skafel, M. G. & Donelan, M. A. 1997 Laboratory measurements of stress modulation by wave groups. *Geophysica* **33**, 9–14

Smedman, A. S., Hogstrom, U., Bergstrom, H., *et al.* 1999 A case study of air–sea interaction during swell conditions. *J. Geophys. Res.* **104**, 25833–25851

Smith, G. G. & Mocke, G. P. 2002 Interaction between breaking/broken waves and infragravity-scale phenomena to control sediment suspension transport in the surf zone. *Marine Geology* **187**, 329–345

Smith, J. 1992 Observed growth of Langmuir circulation. *J. Geophys. Res.* **97**, 5651–5664

Smith, J. 1998 Evolution of Langmuir circulation in a storm. *J. Geophys. Res.* **103**, 12649–12668

Smith, M. H., Park, P. H. & Consterdine, I. E. 1993 Marine aerosol concentrations and estimated fluxes over the sea. *J. Roy. Meteorol. Soc.* **119**, 809–824

Smith, M. J., Poulter, E. M. & McGregor, J. A. 1996 Doppler radar measurements of wave groups and breaking waves. *J. Geophys. Res.* **C101**, 14269–14282

Smith, S. D., Anderson, R. J., Oost, W. A., *et al.* 1992 Sea surface wind stress and drag coefficients: the HEXOS results. *Boundary-Layer Meteorol.* **60**, 109–142

Snodgrass, F. E., Groves, G. W., Hasselmann, K. F., *et al.* 1966 Propagation of ocean swell across the Pacific. *Phil. Trans. R. Soc. Lond.* **259**, 431–497

Snyder, R. L. & Kennedy R. M. 1983 On the formation of whitecaps by a threshold mechanism. Part I: Basic formalism. *J. Phys. Oceanogr.* **13**,1482–1492

Snyder, R. L., Smith, L. & Kennedy R. M. 1983 On the formation of whitecaps by a threshold mechanism. Part I: Field experiment and comparison with theory. *J. Phys. Oceanogr.* **13**, 1505–1518

Soares, C. G., Cherneva, Z. & Antao E. M. 2004 Steepness and asymmetry of the largest waves in storm sea states. *Ocean Eng.* **31**, 1147–1167

Soloviev, A. V., Vershinsky N. V. & Bezverchnii, V. A. 1988 Small-scale turbulence measurements in the thin surface layer of the ocean. *Deep-Sea Res.* **35**, 1859–1874

Soloviev, A. V. & Lukas, R. 2003 Observation of wave-enhanced turbulence in the near-surface layer of the ocean during TOGA COARE. *Deep-Sea Res. I* **50**, 371–395

Soloviev, A. V. & Lukas, R. 2006 *The Near-Surface Layer of the Ocean: Structure, Dynamics and Applications*, Springer, New York, **136**, 365–376

Soloviev, A. V. & Lukas, R. 2010 Effects of bubbles and sea spray on air-sea exchange in hurricane conditions. *Boundary-Layer Meteorol.*, in press

Soloviev, Y. P. & Kudryavtsev, V. N. 2010 Wind-speed undulations over swell: field experiment and interpretation. *Boundary-Layer Meteorol.* **136**, doi:10.1007/s10546-010-9506-z

Song, C. & Sirviente, A. I. 2004 A numerical study of breaking waves. *Phys. Fluids* **16**, 2649–2667

Song, J. & Banner, M. L. 2002 On determining the onset and strength of breaking for deep water waves. Part 1: Unforced irrotational wave groups. *J. Phys. Oceanogr.* **32**, 2541–2558

Srokosz, M. A. 1986 On the probability of breaking in deep water. *J. Phys. Oceanogr.* **16**, 382–385

Stansell, P. & MacFarlane, C. 2002 Experimental investigation of wave breaking criteria based on wave phase speeds. *J. Phys. Oceanogr.* **32**, 1269–1283

Stevens, C. L., Poulter, E. M., Smith, M. J. & McGregor, J. A. 1999 Nonlinear features in wave-resolving microwave radar observations of ocean waves. *IEEE* **24**, 470–480

Stewart, R. W. 1974 The air–sea momentum exchange. *Boundary-Layer Meteorol.* **6**, 151–167

Stiassnie, M. 1984 Note on the modified nonlinear Schrödinger equation for deep water waves. *Water Motion* **6**, 431–433

Stiassnie, M., Agnong, Y. & Janssen, P. A. E. M. 2007 Temporal and spatial growth of wind waves. *J. Phys. Oceanogr.* **37**, 106–114

Stiassnie, M., Regev, A. & Agnon, Y. 2008 Recurrent solution of Alber's equation for random water-wave fields. *J. Fluid Mech.* **598**, 245–266

Stiassnie, M. 2010 Fetch limited growth of wind waves, submitted for publication

Stokes, G. G. 1880 Considerations relative to the greatest height of oscillatory irrotational waves which can be propagated without change of form. In *On the Theory of Oscillatory Waves*, Cambridge University Press, London, England, 225–229

Stolte, S. 1992 Wave breaking characteristics deduced from wave staff measurements. *Forschungsanstalt der Bundeswehr fur Wasserschall- und Geophysik*, Report FB 1992–4, 21p

Stolte, S. 1994 Short-wave measurements by a fixed tower-based and a drifting buoy system. *IEEE J. Oceanic Eng.* **19**, 10–22

Stramska, M. & Petelski, T. 2003 Observations of oceanic whitecaps in the north polar waters of the Atlantic. *J. Geophys. Res.* **C108**, doi:10.1029/2002JC001321, 10p

Su, M.-Y., Bergin, M., Marler, P. & Myrick, R. 1982 Experiments of nonlinear instabilities and evolution of steep gravity-wave trains. *J. Fluid Mech.* **124**, 45–72

Su, M. Y. & Cartmill, J. 1992 Breaking wave statistics during 'SWADE'. In *Breaking Waves. IUTAM Symposium, Sydney, Australia, 1991*, Eds. M. L. Banner, R. H. J. Grimshaw, Springer-Verlag, Berlin, Heidelberg, 161–164

Sullivan, P. P. & McWilliams, J. C. 2010 Dynamics of winds and currents coupled to surface waves. *Ann. Rev. Fluid Mech.* **42**, 19–42

Tayfun, M. A. 1981 Breaking limited wave heights. *J. Waterways, Harbors and Coastal Eng. Div.* **107**, 55–79

Taylor, P. K. & Yelland, M. J. 2001 The dependence of sea surface roughness on the height and steepness of the waves. *J. Phys. Oceanogr.* **31**, 572–590

Teixeira, M. A. C. & Fenton, S. E. 2002 On the distortion of turbulence by a progressive surface wave. *J. Fluid Mech.* **458**, 229–267

Terray, E. A., Donelan, M. A., Agrawal, Y. C., *et al.* 1996 Estimates of kinetic energy dissipation under breaking waves. *J. Phys. Oceanogr.* **26**, 792–807

Terrill, E. J. & Melville, K. W. 2000 A broadband acoustic technique for measuring bubble size distributions: laboratory and shallow water measurements. *J. Atmos. Oceanic Tech.* **17**, 220–239

Terrill, E. J., Melville, W. K. & Stramski, D. 2001 Bubble entrainment by breaking waves and their influence on optical scattering in the upper ocean. *J. Geophys. Res.* **C106**, 16815–16823

The WISE Group: Cavaleri, L., Alves, J.-H. G. M., Ardhuin, F., *et al.* 2007 Wave modelling – The state of the art. *Progr. Oceanogr.* **75**, 603–674

Thornton, E. B. & Guza, R. T. 1983 Transformation of wave height distribution. *J. Geophys. Res.* **88**, 5925–5938

Thorpe, S. A. & Humphries, P. N. 1980 Bubbles and breaking waves. *Nature* **283**, 463–465

Thorpe, S. A. 1982 On the clouds of bubbles formed by breaking wind waves in deep water and their role in air–sea gas transfer. *Phil. Trans. R. Soc. Lond.* **A304**, 155–210

Thorpe, S. A. 1992 Bubble clouds and the dynamics of the upper ocean. *Q. J. R. Meteorol. Soc.* **118**, 1–22

Thorpe, S. A., Bowyer, P. & Woolf, D. K. 1992 Some factors affecting the size distribution of ocean bubbles. *J. Phys. Oceanogr.* **22**, 382–389

Thorpe, S. A. 1995 Dynamical processes of transfer at the sea surface. *Prog. Oceanogr.* **35**, 315–352

Thorpe, S. A. 2004 Langmuir circulation. *Ann. Rev. Fluid Mech.* **36**, 55–79

Thorpe, S. A. 2005 *The Turbulent Ocean*, Cambridge University Press, 458p

Tian, Z., Perlin, M. & Choi, W. 2010 Energy dissipation in two-dimensional unsteady plunging breakers and eddy viscosity model. *J. Fluid Mech.*, doi:10.1017/S0022112010000832, 41p

Ting, C.-H., Babanin, G. V., Chalikov, D. & Hsu, T.-W. 2010 Dependence of drag coefficient on angular spreading of sea waves. *Proc. 32nd Ocean Eng. Conf. in Taiwan, Kelnng, Taiwan, 25–26 Nov. 2010*, Ed. H. K. Chang, National Taiwan Ocean Univ., 37–42

Tkalich, P. & Chan, E. S. 2002 Breaking wind waves as a source of ambient noise. *J. Acoust. Soc. Am.* **112**, 456–463

Toba, Y. 1973 Local balance in the air–sea boundary processes on the spectrum of wind waves. *J. Oceanogr. Soc. Jpn.* **29**, 209–220

Toba, Y. & Chaen, M. 1973 Quantitative expression of the breaking of wind waves on the sea surface. *Rec. Oceanogr. Works Jpn.* **12**, 1–11

Toffoli, A., Babanin, A. V., Onorato, M. & Waseda, T. 2010a The maximum steepness of oceanic waves. *Geophys. Res. Lett.* **37**, L05603, doi:10.1029/2009GL041771, 4p

Toffoli, A., Babanin, A. V., Donelan, M. A., Haus, B. K. & Jeong, D. 2010b Estimate of sea spray concentration with a laser altimeter. *J. Atmos. Oceanic Tech.*, submitted

Tolman, H. L. & Chalikov, D. 1996 Source terms in a third-generation wind wave model. *J. Phys. Oceanogr.* **26**, 2497–2518

Tolman, H. L. 2009 User manual and system documentation of WAVEWATCH-III^{TM}, version 3.14. *Technical Report 276*, NOAA/NWS/NCEP/MMAB

Tracy, B. A. & Resio, D. T. 1982 Theory and calculation of the nonlinear energy transfer between sea waves in deep water. *U.S. Army Engineer Waterways Experiment Station* **Rep. N11**, Vicksburg, U.S.A., 54p

Trizna, D. B., Hansen, J. P., Hwang, P. & Wu, J. 1991 Laboratory studies of radar sea spikes at low grazing angles. *J. Geophys. Res.* **C96**, 12529–12537

Troitskaya, Y. I. & Rybushkina, G. V. 2008 Quasi-linear model of interaction of surface waves with strong and hurricane winds. *Izvestiya, Atmospheric and Oceanic Physics* **44**, 670–694

Trulsen, K. & Dysthe, K. B. 1990 Frequency down-shift through self modulation and breaking. In *Water Wave Kinematics*, Eds. A. Torum, T. Gudmestad, Kluwer, 561–572

Trulsen, K. & Dysthe, K. B. 1992 Action of windstress and breaking on the evolution of a wavetrain. In *Breaking Waves, IUTAM Symposium, Sydney, Australia, 1991*, Eds. M. L. Banner, R. H. J. Grimshaw, Springer-Verlag, Berlin, Heidelberg, 244–249

Trulsen, K. & Dysthe, K. B. 1997 Frequency downshift in three-dimensional wave trains in a deep basin. *J. Fluid Mech.* **352**, 359–373

Trulsen, K., Stansberg, C. T. & Velarde M. G. 1999 Laboratory evidence of three-dimensional frequency downshift of waves in a long tank. *Phys. Fluids* **11**, 235–237

Tsagareli, K. N. 2009 Numerical investigation of wind input and spectral dissipation in evolution of wind waves. *PhD Thesis*, The University of Adelaide, South Australia, 219p

Tsagareli, K. N., Babanin, A. V., Walker, D. J. & Young, I. R. 2010 Numerical investigation of spectral evolution of wind waves. Part 1. Wind input source function. *J. Phys. Oceanogr.* **44**, 656–666

Tulin, M. P., Yao, Y. T. & Wang, P. 1994 The simulation of the deformation and breaking of ocean waves in wave groups. *Proc. 7th Intl. Conf. Behaviour Offshore Struc.*, BOSS'74, Elsevier, Amsterdam, 383–392

Tulin, M. P. 1996 Breaking of ocean waves and downshifting. *Waves and Nonlinear Processes in Hydrodynamics*, Eds. J. Grue, B. Gjevik, J. E. Weber, Kluwer, 177–190

Tulin, M. P. & Li, J. J. 1999 The nonlinear evolution of wind driven, breaking ocean waves: mathematical description. *OEI Tech. Rpt.*, 99–202

Tulin, M. P. & Waseda, T. 1999 Laboratory observations of wave group evolution, including breaking effects. *J. Fluid Mech.* **378**, 197–232

Tulin, M. P. & Landrini, M. 2001 Breaking waves in the ocean and around ships. *Proc. of the Twenty-Third Symposium of Naval Hydrodynamics*, 713–745

Uz, B. M., Donelan, M. A., Hara, T. & Bock, E. J. 2002 Laboratory studies of wind stress over surface waves. *Boundary-Layer Meteorol.* **102**, 301–331

Vakhguelt, A. 2007 Mathematical model of interaction between finite amplitude wave and the liquid particles. *Proc. 18th Australasian Coastal and Ocean Engineering Conf. and 11th Australasian Port and Harbour Conference, 18–20 July 2007, Melbourne, Victoria*, paper 197, 6p

Van der Westhuysen, A. J., Zijlema, M. & Battjes, J. A. 2007 Nonlinear saturation-based whitecapping dissipation in SWAN for deep and shallow water. *Coast. Eng.* **54**, 151–170

Van Dorn, W. G. & Pazan, S. E. 1975 Laboratory investigation of wave breaking. *AOEL rep. no. 71*, Scripps Inst. of Oceanog., Ref. No. 75–21

Van Vledder, G. P. 2002 A subroutine version of the Webb/Resio/Tracy method for the computation of non-linear quadruplet wave–wave interactions in deep and shallow water. *Alkyon Report* **151**, 55p

Van Vledder, G. P. 2006 The WRT method for computation of non-linear four-wave interactions in discrete spectral wave models. *Coast. Eng.* **53**, 223–242

Vazquez, A., Sanchez, R. M., Salinas-Rodriquez, E., Soria, A. & Manasseh, R. 2005 Experimental comparison between acoustic and pressure signals from a bubbling flow. *Exper. Thermal and Fluid Sci.* **30**, 49–57

Vazquez, A., Manasseh, R. & Sanchez R. M. 2008 A look at three measurement techniques for bubble size determination. *Chem. Eng. Sci.* **63**, 5860–5869

Veron, F. & Melville, W. K. 1999 Pulse-to-pulse coherent Doppler measurements of waves and turbulence. *J. Atmos. Oceanic Tech.* **16**, 1580–1596

Volkov, Y. A. 1970 Turbulent flux of momentum and heat in the atmospheric surface layer over a disturbed sea-surface. *Izvestiya, Atmospheric and Oceanic Physics* **6**, 770–774

Walpole, R. E. & Meyers, R. H. 1978 *Probability and Statistics for Scientists and Engineers*, Macmillan, New York, 580p

Wannikhoff, R. & McGillis, W. R. 1999 A cubic relationship between air–sea CO_2 exchange and wind speed. *Geophys. Res. Lett.* **26**, 1889–1892

Waseda, T. & Tulin, M. P. 1999 Experimental study of the stability of deep-water wave trains including breaking effects. *J. Fluid Mech.* **401**, 55–84

Waseda, T., Kinoshita, T. & Tamura, H. 2009a Evolution of a random directional wave and freak wave occurrence. *J. Phys. Oceanogr.* **39**, 621–639

Waseda, T., Kinoshita, T. & Tamura, H. 2009b Interplay of resonant and quasi-resonant interaction of the directional ocean waves. *J. Phys. Oceanogr.*, **39**, 2351–2362

Watanabe, Y. & Tulin, H. 1999 Three-dimensional large-eddy simulation of breaking waves. *Coast. Eng. J. Jpn.* **41**, 281–301

Watson, K. M. & West B. J. 1975 A transport-equation description of nonlinear ocean surface wave interactions. *J. Fluid Mech.* **70**, 815–826

Webb, D. J. 1978 Nonlinear transfer between sea waves. *Deep-Sea Res.* **25**, 279–298

Weissman, M. A., Atakturk, S. S. & Katsaros, K. B. 1984 Detection of breaking events in a wind-generated wave field. *J. Phys. Oceanogr.* **14**, 1608–1619

Wenz, G. M. 1962 Acoustic ambient noise in the ocean: spectra and sources. *J. Acoust. Soc. Am.* **34**, 1936–1956

West, B. J., Brueckner, K. A. & Janda, R. S. 1987 A new numerical method for surface hydrodynamics. *J. Geophys. Res.* **C92**, 11803–11824

Wetzel, L. B. 1990 Electromagnetic scattering from the sea at low grazing angles. In *Surface Waves and Fluxes: Current Theory and Remote Sensing* **2**, Eds. G. L. Geernaert, W. J. Plant, Kluwer Acad., Norwell, Mass., 109–171

Willemsen, J. F. 2002 Deterministic modeling of driving and dissipation for ocean surface gravity waves in two horizontal dimensions. *J. Geophys. Res.* **C107**, C83121, doi:10.1029/2001JC001029, 17p

Williams, G. F. Jr. 1969 Microwave radiometry of the ocean and the possibility of marine wind velocity determination from satellite observations. *J. Geophys. Res.* **74**, 4591–4594

Wilson, J. D., & Makris, N. C. 2006 Ocean acoustic hurricane classification. *J. Acoust. Soc. Am.* **119**, 168–181

Wilson, J. D., & Makris, N. C. 2008 Quantifying hurricane destructive power, wind speed and air–sea material exchange with natural undersea sound. *Geophys. Res. Lett.* **35**, 3177, doi:10.1029/2008GL033200

Woolf, D. K. 1997 Bubbles and their role in gas exchange. In *The Sea Surface and Global Change*, Eds. P. S. Liss, R. A. Duce, Cambridge University Press, UK, 173–205

Wu, C. H., & Nepf, H. M. 2002 Breaking criteria and energy losses for three-dimensional wave breaking. *J. Geophys. Res.* **C107**, 3177, doi:10.1029/2001JC001077, 18p

Wu, J. 1969 Wind stress and surface roughness at the air–sea interface. *J. Geophys. Res.* **74**, 444–455

Wu, J. 1973 Spray in the atmospheric surface layer: Laboratory study. *J. Geophys. Res.* **78**, 511–519

Wu, J. 1979 Oceanic whitecaps and sea state. *J. Phys. Oceanogr.* **9**, 1064–1068

Wu, J. 1988 Variations of whitecap coverage with wind stress and water temperature. *J. Phys. Oceanogr.* **18**, 1448–1453

Wu, J. 1993 Spray in the atmospheric surface layer: Review and analysis of laboratory and oceanic results. *J. Geophys. Res.* **98**, 18221–18227

Wu, J. 2000 Bubbles produced by breaking waves in fresh and salt waters. *J. Phys. Oceanogr.* **30**, 1809–1813

Xu, D., Hwang, P. A. & Wu, J. 1986 Breaking of wind-generated waves. *J. Phys. Oceanogr.* **16**, 2172–2178

Yefimov, V. V. & Khristiforov, G. N. 1971 Spectra and statistical relations between the velocity fluctuations in the upper layer of the sea and surface waves. *Izvestiya, Atmospheric and Oceanic Physics* **7**, 1290–1310

Yefimov, V. V. & Babanin, A. V. 1991 Dispersive relations of envelopes of wind wave groups. *Izvestiya, Atmospheric and Oceanic Physics* **27**, 599–603

Yelland, M. J., Moat, B. I., Taylor, P. K., *et al.* 1998 Wind stress measurements from the open ocean corrected for airflow distortion by the ship. *J. Phys. Oceanogr.* **28**, 1511–1526

Young, I. R. & Banner, M. L. 1992 Numerical experiments on evolution of fetch limited waves. In *Breaking Waves. IUTAM Symposium, Sydney, Australia, 1991*, Eds. M. L. Banner, R. H. J. Grimshaw, Springer-Verlag, Berlin, Heidelberg, 267–275

Young, I. R. & van Vledder, G. P. 1993 A review of the central role of nonlinear interactions in wind–wave evolution. *Phil. Trans. R. Soc. Lond.* **A342**, 505–524

Young, I. R. 1994 On the measurement of directional wave spectra. *Appl. Ocean Res.* **16**, 283–294

Young, I. R. & Holland, G. J. 1996 *Atlas of the Oceans: Wind and Wave Climate*, Pergamon Press, ISBN 0-08-042519-4, 241p

Young, I. R. & Verhagen, L. A. 1996 The growth of fetch limited waves in water of finite depth. Part 1. Total energy and peak frequency. *Coast. Eng.* **29**, 101–121

Young, I. R., Verhagen, L. A. & Khatri, S. K. 1996 The growth of fetch limited waves in water of finite depth. Part 3. Directional spectra. *Coast. Eng.* **29**, 47–78

Young, I. R. 1999 *Wind Generated Ocean Waves*, Elsevier, 288p

Young, I. R. & Babanin, A. V. 2001 Wind wave evolution in finite depth water. *Proc. 14th Australasian Fluid Mech. Conf.*, Adelaide, Australia, 79–86

Young, I. R., Banner, M. L., Donelan, M. A., *et al.* 2005 An integrated study of the wind wave source term balance in finite depth water. *J. Atmos. Oceanic Tech.* **22**, 814–828

Young, I. R. 2006 Directional spectra of hurricane wind waves. *J. Geophys. Res.* **111**, C08020, doi:10.1029/JC003540

Young, I. R. & Babanin, A. V. 2006a Spectral distribution of energy dissipation of wind-generated waves due to dominant wave breaking. *J. Phys. Oceanogr.* **36**, 376–394

Young, I. R. & Babanin, A. V. 2006b The form of the asymptotic depth-limited wind wave frequency spectrum. *J. Geophys. Res.* **C111**, C06031, doi:10.1029/2005JC003398, 15p

Young, I. R. 2010 The form of the asymptotic depth-limited wind wave frequency spectrum. Part III Directional spreading. *Coast. Eng.* **57**, 30–40

Yuan, Y., Tung, C. C. & Huang, N. E. 1986 Statistical characteristics of breaking waves. In *Wave Dynamics and Radio Probing of the Ocean Surface*, Eds. O. M. Phillips, K. Hasselmann, Plenum New York, 265–272

Yuan, Y., Zhang, S. & Han, L. 2008 Statistical model on the surface elevation of waves with breaking. *Sci. China.* **B51**, 759–768

Yuan, Y., Han, L., Hua, F., *et al.* 2009 The statistical theory of breaking entrainment depth and surface whitecap coverage of real sea waves. *J. Phys. Oceanogr.* **39**, 143–161

Yuen, H. C. & Lake, M. 1982 Nonlinear dynamics of deep-water gravity waves. In *Advances in Applied Mechanics* **22**, 67–229

Zakharov, V. E. 1966 The instability of waves in nonlinear dispersive media. *Zhurnal Eksperimental'noi i Teoreticheskoi Fiziki* **51**, 1107–1114 (in Russian)

Zakharov, V. E. & Filonenko, N. N. 1966 The energy spectrum for stochastic oscillation of a fluid's surface. *Dokl. Acad. Nauk SSSR (Transactions of USSR Academy of Sciences – English Translation)* **170**, 1992–1995

Zakharov, V. E. 1967 The instability of waves in nonlinear dispersive media. *Sov. Phys.–JETP (English Translation)* **24**, 740–744

Zakharov, V. E. 1968 Stability of periodic waves of finite amplitude on the surface of deep fluid. *J. Appl. Mech. Tech. Phys.–JETR (English Translation)* **2**, 190–194

Zakharov, M. M. & Zaslavskii, V. E. 1982 Kinetic equation and Kolmogorov's spectra in a weak turbulence theory of wind waves. *Izvestiya Akademii Nauk SSSR. Fizika Atmosferi i Okeana)* **18**, 970–979 (in Russian, English abstract)

Zakharov, V. E., Korotkevich, A. O., Pushkarev, A. & Resio, D. 2007 Coexistence of weak and strong wave turbulence in a swell propagation. *Phys. Rev. Lett.* **99**, doi: 10.1103/PhysRevLett.99.164501

Zakharov, V. E. 2010 Energy balance in a wind-driven sea. *Phys. Sci.*, **T142**, 014052, 14p

Zaslavskii, G. M. & Sharkov, E. A. 1987 Fractal features in breaking wave areas on sea surface. *Doklady Academii Nauk SSSR (Transactions of USSR Academy of Sciences – English Translation)* **294**, 1362–1366

Zhang, J. A., Black, P. G., French, J. R. & Drennan, W. M. 2008 First direct measurements of enthalpy flux in the hurricane boundary layer: The CBLAST results. *Geophys. Res. Lett.* **35**, L14813, doi:10.1029/2008GL034374

Zhao, D. & Toba, Y. 2001 Dependence of whitecap coverage on wind and wind–wave properties. *J. Oceanogr.* **57**, 603–616

Zhao, Q. & Tanimoto, K. 1998 Numerical simulation of breaking waves by large eddy simulation and vof method. *Proc. 26th Int. Conf. Coast. Eng.* **1**, ASCE, 892–905

Zhao, Q., Armfield, S. & Tanimoto, K. 2004 Numerical simulation of breaking waves by multi-scale turbulence model. *Coast. Eng.* **51**, 53–80

Zieger, S., Vinoth, J. & Young, I. R. 2009 Joint calibration of multiplatform altimeter measurements of wind speed and wave height over the past 20 years. *J. Atmos. Ocean. Tech.* **26**, 2549–2564

Zimmermann, C.-A. & Seymour, R. 2002 Detection of breaking in a deep water wave record. *J. Waterway, Port, Coastal, Ocean Eng.* **128**, 72–78

Index

Printed in the United States
By Bookmasters